K	The number of levels of the independent variable
LL	Lower limit of a confidence interval
MS	Mean square; ANOVA term for the variance
N	The number of scores or observations
O	In chi square, the observed frequency
r	Pearson product-moment correlation coefficient
r_s	A correlation coefficient for ranked data; named for Spearman
r^2	The coefficient of determination
S	The standard deviation of a sample; describes the sample
$\hat{s}$	The standard deviation of a sample; estimates σ
$\hat{s}^2$	Variance of a sample; estimates σ^2
$\hat{s}_D$	Standard deviation of a distribution of differences between correlated scores
$s_{\bar{D}}$	Standard error of the difference between correlated means
$s_{\bar{X}_1 - \bar{X}_2}$	Standard error of a difference between means
$s_{\bar{X}}$	Standard error of the mean
SS	Sum of squares; the sum of the squared deviations from the mean
T	Wilcoxon matched-pairs signed-ranks T statistic for correlated samples
t	t test statistic
t_α	Critical value of t; level of significance $= \alpha$
U	Mann-Whitney statistic for independent samples
UL	Upper limit of a confidence interval
X	A score
$\bar{X}$	The mean of a sample
$\bar{\bar{X}}$	The mean of a set of means
X_H	The highest score in a distribution
X_L	The lowest score in a distribution
$\hat{Y}$	The Y value predicted for some X value
$\bar{Y}$	The mean of the Y variable
z	A score expressed in standard deviation units; a standard score
z	Test statistic when the sampling distribution is normal
z_X	A z value for a score on variable X
z_Y	A z value for a score on variable Y

If you answer "YES" to any of these questions . . .

· Do you dive into statistical calculations without a clear idea of why you're doing them?
· Do the concepts and steps behind hypothesis testing confuse you?
· Are you having trouble understanding the difference between one-way and two-way ANOVA?
· Are you finding it difficult to figure out when to use a z-score?

. . . it's time to visit our FREE Online Statistics Workshops

http://psychology.wadsworth.com/workshops/workshops.html

These workshops are a new and different way for students to further clarify and reinforce statistical concepts that may, at first, seem too tough to comprehend. All 10 **Statistics Workshops** are accessible *24 hours a day, 7 days a week*! To get started, simply click on any of the workshop topics that you want to explore!

Seemingly complex, confusing concepts *(yes, even z-scores)* are distilled into straightforward language that you can understand!

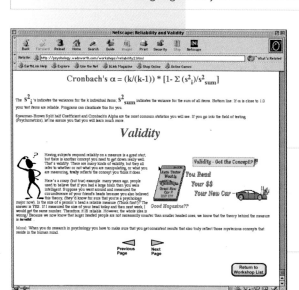

Workshops author Glenn E. Meyer from Trinity University is a fun host, humorous and conversational, as he guides you step-by-step through the concepts that are giving you trouble. Working with Glenn's workshops is like having a patient mentor right by your side!

Online Statistics Workshops . . .
Great ways to reinforce what you learn in the textbook

http://psychology.wadsworth.com/workshops/workshops.html

You'll find that a statistical concept can make much more sense with examples like this one where you can explore *t*-tests in terms of something practical . . . like how much sleep college students normally get!

You'll discover that Chi-Square doesn't seem so inscrutable when you learn that what it really does is compare whether one set of proportions is different from another. And as you progress through each of these workshops, you'll be surprised to find that any apprehension you may have had about the "difficulty" of statistics slipping away.

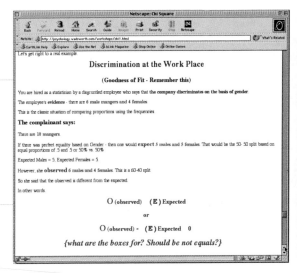

Also available at . . .

http://psychology.wadsworth.com/workshops/workshops.html

10 FREE Research Methods Workshops

Workshops include:

1. What Is Science? Ways of Thinking About the World
2. The Experimental Method (Independent and Dependent Variables)
3. Reliability and Validity
4. Non-Experimental Approaches to Research

5. The Survey Method
6. True Experiments
7. Between vs. Within Designs
8. Getting Ideas for a Study
9. APA style
10. Confounds—Threats to Validity

Additional workshops are coming soon! Watch for them!

seventh edition

BASIC STATISTICS
TALES OF DISTRIBUTIONS

Chris Spatz
Hendrix College

Wadsworth
Thomson Learning

Australia • Canada • Mexico • Singapore • Spain
United Kingdom • United States

Publisher, Psychology: Vicki Knight
Marketing: Jenna Opp
Marketing Communications Team: Margaret Parks, Kelly Millon
Assistant Editor: Jennifer Wilkinson
Manuscript editor: Carol Reitz
Interior Design: E. Kelly Shoemaker

Production Management: New Leaf Publishing Services
Print buyer: Karen Hunt
Cover Design: Carolyn Deacy
Cover Photograph: Matthew Farrugio
Permissions Editor: Bob Kauser
Printing: R. R. Donnelley & Sons/Crawfordsville

Library of Congress Cataloging-in-Publication Data
Spatz, Chris, [date]
    Basic statistics : tales of distributions / Chris Spatz. — 7th ed.
    p. cm.
    Includes index.
    ISBN 0-534-36690-2 (alk. paper)
      1. Statistics. I. Title.
QA276.12.S66 2000
519.5 — dc21   00-039884

Wadsworth/Thomson Learning
10 Davis Drive
Belmont, CA 94002-3098
USA

For more information about our products, contact us:
Thomson Learning Academic Resource Center
1-800-423-0563
http://www.wadsworth.com

International Headquarters
Thomson Learning
International Division
290 Harbor Drive, 2nd Floor
Stamford, CT 06902-7477
USA

UK/Europe/Middle East/South Africa
Thomson Learning
Berkshire House
168-173 High Holborn
London WC1V 7AA
United Kingdom

Asia
Thomson Learning
60 Albert Street, #15-01
Albert Complex
Singapore 189969

Canada
Nelson Thomson Learning
1120 Birchmount Road
Toronto, Ontario M1K 5G4
Canada

the **author**

Chris Spatz is a professor of psychology at Hendrix College in Conway, Arkansas, where he has served twice as chair of the Psychology Department. After completing his undergraduate work at Hendrix College, Spatz earned a Ph.D. from Tulane University in experimental psychology. He subsequently held a two-year National Institutes of Mental Health postdoctoral fellowship in animal behavior at the University of California, Berkeley. Before returning to Hendrix to teach, Spatz held positions at The University of the South and the University of Arkansas at Monticello.

Spatz is married to Thea Siria Spatz and they have three grown children. Two of them work with jet airplanes and one is an artist. They also have four grandchildren. Aside from writing and teaching, Spatz enjoys the outdoors, especially camping and canoeing. He swims several times a week and is active in the United Methodist Church.

With love and affection, this textbook is dedicated to
Thea Siria Spatz, Ed.D., CHES.

brief**contents**

contents

chapter **three**

VARIABILITY   **50**

chapter **four**

COMBINATION STATISTICS   **66**

transition page   79

chapter **five**

CORRELATION AND REGRESSION   **80**

transition page   112

chapter six

THEORETICAL DISTRIBUTIONS INCLUDING THE NORMAL DISTRIBUTION   113

chapter seven

SAMPLES, SAMPLING DISTRIBUTIONS, AND CONFIDENCE INTERVALS   134

transition page

chapter **thirteen**

THE CHI SQUARE DISTRIBUTION   **278**

chapter **fourteen**

NONPARAMETRIC STATISTICS   **299**

chapter **fifteen**
CHOOSING TESTS AND WRITING INTERPRETATIONS   **326**

appendixes

preface

Even if our statistical appetite is far from keen, we all of us should like to know enough to understand, or to withstand, the statistics that are constantly being thrown at us in print or conversation—much of it pretty bad statistics. The only cure for bad statistics is apparently more and better statistics. All in all, it certainly appears that the rudiments of sound statistical sense are coming to be an essential of a liberal education.

—Robert Sessions Woodworth

Basic Statistics: Tales of Distributions, Seventh Edition, is designed for a one-term, introductory course in statistics. My goal has been to produce a book that is comprehensible and complete for students who take only one statistics course and comprehensible and preparatory for students who will take additional statistics courses.

Although detailed directions and examples are given for each statistical procedure, I concentrate heavily on conceptualization and interpretation of the problems. To facilitate this emphasis, some experimental design terms and issues are included along with the traditional statistics topics. In many places, the reader is invited to stop and think or to stop and do an exercise. Some problems simply ask the student to decide which statistical technique is appropriate. Because of this book's emphasis and style, I am confident that it will reinforce an instructor's efforts to promote critical thinking.

Students who work through this book should be able to:

- solve statistical problems
- understand statistical reasoning
- write explanations that are congruent with statistical analyses
- choose proper statistical techniques to analyze data from simple experiments

This textbook is designed to engage students in their study of statistics. Its principal strength is that it is written in an informal style. In addition, many other features facilitate engagement:

- Examples and problems come from different disciplines and from everyday life.

- Problems are distributed throughout the chapter instead of being massed at the end.
- Answers to problems are extensive; almost 90 pages are devoted to answers.
- "Clues to the Future" alert students to concepts that will come up again.
- "Error Detection" notes tell ways to detect and to prevent mistakes.
- Objectives at the beginning of each chapter serve first as an orientation list and later as a review list.
- After each third of the book, there are "What Would You Recommend?" problems.
- There are three glossaries (words, formulas, and symbols).

I have written two ancillary publications to supplement this textbook. For students, a Study Guide provides additional explanations, problems, and a chapter summary. For professors, an Instructor's Manual and Test Bank has teaching suggestions and test items. Test items are also available on disks.

A gentle revolution is going on in the practice of statistics, and this new edition reflects some of the changes. In the past, statistics was moving toward more and more sophisticated ways to test null hypotheses. Now it is moving toward simpler analyses and more emphasis on descriptive statistics. (See *www.apa.org/science/tfsi*.)

The new features of this seventh edition that reflect this attention to descriptive statistics are (1) boxplots, (2) increased emphasis on the effect size index, and (3) the Descriptive Statistics Report, an exercise in reporting several descriptive statistics for a set of data. All three of these topics are included in the new Chapter 4, Combination Statistics.

Also for this seventh edition, I have restored the treatment of confidence intervals about a mean difference (from earlier editions), added an effect size index for chi square, moved instructions for creating a grouped frequency distribution to an appendix, and brought all contemporary data sets up to date. More than 40 problems are new, and many others were rewritten or shortened. Comprehensive problems are identified with a special logo. Supplementary multiple-choice questions and topic workshops are available at the Wadsworth website.

I am pleased to acknowledge all the help I have received from students, colleagues, and Hendrix College. Bob Eslinger produced accurate graphs of the F, t, χ^2, and normal distributions. Rob Nichols wrote a sampling program for me. Roger E. Kirk, my consulting editor for the first six editions, deserves a special thanks for saving me from errors and teaching me some statistics. Carol Reitz started with a manuscript that I considered well written and proceeded to improve it in hundreds of places using her superb copyediting skills. I was very fortunate to once again have Nancy Shammas as my production editor. I especially want to acknowledge James O. Johnston, my friend and former co-author, who contributed to the first three editions of this book.

I also want to acknowledge the help of reviewers for this edition: Dr. Winona Cochran, Bloomsburg University; Dr. Jane Connor, SUNY Binghamton; Dr. Stan G. Danis, California State University — Dominguez Hills; Dr. Kathleen Dillon, Western New England College; Dr. Hank Gorman, Austin College; Dr. Alexandria Guzman, SUNY Binghamton; Dr. William Levine, SUNY Binghamton; Dr. William Lucker, University of Texas at El Paso; Dr. Leroy Matthews, Idaho State University; Dr. William McDaniel, Georgia College & State University; Dr. Sandra McIntire, Rollins College; Dr. Craig Nagoshi, Arizona State University; Dr. Todd Shackelford, Florida Atlantic University; Dr. Bonnie Walker, Rollins College; Dr. Evangeline Wheeler, Towson University; Dr. Robert Woll, Sienna College; and Dr. Hasan Ziaie, Virginia Union University.

I am grateful to the Longman Group UK Ltd., on behalf of the Literary Executor of the late Sir Ronald A. Fisher, F.R.S., and Dr. Frank Yates, F.R.S., for permission to reproduce Tables III, IV, and VII from their book *Statistical Tables for Biological, Agricultural, and Medical Research,* Sixth Edition (1974).

My most important acknowledgment goes to my wife and family, who helped and supported me in many ways over the life of this project.

I've been a teacher for much of my life—first as an older sibling, then as a parent, and now as a professor and grandfather. Education is a first-class task, in my opinion. I hope this book conveys my enthusiasm and my philosophy of teaching. (By the way, if you are a student who is so thorough as to read the whole preface, you should know that I have included phrases and examples in a number of places with *your kind* in mind.)

Chris Spatz

Note: If errors are found in this book, I will post corrections at http://www.hendrix.edu/statistics. My email address is spatz@hendrix.edu.

chapter**one**

INTRODUCTION

OBJECTIVES FOR CHAPTER ONE

Study the text. Work the problems as you come to them. Afterward, you should be able to:

1. Distinguish between descriptive and inferential statistics
2. Define the words *population, sample, parameter, statistic,* and *variable* as these terms are used in statistics
3. Distinguish between quantitative and qualitative variables
4. Identify the lower and upper limits of a quantitative measurement
5. Identify four scales of measurement and give examples of each
6. Distinguish between statistics and experimental design
7. Define some experimental-design terms—*independent variable, dependent variable,* and *extraneous variable*—and identify these variables in the description of an experiment
8. Describe the relationship of statistics to epistemology
9. Identify a few events in the history of statistics

This is a book about statistics that is written for people who do not plan to become statisticians. In fact, I was not trained as a statistician myself; my field in graduate school was psychology. Nevertheless, I studied statistics so I could analyze, understand, and explain the quantitative data from experiments (and also because the courses were required).

Many disciplines use quantitative data. In all of these disciplines, understanding statistics is at least helpful and may even be necessary. Lawyers, for example, sometimes have cases in which statistics can help establish more convincing evidence for their clients. I have assisted lawyers at times, occasionally testifying in court about the proper interpretation of data. (You will analyze and interpret some of these data in later chapters.) Statistics is a powerful method for getting answers from data, and it is sometimes the best way to persuade others that the conclusions are correct.

Whatever your current thoughts about your future as a statistician, I believe you will benefit from this course. When a statistics course is successful, students learn to identify the questions that a set of data can answer, determine the statistical procedures that will provide the answers, carry out the procedures, and then, using plain English and graphs, tell the story the data reveal.

The best way for you to acquire all these skills (especially the part about telling the story) is to use a *generative* approach while you are studying. What should you generate? Generate questions for topics you don't yet understand. When you can, generate answers. If you aren't able to generate an answer, then generate a plan for finding an answer. Decide which sections of this text are more important and generate a summary for them. And, most important, generate answers to the problems scattered throughout the chapter. A generative approach to learning is guaranteed to pay handsome dividends (see Wittrock, 1986). For me, I generate more by writing than by thinking. Is writing also better for you?

Are you uncertain about your arithmetic and algebra skills? Appendix A in the back of this book may help. It consists of a pretest (to see if you need to refresh your memory) and a review (to provide that refresher).

WHAT DO YOU MEAN, "STATISTICS"?

The *Oxford English Dictionary* says that the word *statistics* came into use more than 200 years ago. At that time, *statistics* referred to a country's quantifiable political characteristics — characteristics such as population, taxes, and area. Statistics meant "state numbers." Tables and charts of those numbers turned out to be a very satisfactory way to compare different countries and to make projections about the future. Later, the techniques of using tables and charts proved helpful to people studying trade (economics) and natural phenomena (science). Statistics was spreading.

Today two different techniques are called *statistics.* One technique, **descriptive statistics,**[1] produces a number or a figure that summarizes or describes a set of data. You are already familiar with some descriptive statistics. For example, you know about the arithmetic average, called the **mean.** You have probably known how to compute a mean since elementary school — just add up the numbers and divide the total by the number of entries. As you already know, the mean describes the central tendency of a set of numbers. The basic idea is simple: A descriptive statistic summarizes a set of data with one number or graph. This book will cover about a dozen descriptive statistics; they start in Chapter 2.

The other statistical technique is **inferential statistics.** With inferential statistics, you use measurements from a sample to reach conclusions about a larger, *unmeasured* population. There is, however, a problem with samples. Samples always depend *partly* on the luck of the draw; chance influences what particular measurements you get. If you have the measurements for the entire population, chance doesn't play a part — all the variation in the numbers is "true" variation. But with samples, some of the variation is the true variation in the population and some is just the chance ups and downs that go with a sample. Inferential statistics was developed as a way to measure and reduce the effects of chance that come with sampling.

---

[1] Words and phrases printed in boldface type are defined in Appendix D, Glossary of Words.

Here is a textbook definition: Inferential statistics is a method that takes chance factors into account when samples are used to reach conclusions about populations. Like most textbook definitions, this one condenses many elements into a short sentence. Because the idea of using samples to understand populations is perhaps the *most important concept* in this course, please pay careful attention when elements of inferential statistics are explained.

Inferential statistics has proved to be a very useful method in scientific disciplines. Many other fields use inferential statistics, too, so I have selected examples and problems from a variety of disciplines for this text and its study guide.

Here is an example of the use of inferential statistics in psychology. Today there is a lot of evidence that people remember the tasks they fail to complete better than the tasks they complete. This is known as the *Zeigarnik effect.* Bluma Zeigarnik asked the participants in her experiment to do about 20 tasks, such as work a puzzle, make a clay figure, and construct a box from cardboard.[2] For each participant, half the tasks were interrupted before completion. Later, when the participants were asked to recall the tasks they had worked on, they listed more of the interrupted tasks (about seven) than the completed tasks (about four).

So, should you conclude that interruption improves memory? Not yet. It might be that interruption actually has no effect but that several *chance factors* happened to favor the interrupted tasks in Zeigarnik's particular experiment. One way to meet this objection is to run the experiment again. Similar results would lend support to the conclusion that interruption improves memory. A less expensive way to meet the objection is to use an inferential statistics test.

An inferential statistics test begins with the actual data from the experiment. It ends with a probability — the probability of obtaining data like those actually obtained if it is true that interruption *does not* affect memory. If the probability is very small, then you can conclude that interruption *does* affect memory.[3]

A conclusion based on inferential statistics might be a statement such as: The difference in the memories of the two groups that were treated differently is most likely due to interruption and not to chance because chance *by itself* would rarely produce two samples this different. Probability, as you can see, is an important element of inferential statistics.

STATISTICS: A DYNAMIC DISCIPLINE

A dynamic discipline is a changing discipline sometimes characterized by controversy. Statistics fits the definition. In the past few years new techniques in descriptive statistics have proved helpful. In inferential statistics, new tests have been created. Meanwhile some statisticians say that large numbers of inferential tests are actually harmful and should be abandoned. (For a particularly readable summary of this issue, see

[2] A summary of this study can be found in Ellis (1938). The complete reference and all others in the text are listed in the References section at the back of the book.
[3] If it really is the case that interruption doesn't affect memory, then differences between the samples are the result of chance.

Dillon, 1999.) No doubt, statistics will continue to change. This textbook simply gives you a snapshot view of the field at the start of the century.

Here is a general overview of how statistics is used today. Researchers start with some phenomenon, event, or process that they want to understand better. They translate it into numbers and then manipulate the numbers according to the rules of statistics. Next, they translate the numbers back into the phenomenon, and, finally, they tell the story of their new understanding of the phenomenon, event, or process. Thus, researchers use statistics as a tool to help their understanding.

clue to the future

The first part of this book is devoted to descriptive statistics (Chapters 2–5) and the second part to inferential statistics (Chapters 6–14). The two parts are not independent, though, because to understand and explain inferential statistics you must first understand descriptive statistics.

WHAT'S IN IT FOR ME?

What's in it for me? is a reasonable question. Decide which of the following answers apply to you and then pencil in others that are true for you.

One thing that is in it for you is that, when you have successfully completed this course, you will understand how statistical analyses are used to *make decisions.* You will understand both the elegant beauty and the ugly warts of the process. H. G. Wells, a novelist who wrote about the future, said, "Statistical thinking will one day be as necessary for efficient citizenship as the ability to read and write." So chalk this course up under the heading "general education about decision making, to be used throughout life."

Another result of successfully completing this course is that you will be able to use a tool employed in many disciplines, including psychology, education, medicine, sociology, biology, forestry, economics, and others. Although people involved in these disciplines are interested in different phenomena, they all use statistics. As examples, statistics has been used to determine the effect of reducing cigarette taxes on smoking among teenagers, the safety of a new surgical anesthetic, and the memory of young school-age children for pictures (which is as good as that of college students). Statistics show what diseases have an inheritance factor, how to improve short-term weather forecasts, and why giving intentional walks in baseball is a poor strategy. All these examples come from *Statistics: A Guide to the Unknown,* a book edited by Judith M. Tanur and others (1989). This book, written for those "without special knowledge of statistics," has 29 essays on many specific uses of statistics.

Statistics is used in places that might surprise you. In American history, for example, the authorship of 12 of *The Federalist* papers was disputed for a number of years. (*The Federalist* papers were 85 short essays written under the pseudonym "Publius" and published in New York City newspapers in 1787 and 1788. Written by James Madison, Alexander Hamilton, and John Jay, the essays were designed to persuade the people of the state of New York to ratify the Constitution of the United

States.) To determine authorship, each of the 12 papers was graded with a quantitative *value analysis* in which the importance of such values as national security, a comfortable life, justice, and equality was assessed. The value analysis scores were compared with value analyses of papers known to have been written by Madison and Hamilton (Rokeach, Homant, and Penner, 1970). Another study, by Mosteller and Wallace, analyzed *The Federalist* papers using the frequency of words such as *by* and *to* (reported in Tanur et al., 1989). Both studies concluded that Madison wrote all 12 essays.

Here is an example from law. Rodrigo Partida was convicted of burglary in Hidalgo County, a border county in southern Texas. A grand jury rejected his motion for a new trial. Partida's attorney filed suit, claiming that the grand jury selection process discriminated against Mexican-Americans. In the end (*Castaneda* v. *Partida,* 430 U.S. 482 [1976]), Justice Harry Blackmun of the U.S. Supreme Court wrote, regarding the number of Mexican-Americans on the grand jury, "If the difference between the expected and the observed number is greater than two or three standard deviations, then the hypothesis that the jury drawing was random (is) suspect." In Partida's case the difference was approximately 12 standard deviations, and the Supreme Court ruled that Partida's attorney had presented *prima facie* evidence. (*Prima facie* evidence is so good that one side wins the case unless the other side rebuts the evidence, which in this case did not happen.) *Statistics: A Guide to the Unknown* includes two essays on the use of statistics by lawyers.

I hope that this course will encourage you or even teach you to ask questions about the statistics you hear and read. Consider the questions, one good and one better, asked of a restaurateur in Paris who served delicious rabbit pie. He served rabbit pie even when other restaurateurs could not obtain rabbits. A suspicious customer asked this restaurateur if he was stretching the rabbit supply by adding horse meat.

"Yes, a little," he replied.

"How much horse meat?"

"Oh, it's about 50–50," the restaurateur replied.

The suspicious customer, satisfied, began eating his pie. Another customer, who had learned to be more thorough in asking questions about statistics, said, "50–50? What does *that* mean?"

"One rabbit and one horse" was the reply.

This section began with a question from you: What's in it for me? My answer is that when you learn the basics of statistics, you will understand data better. You will be able to communicate with others who use statistics, and you'll be better able to persuade others. In short, I am sure there will be lots of lasting value for you in this course.

SOME TERMINOLOGY

Like most courses, statistics introduces you to many new words. In statistics, most of the terms are used over and over again. Your best move, when introduced to a new term, is to *stop, read* the definition carefully, and *memorize* it. As the term continues to be used, you will become more and more comfortable with it. Making notes is helpful.

Populations and Samples

A **population** consists of all the scores of some specified group. A **sample** is a subset of a population. The population is the thing of interest, is defined by the investigator, and includes all cases. The following are some populations:

Family incomes of college students in the fall of 1999
Weights of crackers eaten by obese male students
Depression scores of Alaskans
Errors in a maze by rats with brain damage
Gestation times for human beings
Memory scores of human beings[4]

Investigators are always interested in a population. However, you can see from these examples that populations are often so large that not all the members can be studied. The investigator must often resort to measuring a sample that is small enough to be manageable. A sample taken from the population of incomes of families of college students might include only 40 students. From the last population on the list, Zeigarnik used a sample of 164.

Most authors of research articles carefully explain the characteristics of the *samples* they use. Often, however, they do not identify the *population,* leaving that task to the reader. For example, consider a study on a drug therapy program involving a sample of 14 men who are acute schizophrenics at Hospital A. Suppose the report of the study doesn't identify the population. What is the population? Among the possibilities are: all male acute schizophrenics at Hospital A, all male acute schizophrenics, all acute schizophrenics, and all schizophrenics. The answer to the question What is the population? depends on the specifics of a research area, but many researchers generalize generously. For example, for some topics it is reasonable to generalize from the results of a study on rats to "all mammals." In all cases, the reason for gathering data on a sample is to generalize the results to some larger population.

Resorting to the use of samples introduces uncertainty into the conclusions because different samples taken from the same population nearly always differ from one another. Using inferential statistics, however, you may be able to assign a cause to the differences you observe.

Parameters and Statistics

A **parameter** is some numerical (number) or nominal (name) characteristic of a population. An example is the mean IQ score of all first-grade pupils in the United States. A **statistic** is some numerical or nominal characteristic of a sample. The mean IQ score of 50 first-graders is a statistic and so is the observation that all are girls.

A parameter is constant; it does not change unless the population itself changes. The mean of a population is exactly one number. Unfortunately, the parameter often cannot be computed because the population is unmeasurable. So, a statistic is used as

---

[4] I didn't pull these populations out of thin air; they are all populations that researchers have studied. The studies of these populations will be described in this book.

an estimate of the parameter, although, as suggested before, a statistic tends to differ from one sample to another. If you have five samples from the same population, you will probably have five different sample means. In sum, parameters are constant; statistics are variable.

Variables

A **variable** is something that exists in more than one amount or in more than one form. Height and gender are both variables. The variable height is measured using a standard scale of feet and inches. For example, the notation 5′7″ is a numerical way to identify a group of persons who are similar in height. Of course, there are many other groups, each with an identifying number. Gender is also a variable. With gender, there are (usually) only two groups of people. Again, we may identify each group by assigning a number to it, usually 0 for females and 1 for males. All participants represented by 0 have the same gender. I will often refer to numbers like 5′7″ and 0 as *scores* or *test scores.* A score is simply the result of a measurement.

Quantitative Variables

Many of the variables you will work with in this text are **quantitative variables.** When a quantitative variable is measured, the scores tell you something about the amount or degree of the variable. At the very least, a larger score indicates more of the variable than a smaller score does.

Take a closer look at an example of a quantitative variable, IQ. IQ scores come in whole numbers like 94 and 109. However, it is reasonable to think that of two persons who scored 103, one just barely made it and the other scored almost 104. Picture the variable IQ as the straight line in **Figure 1.1.**

Figure 1.1 shows that a score of 103 is used for a range of possible values: the range from 102.5 to 103.5. The number 102.5 is the **lower limit** and 103.5 is the **upper limit** of the score 103. The idea is that IQ can be any value between 102.5 and 103.5 but that all IQ values in that range are rounded to 103.

In a similar way, a score of 12 seconds stands for all the values between 11.5 and 12.5 seconds; 11.5 seconds is the lower limit and 12.5 is the upper limit of a score of 12 seconds.

Sometimes scores are expressed in tenths, hundredths, or thousandths. With a good stopwatch, for example, time can be measured to the nearest hundredth of a second. Scores such as 6.00 and 12.05 result. These scores also have lower and upper limits, and the same reasoning holds: Each score represents a range of possible values. The score 6.00 is the middle of a range of values running from 5.995 to 6.005. The lower and upper limits of 12.05 are 12.045 and 12.055.

FIGURE 1.1 The lower and upper limits of IQ scores of 102, 103, and 104

In summary, to find the lower limit of a score (e.g., 6.1), subtract half the unit of measurement from the score (6.1 − 0.05 = 6.05). To find the upper limit, add half the unit of measurement to the score (6.1 + 0.05 = 6.15). Thus, for the score 6.1 seconds, the lower limit is 6.05 seconds and the upper limit is 6.15 seconds.

The basic idea is that the limits of a score have one more decimal place than the score itself, and the number in this last decimal place is 5. A good way to be sure you understand this idea is to draw a few number lines like the one in Figure 1.1, put on them scores such as 3.0, 3.8, and 5.95, and find the lower and upper limits of each score.

clue to the future

The lower and upper limits of a score will be important in Chapter 2, "Data: Organization, Graphs, and Central Tendency."

Qualitative Variables

Qualitative variables do not have the continuous nature that quantitative variables have. For example, gender is a qualitative variable. The two measurements on this variable, female and male, are different in a qualitative way. Using a 0 and a 1 instead of names doesn't change this fact. Another example of a qualitative variable is political affiliation, which has measurements such as Democrat, Republican, Independent, and Other.

Some qualitative variables have the characteristic of *order*. College year classification is a qualitative variable with ordered measurements of senior, junior, sophomore, and freshman. Another qualitative variable is military rank, which has measurements such as sergeant, corporal, and private.

PROBLEMS AND ANSWERS

At the beginning of this chapter, I urged you to take up the generative approach to learning statistics. Did you? For instance, did you draw those number lines and label them, as suggested several paragraphs back? If not, try it now for the score 0.05. What is its lower limit? If you didn't get 0.045, you need to reread that material. If you still have trouble, get some help from your instructor of from another student who understands the idea. Be active. Do you understand the word *nominal* that was used in the last section? How near is a dictionary? Make it your policy throughout this course never to let anything slip by you.

The generative approach means that you read actively. You will know you are doing this if you reread a paragraph that you did not understand the first time, or wrinkle your brow in concentration, or say to yourself, "Hmm, I really don't understand that; I'll ask my instructor about it tomorrow," or, best of all, nod to yourself, "Oh yes."

From time to time I will interrupt with a set of problems so that you can practice what you have just been reading about. Working these problems correctly is also evidence of active reading. You will find the answers at the end of the book in Appendix F. Here are some suggestions for *efficient* learning.

1. Buy yourself a notebook for statistics. Work all the problems for this course in it because I sometimes refer back to a problem you worked previously. When you make an error, don't erase it — circle it and work the problem correctly below. Seeing your error later serves as a reminder of what not to do on a test. If you find that *I* have made an error, write to me with a reminder of what not to do in the next edition.
2. Never, *never* look at an answer before you have worked the problem (or at least tried twice to work the problem).
3. When you come to a set of problems, work the first one and then immediately check your answer against the answer in the book. If you make an error, find out why you made it — faulty understanding, arithmetic error, or whatever.
4. Don't be satisfied with just doing the math. If a problem asks for an interpretation, write out your interpretation.
5. When you finish a chapter, go back over the problems immediately, reminding yourself of the various techniques you have learned.
6. Use any blank spaces near the end of the book for your special notes and insights.

Now, here is an opportunity to see how actively you have been reading.

PROBLEMS

1. For the quantitative variables that follow, give the lower and upper limits. For the qualitative variables, write "qualitative."
    a. 8, gallons of gasoline
    b. 5, a category of daffodil in a flower show
    c. 10.00, seconds required to run 100 meters
    d. 811.1, the identification of American poetry authors in the Dewey Decimal System
    e. 4.5, millions of dollars
    f. 2.95, tons of grain
2. Write a paragraph that gives the definitions of and the relationships among the terms *population, sample, parameter,* and *statistic.*
3. Classify each statement as being descriptive or inferential statistics.
    a. Based on a sample, the pollster predicted Demosthenes would be elected.
    b. The population of the American colonies in 1776 was 2 million.
    c. Demosthenes received 54.3 percent of the votes cast.
    d. A generative approach to learning works; the 29 students who generated summaries and inferences performed 20 percent better than those who just memorized (Wittrock, 1991).

SCALES OF MEASUREMENT

Numbers mean different things in different situations. Consider three answers that appear to be identical but are not:

What number were you wearing in the race?	"5"
What place did you finish in?	"5"
How many minutes did it take you to finish?	"5"

The three 5s all look the same. However, the three variables (identification number, finish place, and time) are quite different. Because of the differences, each 5 has a different interpretation.

To illustrate this, consider another person whose answers to the same three questions were 10, 10, and 10. If you take the first question by itself and know that the two people had scores of 5 and 10, what can you say? You can say that the first runner was different from the second, but *that is all.* (Think about this until you agree.) On the second question, with scores of 5 and 10, what can you say? You can say that the first runner was faster than the second and, of course, that they are different. Comparing the 5 and 10 on the third question, you can say that the first runner was twice as fast as the second runner (and, of course, was faster and different).

The point of this discussion is to draw the distinction between the *thing* you are interested in and the *number* that stands for the thing. Much of your experience with numbers has been with pure numbers or with the measurement of things such as time, length, and amount. "Four is twice as much as two" is true for the pure numbers themselves and for time, length, and amount, but it is not true for finish places in a race. Fourth place is not twice anything in relation to second place — not twice as *slow* or twice as *far behind* the second runner.

S. S. Stevens's 1946 article is probably the most widely referenced effort to focus attention on these distinctions. He identified four different *scales of measurement,* each of which carries a different set of information. Each scale uses numbers, but the information that can be inferred from the numbers differs. The four scales are nominal, ordinal, interval, and ratio.

The information in the **nominal scale** is quite simple. In the nominal scale, numbers are used simply as names and have no real quantitative value. Numerals on sports uniforms are an example. Thus, 45 is *different* from 32, but that is all you can say. The person represented by 45 is not "more than" the person represented by 32, and certainly it would be meaningless to calculate a mean from the two scores. Examples of nominal variables include psychological diagnoses, personality types, and political parties. The variable called psychological diagnoses consists of a set of categories. A person is classified into one of the categories, which may then be given a number. On a nominal scale, you could even reassign all the numbers and still maintain the original meaning, which is only that *things with different numbers are different.* Of course, all things that are alike have the same number.

A second kind of scale, the **ordinal scale,** has the characteristic of the nominal scale (different numbers mean different things) plus the characteristic of indicating *greater than* or *less than.* In the ordinal scale, the object with the number 3 has less or more of something than the object with the number 5. Finish places in a race are an example of an ordinal scale. The runners finish in rank order, with 1 assigned to the winner, 2 to the runner-up, and so on. Here, 1 means less time than 2. Judgments about anxiety, quality, and recovery often correspond to an ordinal scale. "Much improved," "improved," "no change," and "worse" are levels of an ordinal recovery variable. Ordinal scales are characterized by *rank order*.

The third kind of scale is the **interval scale,** which has the properties of both the nominal and ordinal scales plus the additional property that *intervals between the numbers are equal.* "Equal interval" means that the distance between the things represented by 2 and 3 is the same as the distance between the things represented by 3 and 4. Temperature is measured on an interval scale. The difference in temperature between 10°C and 20°C is the same as the difference between 40°C and 50°C. The Celsius thermometer, like all interval scales, has an arbitrary zero point. On the Celsius thermometer, this zero point is the freezing point of water at sea level. Zero degrees on this scale does not mean the complete absence of heat; it is simply a convenient starting point. With interval data, there is one restriction: You may not make simple ratio statements. You may not say that 100° is twice as hot as 50° or that a person with an IQ of 60 is half as intelligent as a person with an IQ of 120.[5]

The fourth kind of scale, the **ratio scale,** has all the characteristics of the nominal, ordinal, and interval scales plus one other: It has a *true zero point,* which indicates a complete absence of the thing measured. On a ratio scale, zero means "none." Height, weight, and time are measured with ratio scales. Zero height, zero weight, and zero time mean that no amount of these variables is present. With a true zero point, you can make ratio statements such as 16 kilograms is four times heavier than 4 kilograms.[6]

Knowing the distinctions among the four scales of measurement will help you in two tasks in this course. The kind of *descriptive statistics* you can compute from numbers depends, in part, on the scale of measurement the numbers represent. For example, it is senseless to compute a mean of numbers on a nominal scale. Calculating a mean Social Security number, a mean telephone number, or a mean psychological diagnosis is either a joke or evidence of misunderstanding numbers.

Understanding scales of measurement is also important in choosing the kind of *inferential statistic* that is appropriate for a set of data. If the dependent variable (see next section) is a nominal variable, then a chi square analysis is appropriate (Chapter 13). If the dependent variable is a set of ranks (ordinal data), then a nonparametric statistic is required (Chapter 14). Most of the inferential statistics you will study (Chapters 7–12) analyze interval or ratio scale data.

To end this section, I must tell you that it is sometimes difficult to classify some of the variables used in the social and behavioral sciences. Often they appear to fall between the ordinal scale and the interval scale. For example, a score may provide more information than simply rank, but equal intervals cannot be proved. Intelligence test

---

[5] Convert 100°C and 50°C to Fahrenheit ($F = 1.8C + 32$) and suddenly the "twice as much" relationship disappears.
[6] Convert 16 kilograms and 4 kilograms to pounds (1 kg = 2.2 pounds) and the "four times heavier" relationship is maintained.

scores and **Likert scale** scores are examples. In such cases, researchers generally treat the scores as if they were on an interval scale.

STATISTICS AND EXPERIMENTAL DESIGN

Here is a story that will help you distinguish between statistics (applying straight logic) and experimental design (observing what actually happens). This is an excerpt from a delightful book by E. B. White, *The Trumpet of the Swan* (1970, pp. 63–64).

> The fifth-graders were having a lesson in arithmetic, and their teacher, Miss Annie Snug, greeted Sam with a question.
>
> "Sam, if a man can walk three miles in one hour, how many miles can he walk in four hours?"
>
> "It would depend on how tired he got after the first hour," replied Sam. The other pupils roared. Miss Snug rapped for order.
>
> "Sam is quite right," she said. "I never looked at the problem that way before. I always supposed that man could walk twelve miles in four hours, but Sam may be right: that man may not feel so spunky after the first hour. He may drag his feet. He may slow up."
>
> Albert Bigelow raised his hand. "My father knew a man who tried to walk twelve miles, and he died of heart failure," said Albert.
>
> "Goodness!" said the teacher. "I suppose *that* could happen, too."
>
> "Anything can happen in four hours," said Sam. "A man might develop a blister on his heel. Or he might find some berries growing along the road and stop to pick them. That would slow him up even if he wasn't tired or didn't have a blister."
>
> "It would indeed," agreed the teacher. "Well, children, I think we have all learned a great deal about arithmetic this morning, thanks to Sam Beaver."
>
> Everyone had learned how careful you have to be when dealing with figures.

Statistics involves the manipulation of numbers and the conclusions based on those manipulations (Miss Snug). Experimental design deals with all the things that influence the numbers you get (Sam and Albert). This text could have been a "pure" statistics book, from which you would learn to analyze numbers without knowing where they came from or what they referred to. You would learn about statistics, but such a book would be dull, dull, dull. On the other hand, to describe procedures for collecting numbers is to teach experimental design — and this book is for a statistics course. My solution to this conflict is generally to side with Miss Snug but to include some aspects of experimental design throughout the book. Perhaps the information on design that you pick up will encourage you to try your hand at your own data gathering. Here's a start on experimental design.

EXPERIMENTAL DESIGN VARIABLES

The overall task of an experimenter is to discover relationships among variables. Variables are things that vary, and researchers have studied personality, health, gender, anger, caffeine, memory, beliefs, age, skill. (I'm sure you get the picture — almost anything can be a variable.)

Independent and Dependent Variables

A simple experiment has two major variables, the **independent variable** and the **dependent variable.** In the simplest experiment, the researcher *selects* two values of the independent variable for investigation. Values of the independent variable are usually called **levels** and sometimes called **treatments.**

The values of the dependent variable are found by *measuring* or *observing* the participants in the investigation. The purpose of an experiment is to determine whether values of the dependent variable change when the researcher switches from one level of the independent variable to a second level.

The basic idea is that the researcher finds or creates two groups of participants that are similar except for the independent variable. These individuals are measured on the dependent variable. The question is whether the data will allow the experimenter to claim that the values on the dependent variable *depend* on the level of the independent variable.

For the independent variable, the two groups might have been selected because they were already different — in age, gender, personality, and so forth. Alternatively, the experimenter might have produced the difference in the two groups by an experimental manipulation such as creating different amounts of anxiety or providing different levels of practice. The dependent variable might be scores on a personality test, the number of items remembered, or whether or not a participant offered help.

An example might help. Suppose for a moment that as a budding gourmet cook you want to improve your spaghetti sauce. One of your buddies suggests adding marjoram. To investigate, you serve spaghetti sauce at two different gatherings. For one group the sauce is spiced with marjoram; for the other it is not. At both gatherings you count the number of favorable comments about the spaghetti sauce. *Stop reading; identify the independent and the dependent variables.*[7]

The independent variable is marjoram. This variable has two levels, present and absent. The dependent variable is the number of favorable comments, which is a measure of the taste of the sauce.

Extraneous Variables

One of the pitfalls of experiments is that every situation has other variables besides the independent variable that might possibly be responsible for the changes in the dependent variable. These other variables are called **extraneous variables.** In the story of Miss Snug, Sam and Albert noted several extraneous variables that could influence the distance traveled.

Are there any extraneous variables in the spaghetti sauce example? Oh yes, there are many, and just one is enough to raise a suspicion about a conclusion that relates spaghetti sauce taste to marjoram. Extraneous variables include the amount and quality of the other ingredients in the sauce, the spaghetti itself, the "party moods" of the two groups, and how hungry everybody was. If any of these extraneous variables was

---

[7] One strategy is first to identify the dependent variable; you don't know values for the dependent variable until data are gathered. Next, identify the independent variable; you can tell what the values of the independent variable are just from the description of the design.

actually operating, it weakens the claim that a difference in the comments about the sauce is the result of the presence or absence of marjoram.

The simplest way to remove an extraneous variable is to be sure that all participants are equal on that variable. For example, you can ensure that the sauces are the same except for marjoram by mixing up one pot of ingredients, dividing it into two batches, adding marjoram to one batch but not the other, and then cooking. The "party moods" variable can be controlled (equalized) by conducting the taste test in a laboratory. Controlling extraneous variables is a complex topic that is covered in courses that focus on research or experiments.

In many experiments it is impossible or impractical to control all the extraneous variables. Sometimes researchers think they have controlled them all, only to find that they did not. The effect of an uncontrolled extraneous variable is to prevent a simple cause-and-effect conclusion. Even so, if the dependent variable changes when the independent variable changes, then something is going on. In this case, researchers can say that the two variables are *related* but that other variables may play a part, too.

At this point you can test your understanding by generating answers to the questions: What were the independent and dependent variables in the Zeigarnik experiment? How many levels of the independent variable were there?

For the question of how well Zeigarnik controlled extraneous variables, I can give a partial answer. In her experiment, each participant was tested at both levels of the independent variable. One advantage of this technique is that it naturally controls many extraneous variables. Thus, extraneous variables such age and motivation were exactly the same for tasks that were interrupted as for tasks that were not because the same people contributed scores to both levels.

The general question that experiments try to answer is: What effect does the independent variable have on the dependent variable? At various places in the following chapters, I will explain experiments and the statistical analyses of their results using the terms *independent* and *dependent* variables. These explanations will usually assume that all extraneous variables were controlled; that is, you may assume that the experimenter knew how to design the experiment so that changes in the dependent variable could be attributed correctly to changes in the independent variable. However, I will present a few investigations (like the spaghetti sauce example) that I hope you recognize as being so poorly designed that conclusions cannot be drawn about the relationship between the independent variable and dependent variable. Be alert.

Here's my summary of statistics and experimental design. Researchers suspect that there is a relationship between two variables. They design and complete an experiment; that is, they *choose* the levels of the independent variable (treatments), control the extraneous variables, and then *measure* the participants on the dependent variable. The measurements (data) are then analyzed using statistical procedures. Finally, the researcher tells a story that is consistent with the statistical analysis of the experiment.

STATISTICS AND PHILOSOPHY

The previous section directed your attention to the relationship between statistics and experimental design; this section will direct your thoughts to the place of statistics in the grand scheme of things.

The discipline of philosophy has the task of explaining the grand scheme of things and, as you know, many schemes have been advanced. For a scheme to be considered a grand one, it has to propose answers to questions of **epistemology** — that is, answers to questions about the nature of knowledge.

One of the big questions in epistemology is: How do we acquire knowledge? Answers such as *reason* and *experience* have gotten a lot of attention.[8] For those who emphasize the importance of reason, mathematics has been a continuing source of inspiration. In classical mathematics you start with a set of axioms that are assumed to be true. Theorems are thought up and are then proved by giving axioms as reasons. Once a theorem is proved, it can be used as a reason in a proof of other theorems.

Statistics has its foundations in mathematics. Statistics, therefore, is based on reason. As you go about the task of memorizing definitions of terms such as $\overline{X}$, σ, and ΣX, calculating their values from data, and telling the story of what they mean, know deep down that you are using a technique in logical reasoning. Logical reasoning is rationalism, which is one solution to the problem of epistemology. Experimental design is more complex; it includes experience (observation) as well as reasoning.

Let's move from this formal description of philosophy to a more informal one. A very common task of most human beings can be described as *trying to understand.* Statistics has helped many in their search for better understanding, and it is such people who have recommended (or demanded) that statistics be taught in school. A reasonable expectation is that you, too, will find statistics useful in your future efforts to understand and persuade.

Speaking of persuasion, you have probably heard it said, "You can prove anything with statistics." The message conveyed is that a conclusion based on statistics is suspect because statistical methods are unreliable. Well, it just isn't true that statistical methods are unreliable, but it is true that people can misuse statistics (just as any tool can be misused). One of the great advantages of studying statistics is that you get better at recognizing statistics that are used inappropriately.

STATISTICS: THEN AND NOW

Statistics began with counting. The beginning of counting, of course, was prehistory. The origin of the mean is almost as obscure. That statistic was in use by the early 1700s, but no one is credited with its discovery. Graphs, however, began when J. H. Lambert, a Swiss-German scientist and mathematician, and William Playfair, an English political economist, invented and improved graphs in the period 1765 to 1800 (Tufte, 1983).

In 1834 the Royal Statistical Society was formed in London. Just five years later, on November 27, 1839, at 15 Cornhill in Boston, a group of Americans founded the American Statistical Society. Less than 3 months later, for a reason that you can probably figure out, the group changed its name to the American Statistical Association.

According to Walker (1929), the first university course in statistics in the United States was probably "Social Science and Statistics," taught at Columbia University

---

[8] In philosophy, those who emphasize reason are rationalists and those who emphasize experience are empiricists.

in 1880. The professor was a political scientist, and the course was offered in the economics department. In 1887, at the University of Pennsylvania, separate courses in statistics were offered by the departments of psychology and economics. By 1891, Clark University, the University of Michigan, and Yale had been added to the list of schools that taught statistics, and anthropology had been added to the list of departments. Biology was added in 1899 (Harvard) and education in 1900 (Columbia).

You might be interested in when statistics was first taught at your school and in what department. College catalogs are probably your most accessible source of information.

Today statistics is a tool used by many people who are trying to understand or persuade or both. Kirk (1999) says that four groups of statistics users can be identified:

Category 1 — those who understand statistical presentations
Category 2 — those who select and apply statistical procedures to a particular problem
Category 3 — applied statisticians who help others use statistics on a particular problem
Category 4 — mathematical statisticians who develop new statistical techniques and discover new characteristics of old techniques

I hope that, as a result of your studies, you will be well along the path to becoming a category 2 statistician by the end of your statistics course.[9]

HELPFUL FEATURES OF THIS BOOK

At the beginning of this chapter, I mentioned the generative approach to learning. Your active participation is *necessary* if you are to learn statistics. I have tried to organize and write this book in a way that will encourage active participation. Here are some of the features you should find helpful.

Objectives

Each chapter begins with a list of skills the chapter is designed to help you acquire. Read this list of objectives first to find out what you are to learn to do. Then thumb through the chapter and read the headings. Next, study the chapter, working all the problems *as you come to them*. Finally, reread the objectives. Can you meet each one? If so, put a check mark beside that objective.

Problems and Answers

The problems in this text are in small groups within the chapter rather than clumped together at the end. This encourages you to read and work problems, read and work problems, which is a pattern of spaced practice. A spaced practice pattern leads to better performance than a massed practice pattern.

[9] For more information on statistics and careers in statistics, go to the American Statistical Association's website, *www.amstat.org.* Follow the link to Statistics and the Statistics Profession.

The problems come from a variety of disciplines. Because working problems and writing answers are *necessary* to learn statistics, I have used interesting problems and have sometimes written interesting answers for the problems; in some cases, I even put new information in the answers.

Many of the problems are conceptual questions that do not require any arithmetic. Think these through and *write* your answers. Being able to calculate a statistic is almost worthless if you cannot tell in English what it means. Writing shows you whether or not you really understand.

On several occasions, problems or data sets are used more than once, either later in the chapter or even in another chapter. If you do not work the problem when it is first presented, you are likely to be frustrated when it appears again. To alert you, I have put an asterisk (*) beside problems that will be used again.

At the end of many chapters, there is a comprehensive problem marked with a ✳. Working these problems requires knowing most of the material in the chapter. For most students, it is best to work all the problems, but be sure you can work those marked with a ✳.

Sometimes you may find a minor difference between your answer and mine (in the last decimal place, for example). This discrepancy will probably be the result of rounding errors and does not deserve further attention.

Clues to the Future

Often a concept is presented that will be used again in later chapters. These ideas are separated from the rest of the text in a box labeled "Clue to the Future." You have already seen two of these "Clues" in this chapter. Close attention to these concepts will pay dividends later in the course.

Error Detection

I have boxed in, at various points in the book, some ways to detect errors. Some of these "Error Detection" tips will also help your understanding of statistical concepts. Because many of these checks can be made early in a statistical problem, they can prevent the frustrating experience of getting an impossible answer when the error could have been caught in step 2.

Figure and Table References

Sometimes the words *Figure* and *Table* are in boldface print. This means that you should examine the figure or table at that point. Afterward, it will be easy for you to return to your place in the text — just find the word in boldface type.

Websites ▧

At the end of each chapter, you will see the logo above. The websites listed have material appropriate for the chapter.

Transition Pages

At five places in this book, there are major differences between the material you just finished and the material in the next section. At these points I have provided a "Transition Page" that describes the major differences.

Glossaries

This book has three glossaries. Two of them are in the appendixes, and one is printed on the inside front cover of the book.

1. *Glossary of words (Appendix D)*. This is a glossary of statistical words and phrases. When these items are introduced in the text, they appear in boldface type. Often the glossary gives you a more complete definition.
2. *Glossary of formulas (Appendix E)*. The formulas for all the statistical techniques used in the text are printed here, in alphabetical order according to the name of the technique.
3. *Glossary of symbols (inside front cover)*. Statistical symbols are defined here.

Computers, Calculators, and Statistics

Computers, calculators, and pencils and erasers are all tools used at one time or another by statisticians. Any or all of these devices may be part of the course you are taking. Regardless of the calculating aids that you use, however, your task remains the same: to read a problem, decide what statistical procedure to use, apply that procedure using the tools available to you, and write an interpretation of the results. In addition, you should be able to detect any gross errors by comparing the raw data, descriptive statistics, and inferential statistics.

Pencils, calculators, and computers represent, in ascending order, tools that are more and more error-free. Most people who do statistics use computers often. You may or may not be at this point. Remember, though, whether you are using computers or not, your task is principally to understand and describe.

CONCLUDING THOUGHTS FOR THIS INTRODUCTORY CHAPTER

Most students find that this book works well for them as a textbook in their statistics course. Those who keep this book after the course is over often find it a very useful *reference* book. In later courses and after leaving school, they find themselves looking up a definition or reviewing a procedure.[10] I hope that you not only learn from this book but also join those students who keep the book as part of their personal library.

This book is a fairly complete introduction to elementary statistics. There is more to the study of statistics — lots, lots more — but there is a limit to what you can do in one term. Some of you, however, will learn the material in this book and want to know more. If you are such a student, I have provided references in footnotes.

---

[10] This text's index is unusually extensive. If you make margin notes, they will help, too.

I also recommend that when you finish this course (but before any final examination), you study and work the problems in Chapter 15, the last chapter in the book. It is designed to be an overview/integrative chapter.

For me, studying and using statistics have been very satisfying and helpful. I hope you will find statistics worthwhile, too.

PROBLEMS

4. Name the four scales of measurement identified by S. S. Stevens.

5. Give the properties of each of the scales of measurement.

6. Identify which kind of scale is used in each of the following cases.

   a. Geologists have a "hardness scale" for identifying different rocks, called the Mohs' scale. The hardest rock (diamond) has a value of 10 and will scratch all others. The second hardest will scratch all but the diamond, and so on. Talc, with a value of 1, can be scratched by every other rock. (A fingernail, a truly handy field-test instrument, has a value between 2 and 3.)

   b. The volumes of three different cubes are 40, 64, and 65 cubic inches.

   c. Three different highways are identified by their numbers: 40, 64, and 65.

   d. Republicans, Democrats, Independents, and Others are identified on the voters' list with the numbers 1, 2, 3, and 4, respectively.

   e. The winner of the Miss America contest was Miss California; runners-up were Miss Ohio and Miss Pennsylvania.[11]

   f. The prices of the three items are $3.00, $10.00, and $12.00.

   g. She earned three degrees: B.A., M.S., and Ph.D.

7. Undergraduate students conducted the three studies that follow. For each study identify the dependent variable, the independent variable, the number of levels of the independent variable, and the names of the levels of the independent variable.

   a. Becca had students in a statistics class rate a resume, telling them that the person had applied for a position that included teaching statistics at their college. The students rated the resume on a scale of 1 (not qualified) to 10 (extremely qualified). All the students received identical resumes, except that the candidate's first name was Jane on half the resumes and John on the other half.

   b. Michael's participants filled out the Selfism scale, which measures narcissism. (Narcissism is neurotic self-love.) In addition, students were classified as first-born, second-born, and later-born.

   c. Johanna had participants read a description of a crime and "Mr. Anderson," the person convicted of the crime. For some participants, Mr. Anderson was described as a janitor. For others, he was described as a vice president of a large corporation. For still others, no occupation was given.

---

[11] Contest winners have come most frequently from these states, which have had six, six, and five winners, respectively.

After reading the description, participants recommended a jail sentence (in months) for Mr. Anderson.

8. Researchers who are now well known conducted the three classic studies that follow. For each study, identify the dependent variable, the independent variable, and the number and names of the levels of the independent variable. Complete items i and ii.

   a. Theodore Barber hypnotized 25 different people, giving each a series of suggestions. The suggestions included arm rigidity, hallucinations, color blindness, and enhanced memory. Barber counted the number of suggestions that the hypnotized participants complied with (the mean was 4.8). For another 25 people, he simply asked them to achieve the best score they could (but no hypnosis was used). This second group was given the same suggestions, and the number complied with was counted (the mean was 5.1). (See Barber, 1976.)

      i. Identify a nominal variable and a statistic.

      ii. In a sentence, describe what Barber's study shows.

   b. Elizabeth Loftus (1979) had participants view a film clip of a car accident. Afterward, some were asked How fast was the car going? and others were asked How fast was the car going when it passed the barn? (There was no barn in the film.) A week later, Loftus asked the participants Did you see a barn? If the barn had been mentioned earlier, 17 percent said yes; if it had not been mentioned, 3 percent said yes.

      i. Identify a population and a parameter.

      ii. In a sentence, describe what Loftus's study shows.

   c. Stanley Schachter and Larry Gross (1968) gathered data from obese male students for about an hour in the afternoon. At the end of this time, a clock on the wall was correct (5:30 P.M.) for 20 participants, slow (5:00 P.M.) for 20 others, and fast (6:00 P.M.) for 20 more. The actual time, 5:30, was the usual dinnertime for these students. While participants filled out a final questionnaire, Wheat Thins® were freely available. The weight of the crackers each student consumed was measured. The means were: 5:00 group — 20 grams; 5:30 group — 30 grams; 6:00 group — 40 grams.

      i. Identify a ratio scale variable.

      ii. In a sentence, describe what this study shows.

9. There are uncontrolled extraneous variables in the study described here. Name as many as you can. Begin by identifying the independent and dependent variables. An investigator concluded that statistics Textbook A was better than Textbook B, based on comparing two statistics classes. The section 1 class, which met MWF at 10:00 A.M., used Textbook A and was taught by Professor X. The section 2 statistics class, which met for 3 hours on Wednesday evening, used Textbook B and was taught by Professor Y. At the end of the term, all students took the same comprehensive test. The mean score for the Textbook A students was higher than the mean score for the Textbook B students.

10. Read the nine objectives at the beginning of this chapter. Responding to them will help you consolidate what you have learned.

WEBSITE REFERENCES

http://psychology.wadsworth.com

At the end of every chapter you will find this website logo and address. For each chapter, the site has a 12-item multiple-choice test, complete with answers and explanations. In addition to the multiple-choice test, some chapters refer you to another part of the website where workshops explain concepts in that chapter. I suggest that you bookmark this address as a favorite.

For a 12-item multiple-choice test on Chapter 1, go to the address above. Open "Choose Your Quiz" and select Spatz, 7e. Find the index of chapters and click on Chapter 1, *Introduction.*

chapter**two**

DATA: ORGANIZATION, GRAPHS, AND CENTRAL TENDENCY

OBJECTIVES FOR CHAPTER TWO

Study the text. Work the problems as you come to them. Afterward, you should be able to:

1. Arrange data into simple and grouped frequency distributions
2. Describe the characteristics of and explain the information provided in frequency polygons, histograms, bar graphs, and line graphs
3. Name some distributions by looking at their shapes
4. Find the mean, median, and mode of a simple frequency distribution
5. Determine whether a particular central tendency measure is a statistic or a parameter
6. Determine which central tendency measure is most appropriate for a set of data
7. Determine the direction of skew of a frequency distribution
8. Find the mean of a set of means

you have now invested some time, energy, and effort on the preliminaries of elementary statistics. Concepts have been introduced and you have an overview of the course. The chapters that follow immediately are about descriptive statistics. To get an idea of the specific topics, read this chapter title and the next three chapter titles thoughtfully.

To begin your study of descriptive statistics, consider a group of **raw scores.** Raw scores, or raw data, can be obtained in many ways. For example, if you administer a questionnaire to a group of college students, the scores from the questionnaire are raw scores. I will illustrate several of the concepts in this chapter by using the raw scores from a questionnaire administered to 100 representative college students.

TABLE 2.1 Scores of 100 college students on the Personal Control questionnaire

61	55	52	47	49	42	44	50	58	49
52	52	39	55	57	44	53	53	37	46
55	50	45	66	57	51	54	55	48	44
53	70	47	47	52	42	50	56	48	44
51	53	58	62	54	57	46	47	59	41
31	52	55	46	59	47	67	53	46	50
66	49	55	55	58	48	41	52	42	67
52	46	56	61	47	32	57	55	49	52
53	64	45	49	42	54	61	52	37	48
49	56	45	52	55	39	47	61	44	44

If you would like to engage fully in this chapter, take 5 minutes, complete the ten items in the following questionnaire, and calculate your score. (If you read the four chapter titles, I predict that you will answer the ten questions. If you didn't read the chapter titles, it's not too late to read·them. Participating is a step toward more engagement.) Score yourself now before you read further.

The Personal Control (PC) scale was developed over a period of 10 years by Delroy Paulhus at the University of British Columbia. It measures a person's perceived self-competence. (See Paulhus and Van Selst, 1990.) **Table 2.1** is an unorganized collection of the scores of 100 representative college students on the PC scale.

QUESTIONNAIRE

For each item, indicate the extent to which the statement applies to you by using the following scale:

1 = Disagree strongly   5 = Agree slightly
2 = Disagree   6 = Agree
3 = Disagree slightly   7 = Agree strongly
4 = Neither agree nor disagree

_____ 1. I can usually achieve what I want when I work hard for it.
_____ 2. Once I make plans I am almost certain to make them work.
_____ 3. I prefer games involving some luck over games of pure skill.
_____ 4. I can learn almost anything if I set my mind to it.
_____ 5. My major accomplishments are entirely due to my hard work and ability.
_____ 6. I usually do not set goals because I have a hard time following through on them.
_____ 7. Bad luck has sometimes prevented me from achieving things.
_____ 8. Almost anything is possible for me if I really want it.
_____ 9. Most of what will happen in my career is beyond my control.
_____ 10. I find it pointless to keep working on something that is too difficult for me.

Scoring. Add the numbers beside items 1, 2, 4, 5, and 8. On items 3, 6, 7, 9, and 10, reverse the point value (that is, 1 = 7, 2 = 6, 3 = 5, 4 = 4, 5 = 3, 6 = 2, 7 = 1). Add these five values. Finally, add your two sums together to get your final score.

In this chapter you will learn to (1) organize data like those of Table 2.1 into frequency distributions, (2) present the data graphically, and (3) calculate and explain measures of central tendency and determine whether they are statistics or parameters.

SIMPLE FREQUENCY DISTRIBUTIONS

Table 2.1, which is just a jumble of numbers, is not very interesting or informative. (My guess is that you glanced at it and went quickly on.) A much more informative presentation of the 100 scores is an arrangement called a **simple frequency distribution.** Table 2.2 is a simple frequency distribution — an ordered arrangement that shows the frequency of each score. Look at **Table 2.2.** (I would guess that you spent more time on Table 2.2 than on Table 2.1 and that you got more information from it.)

Look again at **Table 2.2.** The PC score column tells the name of the variable that is being measured. The generic name for any variable is X, which is the symbol used in formulas. The Frequency (f) column shows how frequently a score appeared. The tally marks are used when you construct a rough-draft version and are not usually included in the final form. N is the number of scores and is found by summing the num-

TABLE 2.2 Rough-draft form of a simple frequency distribution of Personal Control scores for a representative sample of 100 college students

PC score (X)	Tally marks	Frequency (f)	PC score (X)	Tally marks	Frequency (f)
70	/	1	50	////	4
69		0	49	++++ /	6
68		0	48	////	4
67	//	2	47	++++ //	7
66	//	2	46	++++	5
65		0	45	///	3
64	/	1	44	++++ /	6
63		0	43		0
62	/	1	42	////	4
61	////	4	41	//	2
60		0	40		0
59	//	2	39	//	2
58	///	3	38		0
57	////	4	37	//	2
56	///	3	36		0
55	++++ ////	9	35		0
54	///	3	34		0
53	++++ /	6	33		0
52	++++ ++++	10	32	/	1
51	//	2	31	/	1
					N = 100

TABLE 2.3 Final form of a simple frequency distribution of Personal Control scores for a representative sample of 100 college students

Personal Control score (X)	f	Personal Control score (X)	f
70	1	51	2
67	2	50	4
66	2	49	6
64	1	48	4
62	1	47	7
61	4	46	5
59	2	45	3
58	3	44	6
57	4	42	4
56	3	41	2
55	9	39	2
54	3	37	2
53	6	32	1
52	10	31	1
			$N = 100$

bers in the f column. You will have a lot of opportunities to construct simple frequency distributions, so here are the steps to follow:

1. Find the highest and lowest scores. In Table 2.1 the highest score is 70 and the lowest score is 31.
2. In column form, write in descending order all the numbers from 70 to 31. The heading for this column is the name of the variable being measured.
3. Start with the number in the upper left-hand corner of the unorganized scores (a score of 61 in Table 2.1), draw a line under it, and place a tally mark beside 61 in the frequency distribution.[1]
4. Continue this process with all the unorganized scores.
5. Count the number of tallies by each score and enter the count in a column headed f. Add the numbers in the f column. (If the sum is equal to N, you haven't left out any scores.)

A simple frequency distribution is a common way to present a set of data. It is common because it is useful, and it is useful because you can pick up valuable information with just a glance. For example, the highest and lowest scores are readily apparent. In addition, after some practice with frequency distributions, you can ascertain the general shape or form of the distribution and make an informed guess about central tendency values.

A formal presentation of a simple frequency distribution usually does not include the tally marks and sometimes does not include the zero-frequency scores. **Table 2.3**

[1] Underlining is much better than crossing out. Using the unorganized scores a second time (to check your work or for another analysis) confirms the choice of underlining.

TABLE 2.4 Rough draft form of a grouped frequency
distribution of Personal Control scores ($i = 3$)

PC scores (class interval)	Midpoint (X)	Tally marks	f
69–71	70	/	1
66–68	67	////	4
63–65	64	/	1
60–62	61	⊬⊬	5
57–59	58	⊬⊬ ////	9
54–56	55	⊬⊬ ⊬⊬ ⊬⊬	15
51–53	52	⊬⊬ ⊬⊬ ⊬⊬ ///	18
48–50	49	⊬⊬ ⊬⊬ ////	14
45–47	46	⊬⊬ ⊬⊬ ⊬⊬	15
42–44	43	⊬⊬ ⊬⊬	10
39–41	40	////	4
36–38	37	//	2
33–35	34		0
30–32	31	//	2
			$N = 100$

shows a formal presentation of the data in Table 2.1. A final form is used to present
data to others, such as colleagues, professors, supervisors, or editors.

GROUPED FREQUENCY DISTRIBUTIONS

Most researchers would condense Table 2.3 even more. The result is a **grouped frequency distribution. Table 2.4** is a rough-draft example of such a distribution.[2] Raw
data are usually condensed into a grouped frequency distribution when researchers
want to present the data as a graph or as a table.

In a grouped frequency distribution, scores are grouped into equal-sized ranges
called **class intervals.** In Table 2.4, the entire range of scores, from 70 to 31, is reduced
to 14 class intervals. Each interval covers three scores; the symbol i indicates the size
of the interval. In Table 2.4, $i = 3$. The midpoint of each interval represents all the
scores in that interval. For example, ten students had scores of 42, 43, or 44. The midpoint of the class interval 42–44 is 43.[3] The midpoint, 43, represents all ten scores.
There are no scores in the interval 33–35, but zero-frequency intervals are included in
formal grouped frequency distributions if they are within the range of the distribution.

Class intervals have lower and upper limits, much like simple scores obtained by
measuring a quantitative variable. A class interval of 42–44 has a lower limit of 41.5
and an upper limit of 44.5.

[2] Much like a formal simple frequency distribution, a formal grouped frequency distribution does not include
tally marks.
[3] When the scores are whole numbers, make i an odd number. With i odd, the midpoint of the class interval
is a whole number.

The only difference between grouped frequency distributions and simple frequency distributions is class intervals. The details of establishing class intervals are described in Appendix B. For problems in this chapter, I will give you the class intervals to use.

PROBLEMS

*1. (This is the first of many problems with an asterisk. See the footnote.) The following numbers are the heights in inches of two groups of Americans in their 20s. Choose the group that interests you more and organize the 50 numbers into a simple frequency distribution, using the rough-draft form. For the group you choose, your result will be fairly representative of the whole population of 20–29-year-olds (*Statistical Abstract of the United States: 1999*, 1999).

Women						Men				
64	67	63	65	66		72	65	72	68	70
66	66	62	65	65		69	73	71	69	67
60	72	64	61	65		77	67	72	73	68
69	64	64	66	61		68	64	72	69	69
65	67	65	62	69		70	71	71	70	75
62	65	63	64	61		72	68	62	68	74
66	64	59	63	65		66	70	72	66	71
67	63	66	64	65		69	71	68	73	69
66	67	62	62	63		71	69	69	65	76
64	70	64	67	65		71	73	65	70	70

*2. (The frequency distribution you will construct for this problem is a little different. The "scores" are simply names and thus nominal data.). A political science student covered a voting precinct on election day morning, recording the yard signs (by initial) for the five candidates. She planned to find the relationship between yard signs and actual votes. The five candidates were Attila (A), Bolivar (B), Gandhi (G), Lenin (L), and Mao (M). Construct an appropriate frequency distribution from her observations.

```
G  A  M  M  M  G  G  L  A  G  B  A  A  G  G  B  L  M  M
A  G  G  B  G  L  M  A  A  L  M  G  G  M  G  L  G  A  A
B  L  G  G  A  G  A  M  L  M  G  B  A  G  L  G  M  A
```

*3. You may have heard or read that the normal body temperature (oral) is 98.6°F. The numbers that follow are temperature readings from healthy adults, aged 18–40. Arrange the data into a grouped frequency distribution, using 99.3–99.5 as the highest class interval and 96.3–96.5 as the lowest. (Based on Mackowiak, Wasserman, and Levine, 1992)

98.1	97.5	97.8	96.4	96.9	98.9	99.5	98.6	98.2	98.3
97.9	98.0	97.2	99.1	98.4	98.5	97.4	98.0	97.9	98.3
98.8	99.5	98.7	97.9	97.7	99.2	98.0	98.2	98.3	97.0
99.4	98.9	97.9	97.4	97.8	98.6	98.7	97.9	98.4	98.8

---

* Problems with asterisks are multiple-use problems; they will turn up again. If you do all the problems in a notebook, you will have a handy reference when the problem is referred to again.

*4. An experimenter read 60 related statements to a class. He then asked the students to indicate which of the next 20 statements were among the first 60. Due to the relationships among the concepts in the sentences, many seemed familiar but, in fact, none of the 20 had been read before. The following scores indicate the number (out of 20) that each student had "heard earlier." (See Bransford and Franks, 1971.) Arrange the scores into a simple frequency distribution. Based on this description and your frequency distribution, write a sentence of interpretation.

14	11	10	8	12	13	11	10	16	11
11	9	9	7	14	12	9	10	11	6
13	8	11	11	9	8	13	16	10	11
9	9	8	12	11	10	9	7	10	

GRAPHIC PRESENTATION OF DATA

You have no doubt heard the saying *A picture is worth a thousand words.* When it comes to numbers, a not-yet-well-known saying is *A graph is better than a thousand numbers.* Actually, as long as I am rewriting sayings, I would also like to say *The more numbers you have, the more valuable graphics are.*

Pictures that present statistical data are called graphics. The most common graphic is a graph composed of a horizontal axis (variously called the baseline, X axis, or **abscissa**) and a vertical axis (called the Y axis or **ordinate**). To the right and upward are both positive directions; to the left and downward are both negative directions. Look at **Figure 2.1**

Four different graphs will be described in this section. Frequency polygons and histograms are used to present the frequency of *quantitative* data. Bar graphs are used to present the frequency of *qualitative* data. (Think about the PC scores in Table 2.4 and the yard-sign data in problem 2. Which data set is quantitative and which is qualitative?) Line graphs are the fourth kind of graph described in this section. Line graphs show the relationship between two different variables.

The Frequency Polygon

A **frequency polygon** is used to graph quantitative data. The PC scores in Table 2.4 are quantitative data, so a frequency polygon is appropriate. The result is **Figure 2.2.** An explanation of Figure 2.2 will serve as your introduction to constructing polygons.

Each point of the frequency polygon represents two numbers: the class midpoint directly below it on the X axis and the frequency of that class directly across from it on the Y axis. By looking at the data points in Figure 2.2, you can see that four students are represented by the midpoint 67, zero students by 34, and so on.

The name of the variable being graphed goes on the X axis (Personal Control scores). The X axis is marked off in equal intervals, with each tick mark indicating the midpoint of a class interval. Low scores are on the left. The lowest class interval midpoint, 28, and the highest, 73, have zero frequencies. Frequency polygons are closed at both ends.

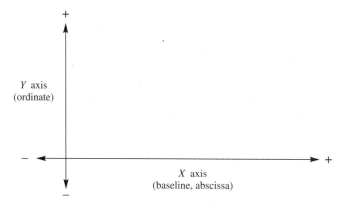

FIGURE 2.1 The elements of a graph

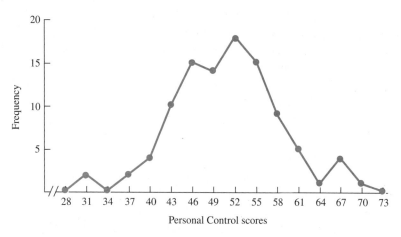

FIGURE 2.2 Frequency polygon of Personal Control scores of 100 college students

To construct graphs with pleasing proportions, it is often necessary to leave out uninformative parts. For example, no students had PC scores from 7 to 29. Rather than including the zero-frequency intervals from 7 to 25, these are replaced with a slash mark on the *X* axis. Excellent graphs "look right." The conventional advice is that graphs look better if the height of the figure is 60 percent to 75 percent of its width. Achieving this proportion usually requires a little planning as you design your graph.

The Histogram

A **histogram** is another graphing technique that is appropriate for quantitative data. **Figure 2.3** shows the PC data of Table 2.4 graphed as a histogram. A histogram is constructed by raising bars from the *X* axis to the appropriate frequencies. The narrow lines or spaces that separate the bars intersect the *X* axis at the lower and upper limits of the class intervals.

FIGURE 2.3 Histogram of Personal Control scores of 100 college students

Here are some considerations for deciding whether to use a frequency polygon or a histogram. On the one hand, if you are displaying two overlapping distributions on the same axes, a frequency polygon is less cluttered than a histogram. On the other hand, it is easier to read frequencies from a histogram, and histograms are the better choice when you are presenting discrete data. (*Discrete data* are quantitative data that do not have intermediate values; the number of children in a family is an example.)

The Bar Graph

A **bar graph** is used to present the frequencies of the categories of a qualitative variable. A conventional bar graph looks exactly like a histogram except for the wider spaces between the bars. The space is usually a signal that a qualitative variable is being graphed. Conventionally, bar graphs also have the name of the variable being graphed on the X axis and frequency on the Y axis.

If an ordinal scale variable is being graphed, the order of the values on the X axis follows the order of the variable. If, however, the variable is a nominal scale variable, then *any* order on the X axis is permissible. Alphabetizing might be best. Other considerations may lead to some other order.

Figure 2.4 is a bar graph of the six most common liberal arts majors among college graduates in 1995–96. The majors are listed on the X axis; the number of graduates is on the Y axis. I ordered the majors from most to fewest graduates, but other orders would be more appropriate in other circumstances. For example, moving the bar for English majors to the first position would be better for a presentation about English majors.

The Line Graph

Perhaps the graph most frequently used by scientists is the **line graph.** A line graph is a picture of the relationship between *two* variables. A point on a line graph represents the value on the Y variable that goes with the corresponding value on the X variable.

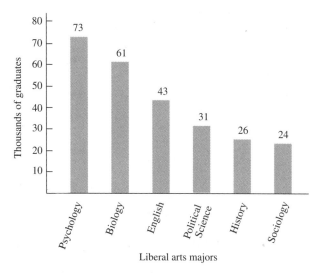

FIGURE 2.4 The six most common liberal arts majors among college graduates for the academic year 1995–96

The point might represent two scores by one person or the mean score of a group of people.

Line graphs are very popular with those who convey information. Peden and Hausmann (2000) found that two-thirds of the graphs in psychology textbooks were line graphs. Likewise, Boehner and Howe (1996) report that line graphs accounted for two-thirds of the data graphs in a random selection of psychology-related journal articles.

Line graphs are used in many different fields to convey the relationship between two variables. If you have a textbook for a course that investigates the relationships among quantitative variables, you might thumb through it and note that many relationships are presented with line graphs. If you would like to learn more about *constructing* graphs (including line graphs), I recommend Henry (1994) and Wallgren et al. (1996).

Figure 2.5 is an example of a line graph. It shows the *serial position effect*. If you practice an ordered list of words until you can repeat it without error, you'll find that you make the most errors on the words just past the middle. The serial position effect applies to any sequence of material that is learned in order. Knowing about the serial position effect, you should devote extra study time to items in the middle.

Other Graphics

There are many other graphics besides those described in this text. Graphics are used increasingly by both those who want to understand and those who want to persuade. One particular champion of graphics is Edward Tufte, a professor at Yale University. He has published three books that demonstrate and celebrate the power that graphics have. The first book, *The Visual Display of Quantitative Information* (1983), gives many examples of this power, including a reprint of "(perhaps) the best statistical

FIGURE 2.5 The serial position effect

graphic ever drawn."[4] Tufte says that, regardless of your field, when you construct a quality graphic, it improves your understanding of the phenomenon you are interested in. So, if you find yourself somewhere on that path between confusion and understanding, you should try to construct a graphic. In the heartfelt words of a sophomore engineering student I know, "Graphs sure do help."

DESCRIBING DISTRIBUTIONS

There are three ways to describe the form or shape of a distribution: words, pictures, and mathematics. In this section you will use the first two ways. The only mathematical method I will cover will be in Chapter 13 on chi square.

Bell-Shaped Distributions

Look at Table 2.4 and its graphic form, Figure 2.2. Notice that the highest frequencies are in the middle of the distribution. The same thing is true in problem 1 of this chapter. These distributions are referred to as *bell-shaped.*

There is a special case of a bell-shaped distribution that you will soon come to know very well. It is called the **normal distribution,** or **normal curve. Figure 2.6** is an illustration of a normal distribution.

Skewed Distributions

In some distributions the scores with the highest frequency are not found in the middle of the graph but near one end. These are called **skewed distributions.** The word *skew* is similar to the word *skewer,* the name of the cooking implement used

[4] I'll give you just a hint about this graphic. A French engineer drew it over a century ago to convey a disastrous military campaign by Napoleon against Russia. Wainer (1984) was also impressed with this graphic, nominating it the "World's Champion Graph."

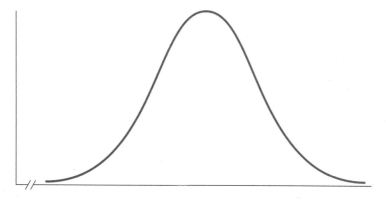

FIGURE 2.6 A normal distribution

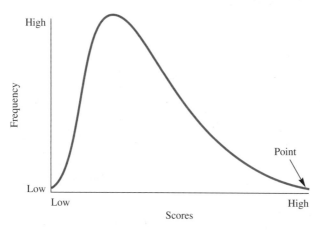

FIGURE 2.7 Example of a positively skewed distribution. Scores with the highest frequencies are concentrated among the low scores.

in making shish kebab. A skewer is thick at one end and pointed at the other (not symmetrical).

    Figure 2.7 is a *positively skewed* curve; the thin point is toward the high scores, and the most frequent scores are the low ones. Note that the point is toward the right — the positive direction.

    Figure 2.8 is a *negatively skewed* curve; the thin point (skinny end) is toward the low scores, and the most frequent scores are the high ones. Note that the point is toward the left — the negative direction.

Other Shapes

Curves of frequency distributions can, of course, take other shapes. Some shapes are so common they have names. Some examples are presented in Figure 2.9.

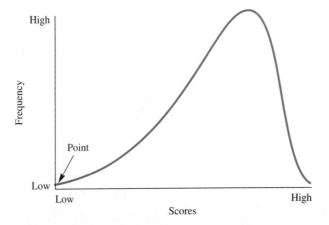

FIGURE 2.8 Example of a negatively skewed distribution. Scores with the highest frequencies are concentrated among the high scores.

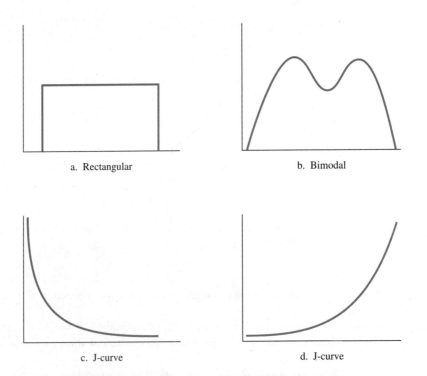

a. Rectangular
b. Bimodal
c. J-curve
d. J-curve

FIGURE 2.9 Examples of rectangular, bimodal, and J-curves

Rectangular distribution. **Figure 2.9a,** a **rectangular distribution** (also called a *uniform distribution*), occurs when the frequency of each value on the X axis is the same. You will see this distribution again in Chapter 6.

Bimodal distribution. A graph of a **bimodal distribution** has two humps in it. In some cases the humps are the same height, as they are in **Figure 2.9b.** To understand "bimodal," you need to know that the mode of a distribution is the score that occurs most frequently. For both distributions and graphs, if two high-frequency scores are separated by scores with lower frequencies, the name *bimodal* is appropriate.

J-curves. Finally, **Figures 2.9c** and **2.9d** are **J-curves.** If you look at Figure 2.9d and use your imagination, the name will make sense to you. The mode of a J-curve is very near one end of the distribution, which results in a very skewed graph.

Any set of measurements of a phenomenon can be arranged into a distribution; scientists and others find them most helpful. Sometimes measurements are presented in a frequency distribution table and sometimes in a graph (and sometimes both ways). In the first chapter I said that by learning statistics you could become more persuasive. Frequency distributions and graphs (but especially graphs) will be valuable to you as you face many audiences over the years.

PROBLEMS

5. Answer the following questions for Figure 2.2.
   a. What is the meaning of the number 55 on the X axis?
   b. What is the meaning of the number 15 on the Y axis?
   c. How many students had scores in the class interval 48–50?

6. For the height data in problem 1 you constructed a frequency distribution. Graph it, being careful to include the zero-frequency scores.

7. The average hourly workweek varies from country to country for industrial workers. In 1995 the figures were: Australia, 38.7; Canada, 38.5; Denmark, 31.5; Japan, 37.8; Korea, 49.2; United Kingdom, 42.1; United States, 41.6. What kind of graph is appropriate for these data? Design a graph and draw it.

*8. Decide whether the following distributions should be graphed as frequency polygons or as bar graphs. Graph both distributions.

a. Class interval	f		**b.** Class interval	f
48–52	1		54–56	3
43–47	1		51–53	7
38–42	2		48–50	15
33–37	4		45–47	14
28–32	5		42–44	11
23–27	7		39–41	8
18–22	10		36–38	7
13–17	12		33–35	4
8–12	6		30–32	5
3–7	2		27–29	2
			24–26	0
			21–23	0
			18–20	1

*9. Determine the direction of the skew for the two curves in problem 8 by examining the curves or the frequency distributions (or both).

10. Look at the frequency distribution that you constructed for the yard-sign observations in problem 2. Which kind of graph should be used to display these data? Graph the distribution.

11. Describe how a line graph is different from a frequency polygon.

12. In the Northern Hemisphere, temperature generally decreases as latitude increases (equator = 0°; North Pole = 90°). The five U.S. cities listed here are all within one degree of 35°N latitude. Construct a line graph relating average temperature for January and elevation. Write a sentence of interpretation.

City	Elevation (feet)	Temperature (°F)
Albuquerque, NM	5000	34
Amarillo, TX	3700	35
Flagstaff, AZ	6900	29
Little Rock, AR	350	39
Oklahoma City, OK	1200	36

13. Without looking at Figures 2.6 and 2.9, sketch a normal distribution, a rectangular distribution, a bimodal distribution, and two J-curves.

14. Is the point of a positively skewed distribution directed toward the right or left?

15. For each frequency distribution listed, tell whether it is positively skewed, negatively skewed, or approximately symmetrical (bell-shaped or rectangular).
    a. Age of all people alive today
    b. Age in months of all first-graders
    c. Number of children in families
    d. Wages in a large manufacturing plant
    e. Age at death of everyone who died last year
    f. Shoe size

MEASURES OF CENTRAL TENDENCY

So far in this chapter you have learned some ways to present scores using frequency distributions and graphs. In this section I will discuss another way to summarize data — calculation of three measures of **central tendency.** Measures of central tendency (sometimes called measures of *central value*) give you one score or measure that represents, or is typical of, an entire group of scores.

Recall from Chapter 1 that statistics come from samples and parameters come from populations. For all three measures of central tendency, the formula for calculating the statistic is the same as that for calculating the parameter.[5] The interpreta-

---

[5] This is not true for the standard deviation, which you will study in Chapter 3.

tions, however, are different. I will use the *mean,* which you are already somewhat familiar with, as an example.

A sample mean, $\overline{X}$ (pronounced "mean" or "ex-bar"), is only one of many possible sample means from a population. Suppose you have a sample from a population and you calculate the sample mean. Because other samples from that same population would probably produce somewhat different $\overline{X}$'s a degree of uncertainty goes with the one $\overline{X}$ you have. However, if you were able to measure an entire population, its mean, μ (pronounced "mew"), would be the only one and would carry no uncertainty with it. Clearly, the parameter μ is more desirable than the statistic $\overline{X}$, but, unfortunately, it is often impossible to measure an entire population. Fortunately, $\overline{X}$ is the best estimator of μ.

The difference in interpretation, then, is that a statistic carries some uncertainty with it; a parameter does not.

The Mean

Here's a very simple example. Suppose a college freshman arrives at school in the fall with a promise of a monthly allowance for spending money. Sure enough, on the first of each month, there is money to spend. However, 3 months into the school term, our student discovers a recurring problem: too much month left at the end of the money.

In pondering this problem, our student realizes that money escapes from his pocket at the Student Center. So, for a 2-week period, he keeps a careful accounting of every cent spent at the center (soft drinks, snacks, video games, coffee, and so forth). His data are presented in **Table 2.5.** You already know how to compute the

TABLE 2.5 Amounts of money spent during a 2-week period at the Student Center

Day	Money spent
1	$ 2.98
2	0.50
3	1.25
4	0.00
5	3.25
6	1.15
7	0.00
8	0.00
9	6.78
10	2.12
11	0.00
12	0.00
13	8.50
14	3.85

$\Sigma = \$30.38$

Your estimate of the mean _____

mean of these numbers, but before you do, eyeball the data and then write down your *estimate* of the mean in the space provided.[6] The formula for the **mean** is

$$\overline{X} = \frac{\Sigma X}{N}$$

where $\overline{X}$ = the mean[7]
$\qquad \Sigma$ = an instruction to add (Σ is uppercase Greek sigma)
$\qquad X$ = a score or observation; ΣX means to add all the X's
$\qquad N$ = number of scores or observations
For the data in Table 2.5,

$$\overline{X} = \frac{\Sigma X}{N} = \frac{\$30.38}{14} = \$2.17$$

These data are for a 2-week period, but our freshman is interested in his expenditures for at least 1 month and, more likely, for many months. Thus, the result is a sample mean and the $\overline{X}$ symbol is appropriate. The amount, $2.17, is an estimate of the amount that our friend spends at the Student Center each day.

Now we come to an important part of any statistical analysis, which is to answer the question So what? Calculating numbers or drawing graphs is a part of almost every statistical problem, but unless you can tell the story of what the numbers and pictures mean, you won't find statistics worthwhile.

The first use you can make of Table 2.5 is to estimate the student's monthly Student Center expenses. This is easy to do. Thirty days times $2.17 is $65.10. Now, let's suppose our student decides that this $65.10 is an important part of the "monthly money problem." The student has three apparent options. The first is to get more money. The second is to spend less at the Student Center. The third is to justify leaving things as they are. For this third option, our student might perform an economic analysis to determine what he gets in return for his $60+ a month. His list might be pretty impressive: an improved social life, occasional information about classes and courses and professors, a borrowed book that was just super, hundreds of calories, and more.

The point of all this is that part of the attack on the student's money problem involved calculating a mean. However, an answer of $2.17 doesn't have much meaning by itself. Interpretations and comparisons are called for.

Characteristics of the mean. Two characteristics of the mean are important for you to know. Both characteristics will come up again later.

First, if the mean of a distribution is subtracted from each score in that distribution and the differences are added, the sum will be zero; that is, $\Sigma(X - \overline{X}) = 0$. The statistic, $X - \overline{X}$, is called a deviation. You might pick a few numbers to play with to *demonstrate* to yourself that $\Sigma(X - \overline{X}) = 0$. (Numbers 1, 2, 3, 4, and 5 are easy to work with.) If you know the rules that govern algebraic operations of summation (Σ) notation, you can *prove* the relationship $\Sigma(X - \overline{X}) = 0$. (See Kirk, 1999, p. 90.)

---

[6] To estimate, begin with the first number as your estimate. If the second number is smaller, reduce your estimate. Add the third number into your consideration, and so on. My estimate was "less than $2.00."
[7] Many scientific journals use a capital M to represent the mean.

TABLE 2.6 Data of Table 2.5 arranged in descending order

X
$8.50
6.78
3.85
3.25   } 7 scores
2.98
2.12
1.25

Median = $1.20

1.15
0.50
0.00
0.00   } 7 scores
0.00
0.00
0.00

Second, the mean is the point about which the sum of the squared deviations is minimized. If the mean is subtracted from each score, each deviation is squared, and all squared deviations are added together, then the resulting sum will be smaller than if any number other than the mean had been used; that is, $\Sigma(X - \overline{X})^2$ is a minimum. This "least squares" characteristic of the mean will come up in Chapters 3 and 5. Again, you can demonstrate this relationship by playing with some numbers.

The Median

The **median** is the *point* that divides a distribution of scores into two parts that are equal in size. To find the median of the Student Center expense data, arrange the daily expenditures from highest to lowest, as shown in **Table 2.6.** Because there are 14 scores, the halfway point, or median, will have seven scores above it and seven scores below it. The seventh score from the bottom is $1.15. The seventh score from the top is $1.25. The median, then, is halfway between these two scores, or $1.20.[8] Remember, the median is a hypothetical *point* in the distribution; it may or may not be an actual score.

What is the interpretation of a median of $1.20? The simplest is that on half the days our student spends less than $1.20 in the Student Center and on the other half he spends more.

What if there had been an odd number of days in the sample? Suppose the student chose to sample half a month, or 15 days. Then the median would be the eighth score. The eighth score has seven scores above and seven below. For example, if an ad-

---

[8] The halfway point between two numbers is the mean of the two numbers. Thus, ($1.15 + $1.25)/2 = $1.20.

ditional day was included, during which $3.12 was spent, the median would be $1.25. If the additional day's expenditure was zero, the median would be $1.15.

The reasoning you just went through can be expressed in formula form. The formula for finding the location of the median is:

$$\text{Median location} = \frac{N + 1}{2}$$

The location may be at an actual score (as in the second example) or at a position between two scores (the first example).

The Mode

The third central tendency statistic is the **mode.** As mentioned earlier, the mode is the most frequently occurring score — the score with the highest frequency. For the Student Center expense data, the mode is $0.00. **Table 2.6** shows the mode most clearly. The zero amount occurred five times, and all other amounts occurred only once.

When a mode is given, it is often accompanied by the percentage of times it occurred. You will probably agree that "The mode was $0.00, which occurred on 36 percent of the days" is more informative than "The mode was $0.00."

FINDING CENTRAL TENDENCY OF SIMPLE FREQUENCY DISTRIBUTIONS

Mean

Table 2.7 is an expanded version of Table 2.3, the frequency distribution of Personal Control (PC) scores. The steps for finding the mean from a simple frequency distribution follow, but first (looking only at the data and not at the summary statistics at the bottom) *estimate* the mean of the PC scores in the space at the bottom of Table 2.7.[9]

The first step in calculating the mean from a simple frequency distribution is to multiply each score in the X column by its corresponding f value, so that all the people who make a particular score are included. Next, sum the fX values and divide the total by N. (N is the sum of the f values.) The result is the mean.[10] In terms of a formula,

$$\mu \text{ or } \overline{X} = \frac{\Sigma fX}{N}$$

For the data in Table 2.7,

$$\mu \text{ or } \overline{X} = \frac{\Sigma fX}{N} = \frac{5100}{100} = 51.00$$

How did your estimate compare to 51.00?

To answer the question of whether this 51.00 is $\overline{X}$ or μ, you need more information. Here's the question to ask: Is there any interest in a group larger than these 100?

---

[9] You can estimate the mean in a very rough fashion by selecting the most frequent score (the mode) or the score in the middle of the list of scores.

[10] When you work with whole numbers, calculating a mean to two decimal places is usually sufficient.

TABLE 2.7 Calculating the mean of the simple frequency distribution of the Personal Control scores

Personal Control score (X)	f	fX	Personal Control score (X)	f	fX
70	1	70	51	2	102
67	2	134	50	4	200
66	2	132	49	6	294
64	1	64	48	4	192
62	1	62	47	7	329
61	4	244	46	5	230
59	2	118	45	3	135
58	3	174	44	6	264
57	4	228	42	4	168
56	3	168	41	2	82
55	9	495	39	2	78
54	3	162	37	2	74
53	6	318	32	1	32
52	10	520	31	1	31
				$\Sigma = 100$	5100

Your estimate of the mean: _____

If the answer is *no*, the 100 scores are a population and $\mu = 51.00$. If the answer is *yes*, the 100 scores are a sample and $\overline{X} = 51.00$.

Median

The formula for finding the location of the median that you used earlier works for a simple frequency distribution, too.

$$\text{Median location} = \frac{N + 1}{2}$$

Thus, for the scores in Table 2.7,

$$\text{Median location} = \frac{N + 1}{2} = \frac{100 + 1}{2} = 50.5$$

To find the 50.5th position, begin adding the frequencies in Table 2.7 from the bottom $(1 + 1 + 2 + 2 + \cdots)$. The total is 49 by the time you include the score of 51. Including 52 would make the total 59 — more than you need. So the median is somewhere among those ten scores of 52. The median is 52.

Suppose you start the quest for the median at the top of the distribution rather than at the bottom. Again, the location of the median is at the 50.5th position in the distribution. To get to 50.5, add the frequencies from the top $(1 + 2 + 2 + \cdots)$. The sum of the frequencies of the scores from 70 down to 53 is 41. The next score, 52, had a frequency of 10. Thus, the 50.5th position is among the scores of 52. The median is 52.

error detection

Calculating the median by starting from the top of the distribution produces the same result as calculating the median by starting from the bottom.

Mode

It is easy to find the mode from a simple frequency distribution. In **Table 2.7,** the score with the highest frequency, 10, is the mode. So 52 is the mode.

PROBLEMS

16. Find the median for the following sets of scores.
   a. 2, 5, 15, 3, 9
   b. 9, 13, 16, 20, 12, 11
   c. 8, 11, 11, 8, 11, 8
17. Which of the following distributions is bimodal?
   a. 10, 12, 9, 11, 14, 9, 16, 9, 13, 20
   b. 21, 17, 6, 19, 23, 19, 12, 19, 16, 7
   c. 14, 18, 16, 28, 14, 14, 17, 18, 18, 6
*18. For the data on memory for statements (problem 4), find the mean, median, and mode. Write a general sentence of interpretation.
*19. Refer back to problem 2, the political scientist's yard-sign data.
   a. Decide which measure of central tendency is appropriate and find it.
   b. Is this central tendency measure a statistic or a parameter?
   c. Write a sentence of interpretation.
20. Refer back to problem 1. Find the mean, median, and mode of the heights of both groups of Americans in their 20s. Work from the answers to problem 1 in Appendix F.
21. Are the data in problem 3 a sample or a population?
22. For distribution a, find the median starting from the bottom. For distribution b, find the median starting from the top. Distribution c can probably be solved by inspection.

a. X	f
15	4
14	3
13	5
12	4
11	2
10	1

b. 4, 0, −1, 2, 1, 0, 3, 1, −2, −2, −1, 2, 1, 3, 0, −2, 1, 2, 0, 1

c. 28, 27, 26, 21, 18, 10

23. Write the two mathematical characteristics of the mean that were covered in this section.

error detection

Eyeballing data is a valuable way to avoid big mistakes. Begin a problem by making a quick estimate of the answers. If the answers you calculate differ from your estimate, wisdom dictates that you reconcile the difference. You have either overlooked something when eyeballing or made a mistake in your computations.

THE MEAN, MEDIAN, AND MODE COMPARED

A common question is: Which measure of central tendency should I use? The general answer is, given a choice, use the mean. Sometimes, however, the data give you no choice. Here are three considerations that limit your choice.

Scale of Measurement

A mean is appropriate for ratio or interval scale data, but not for ordinal or nominal distributions. A median is appropriate for ratio, interval, and ordinal scale data, but not for nominal data. The mode is appropriate for any of the four scales of measurement.

You have already thought through part of this issue in working problem 19. In it you found that the yard-sign names (very literally, a nominal variable) could be characterized with a mode, but it would be impossible to try to add up the names and divide by N or to find the median of the names.

For an ordinal scale such as class standing in college, either a median or a mode makes sense. The median would probably be part of the way through the classification sophomore, and the mode would be freshman.

Skewed Distributions

Even if you have interval or ratio data, using a mean is not recommended if the distribution is severely skewed. The following story demonstrates why the mean gives an erroneous impression for severely skewed distributions. (Note also that the story describes a population of data, not a sample; see if you can tell why.)

The developer of Swampy Acres Retirement Homesites is attempting, with a computer-selected mailing list, to sell the lots in a southern "paradise" to northern buyers. The marks express concern that flooding might occur. The developer reassures them by explaining that the average elevation of the lots is 78.5 feet and that the water level has never exceeded 25 feet in that area. On the average, the developer has told the truth, but this average truth is misleading. Look at the actual lay of the land in **Figure 2.10** and examine the frequency distribution in **Table 2.8.**

The mean elevation, as the developer said, is 78.5 feet; however, only 20 lots, all on a hill, are out of the flood zone. The other 80 lots are, on the average, under water. Using the mean is misleading. The median is a much better measure of central tendency here because it is unaffected by the few extreme lots on the hill. The median elevation is 12.5 feet, well below the high-water mark. (Because our interest is in only this one development, the lot elevations constitute a population of data. There is no interest in generalizing from these data to some larger group.)

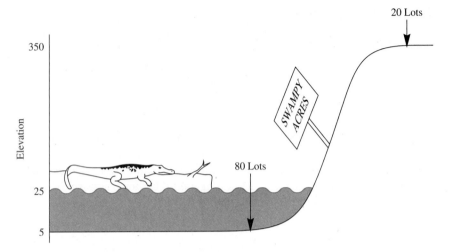

FIGURE 2.10 Elevation of Swampy Acres

TABLE 2.8 Frequency distribution of lot elevations at Swampy Acres

Elevation, in feet (X)	Number of lots (f)	fX
348–352	20	7000
13–17	30	450
8–12	30	300
3–7	20	100
	$\Sigma = 100$	7850

$$\mu = \frac{\Sigma fX}{N} = \frac{7850}{100} = 78.5 \text{ feet}$$

A number of books with engaging titles tell how statistics can be chosen to convey one particular idea rather than another. The grandparent of such books is probably Huff's *How to Lie with Statistics* (1954). Other books are Campbell's *Flaws and Fallacies in Statistical Thinking* (1974) and Runyon's *How Numbers Lie* (1981). All three books are delightfully written.

Open-Ended Categories

There is another instance that requires a median, even though you have interval or ratio data that are symmetrically distributed. This is when the class interval with the highest (or lowest) scores is not limited. In such a case, you do not have a midpoint, and therefore you *cannot* compute a mean. For example, age data are sometimes reported with the highest category as "75 and over"—the mean cannot be computed. Thus, when one or both of the extreme class intervals is not limited, the median is the appropriate measure of central tendency.

DETERMINING SKEWNESS FROM THE MEAN AND MEDIAN

Examining the relationship of the mean to the median is a way of determining the direction of skew of a distribution without drawing a graph. When the *mean is smaller* than the median, there is some amount of negative skew. When the *mean is larger* than the median, there is positive skew. The greater the difference between the mean and the median, the greater the skew. **Figure 2.11** illustrates positively and negatively skewed distributions.

The mean depends on the *size* of every number. The more extreme the number, the greater its influence. With the median, however, an extreme score and a not-extreme score contribute the same weight — just one value. As you might suspect, there are mathematical ways to measure skewness. [See Kirk (1999), p. 129, for one way.]

In summary, choose the mean if it is appropriate. (To follow this advice you must be able to recognize data for which the mean is *not* appropriate.)

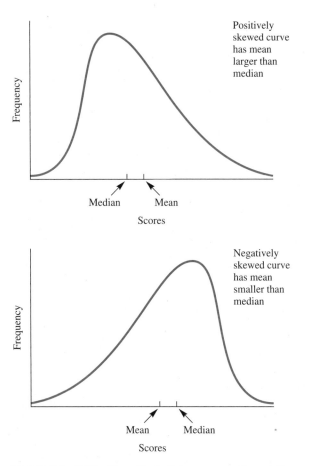

FIGURE 2.11 The effect of skewness on the relative position of the mean and median

THE MEAN OF A SET OF MEANS ($\overline{\overline{X}}$)

Sometimes means are available from several samples taken from the same population. If these means are combined, the mean of the set of means $(\overline{\overline{X}})$ is the best estimate of the population parameter, μ. If every sample has the same N, you can compute $\overline{\overline{X}}$ by adding the means and dividing the sum by the number of means. If the sample means are based on N's of different sizes, however, you cannot use this procedure. Here is a story that illustrates the right way and the wrong way to calculate the mean of a set of means.

At my undergraduate college, a student with a cumulative grade point average (GPA) of 3.25 in the middle of the junior year was eligible to enter a program to "graduate with honors." [In those days (1960) usually fewer than 10 percent of a class had GPAs greater than 3.25.] Discovering this rule after my sophomore year, I decided to figure out if I had a chance to qualify. Calculating a cumulative GPA seemed easy enough to do: Four semesters had produced GPAs of 3.41, 3.63, 3.37, and 2.16. Given another GPA of 3.80, the sum of the five semesters would be 16.37, and dividing by 5 gave an average of 3.27, well above the required 3.25.

Graduating with honors seemed like a great ending for college, so I embarked on a goal-oriented semester — a GPA of 3.80 (a B in German and an A in everything else). And, at the end of the semester I had accomplished the goal. Unfortunately, "graduating with honors" was not to be.

There was a flaw in my method of calculating my cumulative GPA. My method assumed that all of the semesters were equal in weight, that they had all been based on the same number of credit hours. My calculations based on this assumption are shown on the left side of **Table 2.9.**

Unfortunately, all five semesters were not the same; the semester with the GPA of 2.16 was based on 19 hours, rather than the usual 16 or so.[11] Thus, that semester should have been weighted more heavily than semesters with fewer hours.

The right side of Table 2.9 shows the correct way to calculate cumulative GPA. Each semester's GPA is multiplied by its number of credit hours. These products are summed and that total is divided by the sum of the hours. As you can see from the numbers on the right, the actual cumulative GPA was 3.24, not high enough to qualify for the honors program.

More generally, to find the mean of a set of means, multiply each separate mean by its N, add these products together, and divide the total by the sum of the N's. Thus, means of 2.0, 3.0, and 4.0 based on N's of 6, 3, and 2 produce $\overline{\overline{X}} = 2.64$. Do you agree?

clue to the future

The distributions you have worked with in this chapter are **empirical distributions** based on scores that were observed. This chapter and the next three are about these empirical frequency distributions. Starting with Chapter 6 and throughout the rest of the book, you will also use **theoretical distributions** — distributions based on mathematical formulas and logic rather than on actual observations.

---

[11] That semester was more educational than a GPA of 2.16 would indicate. I read a great deal of American literature that spring, but unfortunately, I was not registered for any courses in American literature!

TABLE 2.9 Two methods of calculating the mean of a set of means from five semesters' GPAs; the method on the left is correct only if all semesters have the same number of credit hours

Flawed method	Correct method		
Semester GPAs	Semester GPAs	Credit hours	GPAs × hours
3.41	3.41	17	58
3.63	3.63	16	58
3.37	3.37	19	64
2.16	2.16	19	41
3.80	3.80	16	61
$\Sigma = 16.37$		$\Sigma = 87$	$\Sigma = 282$
$\overline{\overline{X}} = \dfrac{16.37}{5} = 3.27$	$\overline{\overline{X}} = \dfrac{282}{87} = 3.24$		

ESTIMATING ANSWERS

You may have noticed during this chapter that I have been (subtly?) working in some advice: *As your first step, estimate an answer.* Here's why I think you should start by estimating an answer: (1) Taking a few seconds to estimate keeps you from plunging into the numbers before you fully understand the problem. (2) An estimate is especially helpful when you have a calculator or a computer doing the bulk of your number crunching. Although these wonderful machines don't make errors, the people who enter the numbers or give instructions do, occasionally. Your at-the-beginning estimate will help you catch these mistakes. (3) Noticing that an estimate differs from a calculated value gives you a chance to correct an error before anyone else sees it.

It can be exciting to look at an answer, whether it is on paper, a calculator display, or a computer screen, and say, "That *can't* be right!" and then to find out that, sure enough, the displayed answer is wrong. If you develop your ability to estimate answers, I promise that you will sometimes experience this excitement.

A MOMENT TO REFLECT

At this point in your statistics course, I have a question for you: How are you doing? I know that questions like this are better asked by a personable human than by an impersonal textbook. Nevertheless, please give your answer. I hope you can say O.K. or perhaps something better. If your answer isn't O.K. or better, now is the time to make changes. Figure out how you can make things better for yourself and then implement your plan.

PROBLEMS

24. For each of the following situations, tell which measure of central tendency is appropriate and why.

    a. As part of a study on prestige, an investigator sat on a corner in a high-income residential area and classified passing automobiles according to color: black, gray, white, silver, green, and other.

    b. In a study of helping behavior, an investigator pretended to have locked himself out of his car. Passersby who stopped were classified on a scale of 1 to 5 as (1) very helpful, (2) helpful, (3) slightly helpful, (4) neutral, and (5) discourteous.

    c. In a study of household income in a city, the following income categories were established: $0–$10,000, $10,001–$20,000, $20,001–$30,000, $30,001–$40,000, $40,001–$50,000, $50,001–$60,000, and $60,001 and more.

    d. In a study of *per capita* income in a city, the following income categories were established: $0–$10,000, $10,001–$20,000, $20,001–$30,000, $30,001–$40,000, $40,001–$50,000, $50,001–$60,000, $60,001–$70,000, $70,001–$80,000, $80,001–$90,000, and $90,001–$100,000.

    e. First admissions to a state mental hospital for 5 years were classified by disorder: schizophrenic, delusional, anxiety, dissociative, and other.

    f. A teacher gave her class an arithmetic test; most of the children scored in the range 70–79. A few scores were above this, and a few were below.

25. A senior psychology major performed the same experiment on three groups and obtained means of 74, 69, and 75. The groups consisted of 12, 31, and 17 participants, respectively. What is the overall mean for all participants?

26. In problem 9, you determined the direction of the skew by looking at the distributions in problem 8. Verify those judgments now by calculating and comparing the means and medians of each distribution.

27. A 3-year veteran of the local baseball team was calculating his lifetime batting average. The first year he played for half the season and batted .350 (28 hits in 80 at-bats). The second year he had about twice the number of at-bats and his average was .325. The third year, although he played even more regularly, he was in a slump and batted only .275. Adding the three season averages and dividing the total by 3, he found his lifetime batting average to be .317. Is this correct? Explain.

*28. Several statistics are required for this problem. You will use most of the answers again. Problem 3 presented data on oral body temperature. Use those data to construct a simple frequency distribution and find the mean, median, and mode. Using the grouped frequency distribution you created for problem 3, construct a graph. Write a sentence or two about central tendency and a sentence or two about the form of the distribution.

29. Read and respond to the eight objectives at the beginning of the chapter. Generating responses will help you consolidate what you have learned.

☀ Problem marked with this symbol incorporates several techniques from the chapter.

⬛ WEBSITE REFERENCES

http://psychology.wadsworth.com
Open "Choose Your Quiz" and select Spatz, 7e. Find the index of chapters, and click on Chapter 2, *Data: Organization, Graphs, and Central Tendency.*

Open "Choose Your Area," scroll down, and select "Research Methods and Statistics." Click on "Statistics" in the graph. Click on "Student Activities." Click on the Web Page button for "Central Tendency and Variability." This workshop covers materials found in Chapters 2 and 3.

chapter**three**

VARIABILITY

Study the text. Work the problems as you come to them. Afterward, you should be able to:

1. Explain the concept of variability
2. Find the range of a distribution
3. Find the 25th and 75th percentile scores of a distribution
4. Find and interpret the interquartile range of a distribution
5. Distinguish among these concepts: the standard deviation of a population, the standard deviation of a sample used to describe the sample, and the standard deviation of a sample used to estimate the population standard deviation
6. Calculate a standard deviation from grouped or ungrouped data and interpret its meaning
7. Calculate the variance of a distribution

In Chapter 2 you began the study of descriptive statistics. *Frequency distributions* and *graphs* presented all the data. The *mean, median,* and *mode* each gave you one number that captures some central tendency characteristic of the distribution. Chapter 3 will continue with more descriptive statistics. This time you will get one number that expresses the **variability** of a distribution.

The value of knowing about variability is illustrated by the story of two brothers who went water skiing on Christmas Day (the temperature was about 35°F). On the *average* each skier finished his turn 5 feet from the shoreline (where one may step off the ski into only 1 foot of very cold water). This bland average of the two actual stopping places, however, does not convey the excitement of the day.

The first brother, determined to avoid the cold water, held onto the towrope too long. Scraped and bruised, he finally stopped rolling at a spot 35 feet up on the rocky shore. The second brother, determined not to share the same fate, released the towrope too soon. Although he swam the 45 feet to shore very quickly, his lips were very blue. No, the average stopping place of 5 feet from the shoreline doesn't capture the excitement of the day. To get the full story, you should ask about variability.

Here are some other situations in which knowing the average leaves you without enough information.

1. You are an elementary school teacher who has been assigned a class of fifth-graders whose mean IQ is 115, well above the IQ of 100 that is the average for the general population. Because children with IQs of 115 can handle more complex, abstract material, you plan many sophisticated projects for the year.

But consider variability: *If the variability of the IQs in the class is small, your projects will probably succeed. If the variability is large, the projects will be too complicated for some, but for others, even these projects will not be challenging enough.*

2. Suppose that your temperature, taken with a thermometer under your tongue, is 97.5°F. You begin to worry. This is below even the average of 98.2°F that you learned in the previous chapter (based on Mackowiak, Wasserman, and Levine, 1992).

But consider variability: *There is variability around that mean of 98.2°F. Is 97.5°F below the mean by just a little or by a lot? Measuring variability is necessary if you are to answer this question.*

3. Having graduated from college, you are considering two offers of employment, one in sales and the other in management. The pay is about the same for both. After checking out the statistics for salespersons and managers at the library, you find that those who have been working for 5 years in each type of job also have similar averages. You conclude that the pay for the two occupations is equal.

But consider variability: *Pay is more variable for those in sales than for those in management. Some in sales make much more than the average and some make much less, whereas the pay of those in management is clustered together. Your reaction to this difference in variability might help you choose.*

This chapter is about statistics and parameters that measure the variability of a distribution. The *range* is the first measure I will describe; the second is the *interquartile range*. The third, the *standard deviation,* is the most important. Most of this chapter will be about the standard deviation. A fourth way to measure variability is with the *variance.*

THE RANGE

The **range** of a quantitative variable is the highest score minus the lowest score. The range of the Personal Control scores you worked with in the previous chapter (Table 2.2) was 39. The highest score was 70; the lowest was 31. The range is therefore 70 − 31 = 39.

In Chapter 2 you worked with oral body temperatures. (See problem 3 and its answer.) The highest temperature was 99.5°F; the lowest was 96.4°F. The range is 3.1°F.

In manufacturing, the range is used in some quality-control procedures. From a small sample of whatever is being manufactured, inspectors may calculate a range and compare it to an expected figure. A large range means there is too much variability in the process and adjustments are called for (see Chapter 6 of Berger and Hart, 1986).

Because the range depends on only the two *extreme* scores, you can probably imagine two very different distributions that have the same mean and the same range. (Go ahead, imagine them.) If you are able to dream up such distributions, you realize that having the same mean and range does not guarantee that the shapes (form) of two frequency polygons are identical.

The range is a quickly calculated, easy-to-understand statistic that is just fine in some situations. Other measures work better in other cases.

INTERQUARTILE RANGE

The next measure of variability, the **interquartile range,** tells the range of scores that make up the middle 50 percent of the distribution. It is an important element of *box-plots,* which you will study in Chapter 4. (A boxplot is a graphic that conveys the data of a distribution and some of its statistical characteristics with *one picture.*)

To find the interquartile range, you need the 25th percentile score and the 75th percentile score. You may already be familiar with the concept of percentile scores. The 10th percentile score has 10 percent of the distribution below it; it is near the bottom. The 95th percentile score is near the top; 95 percent of the distribution has scores that are smaller. The 50th percentile divides the distribution into equal halves.[1]

Determining percentiles is something you have already practiced. In Chapter 2 you calculated several medians. The median is the point that divides a distribution into equal halves. Thus, the median is the 50th percentile. Calculating the 25th and 75th percentile scores involves the same kind of thinking you used in finding the median.

The 25th percentile score is the one that has 25 percent of the scores below it. Look at **Table 3.1,** which shows a frequency distribution of 40 scores. To find the 25th percentile score, multiply 0.25 times N ($0.25 \times 40 = 10$). When $N = 40$, the 10th score *up* from the *bottom* is the 25th percentile score. You can see in Table 3.1 that there are 5 scores of 23 or lower. The 10th score is among the 7 scores of 24. Thus, the 25th percentile is a score of 24.

The 75th percent score has 75 percent of the scores below it. The easiest way to find it is to work from the top, using the same multiplication procedure, 0.25 times N ($0.25 \times 40 = 10$). The 75th percentile score is the 10th score *down* from the *top* of the distribution. In Table 3.1 there are 9 scores of 29 or higher. The 10th score is among the 4 scores of 28, so the 75th percentile score is 28.

The interquartile range is simply the 75th percentile minus the 25th percentile:

IQR = 75th percentile − 25th percentile

where IQR = interquartile range. Thus, for the distribution in Table 3.1, IQR = 28 − 24 = 4. The interpretation is that the middle 50 percent of the scores have values from 24 to 28.

---

[1] For more details on percentile scores, see Runyon, Coleman, and Pittenger (2000, pp. 59–61).

TABLE 3.1 Finding the 25th and 75th percentiles

Score	f	
31	1	⎫
30	3	⎬ 9 scores
29	5	⎭
28	4	
27	5	
26	4	
25	6	
24	7	
23	3	⎫
22	2	⎬ 5 scores
	N = 40	

[handwritten: 2⟍8⟋ − 22 = 9 ; −22 ; 9 ; IQR = 4]

PROBLEMS

1. Find the range for the two distributions.
   a. 17, 5, 1, 1
   b. 0.45, 0.30, 0.30
*2. Find the interquartile range of the Personal Control scores in Table 2.7. Write a sentence of interpretation. *[handwritten: Range = 39   I = 9]*
*3. Find the interquartile range of the heights of both groups of those 20–29-year-old Americans. Use the frequency distributions you constructed for problem 1 in Chapter 2. (These distributions may be in your notebook of answers; they are also in Appendix F.)

THE STANDARD DEVIATION

When it comes to describing the variability of a distribution, the most widely used statistic is the **standard deviation.** It is popular because it is useful. Once you understand standard deviations, you can express quantitatively the difference between the two distributions you imagined previously in the section on the range. (You did do the imagining, didn't you?)

In addition, the standard deviation (or its cousin statistics) will turn up in every chapter after this one. Of course, in order to understand, you'll have to read all the material, work the problems, and do the interpretations. (What you get for all this work is a lifetime of understanding the most popular yardstick of variability.)

---

* Problems with an asterisk are multiple-use problems. Your answer and your understanding of an asterisked problem will help when the data set is presented again.

TABLE 3.2   Symbols, purposes, and descriptions of three standard deviations

Symbol	Purpose	Description
σ	Measure a population's variability	Lowercase Greek *sigma* is the symbol for the standard deviation of a population. σ is a parameter and is used to *describe* variability when a population of data is available.
$\hat{s}$	Estimate a population's variability	Lowercase $\hat{s}$ (s-hat) is an *estimate* of σ, in the same way that $\overline{X}$ is an estimate of μ. Parameters are rarely known, so often the best you can do is draw a sample from the population and use statistics such as $\overline{X}$ and $\hat{s}$ as estimates of parameters such as μ and σ. The variability statistic you will use most often in this book is $\hat{s}$.
S	Measure a sample's variability	There are occasions when you want to *describe* the variability of a sample but have no interest in estimating σ. In such cases, a capital S is the statistic to use.

Your principal task in this section is to learn the distinctions among three different standard deviations. Which standard deviation you use in a particular situation will be determined by your purpose. Table 3.2 lists symbols, purposes, and descriptions. It will be worth your time to study **Table 3.2** thoroughly now.

Distinguishing among these three standard deviations is sometimes a problem for beginning students. Be alert in situations where a standard deviation is used. With each situation, you will acquire more understanding. I will first discuss the calculation of σ and S (which are computed in the same way except for the use of μ and $\overline{X}$, respectively, in the formulas) and then deal with $\hat{s}$.

THE STANDARD DEVIATION AS A DESCRIPTIVE INDEX OF VARIABILITY

Both σ and S are used to *describe* the variability of a set of data. The symbol σ is a parameter of a population; S is a statistic of a sample. Both are calculated with similar formulas. I will show you two ways to arrange the arithmetic for these formulas, the deviation-score formula and the raw-score formula.

You can best learn what a standard deviation is actually measuring by working through the steps of the *deviation-score* formula. *The raw-score* formula, however, is quicker (and sometimes more accurate, depending on rounding). Algebraically, the two formulas are identical.

My suggestion is that you learn both methods. By the time you get to the end of the chapter, you should have both the understanding that the deviation-score formula produces and the efficiency that the raw-score formula gives. After that, you may want to use a calculator or computer to do the arithmetic for you.

Deviation Scores

A **deviation score** is a raw score minus the mean, either $X - \overline{X}$ or $X - \mu$. It is simply the difference between a score in the distribution and the mean of that distribution. Raw scores that are greater than the mean have positive deviation scores, raw scores that are less than the mean have negative deviation scores, and raw scores that are equal to the mean have a deviation score of zero.

Table 3.3 provides an illustration of how to compute deviation scores for a small population of data. In Table 3.3, I first computed the mean, 8, and then subtracted it from each score. The result is deviation scores, which appear in the right-hand column.

A deviation score tells you the number of points that a particular score deviates from, or differs from, the mean. In Table 3.3, the $X - \mu$ value for Selene, 6, tells you that she scored six points above the mean. Caitlin, 0, scored at the mean, and Ian, −5, scored five points below the mean.

error detection

Notice that the sum of the deviation scores is always zero. Add the deviation scores; if the sum is not zero, you have made an error. You have studied this concept before. In the last chapter you learned that $\Sigma(X - \overline{X}) = 0$.

Computing σ and S Using Deviation Scores

The deviation-score formula for computing the standard deviation as a descriptive index is

$$\sigma = \sqrt{\frac{\Sigma(X - \mu)^2}{N}} \quad \text{or} \quad S = \sqrt{\frac{\Sigma(X - \overline{X})^2}{N}}$$

where σ = standard deviation of a population
S = standard deviation of a sample
N = number of scores (same as the number of deviations)

T A B L E 3.3 The computation of deviation scores from raw scores

Name	Score	$X - \mu$	Deviation score
Selene	14	14 − 8	6
Kevin	10	10 − 8	2
Caitlin	8	8 − 8	0
Logan	5	5 − 8	−3
Ian	3	3 − 8	−5
	$\Sigma X = 40$		$\Sigma(X - \overline{X}) = 0$

$$\mu = \frac{\Sigma X}{N} = \frac{40}{5} = 8$$

$\Sigma(X - \overline{X})^2$ is shorthand notation that tells you to find the deviation score for each raw score, square the deviation score, and then add all the squares together. This sequence illustrates one of the rules for working with summation notation: Perform the operations within the parentheses first.

How can a standard deviation add to your understanding of a phenomenon? Let's take the sales of Girl Scout cookies. Table 3.4 presents some imaginary (but true-to-life) data on boxes of cookies sold by six Girl Scouts. I'll use these data to illustrate the calculation of the standard deviation. Let's define these six sales reports as a population; thus, the standard deviation is σ.

The numbers of boxes sold are listed in the X column of **Table 3.4.** The rest of the arithmetic needed for calculating σ is also given. To compute σ by the deviation-score formula, first find the mean.[2] Obtain a deviation score for each raw score by subtracting the mean from the raw score. Square each deviation score and sum the squares to obtain $\Sigma(X - \mu)^2$. Divide $\Sigma(X - \mu)^2$ by N. Take the square root. The result is $\sigma = 8.66$ boxes.

Now, what does $\sigma = 8.66$ boxes mean? How does it help your understanding? The 8.66 boxes is a measure of the variability in the number of boxes the six Girl Scouts sold. If σ was zero, you would know that each girl sold the same number of boxes. The closer σ is to zero, the more confidence you can have in predicting that the number of boxes any girl sold was equal to the mean of the group. Conversely, the further σ is from zero, the less confidence you have. With $\sigma = 8.66$ and $\mu = 10$, you know that the girls varied a great deal in cookie sales.

I realize that, so far, my interpretation of the standard deviation has not given you any more information than the range does. (A range of zero means that each girl sold the same number of boxes, and so forth.) The range, however, has no additional information to give you — the standard deviation does, as you will see.

Now look again at **Table 3.4** and the formula for σ. Notice what is happening. The mean is subtracted from each score. This difference, whether positive or negative, is squared and these squared differences are added together. This sum is divided by N and the square root is found. Every score in the distribution contributes to the final answer.

Notice the contribution made by a score like 28, which is far from the mean; its contribution to $\Sigma(X - \mu)^2$ is large. This makes sense because the standard deviation is a yardstick of variability. Scores that are far from the mean cause the standard deviation to be greater. Take a moment to think through the contribution to the standard deviation made by a score near the mean.[3]

error detection

All standard deviations are positive numbers. If you find yourself trying to take the square root of a negative number, you've made an error.

---

[2] For convenience I arranged the data so the mean is an integer. If you are confronted with decimals, it usually works to carry three decimals in the problem and then round the final answer to two decimal places.

[3] If you play with a formula, you will become more comfortable with it and understand it better. Make up a small set of numbers and calculate a standard deviation. Change one of the numbers, or add a number, or leave out a number. See what happens. Besides teaching yourself about standard deviations, you may learn more efficient ways to use your calculator.

TABLE 3.4 Using the deviation-score formula to compute σ for the cookie sales of a population of six Girl Scouts

Boxes of cookies X		Deviation scores $X - \mu$	$(X - \mu)^2$
28	$-10 =$	18	324
11	-10	1	1
10	-10	0	0
5	-10	-5	25
4		-6	36
2		-8	64

sum of $\Sigma X = 60$ $\Sigma(X - \mu) = 0$ $\Sigma(X - \mu)^2 = 450$

mean $\mu = \dfrac{\Sigma X}{N} = \dfrac{60}{6} = 10$

of

$\sigma = \sqrt{\dfrac{\Sigma(X - \mu)^2}{N}} = \sqrt{\dfrac{450}{6}} = \sqrt{75} = 8.66 \text{ boxes}$

PROBLEMS

4. Give the symbol and purpose of each of the three standard deviations.
5. Using the deviation-score method, compute S for the three distributions.
   *a. 7, 6, 5, 2
   b. 14, 11, 10, 8, 8
   *c. 107, 106, 105, 102
6. Compare the standard deviation of problem 5a with that of problem 5c. What conclusion can you draw about the effect of the size of the numbers on the standard deviation?
7. Does the size of the numbers in a distribution have any effect on the mean?
8. The temperatures listed are averages for March, June, September, and December. Calculate the mean and standard deviation for each city. Summarize your results in a sentence.

San Francisco, CA	54°F	59°F	62°F	52°F
Albuquerque, NM	46°F	75°F	70°F	36°F

9. No computation is needed for this one; just eyeball the data. Determine whether the first distribution is more variable than the second or whether the variabilities in the two distributions are equal.
   a. 1, 2, 4, 1, 3   and   9, 7, 3, 1, 0
   b. 9, 10, 12, 11   and   4, 5, 7, 6
   c. 1, 3, 9, 6, 7   and   14, 15, 14, 13, 14
   d. 114, 113, 114, 112, 113   and   14, 13, 14, 12, 13
   e. 8, 4, 6, 3, 5   and   4, 5, 7, 6, 15

Computing σ and S by the Raw-Score Method

The deviation-score formula helps you understand what is actually going on when you calculate a standard deviation. It is the formula to use until you do understand. Unfortunately, except in textbook examples, the deviation-score formula almost always has you working with decimal values. The raw-score formula, which is algebraically equivalent, involves far fewer decimals. It also produces answers more quickly, especially if you are working with a calculator.

The raw-score formula is

$$\sigma \text{ or } S = \sqrt{\frac{\Sigma X^2 - \frac{(\Sigma X)^2}{N}}{N}}$$

where
ΣX^2 = sum of the squared scores
$(\Sigma X)^2$ = square of the sum of the raw scores
N = number of scores

Although the raw-score formula may appear more forbidding, it is actually easier to use than the deviation-score formula because you don't have to compute deviation scores. The numbers you work with will be larger, but your calculator won't mind.

Table 3.5 shows the steps for calculating σ or S by the raw-score formula. The data are for boxes of cookies sold by the six Girl Scouts. You can put the arithmetic of Table 3.5 into words: Square the sum of the values in the X column and divide the total by N. Subtract this quotient from the sum of the values in the X^2 column. Divide this difference by N and find the square root. The result is σ or S. Notice that the value of σ in Table 3.5 is the same as the one you calculated in Table 3.4. In this case, the mean is an integer, so the deviation scores introduced no rounding errors.[4]

ΣX^2 and $(\Sigma X)^2$: Did you notice the difference in these two terms when you were working with the data in Table 3.5? If so, congratulations. You cannot calculate a standard deviation correctly unless you understand the difference. Reexamine Table 3.5 if you aren't sure of the difference between ΣX^2 and $(\Sigma X)^2$. Be alert for these two sums in the problems that are coming up.

error detection

The range is usually two to five times greater than the standard deviation when $N =$ 100 or less. The range (which can be calculated quickly) will tell you if you made any large errors in calculating a standard deviation.

---

[4] Some textbooks and statisticians prefer the algebraically equivalent formula

$$\sigma \text{ or } S = \sqrt{\frac{N\Sigma X^2 - (\Sigma X)^2}{N^2}}$$

I am using the formula in the text because the same form, or parts of it, will be used in other procedures. Yet another formula is often used in the field of testing. To use this formula, you must first calculate the mean:

$$\sigma = \sqrt{\frac{\Sigma X^2}{N} - \mu^2} \quad \text{and} \quad S = \sqrt{\frac{\Sigma X^2}{N} - \bar{X}^2}$$

All these arrangements of the arithmetic are algebraically equivalent.

TABLE 3.5 Using the raw-score formula to compute σ for the cookie sales of a population of six Girl Scouts

Boxes of cookies	
X	X^2
28	784
11	121
10	100
5	25
4	16
2	4
$\Sigma X = 60$	$\Sigma X^2 = 1050$

Note: $(\Sigma X)^2 = (60)^2 = 3600$

$$\sigma = \sqrt{\frac{\Sigma X^2 - \frac{(\Sigma X)^2}{N}}{N}} = \sqrt{\frac{1050 - \frac{(60)^2}{6}}{6}} = \sqrt{\frac{1050 - 600}{6}} = \sqrt{\frac{450}{6}} = \sqrt{75} = 8.66 \text{ boxes}$$

PROBLEMS

*10. Look at the following two sample distributions. Without calculating (just look at the data), decide which one has the larger standard deviation. Guess the size of S for each distribution. (You may wish to calculate the range before you decide.) Finally, make a choice between the deviation-score and the raw-score formulas and compute S for each distribution. Compare your computation with your guess.
  a. 5, 4, 3, 2, 1, 0   **b.** 5, 5, 5, 0, 0, 0

11. For each of the distributions in problem 10, divide the range by the standard deviation. Is the result between 2 and 5?

12. By now you can look at the following two distributions and see that a is more variable than b. The difference in the two distributions is in the lowest scores (2 and 6). Calculate σ for each distribution, using the raw-score formula. Notice the difference in σ produced by the change in just one score.
  a. 9, 8, 8, 7, 2   **b.** 9, 8, 8, 7, 6

$\hat{s}$ AS AN ESTIMATE OF σ

Remember that $\hat{s}$ is the principal statistic you will learn in this chapter. It will be used again and again throughout the rest of this text.

As I explained in Chapters 1 and 2, the purpose of using a sample is usually to find out something about a population; that is, a statistic from a sample is used to estimate a parameter of the population. It is both obvious and unfortunate that samples from the same population are often slightly different, yielding different statistics. Which of the statistics is closest in value to the parameter? There is no way to know other than to measure the entire population, which is usually impossible. Statisticians say that the best you can do is to calculate the statistic in such a way that, *on the average,* its value is equal to the parameter. In the language of the mathematical statistician, you want a statistic that is an "unbiased estimator" of the corresponding population parameter.

If you have sample data and you want to calculate an estimate of σ, you should use the statistic $\hat{s}$:

$$\hat{s} = \sqrt{\frac{\Sigma(X - \overline{X})^2}{N - 1}}$$

Statisticians often add a "hat" to a symbol to indicate that something is being estimated. Note that the difference between $\hat{s}$ and σ is that $\hat{s}$ has $N - 1$ in the denominator, whereas σ has N.

This issue of dividing by N or by $N - 1$ sometimes leaves students shrugging their shoulders and muttering, "OK, I'll memorize it and do it however you want." I would like to explain, however, why, when you have sample data and want to estimate σ, you use $N - 1$ in the denominator.

Because the formula for σ is

$$\sigma = \sqrt{\frac{\Sigma(X - \mu)^2}{N}}$$

it would seem logical just to calculate $\overline{X}$ from the sample data, substitute $\overline{X}$ for μ in the numerator of the formula, and calculate an answer. This solution will, more often than not, produce a numerator that is *too small* [as compared to the value of $\Sigma(X - \mu)^2$].

To explain this surprising state of affairs, remember from Chapter 2 (page 39) a characteristic of the mean: For any set of scores, the expression $\Sigma(X - \overline{X})^2$ is minimized. That is, for any set of scores, this sum is smaller when $\overline{X}$ is used than it is if some other number (either larger or smaller) is used in place of $\overline{X}$.

Thus, for a sample of scores, substituting $\overline{X}$ for μ gives you a numerator that is minimized. However, what you want is a value that is the same as $\Sigma(X - \mu)^2$. Now, if the value of $\overline{X}$ is at all different from μ, then the minimized value you get using $\Sigma(X - \overline{X})^2$ will be too small.

The solution that statisticians have adopted for this underestimation problem is to use $\overline{X}$ and then to divide the too-small numerator by a smaller denominator — namely, $N - 1$. This results in a statistic that is a much better estimator of σ.[5]

You just finished four dense paragraphs — lots of ideas per square inch. You may understand it already, but if you don't, take 10 or 15 minutes to reread, do the exercise in footnote 5, and think.

---

[5] To illustrate this for yourself, use a small population of scores and do some calculations. For a population with scores of 1, 2, and 3, $\sigma = 0.82$. With $N = 2$ there are three samples in the population. For each sample, calculate the standard deviation using $\overline{X}$ and N in the denominator. Find the mean of these three standard deviations. Now, for each of the three samples, calculate the standard deviation using $N - 1$ in the denominator and find the mean of these three. Compare the two means to the σ that you want to estimate.

Here are two technical points that you should know if you plan to learn more about statistics after this course. First, even with $N - 1$ in the denominator, $\hat{s}$ is not an unbiased estimator of σ. Because the bias is not very serious, however, $\hat{s}$ is used as the estimator of σ. Second, $\hat{s}^2$ (see next section) *is* an unbiased estimator of σ^2.

Note also that as N gets larger, the subtraction of 1 from N has less and less effect on the size of the estimate of variability. This makes sense because the larger the sample size is, the closer $\overline{X}$ will be to μ, on average.

One new task comes with the introduction of $\hat{s}$: It is the decision whether to calculate σ, S, or $\hat{s}$ for a given set of data. Your choice will be based on your purpose. If your purpose is to estimate the variability of a population using data from a sample, calculate $\hat{s}$. (This purpose is common in inferential statistics.) If your purpose is to describe the variability of a sample or a population, use S or σ.

Calculating $\hat{s}$

A raw-score formula is recommended for calculating $\hat{s}$:

$$\hat{s} = \sqrt{\frac{\Sigma X^2 - \dfrac{(\Sigma X)^2}{N}}{N - 1}}$$

This is almost the same as the raw-score formula for σ (except for $N - 1$ in the denominator). This raw-score formula is the one you will probably use for your own data.[6]

Sometimes you may need to calculate a standard deviation for data already arranged in a frequency distribution. For a simple frequency distribution or a grouped frequency distribution, the formula is

$$\hat{s} = \sqrt{\frac{\Sigma fX^2 - \dfrac{(\Sigma fX)^2}{N}}{N - 1}}$$

where f is the number of scores in an interval.

Here are some data to illustrate the calculation of $\hat{s}$ both for ungrouped raw scores and for a frequency distribution. Consider puberty. As you know, females reach puberty earlier than males (about 2 years earlier on the average). Is there any difference between the genders in the *variability* of reaching this developmental milestone? Comparing standard deviations will give you an answer.

If you have only a sample of ages for each sex and your interest is in all females and all males, $\hat{s}$ is the appropriate standard deviation. **Table 3.6** shows the calculation of $\hat{s}$ for the females. Work through the calculations. The standard deviation is 2.19 years.

Data for the ages at which males reach puberty are presented in a simple frequency distribution in **Table 3.7**. Work through these calculations, noting that grouping causes two additional columns of calculations. For males, $\hat{s}$ is 1.44 years.

Thus, based on sample data, you can conclude that there is more variability among females in the age of reaching puberty than there is among males. (Inciden-

---

[6] Calculators with standard deviation functions differ. Some use N in the denominator, some use $N - 1$, and some have both standard deviations. You will have to check yours to see how it is programmed.

TABLE 3.6 Calculation of $\hat{s}$ for ages at which females reach puberty (ungrouped raw-score method)

Age (X)		X^2
17	4.25	289
15	2.25	225
13	0.25	169
12	−0.75	144
12	−0.75	144
11	−1.75	121
11	−1.75	121
11	−1.75	121

$\Sigma X = 102 \qquad \Sigma X^2 = 1334 \qquad (\Sigma X)^2 = 10{,}404$

$\overline{X} = 12.75$ years

$$\hat{s} = \sqrt{\dfrac{\Sigma X^2 - \dfrac{(\Sigma X)^2}{N}}{N-1}} = \sqrt{\dfrac{1334 - \dfrac{(102)^2}{8}}{7}} = \sqrt{\dfrac{1334 - 1300.50}{7}} = 2.19 \text{ years}$$

TABLE 3.7 Calculation of $\hat{s}$ for ages at which males reach puberty (simple frequency distribution)

Age (X)	f	fX	fX^2
18	1	18	324
17	1	17	289
16	2	32	512
15	4	60	900
14	5	70	980
13	3	39	507
	$N = 16$	$\Sigma fX = 236$	$\Sigma fX^2 = 3512$

$\overline{X} = 14.75$ years

$$\hat{s} = \sqrt{\dfrac{\Sigma fX^2 - \dfrac{(\Sigma fX)^2}{N}}{N-1}} = \sqrt{\dfrac{3512 - \dfrac{(236)^2}{16}}{15}} = \sqrt{\dfrac{3512 - 3481}{15}} = 1.44 \text{ years}$$

tally, you would be correct in your conclusion — I chose the numbers so they would produce results that are similar to population figures.)

Here are three final points about working with simple and grouped frequency distributions. Recognize that ΣfX^2 is found by squaring X, multiplying by f, and then summing. $(\Sigma fX)^2$ is found by multiplying f by X, summing, and then squaring. For grouped frequency distributions, use the midpoints of the class intervals as the X values.

For both grouped and simple frequency distributions, most calculators with memory will give you ΣfX and ΣfX^2 if you properly key in X and f for each line of the distribution. Procedures differ depending on the brand. The time you invest in learning (even if you have to read the instructions) will be repaid several times over in future chapters (not to mention the satisfying feeling you will get).

clue to the future

You will be glad to know that in your efforts to calculate a standard deviation, you have produced two other useful statistics along the way. Each has a name and they will turn up again in future chapters. The number that you took the square root of to get the standard deviation is called the **variance** (also called a *mean square*). The expression in the numerator of the standard deviation, $\Sigma(X - \overline{X})^2$, is called the **sum of squares.**

THE VARIANCE

The last step in calculating a standard deviation is to find a square root. The number you take the square root of is the **variance.** The symbols for the variance are σ^2 (population variance) and $\hat{s}^2$ (sample variance used to estimate the population variance). By formula,

$$\sigma^2 = \frac{\Sigma(X - \mu)^2}{N} \quad \text{and} \quad \hat{s}^2 = \frac{\Sigma(X - \overline{X})^2}{N - 1} \quad \text{or} \quad \hat{s}^2 = \frac{\Sigma X^2 - \dfrac{(\Sigma X)^2}{N}}{N - 1}$$

The difference between σ^2 and $\hat{s}^2$ is the term in the denominator. The population variance uses N and the sample variance uses $N - 1$.[7] The variance is not very useful as a *descriptive* statistic. It is, however, of enormous importance in inferential statistics, especially in a technique called the analysis of variance (Chapters 10, 11, and 12).

PROBLEMS

13. Here is an interpretation problem. Remember the student who recorded money spent at the Student Center every day for 14 days (pages 37–40)? The mean was $2.17 per day. Suppose the student wanted to reduce his Student Center spending. Write a sentence of advice if $\hat{s}$ was $0.02 and a second sentence if $\hat{s}$ was $2.50.

14. Describe in words the relationship between the variance and the standard deviation.

15. A researcher had a sample of scores from the freshman class on a test that measured attitudes toward authority. She wished to estimate the standard deviation of the scores of the entire freshman class. (She had data from 20 years ago and she believed that current students were more homogeneous than students in the past.) Given the following summary statistics, calculate the proper standard deviation and variance:

$$N = 21 \qquad \Sigma X = 304 \qquad \Sigma X^2 = 5064$$

---

[7] Many calculators have a variance key. One kind uses $N - 1$ to calculate the standard deviation and N to calculate the variance. With a square and a square root function key, this arrangement covers σ, σ^2, $\hat{s}$, and $\hat{s}^2$.

16. Here are those data on the heights of the Americans in their 20s that you have been working with. For each group, calculate $\hat{s}$. Write a sentence of interpretation.

Females			Males	
Height (in.)	f		Height (in.)	f
72	1		77	1
70	1		76	1
69	2		75	1
67	5		74	1
66	7		73	4
65	10		72	6
64	9		71	7
63	5		70	6
62	5		69	8
61	3		68	6
60	1		67	2
59	1		66	2
			65	3
			64	1
			62	1

17. A high school English teacher measured the attitudes of 11th-grade students toward poetry. After a 9-week unit on poetry, she measured the students' attitudes again. She was disappointed to find that the mean change was zero. Following are some representative scores. (High scores represent more favorable attitudes.) Calculate $\hat{s}$ for both before and after, and write a conclusion based on the standard deviations.

Before	7	5	3	5	5	4	5	6
After	9	8	2	1	8	9	1	2

18. In manufacturing, engineers strive for consistency. The following data are the errors, in millimeters, in giant doodads manufactured by two different processes. Choose S or $\hat{s}$ and determine which process produces the more consistent doodads.

Process A	0	1	−2	0	−2	3
Process B	1	−2	−1	1	−1	2

19. Estimate the population standard deviation for oral body temperature using data you worked with in Chapter 2, problem 28. You may find ΣfX and ΣfX^2 by working from the answer to problem 28, or, if you understand what you need to do to find these values from that table, you can use $\Sigma fX = 3928.0$ and $\Sigma fX^2 = 385,749.24$.

20. Reread the description of the classic study by Bransford and Franks (problem 4 in Chapter 2, page 28). Using the answers you compiled on those data, find the range and the interquartile range. After considering the reason that Bransford and Franks would gather such data, calculate an appropriate standard deviation and variance.

21. Return to the objectives at the beginning of the chapter. Can you do each one?

 ## WEBSITE REFERENCES

http://psychology.wadsworth.com

Open "Choose Your Quiz" and select Spatz, 7e. Find the index of chapters, and click on Chapter 3, *Variability.*

Open "Choose Your Area," scroll down, and select "Research Methods and Statistics." Click on "Statistics" in the graph. Click on "Student Activities." Click on the Web Page button for "Central Tendency and Variability." This workshop covers materials found in Chapters 2 and 3.

chapter **four**

COMBINATION STATISTICS

OBJECTIVES FOR CHAPTER FOUR

Study the text. Work the problems as you come to them. Afterward, you should be able to:

1. Use *z* scores to compare two scores in one distribution
2. Use *z* scores to compare scores in one distribution with scores in a second distribution
3. Construct and interpret boxplots
4. Calculate an effect size index and describe it in words
5. Compile descriptive statistics and an explanation into a Descriptive Statistics Report

his chapter will introduce you to four statistical techniques that have two things in common. First, all four are descriptive statistics. Second, each one combines two or more statistical components. Fortunately, the concepts are ones you have studied in the preceding two chapters. These four statistical techniques should prove quite help- ful to you as you develop your own ability to understand data.

The first statistic, the *z score,* combines a raw score with the mean and standard deviation of a distribution in a way that allows you to know the relative standing of the raw score in the distribution. A *z*-score description works regardless of the kind of scores you are using or the shape of the distribution.

The second section of this chapter will cover *boxplots.* A boxplot is a graphic with information on one variable, much like a frequency polygon. It conveys much more information, however. A boxplot gives you the mean, median, range, interquartile range, and skew of the distribution with just one picture.

The next combination statistic presented in this chapter is the *effect size index,* symbolized with *d.* An effect size index gives you a way to describe the size of the dif- ference between two distributions. If you know that an independent variable makes a difference in the scores on the dependent variable, then *d* tells you how much of a dif- ference the independent variable makes.

The final section of this chapter does not introduce any new statistics. It shows you how to put together the ones you have learned into a Descriptive Statistics Report. A Descriptive Statistics Report is an organized collection of descriptive statistics that helps a reader understand the data that were gathered.

Probably all college graduates are familiar with measures of central tendency; most are familiar with measures of variability. However, only those with a good education in quantitative thinking are familiar with all four techniques presented in this chapter. So, learn this material. You will then understand more than most and you will be better equipped to help others.

USING z SCORES TO DESCRIBE INDIVIDUALS

Suppose one of your friends says he got a 95 on a math exam. What does that tell you about his mathematical ability? From your previous experience with tests, 95 may seem like a pretty good score. This conclusion, however, depends on a couple of assumptions, and without those assumptions, a score of 95 is *meaningless*. Let's return to the conversation with your friend.

After you say, "95! Congratulations," suppose he tells you that 200 points were possible. Now a score of 95 seems like something to hide. "My condolences," you say. But then he tells you that the highest score on that difficult exam was 101. Now 95 has regained respectability and you chortle, "Well, all right!" In response, he shakes his head and tells you that the mean score was 98. The 95 takes a nose dive. As a final blow, you find out that 95 was the lowest score, that nobody scored worse than your friend. With your hand on his shoulder, you cut off further discussion of the test with, "Come on, I'll buy you an ice cream cone."

This example illustrates that the score of 95 acquires meaning only when it is compared with the rest of the test scores. In particular, a score gets its meaning from *its relation to the mean and the variability of other scores in the distribution.* A **z score** is a mathematical way to modify an individual raw score so that the result conveys the score's relationship to the mean and standard deviation of its fellow scores. The formula is

$$z = \frac{X - \overline{X}}{S}$$

Remember that $X - \overline{X}$ is an acquaintance of yours, the deviation score. A z score describes the relation of X to $\overline{X}$ with respect to the variability of the distribution. For example, if you know that a score (X) is 5 units from the mean $(X - \overline{X} = 5)$, you know only that the score is better than average, but you have no idea how far above average it is. If the distribution has a range of 10 units and $\overline{X} = 50$, then an X of 55 is a very high score. On the other hand, if the distribution has a range of 100 units, an X of 55 is barely above average. Look at **Figure 4.1,** which is a picture of the ideas in this paragraph.

Thus, to know a score's position in a distribution, the variability of the distribution must be taken into account. The way to do this is to divide $X - \overline{X}$ by a unit that

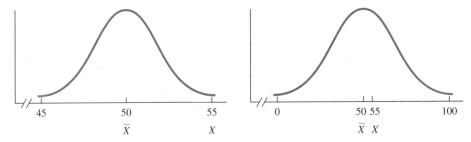

F I G U R E 4.1 A comparison of $X - \overline{X} = 5$ in two distributions with different standard deviations

measures variability, the standard deviation. The result is a deviation score per unit of standard deviation.[1] A z score is sometimes referred to as a **standard score** because it is a deviation score expressed in standard deviation units.

Any distribution of raw scores can be converted into a distribution of z scores; for each raw score, there is one z score. Positive z scores represent raw scores that are greater than the mean; negative z scores go with raw scores that are less than the mean. In both cases, the absolute value of the z score tells the number of standard deviations the score is from the mean.

Converting a raw score to a z score gives you a number that indicates the raw score's relative position in the distribution. If *two* raw scores are converted to z scores, you will know their positions relative to *each other* as well as to the distribution. Finally, z scores are also used to compare two scores from different distributions, even when the scores are measuring different things. (If this seems like trying to compare apples and oranges, see problem 5.)

In the general psychology course I took as a freshman, the professor returned tests with a z score rather than a percentage score. This z score was the key to figuring out your grade. A z score of $+1.50$ or higher was an *A*, and -1.50 or lower was an *F*. (z scores between $+.50$ and $+1.50$ received *B*'s. If you assume the professor practiced symmetry, you can figure out the rest of the grading scale.)

Table 4.1 lists the raw scores (percentage correct) and z scores for four of the many students who took two of the tests in that class. Consider the first student, Kris, who scored 76 on both tests. The two 76s appear to be the same, but the z scores show that they are not. The first 76 was a high A and the second 76 was a high F. The second student, Robin, appears to have improved on the second test, going from 54 to 86, but in fact, the scores were the test averages both times (a grade of C). Robin stayed the same. Marty also appears to have improved if you examine only the raw scores. However, the z scores reveal that Marty did *worse* on the second test. Finally, comparing Terry's and Robin's raw scores, you can see that Terry scored 4 points higher

---

[1] This technique is based on the same idea that percents are based on. For example, 24 instances one time and 24 instances a second time appear to be the same. But don't stop with appearances. Ask for the additional information that you need, which is, "24 out of how many events?" An answer of "50 the first time and 200 the second" allows you to convert both 24s to "per centum" (per one hundred). Clearly, 48 percent and 12 percent are different, even though both are based on 24 instances.

TABLE 4.1 The z scores of selected students on two 100-point tests in general psychology

Student	Test 1		Test 2	
	Raw score	z score	Raw score	z score
Kris	76	+2.20	76	−1.67
Robin	54	.00	86	.00
Marty	58	+.40	82	−.67
Terry	58	+.40	90	+.67
	Test 1: $\bar{X} = 54$		Test 2: $\bar{X} = 86$	
	$S = 10$		$S = 6$	

than Robin on each test, but the z scores show that Terry's improvement was greater on the second test.

The reason for these surprising comparisons is that the means and standard deviations are so different for the two tests. Perhaps the second test was easier, or the material was more motivating to students, or the students studied more. Maybe the teacher was better. Perhaps all of these reasons were true.

To summarize, z scores give you a way to compare raw scores. The basis of the comparison is the distribution itself rather than some external standard (such as a grading scale of 90–80–70–60 percent for As, Bs, and so on).

Now, z is used as both a descriptive statistic and an inferential statistic. As a descriptive statistic, its range is limited. For a distribution of 100 or so scores, the z scores might range from approximately −3 to +3. For many distributions, especially when N is small, the range is more narrow.

As an inferential statistic, z values are not limited to ±3. To illustrate, a z score test is used in Chapter 14 to help decide whether two populations are different. The value of z depends heavily on how different the two populations actually are. Thus, values much greater than 3 occur when z is used as an inferential statistic.

clue to the future

z scores will turn up often as you study statistics. They are prominent in this book in Chapters 4, 6, 7, and 14.

PROBLEMS

1. The mean of any distribution has a z score equal to what value?
2. What conclusion can you reach about Σz?
3. Under what conditions would you prefer that your personal z score be negative rather than positive?

4. Hattie and Missy, twin sisters, were intense competitors, but they never competed against each other. Hattie specialized in long-distance running and Missy was an excellent sprint swimmer. As you can see from the distributions in the table, each was the best in her event. Take the analysis one step further and use z scores to determine who is the more outstanding twin. You might start by looking at the data and making an estimate.

10K runners	Time (min)	50M swimmers	Time (sec)
Hattie	37	Missy	24
Dott	39	Ta-Li	26
Liz	40	Terri	27
Marette	42	Betty	28

5. Tobe grows apples and Zeke grows oranges. In the local orchards the mean weight of apples is 5 ounces, with $S = 1.0$ ounce. For oranges the mean weight is 6 ounces, with $S = 1.2$ ounces. At harvest time, each enters his largest specimen in the Warwick County Fair. Tobe's apple weighed 9 ounces and Zeke's orange weighed 10 ounces. This particular year Tobe was ill on the day of judgment, so he sent his friend Hamlet to inquire who had won. Adopt the role of judge and use z scores to determine the winner. Hamlet's query to you is: "Tobe, or not Tobe; that is the question."

6. Ableson's anthropology professor announced that the poorest exam grade for each student would be dropped. Ableson scored 79 on the first anthropology exam. The mean was 67 and the standard deviation 4. On the second exam, he made 125. The class mean was 105 and the standard deviation 15. On the third exam, the mean was 45 and the standard deviation 3. Ableson got 51. Which test should be dropped?

BOXPLOTS

So far you have studied three different characteristics of distributions: central tendency, variability, and form. A **boxplot** is a way to present all three characteristics with one graphic.

What information do you get from a boxplot? You get two measures of central tendency, two measures of variability, and a way to estimate skewness. Skewness, of course, is a description of the form of the distribution.

Figure 4.2 is a boxplot of the Personal Control scores you first encountered in Chapter 2. The older name for this graphic is a "box-and-whisker plot," and as you can see, it consists of a *box* and *whiskers* (and a *line* and a *dot*). The horizontal line shows the variable being measured and its values.

Variability Both the box and the whiskers in Figure 4.2 tell you about variability. The *box* covers the interquartile range, which you studied in Chapter 3. The left-hand

Personal Control scores

FIGURE 4.2 Boxplot of Personal Control scores

end of the box is the 25th percentile score, and the right-hand end is the 75th percentile score. You can estimate these scores from the horizontal axis. The horizontal width of the box is the interquartile range.

The *whiskers* extend to the extreme scores in the distribution. Thus, the whiskers give you a picture of the range. Again, reading from the scale in Figure 4.2, you can estimate the highest score as 70 (which is correct) and the lowest as 30 (it is actually 31).

Central Tendency Besides the two measures of variability, you get two measures of central tendency. The vertical line inside the box is the median. The dot is the mean. In Figure 4.2 you can estimate the mean as slightly greater than 50 (it is 51.00) and the median as slightly greater than the mean (51.55).

Skew Finally, both the relationship of the mean to the median and any difference in the lengths of the whiskers help you determine the skew of the distribution. You already know that when the mean is less than the median, the skew is negative; when the mean is greater than the median, the skew is positive. This relationship is readily apparent in a boxplot.

In addition to the mean/median relationship, skew is indicated by whiskers that differ in length. The longer whisker tells you the direction of the skew. Thus, if the longer whisker is over the higher scores, the skew is positive; if the longer whisker is over the lower scores, the skew is negative.

Given these two ways to determine skew, please describe the form of the distribution of PC scores in **Figure 4.2.**

The conclusion "approximately symmetrical" is appropriate. (Slightly negatively skewed is not wrong either.)

Interpreting Boxplots

Look at **Figure 4.3,** which shows the boxplots of four distributions. Eight questions follow. See how many you get right. If you get all of them right, consider skipping the explanations that follow the questions.

> *Questions*
> 1. Which distribution has positive skew?
> 2. Which distribution is the most compact?
> 3. Which distribution has a mean closest to 40?
> 4. Which distribution is most symmetrical?
> 5. Which distribution has a median closest to 50?
> 6. Which distribution is most negatively skewed?

Distribution

FIGURE 4.3 Boxplots of four distributions

7. Which distribution has the largest range?
8. Which two distributions are the most different?

Answers
1. *Positive skew:* Distribution D. The mean is greater than the median, and the high-score whisker is longer than the low-score whisker.
2. *Most compact:* Distribution C. The range is less than that of any other distribution.
3. *Mean closest to 40:* Distribution D.
4. *Most symmetrical:* Distribution A. The mean and median are about the same, and the whiskers are about the same length.
5. *Median closest to 50:* Distribution B.
6. *Most negative skew:* Distribution B. The difference between the mean and median is greater in B than in C, and the difference in whisker length is greater in B than in C.
7. *Largest range:* Distribution A.
8. *Two most different distributions:* Distributions C and D are the most different. There is almost no overlap of scores. The means are the farthest apart and the medians are the farthest apart.

Boxplots have proved to be quite versatile. The basic boxplot can be modified so that it presents information other than the basics. As just one example, for some data it is helpful to note the outliers. **Outliers** are extreme scores, which are presented as asterisks or dots out beyond the whiskers. Thus, when you encounter a boxplot, you may have to consult the legend or the caption to find out what information is being conveyed. I might add that sometimes boxplots are oriented vertically rather than horizontally.

error detection box

Gross errors from misrecording data are often apparent from a boxplot. Detecting errors is easiest when you are using a computer program to generate the boxplot and you are familiar with boxplot interpretation.

PROBLEMS

7. Tell the story (mean, median, range, interquartile range, and form) of each of the three boxplots in the figure.

8. Which two of the boxplots in the figure are most different from each other?
9. You have already created all the elements for boxplots of the heights of 20–29-year-old women and men. (See problems 1 and 20 of Chapter 2 and problem 3 of Chapter 3.) Draw boxplots of the two groups using one horizontal axis.
10. Create a boxplot (but without the mean) of the oral body temperature data based on Mackowiak, Wasserman, and Levine (1992). Find the statistics you need from the simple frequency distribution answer for problem 28 of Chapter 2 (Appendix F).

EFFECT SIZE INDEX

Does the statement, "On average, women's verbal scores are higher than men's" leave you with a question? Perhaps the question is How much higher?

An **effect size index** is the statistician's way to answer the question: How much difference is there? Here are some statements that don't tell you how different the two groups are.

■ Men are taller than women.
■ First-born children score higher than second-born children on tests of cognitive ability.
■ Interrupted tasks are remembered better than uninterrupted tasks.

For the first statement, this additional information is probably satisfactory: Men, on average, are about 5 inches taller than women. But, how satisfactory is it to know that the mean difference in cognitive ability scores of first-born and second-born children is 13 points or that the difference in recall of interrupted and uninterrupted tasks is 3 tasks? Something else is needed.[2]

---

[2] What is needed is information about averages and variability. This is the same information you needed when your friend told you he scored 95 on a math exam.

An *effect size index* gives you a mathematical way to answer the question How much more? It works whether you are already familiar with the scale of measurement (such as inches) or you have never heard of the scale before (cognitive ability and re-call scores).

The effect size index is symbolized by *d*, where

$$d = \frac{|\mu_1 - \mu_2|}{\sigma}$$

To calculate *d*, you must estimate the parameters with statistics. Samples from the μ_1 population and from the μ_2 population produce $\overline{X}_1$ and $\overline{X}_2$. Calculating an estimate of σ requires knowing about degrees of freedom, a concept that is better introduced in connection with hypothesis testing (Chapter 8 and chapters that follow). So, at this point you will have to be content to have σ given to you.

Because the difference between means is an absolute value, *d* is always expressed as a positive number. Thus, *d* varies from zero upward. You can easily see that if the difference in *means* is zero, then *d* = 0. Also, depending on σ, a specific difference be-tween the two means might produce a small *d* or a large *d*.

Many areas of statistics (and life) deal with variables that start at zero and range upward. Typically, conventions develop that divide the continuum into categories. For *d*, Jacob Cohen (1969) proposed the conventions that follow.[3] They seem to be uni-versally accepted.

Small effect   *d* = .20
Medium effect   *d* = .50
Large effect   *d* = .80

To illustrate, let's take the heights of women and men. Just intuitively, what ad-jective would you use to describe the different heights of women and men? A small difference? A large difference?

Well, it certainly is an obvious difference, one that everyone sees. Let's find the effect size index for the difference in heights of women and men. You already have estimates of μ_{women} and μ_{men} from your work on problem 20, Chapter 2: $\overline{X}_{\text{women}} = $ 64.6 inches and $\overline{X}_{\text{men}} = 69.8$ inches. For this problem, $\sigma = 2.8$ inches. Thus,

$$d = \frac{|\mu_1 - \mu_2|}{\sigma} = \frac{|\overline{X}_1 - \overline{X}_2|}{\sigma} = \frac{|64.6 - 69.8|}{2.8} = |-1.86| = 1.86$$

The interpretation of *d* = 1.86 is that the difference in heights of women and men is just huge, more than twice the size of a difference that would be judged "large." So, here is a reference point for you for effect size indexes. If a difference is so great that everyone is aware of it, then the effect size index, *d*, will be greater than what is con-sidered large.

Let's take another example, the one at the beginning of this section. What is the effect size index for the difference in verbal ability between women and men? To an-

---

[3] For an easily accessible source of Cohen's reasoning in proposing these conventions, see Cohen (1992, p. 99).

swer this question, I referred to Hedges and Nowell (1995). Based on their analysis (six studies with very large representative samples with a total $N > 150,000$), $\overline{X}_{\text{women}} = 513$, $\overline{X}_{\text{men}} = 503$, and $\sigma = 110$, you can find[4]

$$d = \frac{|\mu_1 - \mu_2|}{\sigma} = \frac{|\overline{X}_1 - \overline{X}_2|}{\sigma} = \frac{|513 - 503|}{110} = .09$$

A d value of .09 is less than half the size of a value considered "small." Thus, you can say that although the average verbal ability of women is better than that of men, the difference is very small.

A common question is: How do you interpret d values that are intermediate between .20, .50, and .80? The answer is to use modifiers. So far, I've used the terms "huge" and "half the size of small." Phrases such as "somewhat larger than" and "intermediate between" are often useful.

THE DESCRIPTIVE STATISTICS REPORT

Remember that by learning statistics, you can (1) understand data better yourself, (2) communicate with others who use statistics, and (3) persuade others. In this section you will combine many of this chapter's techniques into a Descriptive Statistics Report. A Descriptive Statistics Report gives you a fairly complete story of the data.[5] Thus, it will help with all three benefits of knowing statistics.

The most interesting Descriptive Statistics Reports are those that compare two or more distributions of scores. To complete a Descriptive Statistics Report of two groups, (1) construct boxplots, (2) find the effect size index, and (3) tell the story that the data reveal. As for telling the story, these points should be covered:

■ Form of the distributions
■ Central tendency
■ Overlap of the two distributions
■ Interpretation of the effect size index

Of course, you should arrange the points so that your story is told well.

To illustrate, let's return to the heights of the 20–29-year-old women and men. **Table 4.2** shows the statistics needed for a Descriptive Statistics Report. Most of the statistics are used to construct boxplots (a task you did in problem 9). I wrote and revised (and revised) an interpretation. The result is **Table 4.3,** a Descriptive Statistics Report for the height data.

I will stop with just this one example of a Descriptive Statistics Report. You have studied all the report's elements in this chapter and the previous two. The best way to learn and understand is to create reports of your own. Thus, problems follow shortly.

[4] I created the means and standard deviation, which are similar to SAT scores, so the result would mirror the conclusions of Hedges and Nowell (1995).
[5] An even more complete report would contain inferential statistics and their interpretation.

T A B L E 4.2 Descriptive statistics needed for a Descriptive
Statistics Report

	Heights of 20–29-Year-Old Americans	
	Women (in.)	Men (in.)
Mean	64.6	69.8
Median	65	70
Minimum	59	62
Maximum	72	77
25th percentile score	63	68
75th percentile score	66	72
Effect size index	1.86	

error detection

The element of a Descriptive Statistics Report that students have the most trouble with is the median, which is needed for the boxplot. If you aren't sure of yourself about medians, review text examples in Chapter 2 and the problems you worked there.

For the first (but not the last) time, I want to call your attention to the subtitle of this book: *Tales of Distributions*. A Descriptive Statistics Report is a tale of a distribution. What on earth would be the purpose of such stories?

The purpose might be to better understand a scientific phenomenon that you are intensely interested in. The purpose might be to explain to your boss the changes that

T A B L E 4.3 A Descriptive Statistics Report on the Heights of Women and Men

The graph shows boxplots of heights of women and men. The difference in means produces an effect size index of 1.86.

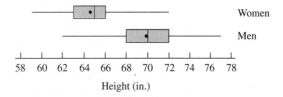

The mean height of women is 64.6 inches; the median is 65 inches. The mean height of men is 69.8 inches; the median is 70 inches. The two genders differ by about 5 inches. Although the two distributions overlap, the middle 50 percent of the men are all taller than the middle 50 percent of the women. This difference in the two distributions is reflected by an effect size index of 1.86, a value that is very large. (A value of 0.80 is traditionally considered large.) The heights for both women and men are distributed symmetrically.

are taking place in your industry; perhaps it might be to convince a quantitative-minded customer to place a big order with you. Your purpose might be to better understand a set of reports on a troubled child (perhaps your own). At this point in your efforts to educate yourself in statistics, you have a good start toward being able to tell the tale of a distribution of data. Congratulations! And, oh yes, here are some more problems so you can get better.

PROBLEMS

11. Find the effect size indexes for the three sets of data in the table. Write an interpretation for each d value.

	Group 1 mean	Group 2 mean	Standard deviation
a.	14	12	4
b.	10	19	10
c.	10	19	30

12. In hundreds of studies with many thousands of participants, first-born children scored better than second-born children on tests of cognitive ability. Representative data from Zajonc and Bargh (1980) provide mean scores of 469 for first-born and 456 for second-born. For this measure of cognitive ability, $\sigma = 110$. Find d and write a sentence of interpretation.

13. Having *the biggest* is a game played far and wide for fun, fortune, and fame (such as being listed in *Guinness World Records*). For example, the biggest cabbage was grown by Bernard Lavery of Rhonnda, Wales. It weighed 124 pounds. The biggest pumpkin was grown by Gary Burke of Simcoe, Ontario, and weighed 1092 pounds (*Guinness World Records 2000*, 2000). Assume that the scores below are representative of contest cabbages and pumpkins. Use z scores to determine the *BIG* winner between Lavery and Burke.

Cabbages	Pumpkins
110	1000
100	900
90	800

14. Is psychotherapy effective? This first-class question has been investigated by many researchers. The data that follow were constructed to mirror the classic findings of Smith and Glass (1977), who analyzed 375 studies.

The *psychotherapy* group received treatment during the study; the *control* group did not. A participant's psychological health score at the beginning of the study was subtracted from the same participant's score at the end of the study, giving the score listed in the table. Thus, each score provides a measure of the *change* in psychological health for an individual. (A negative score means that the person was worse at the end of the study.) For these change scores, $\sigma = 10$. Create a Descriptive Statistics Report.

Psychotherapy	Control	Psychotherapy	Control
7	9	13	−3
11	3	−15	22
0	13	10	−7
13	1	5	10
−5	4	9	−12
25	3	15	21
−10	18	10	−2
34	−22	28	5
7	0	−2	2
18	−9	23	4

15. It's time to check yourself. Turn back to the beginning of this chapter and read the objectives. Can you meet each one?

 WEBSITE REFERENCES

http://psychology.wadsworth.com
    Open "Choose Your Quiz" and select Spatz, 7e. Find the index of chapters, and click on Chapter 4, *Combination Statistics.*

transition page

So far in this exposition of the wonders of statistics, the examples and problems have been about only one variable. The variable might have been Personal Control scores, heights, money, or apples, but your statistic described that one variable. One-variable data are *univariate* data.

Another kind of statistical analysis is about two variables (*bivariate* data) and the relationship between them. For example, What is the relationship between a person's verbal ability and mathematical ability? You will find out in Chapter 5. Other pairs of variables that might be related follow. By the time you finish the next chapter you will know whether or not they are related and to *what degree*.

Height of daughters and height of their fathers
Amount of motivation and quality of performance
Church membership and homicide
Unemployment and traffic deaths
Income level and probability of being diagnosed as psychotic

Chapter 5, Correlation and Regression, will explain two descriptive statistics. Correlation is a method that is used to determine the *degree* of relationship between two variables when you have bivariate data. Regression is a method that is used to *predict* scores when you have bivariate data.

chapter **five**

CORRELATION AND REGRESSION

Correlation and regression: My guess is that you have some understanding of the concept of correlation and that you are less comfortable with the word *regression*. Speculation aside, correlation is simpler.

Correlation is a statistical technique that describes the degree of relationship between two variables. Regression is more complex. In this chapter you will use the regression technique for two tasks, *drawing* the line that best fits the data and *predicting* a person's score on one variable when you know that person's score on a second, correlated variable. Regression has other, more sophisticated uses, but you will have to put those off until you study more advanced statistics.

The ideas identified by the terms *correlation* and *regression* were developed by Sir Francis Galton in England a little over 100 years ago. Galton was a genius (he could read at age 3) who had an amazing variety of interests, many of which he actively pursued during his 89 years. He once listed his occupation as "private gentleman," which meant that he had inherited money and did not have to work at a job. Lazy, however, he was not. Galton traveled widely and wrote prodigiously (16 books and more than 200 articles).

From an early age, Galton was enchanted with counting and quantification. Among the many things he tried to quantify were weather, individuals, beauty, characteristics of criminals, boringness of lectures, and effectiveness of prayers. Often he was successful. For example, it was Galton who discovered that atmospheric pressure highs produce clockwise winds around a calm center, and one of his efforts at quantifying individuals resulted in a book on fingerprints. Because it worked so well for him, Galton actively promoted the philosophy of **quantification,** the idea that you can understand a phenomenon much better if you translate its essential parts into numbers.

Now, as you well know, simply measuring some interesting phenomenon such as a breeze and getting a number like 12 doesn't leave you sighing the sigh of the newly enlightened. Researchers say that such satisfying sighs come when your measurements confirm your expectation that two variables are related. Francis Galton was a master at identifying variables that were related.

Many of the variables that interested Galton were in the field of heredity. Although it was common in the 19th century to comment on physical similarities within a family (height and facial characteristics, for example), Galton thought that psychological characteristics, too, tended to run in families. Specifically, he thought that characteristics such as genius, musical talent, sensory acuity, and quickness had a hereditary basis. [A clue to the position that Galton took on this matter is that he and his illustrious cousin, Charles Darwin, shared an illustrious grandfather, Erasmus Darwin (although they had no grandmother in common).] Galton's 1869 book, *Hereditary Genius,* listed many families and their famous members.

Galton wasn't satisfied with the list in that early book; he wanted to express the relationships in quantitative terms. To get quantitative data, he established an anthropometric (people-measuring) laboratory at a health exposition (a fair) and later at the South Kensington Museum in London. Approximately 17,000 people who stopped at a booth paid three pence to be measured. They left with self-knowledge; Galton left with quantitative data and a pocketful of coins. In today's research the investigator usually pays the participants. For us researchers, the old days were better in some ways. (For one summary of Galton's results, see Johnson et al., 1985.)

Galton's most important legacy is probably his invention of the concepts of correlation and regression. Using correlation, you can express the degree of relationship between any two paired variables. (The relationship between the height of fathers and the height of their adult sons was Galton's classic example.)

Galton was not enough of a mathematician to work out the theory and formulas for his concepts. This task fell to Galton's friend and protégé, Karl Pearson, Professor of Applied Mathematics and Mechanics at University College in London.[1] Pearson's 1896 *product-moment correlation coefficient* and other **correlation coefficients** that he and his students developed were quickly adopted by researchers in many fields and

---

[1] I have some biographical information on Pearson in Chapter 13, the chapter on chi square. Chi square is another statistical invention of Karl Pearson.

are widely used today in psychology, sociology, education, political science, the biological sciences, and other areas.

Finally, although Galton and Pearson became famous for their data gathering and statistics, this was not what they had set out to do. For both, the principal goal was to develop recommendations that would improve the human condition. Making recommendations required a better understanding of heredity and evolution, and statistics was simply the best way to arrive at this better understanding.

In 1889 Galton described how valuable statistics are (and also let us in on his emotional feelings about statistics):

> Some people hate the very name of statistics, but I find them full of beauty and interest. . . . Their power of dealing with complicated phenomena is extraordinary. They are the only tools by which an opening can be cut through the formidable thicket of difficulties that bars the path of those who pursue the Science of [Human Beings].[2]

My plan in this chapter is for you to read about bivariate distributions (necessary for both correlation and regression), to learn to compute and interpret Pearson product-moment correlation coefficients, and to use the regression technique to draw a best-fitting straight line and predict outcomes.

BIVARIATE DISTRIBUTIONS

In the chapters on central tendency and variability, you worked with one variable at a time (**univariate distributions**). Height, time, test scores, and errors all received your attention. If you look back at those problems, you'll find a string of numbers under one heading (see, for example, Table 2.6). Compare those distributions with the one in **Table 5.1.** In Table 5.1 there are scores under the variable *Humor test* and other scores under a second variable, *Intelligence test.* You could find the mean and standard deviation of either of these variables. The characteristic of the data in this table that makes it a **bivariate distribution** is that the scores on the two variables are paired. The 50 and the 8 go together; the 20 and the 4 go together. They are paired, of course, because the same person made the two scores. As you will see, there are also other reasons for pairing scores. All in all, bivariate distributions are fairly common.

The essential idea of a bivariate distribution (which is required for correlation and regression techniques) is that two variables have values that are paired for some logical reason. A bivariate distribution may show positive correlation, negative correlation, or zero correlation.

POSITIVE CORRELATION

In the case of a *positive correlation* between two variables, high measurements on one variable tend to be associated with high measurements on the other variable, and low measurements on one variable with low measurements on the other. This is the case

---

[2] For a short biography of Galton, I recommend David (1968).

TABLE 5.1 A bivariate distribution of scores on two tests taken by the same individuals

	Humor test X variable	Intelligence test Y variable
Larry	50	8
Shep	40	9
Curly	30	5
Moe	20	4

TABLE 5.2 Manufactured data on two variables: heights of fathers and their sons*

Father	Height (in.) X	Height (in.) Son	Y
Michael Smith	74	Mike, Jr.	74
Christopher Johnson	72	Chris, Jr.	72
Matthew Williams	70	Matt, Jr.	70
Joshua Brown	68	Josh, Jr.	68
Andrew Jones	66	Andy, Jr.	66
James Miller	64	Jim, Jr.	64

* The first names are, in order, the six most common in the United States. Rounding out the top ten are John, Nicholas, Justin, and David. The surnames are also the six most common. Completing this top ten are Davis, Wilson, Anderson, and Taylor (Dunkling, 1993).

for the manufactured data in **Table 5.2.** Tall fathers tend to have sons who grow up to be tall men. Short fathers tend to have sons who grow up to be short men. If the true relationship were so undeviating that every son grew to be exactly his father's height (as in the manufactured data), the correlation would be perfect, and the correlation coefficient would be 1.00. A graph plotting this relationship would look like **Figure 5.1.** If such were the case (which, of course, is ridiculous; mothers and environments have their say, too), then it would be possible to predict without error the adult height of an unborn son simply by measuring his father.

Examine Figure 5.1 carefully. Each point represents a pair of scores, the height of a father and the height of his son. Such an array of points is called a **scatterplot,** or scatter diagram. I will use scatterplots throughout this chapter. Incidentally, it was when Galton cast his data as a scatterplot graph that the idea of a co-relationship began to become clear to him.

The line that runs through the points in Figure 5.1 (and in Figures 5.2, 5.3, and 5.5) is called a **regression line.** It is a "line of best fit." When there is perfect correlation ($r = 1.00$), all points fall exactly on the regression line. It is Galton's use of the term *regression* that gives the symbol r for correlation.[3]

[3] The term *regression* can be confusing because it has two separate meanings. As you already know, it is a *statistical method* that allows you to draw the line of best fit and to make predictions with bivariate data. Re-

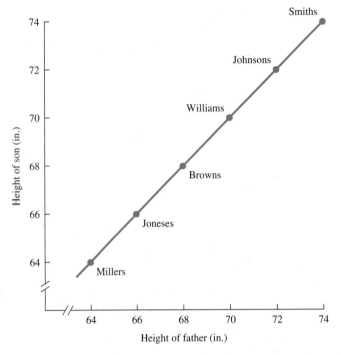

FIGURE 5.1 A scatterplot with a perfect positive correlation (*r* = 1.00)

In the past generation or so, with changes in nongenetic factors such as nutrition and medical care, sons tend to be somewhat taller than their fathers, except for extremely tall fathers. If every son grew to be exactly 2 inches taller than his father (or 1 inch or 6 inches, or even 5 inches shorter), the correlation would still be perfect, and the coefficient would again be 1.00. **Figure 5.2** demonstrates this point: You can have a perfect correlation even if the paired numbers aren't the same. The only requirement for perfect correlation is that the *differences* between pairs of scores all be the same. If they are the same, then all the points of a scatterplot lie on the regression line, correlation is perfect, and an exact prediction can be made.

Of course, people cannot predict their sons' heights precisely. The correlation is not perfect and the points do not all fall on the regression line. As Galton found, however, there is some positive relationship; the correlation coefficient is about .50. The points do tend to cluster around the regression line.

---

gression also refers to a *phenomenon* that occurs when a select group is tested a second time. Suppose that from a large sample of scores, you select a subsample of extreme scores for retesting (those who did either very well or very poorly the first time). On the retest, the mean of the subsample tends to be closer to the larger group's mean than it was on the first test. Galton found that the mean height of sons of extremely tall men was less than the mean height of their fathers and also that the mean height of sons of extremely short men was greater than the mean height of their fathers. In both cases, the sons' mean is closer to the population mean. Because the mean of an extreme group, when measured a second time, tended to regress toward the population mean, Galton named this phenomenon *regression*. Unfortunately, he then used the same word for one of the statistical techniques he was developing.

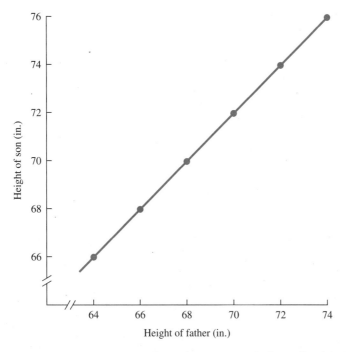

FIGURE 5.2 A scatterplot with every son 2 inches taller than his father ($r = 1.00$)

Here is a new problem. In your academic career you have taken an untold number of aptitude and achievement tests. For several of these tests, separate scores have been computed for verbal aptitude and mathematics aptitude. In general, what is the relationship between verbal aptitude and math aptitude? That is, are people who are good in one also good in the other, or are they poor in the other, or is there no relationship? Stop for a moment and compose your own answer to that question.

As you may have suspected, the next graph is a scatterplot of data that will begin to answer my question. **Figure 5.3** shows the scores of eight high school seniors who took the Scholastic Aptitude Test (SAT). The SAT verbal scores are on the X axis, and the SAT math scores are on the Y axis. As you can see in Figure 5.3, there is a positive relationship, though not a perfect one. As verbal scores vary upward, mathematics scores *tend* to vary upward. If the score on one is high, the other score tends to be high, and if one is low, the other tends to be low. Later in this chapter you'll learn to calculate the precise *degree* of relationship, and because there is a relationship, you can use regression to predict students' math scores if you know their verbal score.

(Examining Figure 5.3, you might complain that the graph is oddly shaped; it violates the 60 to 75 percent rule for the Y axis and all the data points are stuck up in one corner. It looks ungainly, but I was in a dilemma, which I'll explain later in the chapter.)

Here is another bivariate distribution for you to think about: traffic death rates (per 100,000) and unemployment over a 40-year period. Do you think that the relationship is positive like that of verbal and math aptitude scores, or that there is no relationship, or that the relationship is negative?

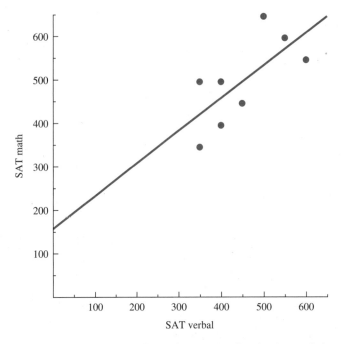

F I G U R E 5.3 Scatterplot and regression line for SAT verbal and SAT math scores for eight high school students (*r* = .69)

NEGATIVE CORRELATION

As you have probably figured out from the heading, the answer to the preceding question is —*negative*. **Figure 5.4** is a scatterplot of unemployment and traffic death rates in the United States based on Wilde (1991).[4] High rates of unemployment are associated with low rates of traffic fatalities, and low rates of unemployment are associated with high traffic fatality rates. The explanation is that people who are unemployed travel less and are more cautious.

When a correlation is negative, increases in one variable are accompanied by decreases in the other variable (an inverse relationship). With negative correlation, the regression line goes from the upper left corner of the graph to the lower right corner. As you may recall from algebra, such lines have a negative slope.

Some other examples of variables with negative correlation are highway driving speed and gas mileage, daily rain and daily sunshine, and grouchiness and friendships.

As was the case with perfect positive correlation, there is such a thing as perfect negative correlation. In cases of perfect negative correlation also, all the data points of the scatterplot fall on the regression line.

Although some correlation coefficients are positive and some are negative, one is *not more valuable than the other*. The algebraic sign simply tells you the *direction* of the relationship (which is important when you are describing how the variables are related). The absolute size of *r*, however, tells you the *degree* of the relationship. A

---

[4] The correlation coefficient is −.68. For six other industrialized countries, the *r*'s ranged from −.69 to −.88.

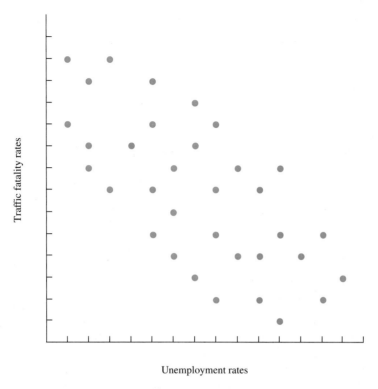

FIGURE 5.4 Unemployment and traffic fatalities for 40 years

strong relationship (either positive or negative) is usually more valuable than a weaker one.

ZERO CORRELATION

A *zero correlation* means there is no linear relationship between two variables. High and low scores on the two variables are not associated in any predictable manner.

 The 50 American states have different murder rates per 100,000. [Recent data show Louisiana with the highest and North Dakota with the lowest (*Statistical Abstract of the United States, 1999*, 1999).] The states also differ in per capita church membership. (Utah has the highest and Nevada has the lowest.) The correlation coefficient between murder and church membership is .06, which is practically zero.

 Figure 5.5 shows a scatterplot that produces a zero correlation coefficient. When $r = 0$, the regression line is horizontal at a height of $\overline{Y}$. This makes sense; if $r = 0$, then your best estimate of Y for any value of X is $\overline{Y}$.

clue to the future

Correlation will come up again in future chapters. If there is correlation between two sets of measurements and you want to compare the *mean* of one set with the mean of the other set, you can make a more precise comparison if you take the correlation into account (used in Chapters 9, 12, and 14).

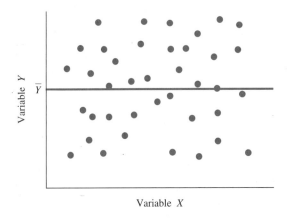

Variable Y

$\overline{Y}$

Variable X

FIGURE 5.5 Scatterplot that produces a zero correlation

PROBLEMS

1. What are the characteristics of a bivariate distribution?
2. What is meant by the statement "Variable X and variable Y are correlated"?
3. Tell how X and Y vary in positive correlation. In negative correlation.
4. Can the following variables be correlated and, if so, would you expect the correlation to be positive or negative?
   a. Height and weight of a group of adults
   b. Height of oak trees and height of pine trees
   c. Average daily temperature and cost of heating a home
   d. IQ and reading comprehension
   e. The quiz 1 scores and the quiz 2 scores of students in section 1 and section 2 of General Biology
   f. The section 1 scores and the section 2 scores of students in General Biology on quiz 1

THE CORRELATION COEFFICIENT

A correlation coefficient provides a quantitative way to express the *degree* of relationship that exists between two variables. The definition formula is

$$r = \frac{\Sigma(z_X z_Y)}{N}$$

where r = Pearson product-moment correlation coefficient
             z_X = a z score for variable X
             z_Y = the corresponding z score for variable Y
            N = number of pairs of X and Y values

Think through this formula and discover what happens when high scores on one variable are paired with high scores on the other variable (positive correlation). The large positive z scores are paired and the large negative z scores are paired. In each case, the multiplication produces large *positive* products, which, when added together, make a large positive numerator. The result is a large positive value of r. Think through for yourself what happens in the formula when there is a negative correlation and when there is a zero correlation.

Though valuable for definitional purposes, the z-score formula is not easy to use when you are *calculating* a value for r. Fortunately, however, there are many equivalent formulas for r. I'll describe two and give you some guidance on when to use each.

Computational Formulas for r

One formula for computing r is referred to as the *blanched* formula; a second one is called the *raw-score* formula. The blanched formula requires that you first find means and standard deviations and then calculate r. For the raw-score formula, you enter the scores directly into the formula. The blanched formula gives you a better feel for the data because you know means and standard deviations. If your calculator produces these statistics (but not r itself), using the blanched formula is quicker. The raw-score formula, on the other hand, doesn't have intermediate steps. If all you want is r, then the raw-score formula is the quicker method.

Blanched formula. Because researchers often use means and standard deviations when telling the story of the data, this formula is used by many:

$$r = \frac{\dfrac{\Sigma XY}{N} - (\overline{X})(\overline{Y})}{(S_X)(S_Y)}$$

where X and Y are paired observations
XY = product of each X value multiplied by its paired Y value
$\overline{X}$ = mean of variable X
$\overline{Y}$ = mean of variable Y
S_X = standard deviation of variable X
S_Y = standard deviation of variable Y
N = number of pairs of observations

One of the terms in the formula is new and a second one has a new meaning. The term with a new meaning is N. In correlation problems, N means the number of *pairs* of scores. The new expression, ΣXY, instructs you to multiply each of the X values by its paired Y value and then sum those products. Do *not* try to obtain ΣXY by multiplying $(\Sigma X)(\Sigma Y)$. *It won't work.* The following table demonstrates this fact. Work through it.

X	Y	XY
2	5	10
3	4	12
Sum 5	9	22

$$\Sigma XY = 22$$
$$(\Sigma X)(\Sigma Y) = (5)(9) = 45$$

The term ΣXY is called "the sum of the cross-products." Some form of the sum of the cross-products is necessary for computing a Pearson r.

As for which variable to call X and which to call Y, it doesn't make any difference for a correlation coefficient. It does, however, make a big difference in the regression problems you will work later in this chapter.

Table 5.3 demonstrates the steps you use to compute r by the blanched procedure. The data are those that were used to draw Figure 5.3, the scatterplot of the SAT verbal and SAT math scores for the eight students. I made up the numbers so that they would produce the same correlation coefficient as reported by Wallace (1972). Work through the numbers in Table 5.3, paying careful attention to the calculation of ΣXY.

Raw-score formula. With the raw-score formula, you calculate r from the raw scores without computing means and standard deviations. The formula is

$$r = \frac{N\Sigma XY - (\Sigma X)(\Sigma Y)}{\sqrt{[N\Sigma X^2 - (\Sigma X)^2][N\Sigma Y^2 - (\Sigma Y)^2]}}$$

You have already learned what all the terms of this formula mean. Remember that N is the number of *pairs* of values.

TABLE 5.3 Calculation of r for SAT verbal and SAT math aptitude scores by the blanched formula

Student	SAT verbal X	SAT math Y	X^2	Y^2	XY
1	350	350	122,500	122,500	122,500
2	350	500	122,500	250,000	175,000
3	400	400	160,000	160,000	160,000
4	400	500	160,000	250,000	200,000
5	450	450	202,500	202,500	202,500
6	500	650	250,000	422,500	325,000
7	550	600	302,500	360,000	330,000
8	600	550	360,000	302,500	330,000
	3600	4000	1,680,000	2,070,000	1,845,000

$$\bar{X} = \frac{\Sigma X}{N} = \frac{3600}{8} = 450 \qquad \bar{Y} = \frac{\Sigma Y}{N} = \frac{4000}{8} = 500$$

$$S_X = \sqrt{\frac{\Sigma X^2 - \frac{(\Sigma X)^2}{N}}{N}} = \sqrt{\frac{1,680,000 - \frac{(3600)^2}{8}}{8}} = 86.60$$

$$S_Y = \sqrt{\frac{\Sigma Y^2 - \frac{(\Sigma Y)^2}{N}}{N}} = \sqrt{\frac{2,070,000 - \frac{(4000)^2}{8}}{8}} = 93.54$$

$$r = \frac{\frac{\Sigma XY}{N} - (\bar{X})(\bar{Y})}{S_X S_Y} = \frac{\frac{1,845,000}{8} - (450)(500)}{(86.60)(93.54)} = .69$$

All calculators with two or more memory storage registers — and some with one — allow you to accumulate simultaneously the values for ΣX and ΣX^2. After doing so, compute values for ΣY and ΣY^2 simultaneously. This leaves only ΣXY for you to compute.

Many calculators have a built-in function for r. When you enter X and Y values and press the r key, the coefficient is displayed. If you have such a calculator, I recommend that you use this labor-saving device *after* you have used the computation formulas a number of times. Working directly with terms like ΣXY leads to an understanding of what goes into r.

If you calculate sums that reach above the millions, your calculator may switch into scientific notation. A display such as 3.234234 08 might appear. To convert this number back to familiar notation, just move the decimal point to the right the number of places indicated by the number on the right. Thus, 3.234234 08 becomes 323,423,400. The display 1.23456789 12 becomes 1,234,567,890,000.

Table 5.4 illustrates the use of the raw-score procedure for computing r. The data are the same as those used in Table 5.3, which demonstrates that the value of r is the same for both methods.

The final step in any statistics problem is interpretation. What story goes with a correlation coefficient of .69 between SAT verbal scores and SAT math scores? An r of .69 is a fairly substantial correlation coefficient. Students who have high SAT verbal scores certainly *tend* to have high SAT math scores. Note, however, that if the correlation had been near zero, you could say that the two abilities are unrelated. If the coefficient had been strong and negative, you could say, Good in one, poor in the other.

Correlation coefficients should be based on an "adequate" number of pairs of observations. As a general rule of thumb, "adequate" means 30 or more. My SAT example, however, had an N of 8, and most of the problems have fewer than 30 pairs. Small-N problems allow you to spend your time on interpretation and understanding rather than on "number crunching." In Chapter 8 you will learn the reasoning behind my admonition that N be adequate.

You may have noticed that r is a sample statistic; there are no σ_X, σ_Y, μ_X, or μ_Y symbols in this chapter. Of course, if a population of data is available, the parameter may be computed. The procedure is the same as for sample data. The symbol for this parameter is ρ (the Greek letter rho). Unfortunately, ρ was used by early writers as

T A B L E 5.4 Calculation of r for SAT verbal and SAT math aptitude scores by the raw-score formula

$\Sigma X = 3600 \quad \Sigma Y = 4000 \quad \Sigma X^2 = 1{,}680{,}000 \quad \Sigma Y^2 = 2{,}070{,}000 \quad \Sigma XY = 1{,}845{,}000$

$$r = \frac{N\Sigma XY - (\Sigma X)(\Sigma Y)}{\sqrt{[N\Sigma X^2 - (\Sigma X)^2][N\Sigma Y^2 - (\Sigma Y)^2]}}$$

$$= \frac{(8)(1{,}845{,}000) - (3600)(4000)}{\sqrt{[(8)(1{,}680{,}000) - (3600)^2][(8)(2{,}070{,}000) - (4000)^2]}}$$

$$= \frac{360{,}000}{518{,}459} = .69$$

the symbol for another kind of correlation coefficient (Spearman's), which you will learn about in Chapter 14. So, be alert in your outside reading.

error detection

The Pearson correlation coefficient ranges between -1.00 and $+1.00$. Values less than -1.00 or greater than $+1.00$ indicate that you have made an error.

PROBLEMS

*5. This problem is based on data published in 1903 by Karl Pearson and Alice Lee. In the original article, 1376 pairs of father–daughter heights were analyzed. The scores here produce the same means and the same correlation coefficient that Pearson and Lee obtained. For these data, draw a scatterplot and calculate r by both the raw-score formula and the blanched formula.

Father's height, X (in.)	69	68	67	65	63	73
Daughter's height, Y (in.)	62	65	64	63	58	63

*6. This problem is based on actual data also. By now you know how to obtain the summary values for ΣX, ΣX^2, ΣY, ΣY^2, and ΣXY, so I am providing them. The raw data are scores on two different tests of self-esteem. Participants were seventh-grade students. Summary values: $N = 38$, $\Sigma X = 1755$, $\Sigma Y = 1140$, $\Sigma X^2 = 87,373$, $\Sigma Y^2 = 37,592$, $\Sigma XY = 55,300$. Compute r.

*7. The X variable in this problem is the population in millions of the states in the United States. The Y variable is expenditure per pupil in public schools for the states. Summary values: $N = 50$, $\Sigma X = 202$, $\Sigma Y = 41,048$, $\Sigma X^2 = 1740$, $\Sigma Y^2 = 35,451,830$, $\Sigma XY = 175,711$. Calculate r using either method.

SCATTERPLOTS

You already know something about scatterplots — what their elements are and what they look like when $r = 1.00$, $r = .00$, and $r = -1.00$. In this section I will illustrate some intermediate cases and reiterate my philosophy about the value of pictures.

Figure 5.6 shows scatterplots of data that have correlation coefficients of .20, .40, $-.60$, .80, .90, and $-.95$. If you draw an envelope around the points in a scatterplot, the picture that the data present becomes clearer. The thinner the envelope, the higher the correlation. To say this in more mathematical language, the closer the points are clustered around the regression line, the higher the correlation.

Pictures help you understand, and scatterplots are easy to construct. Although the plots require some time, the benefits are worth it. If I had given you the raw data on the populations of states and their school expenditures (see problem 7) and you had drawn a scatterplot, you would have plotted one point in the upper left-hand corner of the graph (a place where data points contribute to a *negative r*). Very likely you would have raised your eyebrows at the unusual point and wondered what state it

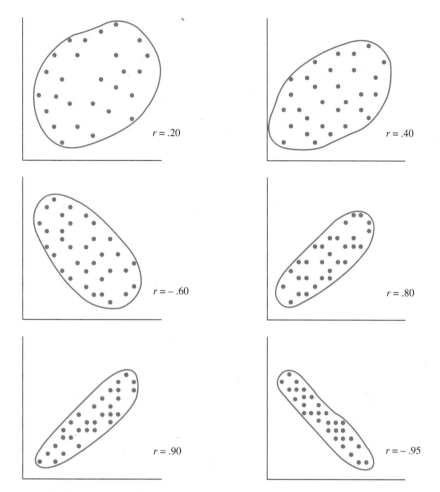

FIGURE 5.6 Scatterplots of data in which $r = .20, .40, -.60, .80, .90,$ and $-.95$

was.[5] So this is a paragraph that encourages you to construct scatterplots — they're worth it.

INTERPRETATIONS OF *r*

The basic and simple interpretation of *r* is probably familiar to you at this point. A correlation coefficient measures the degree of linear relationship between two variables of a bivariate distribution. Fortunately for those eager to understand, correlation coefficients can provide even more information. However, the interpretation of *r* can be a tricky business; there are several common errors. Over the next few pages you

[5] Alaska

will learn additional interpretations and uses of correlation coefficients. I'll also point out some common errors in interpreting correlation coefficients.

Effect Size Index for *r*

What qualifies as a large correlation coefficient? What is small? You will remember that you dealt with similar questions in Chapter 4 when you studied the *effect size index*. In that situation, you had two sample means that were different. The question was: Is the difference big? You will recall that Jacob Cohen proposed that the question be answered by calculating an effect size index (d) and that d values of .20, .50, and .80 were designated as small, medium, and large, respectively. In a similar way, Cohen (1969) addressed the question of calculating an effect size index for correlation coefficients.

You will be delighted with Cohen's calculation of an effect size index for r: the value is r. The remaining question is: What is small, medium, and large? Based on both experience with correlation coefficients and logic, Cohen proposed the following designations:

Small $r = .10$
Medium $r = .30$
Large $r = .50$

Using Cohen's guidelines, you can see that the $r = .69$ between SAT verbal and SAT mathematics scores is a very large correlation. The correlation between the heights of fathers and their daughters is large (.51, problem 5). The correlation between a state's population and its expenditure per pupil is almost medium sized (.25, problem 7).

Coefficient of Determination

The correlation coefficient is also the basis of the **coefficient of determination,** which tells the proportion of variance that two variables in a bivariate distribution have in common. The coefficient of determination is calculated by squaring r; it is always a positive value between 0 and 1:

Coefficient of determination $= r^2$

Look back at Table 5.2, the heights that produced $r = 1.00$. There is variation among the fathers' heights as well as among the sons' heights. How much of the variation among the sons' heights is associated with the variation in the fathers' heights? All of it! That is, the variation among the sons' heights (74, 72, 70, and so on) is exactly the same variation that is seen among their fathers' heights (74, 72, 70, and so on). In the same way, the variation among the sons' heights in Figure 5.2 (76, 74, 72, and so on) is the same *variation* that is seen among their fathers' heights (74, 72, 70, and so on). For Table 5.2 and Figure 5.2, $r = 1.00$ and $r^2 = 1.00$.

Now look at Table 5.3, the SAT verbal and SAT math scores. There is variation among the SAT verbal scores as well as among the SAT math scores. How much of the variation among the SAT math scores is associated with the variation among the SAT verbal scores? Some of it. That is, the variation among the SAT math scores (350, 500,

FIGURE 5.7 Two separate illustrations of common variance

400, and so on) is only partly reflected in the variation among the SAT verbal scores (350, 350, 400, and so on). The *proportion* of variance in the SAT math scores that is associated with the variance in the SAT verbal scores is r^2. In this case, $(.69)^2 = .48$.

Think for a moment about the many factors that influence the scores in Table 5.3. Some factors influence both scores — factors such as motivation, mental sharpness on test day, and, of course, the big one: general intellectual ability. Other factors influence one test but not the other — factors such as anxiety about math tests, chance successes and errors, and, of course, the big ones: specific verbal knowledge and specific math knowledge.

What a coefficient of determination of .48 tells you is that 48 percent of the variance in the two sets of scores is **common variance.** However, 52 percent of the variance is independent variance — that is, variance in one test that is not associated with variance in the other test.

Here is another example. The correlation of first-term college grade point averages (GPAs) with academic aptitude test scores is about .50. The coefficient of determination is .25. This means that of all that variation in GPAs (from flunking out to 4.0), 25 percent is associated with aptitude scores. The rest of the variance (75 percent) is related to other factors. As examples of other factors that influence GPA (for good or for ill) consider health, roommates, new relationships, and financial situation. Academic aptitude tests cannot predict the variation that these factors will produce.

Common variance is often illustrated by two overlapping circles, each of which represents the total variance of one variable. The overlapping portion is the amount of common variance. The left half of **Figure 5.7,** for example, shows these circles for the GPA–college aptitude test scores and the right half shows the SAT verbal–SAT math data.

Note what a big difference there is between a correlation of .69 and one of .50 when they are interpreted in terms of common variance. Although .69 and .50 seem fairly close, an *r* of .69 predicts almost twice the amount of variance that an *r* of .50 predicts: 48 percent to 25 percent. By the way, common variance is the way professional statisticians interpret correlation coefficients.[6]

---

[6] If you are moderately skilled in algebra and would like to understand more about common variance, I recommend that you finish the material in this chapter on regression and then work through the presentation of Minium, King, and Bear (1993, pp. 206–208).

USES OF *r*

Reliability

One important use of correlation coefficients is to assess the **reliability** of devices that measure things. A device (a test is an example) is reliable if the score you get is not subject to chance variation.

One way to assess chance variation is to measure a number of individuals and then measure them a second time. If the device that measures is not influenced by chance, then you get the same measurements both times. If the second measurement is *exactly* the same as the first for each individual, it is easy to conclude that the measuring device is perfectly reliable — that chance has nothing to do with the score you get. However, if the measurements are not *exactly* the same, you experience some uncertainty. Fortunately, a correlation coefficient tells you the *degree* of agreement between the test and the retest scores. High correlation coefficients mean lots of agreement and therefore high reliability; low coefficients mean lots of disagreement and therefore low reliability. But what size coefficient indicates reliability? The rule of thumb is that an *r* of .80 or higher is expected for social science measurements.

As an example, in Galton's data, 435 adults' heights were measured twice. The correlation was .98. It is not surprising that Galton's method of measuring height was very reliable. The correlation, however, for "highest audible tone," a measure of pitch perception, was only .28 for 349 subjects who were tested a second time within a year (Johnson et al., 1985). One of two interpretations is possible. Either people's ability to hear high sounds changes up and down during a year, or the test was not reliable. In this case, the test was not reliable. Two possible explanations for this lack of reliability are (1) the test environment was not as quiet from one time to the next and (2) the instruments the researchers used were not calibrated exactly the same on both tests.

This section is concerned with the reliability of a *measuring instrument,* which involves measuring the same variable twice. The reliability of a *relationship* between two different variables is a different question and will be covered in Chapter 8. When you ask whether a relationship is *reliable,* the .80 rule of thumb does not apply.

Correlation—Evidence for Causation?

A high correlation coefficient does *not* give you the kind of evidence that allows you to make cause-and-effect statements. Therefore, don't do it. Ever.

Jumping to a cause-and-effect conclusion is a cognitively easy leap for humans. For example, Shedler and Block (1990) found that among a sample of 18-year-olds whose marijuana use ranged from abstinence to once a month, there was a positive correlation between use and psychological health. Is this evidence that occasional drug use promotes psychological health?

Because Shedler and Block had followed the sample from age 3 on, they knew about a third variable, the quality of the *parenting* that the 18-year-olds had received. Not surprisingly, parents who were responsive, accepting, and patient and who valued originality had children who were psychologically healthy. In addition, these same children as 18-year-olds had used marijuana on occasion. Thus, two variables — drug use and parenting style — were each correlated with psychological health. Shedler and Block concluded that psychological health and adolescent drug use were both traceable to quality of parenting. (This research also included a sample of frequent users, who were not psychologically healthy and who had been raised with a parenting style not characterized by the adjectives above.)

Of course, if you have a sizable correlation coefficient, it *may* be the result of a cause-and-effect relationship between the two variables. For example, early statements about cigarette smoking causing lung cancer were based on simple correlational data. Persons with lung cancer were often heavy smokers. Also, national comparisons indicated a relationship (see problem 13). However, as careful thinkers (and the cigarette companies) pointed out, both cancer and smoking might have been caused by a third variable; stress was often suggested. That is, stress caused cancer and stress also caused people to smoke. Thus, cancer rates and smoking rates were related (a high correlation), but one did not cause the other; both were caused by a third variable. What was required to establish a cause-and-effect relationship was data from controlled experiments, not correlational data. Experimental data, complete with control groups, finally established the cause-and-effect relationship between cigarette smoking and lung cancer. (Controlled experiments will be discussed in Chapter 9, Hypothesis Testing, Effect Size, and Confidence Intervals: Two-Sample Designs.)

To summarize this section using the language of logic: A sizable correlation is a necessary but not a sufficient condition for establishing causality.

PROBLEMS

8. Estimate the correlation coefficients for these scatterplots.

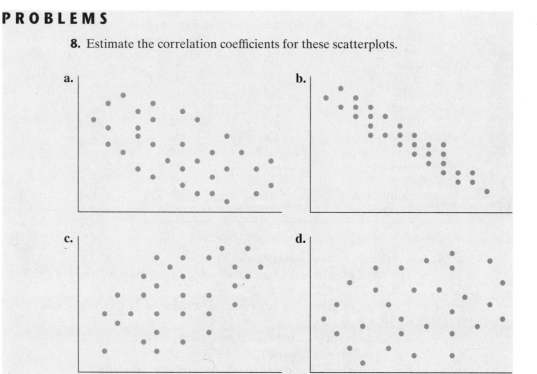

9. You found a correlation of .57 between two measures of self-esteem. What is the coefficient of determination, and what does it mean?

10. You found a correlation of .25 between the populations of American states and their expenditures per pupil in public schools. Calculate the coefficient of determination and write an interpretation.

11. Here are some summary statistics that you have seen before. Can you find the correlation coefficient?

	Height of women (in.)	Height of men (in.)
ΣX	3230	3490
ΣX^2	208,966	244,050
$N = 50$ pairs		

12. What percent of variance in common do two variables have if their correlation is .10? What if the correlation is increased to .40?

13. For each of 11 countries, the table gives the cigarette consumption per capita in 1930 and the male death rate from lung cancer 20 years later in 1950 (Doll, 1955; reprinted in Tufte, 1983). Calculate a Pearson r and write a statement telling what the data show.

Country	Per capita cigarette consumption	Male death rate (per million)
Iceland	217	59
Norway	250	91
Sweden	308	113
Denmark	370	167
Australia	455	172
Holland	458	243
Canada	505	150
Switzerland	542	250
Finland	1112	352
Great Britain	1147	467
United States	1283	191

14. Interpret each of these statements.
  a. The correlation between vocational-interest scores at age 20 and at age 40 for the 150 participants was .70
  b. The correlation between intelligence test scores of identical twins raised together is .86
  c. A correlation of $-.30$ between IQ and family size
  d. $r = .22$ between height and IQ for 20-year-old men
  e. $r = -.83$ between income level and probability of psychosis

STRONG RELATIONSHIPS BUT LOW CORRELATIONS

One good thing about understanding something is that you come to know what's going on beneath the surface. Knowing the inner workings, you can judge whether the surface appearance is to be trusted or not. You are about to learn two of the "inner workings" of correlation. These will help you evaluate the meaning of low correlations. Low correlations do not always mean there is no relationship between two variables.[7]

Nonlinearity

For r to be a meaningful statistic, the best-fitting line through the scatterplot of points must be a *straight line*. If a curved regression line fits the data better than a straight line, r is low, which will not reflect the true degree of relationship between the two variables.

Figure 5.8 is an example of a situation in which r is inappropriate because the best-fitting line is curved. The X variable is arousal and the Y variable is efficiency of performance. At low levels of arousal (sleepy, for example), performance is not very good. Likewise, at very high levels of arousal (agitation, for example), people don't perform well. In the middle range, however, there is a degree of arousal that is optimum; performance is best at moderate levels of arousal.

·In Figure 5.8 there is obviously a strong relationship between arousal and performance, but r for the distribution is −.10, a value that indicates a very weak relationship. The product-moment correlation coefficient is just not useful for measuring the strength of curved relationships. Special nonlinear correlation techniques for such relationships have been developed and are described in texts such as Howell (1997, chap. 10) and Rosenthal and Rosnow (1991, chap. 14).

Truncated Range

Besides nonlinearity, a second situation can give a low Pearson correlation coefficient even though there is a strong relationship between the two variables. Spuriously low r values can occur when the range of scores in the sample is much smaller than the range of scores in the population (a **truncated range**).

What is the relationship between GRE scores and grades in graduate school?[8] The relationship graphed in **Figure 5.9** is based on the study by Sternberg and Williams (1997). These data look like a snowstorm; they lead to the conclusion that there is little relationship between GRE scores and graduate school grades.

However, students in graduate school do not represent the full range of GRE scores; those with low GRE scores are not included. What effect does this restriction of the range have? You can get an answer to this question by looking at **Figure 5.10,** which shows a hypothetical scatterplot of data for the full range of GRE scores. This

[7] Correlations that do not reflect the true degree of the relationship are said to be *spuriously* low or *spuriously* high.

[8] The Graduate Record Examination (GRE) is used by many graduate schools to help select which students to admit.

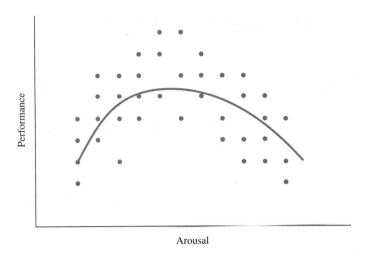

FIGURE 5.8 Generalized relationship between arousal and efficiency of performance

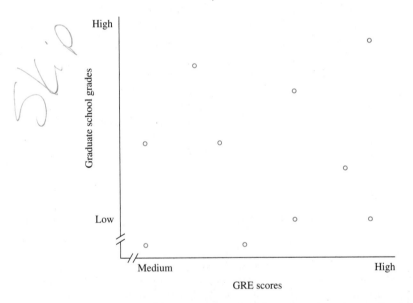

FIGURE 5.9 Scatterplot of GRE scores and graduate school grades in one school

scatterplot shows a moderate relationship. So, unless you recognized that your sample of graduate students truncated the range of GRE scores, you might be tempted to dismiss GRE scores as "worthless."

OTHER KINDS OF CORRELATION COEFFICIENTS

The kind of correlation coefficient you have been learning about — the Pearson product-moment correlation coefficient — is appropriate for measuring the degree of the relationship between two linearly related, continuous variables. Sometimes, however,

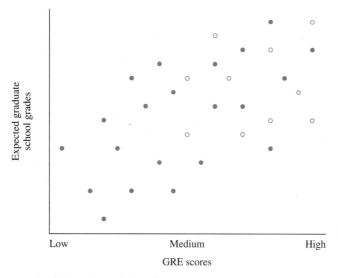

FIGURE 5.10 Hypothetical scatterplot of GRE scores and expected graduate school grades for the population

the data do not consist of two linearly related, continuous variables. What follows is a description of five other situations. In each case, you can express the degree of relationship in the data with a correlation coefficient — but not a Pearson product-moment correlation coefficient. Fortunately, these correlation coefficients can be interpreted much like Pearson product-moment coefficients.

1. If one of the variables is **dichotomous** (has only two values), then a biserial correlation (r_b) or a point-biserial correlation (r_{pb}) is appropriate. Variables such as height (measured as simply tall or short) and gender (male or female) are examples of dichotomous variables.

2. If the relationship between two variables is curved rather than linear, then the correlation ratio *eta* (η) gives the degree of association.

3. Several variables can be combined, and the resulting combination can be correlated with one variable. With this technique, called **multiple correlation,** a more precise prediction can be made. Performance in school can be predicted better by using several measures of a person rather than one.

4. A technique called **partial correlation** allows you to separate or partial out the effects of one variable from the correlation of two variables. For example, if you want to know the true correlation between achievement test scores in two school subjects, it is probably necessary to partial out the effects of intelligence because IQ and achievement are correlated.

5. When the data are ranks rather than scores from a continuous variable, the proper coefficient is the Spearman *r*, which will be covered in Chapter 14.

These and other correlational techniques are discussed in intermediate-level textbooks. (See Howell, 1997, or Rosenthal and Rosnow, 1991.)

PROBLEMS

15. The correlation between number of older siblings and degree of acceptance of personal responsibility for one's own successes and failures is −.37.
  a. Interpret this correlation.
  b. What can you say about the cause of this correlation?
  c. What practical significance might it have?

16. A correlation of .97 is found between anxiety and neuroticism.
  a. Interpret the meaning of this *r*.
  b. Are you justified in assuming that being neurotic is responsible in large part for manifestations of anxiety?

17. Examine the following data, make a scatterplot, and compute *r* if appropriate.

Serial position	1	2	3	4	5	6	7	8
Errors	2	5	6	9	13	10	6	4

LINEAR REGRESSION

You are now prepared to learn how to make predictions. The technique you will use is called *linear regression.* As you will see, what you have learned about correlation (the degree of relationship between variables) will be most helpful when you interpret the results of a regression analysis (a formula that allows predictions).

A few sections ago I said that the correlation between college entrance examination scores and first-semester grade point averages is about .50. Thus, you can predict that those who score high on the entrance examination are more likely to succeed as freshmen than those who score low. All this is rather general, though. Usually you want to predict a *specific* grade point for a *specific* applicant. For example, if you were in charge of admissions at Collegiate U., you would want to know the entrance examination score that predicts a GPA of 2.00, the minimum required for graduation. To make specific predictions, you must do a regression analysis.

Linear regression is a technique that uses the data to write an equation for a straight line. This equation is then used to make predictions. I'll begin with some background on making predictions from equations.

MAKING PREDICTIONS FROM A LINEAR EQUATION

You are used to making predictions; some you make with a great deal of confidence. "If I get my average up to 80 percent, I'll get a B in this course." "If I spend $15 plus tax on this compact disc, I won't have enough left from my $20 to go to a movie."

Often predictions are based on an assumption that the relationship between two variables is linear, that a straight line will tell the story exactly. Frequently, this assumption is quite justified. For the preceding short-term economics problem, imagine a set of axes with "Amount spent" on the *X* axis ($0 to $20) and "Amount left" on the *Y* axis ($0 to $20). A straight line that connects the two $20 marks on the axes tells the whole story. (This is a line with a negative slope that is inclined 45 degrees from horizontal.) Draw this picture. Can you also write the equation for this graph?

Part of your education in algebra was about straight lines. You may recall that the slope-intercept formula for a straight line is

$$Y = mX + b$$

where Y and X are variables representing scores on the Y and X axes
m = slope of the line (a constant)
b = intercept (intersection) of the line with the Y axis (a constant)

Here are some reminders about the *slopes* of lines. The slope of a line is positive if the highest point on the line is to the right of the lowest point. If the highest point is to the left of the lowest point, the slope is negative. Horizontal lines have a slope equal to zero. Lines that are almost vertical have slopes that are either large positive numbers or large negative numbers.

Now, back to the slope-intercept formula. If you are given $m = 3$ and $b = 6$, the formula becomes $Y = 3X + 6$. Now, if you know a value of X, you can easily find the value of Y. If $X = 4$, then $Y = $ __?

The unsolved problem (so far) is how to go from the general formula, $Y = mX + b$, to a specific formula like $Y = 3X + 6$ — that is, how to find the values for m and b.

A common solution involves a rule and a little algebra. The rule is that if a point lies on the line, then the coordinates of the point satisfy the equation of the line. To use this rule to find an equation, you must know the coordinates for two points. Suppose you are told that the point where $X = 5$ and $Y = 8$ is on the line. This point, represented as (5, 8), produces $8 = 5m + b$ when the coordinates are substituted into the general equation. If you are given a second point that lies on the line, you will get another equation with m and b as unknowns. Now you have two equations with two unknowns and you can solve for each in turn, giving you values for m and b. If you would like to check your understanding of this on a simple problem, figure out the formula for the line that tells the story of the $20 problem. For simplicity, use the two points where the line crosses the axes, (0, 20) and (20, 0).

Once you have assumed or have satisfied yourself that a relationship is linear, *any* two points will allow you to draw the line that tells the story of the relationship.

THE REGRESSION EQUATION—A LINE OF BEST FIT

With this background in place, you are in a position to appreciate the problem Karl Pearson faced at the end of the 19th century when he looked at father–daughter height data like those in problem 5. Look at your scatterplot of those data and at Figure 5.4. It is reasonable to assume that the relationship is linear, but which two points should be used to write the equation for the line? The equation will depend on which two particular points you choose to plug into the formula. The two-point approach produces dozens of different lines. Which one is the best line? What solutions come to your mind?

One solution is to find the mean Y value for each of the X values. Connect those means with straight lines, and then choose two points that make the regression line fall within this narrower range of the means. Pearson's solution, which statistics embraced, is to use a more mathematically sophisticated version of this idea. It is called the **least squares** method. Look at **Figure 5.11,** which shows the heights of two daughters from Pearson and Lee's 1903 data, both of whom had 5'9" fathers. Based on all the data,

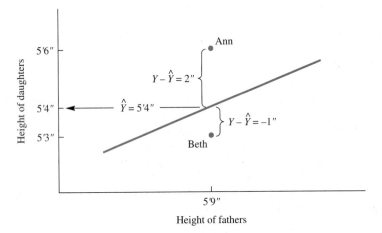

FIGURE 5.11 Hypothetical case of two daughters with a father 5′9″ tall

I drew a regression line using the least squares method. As you can see, 5′4″ is the height predicted for a 5′9″ father. (Predicted values are symbolized by $\hat{Y}$ and pronounced "Y-hat" or "Y-predicted.") The actual data show that Ann was 5′6″ and Beth was 5′3″ (5′6″ and 5′3″ are data points, symbolized by Y). You can imagine that if the other daughters' heights were plotted on the graph, most of them would not fall on the regression line either. With all this error, why use the least squares method for drawing the line?

The answer is that a least squares regression line *minimizes* error. Calculating error is easy. For each prediction,

$$\text{Error} = Y - \hat{Y}$$

Using the two daughters in our example, you get

Ann:	error $= Y - \hat{Y} = 66 - 64 = 2$
Beth:	error $= Y - \hat{Y} = 63 - 64 = -1$

In a similar way, there is an error for each person on the scatterplot (though for some the error is zero). The least squares method creates a straight line such that the *sum of the squares* of the errors is a minimum. In symbol form, $\Sigma(Y - \hat{Y})^2$ is a minimum for a straight line calculated by the least squares method.

The least squares method produces numerical values for the *slope* and the *intercept*. With a slope and an intercept, you can write the equation for a straight line; this line is the one that *best* fits the data.

One more transitional point is necessary. In the language of algebra, the idea of a straight line is expressed as $Y = mX + b$. In the language of statistics, exactly the same idea is expressed as $Y = a + bX$. Y and X are the same in the two formulas, but different letters are used for the slope and the intercept and the order of the terms is different. Unfortunately, the terminology is well established in both fields. However, the mental translation required of students doesn't cause many problems. Thus, in statistics b is the slope of the line and a is the intercept of the line with the Y axis.

In statistics, the **regression equation** is

$$\hat{Y} = a + bX$$

where $\hat{Y}$ = Y value predicted from a particular X value
a = point at which the regression line intersects the Y axis
b = slope of the regression line
X = X value for which you wish to predict a Y value

In a correlation problem, the symbol Y can be assigned to either variable, but in a regression equation, Y *is assigned to the variable you wish to predict.*

The Regression Coefficients

To use the equation $\hat{Y} = a + bX$ you need the values for a and b, which are called **regression coefficients.** Values for a and b can be calculated from any bivariate set of data. The arithmetic for these calculations comes from the least squares method of line fitting.

If you have already computed r and the standard deviations for both X and Y, then b can be obtained by the formula

$$b = r\frac{S_Y}{S_X}$$

where r = correlation coefficient for X and Y
S_Y = standard deviation of the Y variable
S_X = standard deviation of the X variable

Is it clear to you that for positive correlations, b will be a positive number? For negative correlations, b will be negative.[9]

You can compute a from the formula

$$a = \overline{Y} - b\overline{X}$$

where $\overline{Y}$ = mean of the Y scores
b = regression coefficient computed previously
$\overline{X}$ = mean of the X scores

The regression coefficients are illustrated in **Figure 5.12.** The coefficient a is the point where the regression line crosses the Y axis (the Y intercept, which is 4.00 in Figure 5.12). The coefficient b is the numerical value of the slope of the regression line. To independently determine the slope of a line from a graph, divide the vertical distance the line rises by the horizontal distance the line covers. In Figure 5.12 the line DE (vertical rise of the line FD) is half the length of FE (the horizontal distance of FD). Thus, the slope of the regression line is 0.50 ($DE/FE = b = 0.50$). Put another way, the value of Y increases one-half point for every one-point increase in X.

[9] The value of b can sometimes be obtained more easily by the formula

$$b = \frac{N\Sigma XY - (\Sigma X)(\Sigma Y)}{N\Sigma X^2 - (\Sigma X)^2}$$

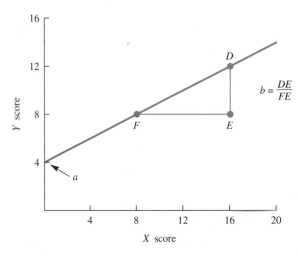

FIGURE 5.12 The regression coefficients *a* and *b*

Writing an SAT Regression Equation

To illustrate the construction of a regression equation, I'll use the SAT data, choosing to make predictions about math aptitude scores (Y) from verbal aptitude scores (X). The descriptive statistics needed from the data are

$$r = .69 \qquad S_X = 86.60 \qquad S_Y = 93.54 \qquad \overline{X} = 450 \qquad \overline{Y} = 500$$

The formula for *b* gives

$$b = r\frac{S_Y}{S_X} = (.69)\frac{93.54}{86.60} = (.69)(1.08) = 0.75$$

The formula for *a* gives

$$a = \overline{Y} - b\overline{X} = 500 - (0.75)(450) = 500 - 337.50 = 162.50$$

The *b* coefficient tells you that when math aptitude increases by three-fourths of a point, verbal aptitude increases by one point. The *a* coefficient tells you that the regression line intersects the Y axis at 162.50. Entering these regression coefficient values into the regression equation produces a formula that predicts math achievement from verbal achievement:

$$\hat{Y} = 162.50 + 0.75X$$

Drawing an SAT Regression Line

To draw a regression line on a scatterplot, you need a straightedge and two points that are on the line. Any two points will do. I'll demonstrate with **Figure 5.13,** which is a redrawing of the SAT data. One point that is always on the regression line for any bivariate data is $(\overline{X}, \overline{Y})$. Thus, for the SAT data, the two means (450, 500) identify a

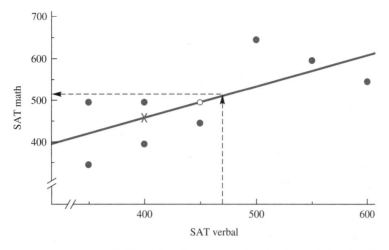

FIGURE 5.13 Scatterplot and regression line for the verbal and math aptitude data (Table 5.3)

point, which is marked on Figure 5.13 with an open circle. A second point can be found by substituting an X value of your choosing into the formula in the previous section. By choosing X as 400, the arithmetic can be done almost at a glance:

$$\hat{Y} = 162.50 + 0.75X = 162.50 + 300 = 462.50$$

The second point is (400, 462.50), which is marked on Figure 5.13 with an **X**. Finally, simply line up the straightedge on the two points and extend the line in both directions. Notice that the line crosses the Y axis at 400, which may surprise you because $a = 162.5$.

The Appearance of Regression Lines

A word of caution is in order about appearances. The *appearance* of the slope of a line drawn on a scatterplot or other graph depends heavily on the units chosen for the X and Y axes. Look at **Figure 5.14.** Although the two lines appear different, $b = 1.00$ for both. They appear different because the space allotted to each Y unit in the right graph is half that allotted to each Y unit in the left graph.

I can now explain the dilemma I faced when I composed the "ungainly" **Figure 5.3** for you. The graph is ungainly because it is square (100 X units is the same length as 100 Y units) and because both axes start at zero (even though the lowest score is 350). I composed it the way I did because I wanted the regression line to cross the Y axis at a (note that it does) and because I wanted its slope to *appear* equal to b (note that it does).

The more attractive **Figure 5.13** is a scatterplot of the same data as those used in Figure 5.3. The difference is that Figure 5.13 has breaks in the axes and 100 Y units is about one-third the length of 100 X units. I'm sure by this point you have the message — you cannot necessarily determine a and b by looking at a graph.

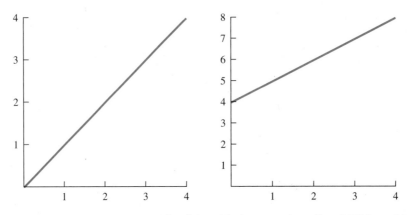

FIGURE 5.14 Two regression lines with the same slope ($b = 1.00$) but with different appearances. The difference is caused by the units chosen for the Y axis.

Predicting a Y Score—Finding $\hat{Y}$

There are two methods you may use to find $\hat{Y}$ for a particular X score. For a rough estimate, you can use an accurately drawn regression line. From the X score, draw a vertical line up to the regression line. Then draw a horizontal line to the vertical axis. The Y score there is $\hat{Y}$. This graphical method is demonstrated by the broken lines in **Figure 5.13** for a verbal aptitude score of 470. The projection to the Y axis shows a predicted math aptitude score of just over 500.

The precise way to find $\hat{Y}$ is to use the regression equation. To predict a math aptitude score for a verbal aptitude score of 470, the regression equation is

$$\hat{Y} = a + bX$$
$$= 162.50 + (0.75)(470)$$
$$= 162.50 + 352.50$$
$$= 515$$

Thus, for an X score of 470, the predicted Y score is 515.

You can also find $\hat{Y}$ values from summary data without calculating a and b by using the following formula:

$$\hat{Y} = r\frac{S_Y}{S_X}(X - \overline{X}) + \overline{Y}$$

For the verbal aptitude score of 470, the math aptitude score is

$$\hat{Y} = (.69)(1.08)(20) + 500 = 14.90 + 500 = 515$$

This is the same value as before.

Now you know how to make predictions. Predictions, however, are cheap; anyone can make them. Respect accrues only when your predictions come true. So far, I have dealt with accuracy by simply pointing out that when r is high, accuracy is high, and when r is low, you cannot put much faith in your predicted values of $\hat{Y}$.

To actually *measure* the accuracy of predictions made from a regression analysis, you need the **standard error of estimate.** This statistic is discussed in most intermediate-level textbooks and in textbooks on testing. After you have covered the material in Chapters 6 and 7 of this book, you will have the background necessary to understand the standard error of estimate.

Finally, a note of caution: There are *two* regression lines for every bivariate distribution. One is a regression line for Y on X, which you have been computing in this chapter. The other is a regression line for X on Y. To make predictions with formulas in this chapter, assign Y to the variable you want to predict and X to the variable that provides the "given."

PROBLEMS

18. In problem 5, the father–daughter height data, you found $r = .513$.
  a. Compute the regression coefficients a and b.
  b. Use your scatterplot from problem 5 and draw the regression line.

19. In problem 6, the data on self-esteem tests, you computed r.
  a. Compute a and b.
  b. What Y score would you predict for a child who scores 42 on X?

 20. Regression is a technique that economists and businesspeople rely on heavily. Think about the relationship between advertising expenditures and sales. Use the data in the table, which are based on national statistics.
  a. Find r.
  b. Write the regression equation.
  c. Plot the regression line on the scatterplot.
  d. Predict sales for an advertising expenditure of $10,000.
  e. Write a sentence that tells whether any confidence at all can be put in this prediction.

Advertising, X ($ thousands)	Sales, Y ($ thousands)
3	70
4	120
3	110
5	100
6	140
5	120
4	100

21. The correlation between Stanford-Binet IQ scores and Wechsler Adult Intelligence Scale (WAIS) IQs is about .80. Both tests have a mean of 100. The standard deviation of the Stanford-Binet is 16. For the WAIS, $S = 15$. What WAIS IQ do you predict for a person who scores 65 on the Stanford-Binet?

(An IQ score of 70 has been used by some schools as a cutoff point between regular classes and special education classes.)

22. From 1990 through 1995, more than 1 million people a year graduated from college with a bachelor's degree. Using the following data from the *Statistical Abstract of the United States,* predict the number that graduate in the year 2000. Carefully choose which variable to call *X* and which to call *Y*.

Year	'90	'91	'92	'93	'94	'95
Graduates (100,000s)	10.5	10.9	11.4	11.7	12.0	11.6

23. Once again, look over the objectives at the beginning of the chapter. Can you do them?

24. Now it is time for *integrative* work on the descriptive statistics you have studied in Chapters 2–5. Choose one of the two options that follow.

    a. Write an essay on descriptive statistics. Start by jotting down from memory things you could include. Review the three chapters, adding to your list additional facts or other considerations. Draft the essay. Rest. Revise it.

    b. Construct a table that summarizes the descriptive statistics in Chapters 2–5. List the techniques in the first column. Across the top of the table list topics that distinguish among the techniques — topics such as purpose, formula, and so forth. Fill in the table.

    Whether you choose option a or b, save your answer for that time in the future when you are reviewing what you are learning in this course (final exam time?).

WEBSITE REFERENCES

http://psychology.wadsworth.com

Open "Choose Your Quiz" and select Spatz, 7e. Find the index of chapters, and click on Chapter 5, *Correlation and Regression.*

Open "Choose Your Area," scroll down, and select "Research Methods and Statistics." Click on "Statistics" in the graph. Click on "Student Activities." Click on the Web Page button for "Correlation." This workshop includes additional concepts that you will cover in chapters after Chapter 5.

WHAT WOULD YOU RECOMMEND? chapters one–five

At this point in the text (and at two later points) there is a set of *What would you recommend?* problems. These problems will help you review and integrate your knowledge. You are to choose an appropriate statistic from among several you have learned in the first five chapters. For each problem that follows, recommend a statistic that will answer the question and note why you recommend that statistic.

a. Registration figures for the American Kennel Club show which dog breeds are common and which are uncommon. For a frequency distribution for all breeds, what central tendency statistic is appropriate?

b. One person is the best golfer in a group. Another person in the group is the best at bowling. What statistical technique allows you to say that one of the two is better than the other?

c. Tuition at Almamater U. has gone up each of the past 5 years. How can I predict what it will be in 25 years when my child enrolls?

d. Each of the American states has a certain number of miles of ocean coastline (ranging from 0 to 6640 miles). Consider a frequency distribution of these 50 scores. What central tendency statistic is appropriate for this distribution (and why)? What measure of variability do you recommend?

e. Many jobs require judgments of monetary value. Later a sale price establishes an actual value. Suppose two job applicants made judgments for 30 items. After the items sold, an analysis showed that the average error for each candidate was zero. How can the data be reanalyzed to allow a choice between the two applicants?

f. Suppose you study some new, relatively meaningless material until you know it all. If you are tested 40 minutes later, you recall 85 percent; 4 hours later, 70 percent; 4 days later, 55 percent; and 4 weeks later, 40 percent. How can you express the relationship between time and memory?

g. A table shows the ages and the number of voters in the year 2000. The age categories start with "18–20" and end with "65 and over." What statistic can be calculated to best describe the age of the typical voter?

h. For a class of 40 students, the study time for the first test ranged from 30 minutes to 6 hours. The grades ranged from a low of 48 to a high of 98. What statistic describes how the variable *Study time* is related to the variable *Grade?*

transition page

You are now through with the part of the book that is clearly about descriptive statistics. You should be able to describe a set of data by using a graph, a few choice words, and numbers such as a mean, a standard deviation, and (if appropriate) a correlation coefficient.

The next chapter will serve as a transition between descriptive and inferential statistics. All the problems you will work will give you answers that describe something about a person, score, or group of people or scores. However, the ideas about probability and theoretical distributions that you will use to work these problems are essential elements of inferential statistics.

So, the transition this time is to concepts that will prepare you to plunge into material on inferential statistics. As you will see rather quickly, most of the descriptive statistics that you have been learning are very useful for inferential statistics.

chapter**six**

THEORETICAL DISTRIBUTIONS INCLUDING THE NORMAL DISTRIBUTION

OBJECTIVES FOR CHAPTER SIX

Study the text. Work the problems as you come to them. Afterward, you should be able to:

1. Distinguish between a theoretical and an empirical distribution
2. Distinguish between theoretical and empirical probability
3. Predict the probability of certain events from your knowledge of the theoretical distribution of those events
4. List the characteristics of the normal distribution
5. Find the proportion of a normal distribution that lies between two scores
6. Find the scores between which a certain proportion of a normal distribution falls
7. Find the number of scores associated with a particular proportion of a normal distribution

T his chapter has more figures than any other chapter, almost one per page. The reason for all these figures is that they are the best way I know to convey to you ideas about theoretical distributions and probability. So, please examine these figures carefully, making sure you understand what each part means. When you are working problems, drawing your own pictures is a big help.

I'll begin by distinguishing between empirical distributions and theoretical distributions. In Chapter 2 you learned to arrange scores in frequency distributions. The scores you worked with were selected because they were representative of scores from actual research. Distributions of such observed scores are **empirical distributions.**

This chapter will be about theoretical distributions. Like the empirical distributions in Chapter 2, a theoretical distribution is a presentation of all the scores, usually presented as a graph. **Theoretical distributions,** however, are based on mathematical formulas and logic rather than on empirical observations.

Theoretical distributions are used in statistics to determine probabilities. When there is a correspondence between an empirical distribution and a theoretical distribution, you can use the theoretical distribution to arrive at *probabilities* about future empirical events. Probabilities, as you know, are quite helpful in reaching decisions.

This chapter will cover three theoretical distributions: rectangular, binomial, and normal. Rectangular and binomial distributions will be used to illustrate probability more fully and to establish some points that are true for all theoretical distributions. The third distribution, the normal distribution, will occupy the bulk of your time and attention in this chapter.

PROBABILITY

You are already somewhat familiar with the concept of probability. You know, for example, that probability values range from .00 (there is no possibility that an event will occur) to 1.00 (the event is certain to happen).

In statistics, events are sometimes referred to as "successes" or "failures." To calculate the probability of a success using the *theoretical* approach, you first enumerate all the ways a *success* can occur. Then you enumerate all the *events* that can occur (whether successes or failures). Finally, you form a ratio with successes on top (the numerator) and total events on the bottom (the denominator). This fraction, changed to a decimal, is the theoretical probability of a success.

For example, with coin flipping, the theoretical probability of "head" is .50. A head is a success and it can occur in only one way. The total number of possible outcomes is two (head and tail), and the ratio $\frac{1}{2}$ is .50. In a similar way, the probability of rolling a six on a die is $\frac{1}{6}$ = .167. For playing cards, the probability of drawing a jack is $\frac{4}{52}$ = .077.

The *empirical* approach to finding probability involves observing actual events, some of which are successes and some of which are failures. The ratio of successes to total events produces a probability, a decimal number between .00 and 1.00. To find an empirical probability, you use observations rather than logic to get the numbers.[1]

What is the probability of particular college majors? Remember the data in Figure 2.4, which showed college majors? The probability question can be answered by processing numbers from that figure. Here's how. Choose the major that you are interested in and label the frequency of that major as "number of successes." Divide that number by 1,165,000, which is the total number of baccalaureate degrees granted in 1995–96. The figure you get answers the probability question. If the major in question is sociology, then 24,000/1,165,000 = .02 is the answer. For English, 43,000/1,165,000 = .04 is the answer.[2] Now, here's a question for you to answer for yourself. Were these probabilities determined theoretically or empirically?

---

[1] The empirical probability approach is sometimes called the relative frequency approach.

[2] If I missed doing the arithmetic for the major *you* are interested in, I hope you'll do it for yourself.

The rest of this chapter will emphasize theoretical distributions and theoretical probability. You will work with coins and cards next, but before you are finished, I promise you a much wider variety of applications.

A RECTANGULAR DISTRIBUTION

To show you the relationship between theoretical distributions and theoretical probabilities, I'll use an example of a theoretical distribution that you may be familiar with. **Figure 6.1** is a histogram that shows the distribution of types of cards in an ordinary deck of playing cards. There are 13 kinds of cards, and the frequency of each card is 4. This theoretical curve is a **rectangular distribution.** (The line that encloses a frequency polygon is called a *curve,* even if it is straight.) The number in the area above each card is the probability of obtaining that card in a chance draw from the deck. That theoretical probability (.077) was obtained by dividing the number of cards that represent the event (4) by the total number of cards (52).

Probabilities are often stated as "chances in a hundred." The expression $p = .077$ means that there are 7.7 chances in 100 of the event in question occurring. Thus, from Figure 6.1 you can tell at a glance that there are 7.7 chances in 100 of drawing an ace from a deck of cards. This knowledge might be helpful in some card games.

With this theoretical distribution, you can determine other probabilities. Suppose you want to know your chances of drawing a face card or a 10. These are the shaded events in Figure 6.1. Simply add the probabilities associated with a 10, jack, queen, and king. Thus, $.077 + .077 + .077 + .077 = .308$. This knowledge might be helpful in a game of blackjack, in which a face card or a 10 is an important event (and may even signal "success").

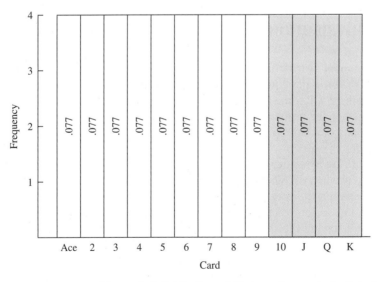

FIGURE 6.1 Theoretical distribution of 52 draws from a deck of playing cards

In Figure 6.1 there are 13 kinds of events, each with a probability of .077. It is not surprising that when you add up all the events [(13)(.077)], the result is 1.00. In addition to the probabilities adding up to 1.00, *the areas add up to 1.00.* That is, by conventional agreement, the *area* under the curve is taken to be 1.00. With this arrangement, any statement about area is also a statement about probability. (If you like to verify things for yourself, you'll find that each slender rectangle has an *area* that is .077 of the area under the curve.) Of the total area under the curve, the proportion that signifies ace is .077, and that is also the probability of drawing an ace from the deck.[3]

clue to the future

The probability of an event or a group of events corresponds to the *area* of the theoretical distribution associated with the event or group of events. This idea will be used throughout this book.

PROBLEMS

1. What is the probability of drawing a card that falls between 3 and jack, excluding both?

2. If you drew a card at random, recorded the result, and replaced the card in the deck, how many 7s would you expect in 52 draws?

3. What is the probability of drawing a card that is higher than a jack *or* lower than a 3?

4. If you made 78 draws from a deck, replacing each card, how many 5s and 6s would you expect?

A BINOMIAL DISTRIBUTION

The **binomial distribution** is another example of a theoretical distribution. Suppose you take three new quarters and toss them into the air. What is the probability that all three will come up heads? As you may know, the answer is found by multiplying together the probabilities of each of the independent events. For each coin, the probability of a head is $\frac{1}{2}$, so the probability that all three will be heads is $(\frac{1}{2})(\frac{1}{2})(\frac{1}{2}) = \frac{1}{8} = .1250$.

Here are two other questions about tossing those three coins. What is the probability of two heads? What is the probability of one head or zero heads? You could answer these questions easily if you had a theoretical distribution of the probabilities. Here's how to construct one. Start by listing, as in **Table 6.1,** the eight possible outcomes of tossing the three quarters into the air. Each of these eight outcomes is

---

[3] In gambling situations, uncertainty is commonly expressed as odds. The expression "odds of 5:1" means that there are five ways to fail and one way to succeed; 3:2 means three ways to fail and two ways to succeed. The odds of drawing an ace are 12:1. To convert odds to a probability of success, divide the second number by the sum of the two numbers.

TABLE 6.1 All possible outcomes when three coins are tossed

Outcomes	Number of heads	Probability of outcome
Heads, heads, heads	3	.1250
Heads, heads, tails	2	.1250
Heads, tails, heads	2	.1250
Tails, heads, heads	2	.1250
Heads, tails, tails	1	.1250
Tails, heads, tails	1	.1250
Tails, tails, heads	1	.1250
Tails, tails, tails	0	.1250

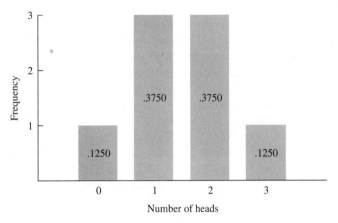

FIGURE 6.2 A theoretical binomial distribution showing the number of heads when three coins are tossed

equally likely, so the probability for any one of them is $\frac{1}{8}$ = .1250. There are three outcomes in which two heads appear, so the probability of two heads is .1250 + .1250 + .1250 = .3750. This is the answer to the first question. Based on Table 6.1, I constructed **Figure 6.2,** which is the theoretical distribution of probabilities you need. You can use it to answer problems 5 and 6, which follow.[4]

PROBLEMS

5. If you toss three coins into the air, what is the probability of a success if success is (a) either one head or two heads? (b) all heads or all tails?

6. If you throw the three coins into the air 16 times, how many times would you expect to find zero heads?

---

[4] The binomial distribution is discussed more fully by Moore and McCabe (1993), Howell (1997), and Minium, Clarke, and Coladarci (1999).

COMPARISON OF THEORETICAL AND EMPIRICAL DISTRIBUTIONS

I have carefully called Figures 6.1 and 6.2 *theoretical* distributions. A theoretical distribution may not reflect *exactly* what would happen if you drew cards from an actual deck of playing cards or tossed quarters into the air. Actual results could be influenced by lost or sticky cards, sleight of hand, unbalanced coins, or chance deviations. Now let's turn to the empirical question of what a frequency distribution of actual draws from a deck of playing cards looks like. **Figure 6.3** is a histogram based on 52 draws from a used deck shuffled once before each draw.

As you can see, Figure 6.3 is not exactly like Figure 6.1. In this case, the differences between the two distributions are due to chance or worn cards and not to lost cards or sleight of hand (at least not conscious sleight of hand). Of course, if I made 52 more draws from the deck and constructed a new histogram, the picture would probably be different from both Figures 6.3 and 6.1. However, if I continued, drawing 520 or 5200 or 52,000 times,[5] and only chance was at work, the curve would be practically flat on the top; that is, the empirical curve would look like the theoretical curve.

The major point here is that a theoretical curve represents the "best estimate" of how the events would actually occur. As with all estimates, a theoretical curve may produce predictions that vary from actual observations, but in the world of real events, it is better than any other estimate.

In summary, then, a theoretical distribution is one based on logic and mathematics rather than on observations. It shows you the probability of each event that is part of the distribution. When it is similar to an empirical distribution, the probability figures obtained from the theoretical distribution are accurate predictors of actual events.

THE NORMAL DISTRIBUTION

One theoretical distribution that has proved to be extremely valuable is the **normal distribution.** In about 1800, Carl Friedrich Gauss (1777–1855) worked out the mathematics of the curve and promoted it as a way to represent random error in astronomy observations (Stewart, 1977). Because the "Gaussian" curve was such an accurate picture of the effects of random variation, early writers referred to the curve as the *law* of error.[6] At the end of the 19th century, Karl Pearson, as part of his efforts to standardize the terminology in statistics, coined the name *normal distribution*.[7] Pearson probably chose the word *normal* because a wide variety of measurements were distributed in this bell-shaped fashion.

One of the early promoters of the normal curve was Adolphe Quetelet (Ka'-tle) (1796–1874), a Belgian who showed that many social and biological measurements are distributed normally. Quetelet, who knew about the "law of error" from his work

---

[5] Statisticians describe this extensive sampling as "the long run."

[6] In this context, *error* means random variation. Used in this manner, the term *error* is an important part of today's statistical terminology.

[7] Pearson also chose the term *standard deviation* for the statistic that you know by that name. (See Porter, 1986.)

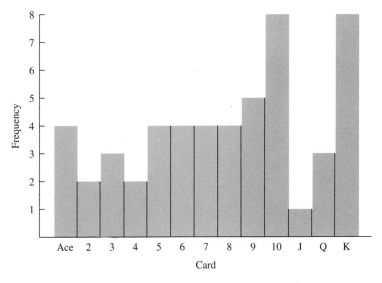

F I G U R E 6.3 Empirical frequency distribution of 52 draws from a deck of playing cards

as an astronomer, presented tables showing the correspondence between measurements such as height and chest size and the normal curve. During the 19th century Quetelet promoted statistics and was widely influential (Porter, 1986). For example, Florence Nightingale, a pioneer in using statistical analyses to improve health care, said that Quetelet was the founder of "the most important science in the whole earth" (Cohen, 1984). Also Quetelet's work suggested to Francis Galton that genius could be treated mathematically, an idea that led to the concept of correlation.[8]

Although many measurements are distributed approximately normally, it is not the case that data "should" be distributed normally. This unwarranted conclusion has been reached by some scientists in the past.

Finally, the theoretical normal curve has an important place in statistical theory. This importance is quite separate from the fact that empirical frequency distributions often correspond closely to the normal curve.

Description of the Normal Distribution

Figure 6.4 is a normal distribution. It is a bell-shaped, symmetrical, theoretical distribution based on a mathematical formula rather than on any empirical observations. (Even so, if you peek ahead to Figures 6.7, 6.8, and 6.9, you will see that empirical curves often look similar to this theoretical distribution.) When the theoretical curve is drawn, the *Y* axis is sometimes omitted. On the *X* axis, *z* scores are used as the unit of measurement for the standardized normal curve, where

[8] Quetelet qualifies as a famous person: A statue was erected in his honor in Brussels, he was the first foreign member of the American Statistical Association, and the Belgian government commemorated the centennial of his death with a postage stamp (1974). You can read a short intellectual biography of Quetelet by Landau and Lazarfeld (1968) in the *International Encyclopedia of the Social Sciences*.

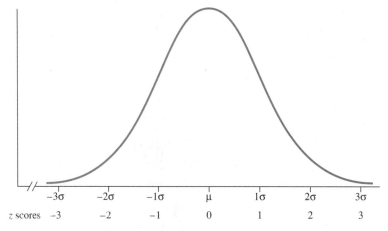

z scores   −3   −2   −1   0   1   2   3

FIGURE 6.4 **The normal distribution**

$$z = \frac{X - \mu}{\sigma}$$

where X = a raw score
μ = the mean of the distribution
σ = the standard deviation of the distribution

I will discuss the standardized curve first and later add the raw scores to the X axis.

There are several other things to note about the normal distribution. The mean, the median, and the mode are the same score — the score on the X axis where the curve peaks. If a line is drawn from the peak to the mean score on the X axis, the area under the curve to the left of the line is half the total area — 50 percent — leaving half the area to the right of the line. The tails of the curve are **asymptotic** to the X axis; that is, they never actually cross the axis but continue in both directions indefinitely, with the distance between the curve and the X axis becoming less and less. Although, in theory, the curve never ends, it is convenient to think of (and to draw) the curve as extending from -3σ to $+3\sigma$. (The *table* for the normal curve in Appendix C, however, covers the area from -4σ to $+4\sigma$.)

Another point about the normal distribution is that the two inflection points in the curve are at exactly -1σ and $+1\sigma$. The **inflection points** are where the curve is the steepest — that is, from where the curve changes from bending upward to bending over. (See the points above -1σ and $+1\sigma$ on Figure 6.4 and think of walking up, over, and down a bell-shaped hill.)

To end this introductory section, here's a caution about the word *normal*. The antonym for *normal* is *abnormal*. *Curves that are not normal distributions, however, are definitely not abnormal.* There is nothing uniquely desirable about the normal distribution. Many nonnormal distributions are also useful to statisticians. Figure 6.1 is an example. It isn't a normal distribution, but it can be very useful. Figure 6.5 shows what numbers were picked when an instructor asked introductory psychology students to pick a number between 1 and 10. **Figure 6.5** is a bimodal distribution with modes at 3 and 7. It isn't a normal distribution, but it will prove useful later in this book.

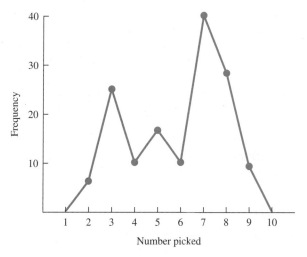

FIGURE 6.5 Frequency distribution of choices of numbers between 1 and 10

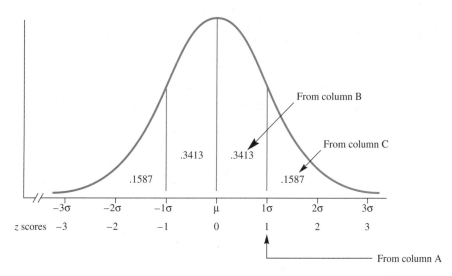

FIGURE 6.6 The normal distribution showing the probabilities of certain *z* scores

The Normal Distribution Table

The theoretical normal distribution is used to determine the probability of an event, just as Figure 6.1 was. **Figure 6.6** is a picture of the normal curve, showing the probabilities associated with certain areas. The figure shows that the probability of an event with a *z* score between 0 and 1.00 is .3413. For events with *z* scores of 1.00 or larger, the probability is .1587. These probability figures were obtained from Table C in Appendix C. Turn to **Table C** now and insert a bookmark there. Table C is arranged so that you can begin with a *z* score (column A) and find the following:

1. The area between the mean and the *z* score (column B)
2. The area beyond the *z* score (column C)

In column A find the z score of 1.00. The proportion of the curve between the mean and a z score of 1.00 is .3413. The proportion beyond the z score of 1.00 is .1587. Because the normal curve is symmetrical and because the area under the entire curve is 1.00, the sum of .3413 and .1587 will make sense to you. Also, because the curve is symmetrical, these same proportions hold for $z = -1.00$. Thus, all the proportions in Figure 6.6 were derived by finding the proportions associated with a z value of 1.00 in Table C. Don't just read this paragraph; do it. Understanding the normal curve *now* will pay you dividends throughout the book.

Notice that the proportions in Table C are carried to four decimal places and that I used all of them. This is customary practice in dealing with the normal curve because you often want two decimal places when a proportion is converted to a percentage.

PROBLEMS

7. In Chapter 4 you read of a professor who gave A's to students with z scores of +1.50 or higher.
   a. What proportion of a class would be expected to make A's?
   b. What assumption must you make to arrive at the proportion you found in part a?

8. What proportion of the normal distribution is found in the following areas?
   a. Between the mean and $z = .21$
   b. Beyond $z = .55$
   c. Between the mean and $z = -2.01$

9. Is the distribution in Figure 6.5 theoretical or empirical?

As I've already mentioned, many empirical distributions are approximately normally distributed. **Figure 6.7** shows a set of 261 IQ scores, **Figure 6.8** shows the diam-

FIGURE 6.7 Frequency distribution of IQ scores of 261 fifth-grade students (unpublished data from J. O. Johnston)

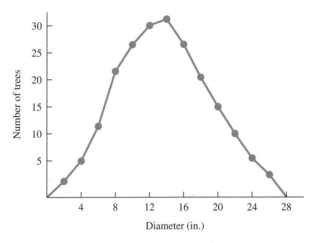

FIGURE 6.8 Frequency distribution of diameters of 100-year-old ponderosa pine trees on 1 acre, N = 199 (Forbs and Meyer, 1955)

FIGURE 6.9 Frequency distribution of hourly wage rates of union truck drivers on July 1, 1944, N = 185,822 (U.S. Bureau of Labor Statistics, December 1944)

eter of 199 ponderosa pine trees, and **Figure 6.9** shows the hourly wage rates of 185,822 union truck drivers in the middle of the last century (1944). As you can see, these distributions from diverse fields are similar to Figure 6.4, the theoretical normal distribution. Please note that all of these empirical distributions are based on a "large" number of observations. Usually more than 100 observations are required for the curve to fill out nicely.

In this section I made two statistical points: first that Table C can be used to determine areas (proportions) of a normal distribution, and second that many empirical distributions are approximately normally distributed.

Converting Empirical Distributions to the Standard Normal Distribution

The point of this section is that any normally distributed empirical distribution can be made to correspond to the theoretical distribution in Table C by using z scores. If the raw scores of an empirical distribution are converted to z scores, the mean of the z scores will be 0 and the standard deviation will be 1. A mean of 0 and a standard deviation of 1 are the statistics of the theoretical normal distribution (which is also called the standardized normal distribution).

With a correspondence established between the z scores from an empirical distribution and the z scores of the normal curve, you can determine the probabilities of empirical events such as IQ scores, tree diameters, and hourly wages. With this statistical technique, you can find the probabilities of *any* empirical events that are distributed normally.

In the following examples, the normal curve will be used with IQ scores. IQ scores are distributed normally (Figure 6.7 is evidence for that) with a theoretical mean of 100 and a standard deviation of 15.[9]

PROBLEM

10. Calculate the z scores for IQ scores of 55, 110, 103, and 100.

Finding the Proportion of a Population That Has Scores of a Particular Size or Greater

Suppose you are faced with finding out what proportion of the population has an IQ of 120 or higher. Begin by sketching a normal curve (either in the margin or on a separate paper). Note on the baseline the positions of IQs of 100 and 120. What is your estimate of the proportion with IQs of 120 or higher?

Look at **Figure 6.10.** It is a more formal version of your sketch, giving additional IQ scores on the X axis. The proportion of the population with IQs of 120 or higher is shaded. The z score that corresponds with an IQ of 120 is

$$z = \frac{120 - 100}{15} = \frac{20}{15} = 1.33$$

Table C shows that the proportion beyond $z = 1.33$ is .0918. Thus, you expect a proportion of .0918 or 9.18 percent of the population to have an IQ of 120 or higher. *Because the size of an area under the curve is also a probability statement about the events in that area, there are 9.18 chances in 100 that any randomly selected person will have an IQ of 120 or above.* **Figure 6.11** shows the proportions just determined.

Table C gives the proportions of the normal curve for positive z scores only. However, because the distribution is symmetrical, knowing that .0918 of the population has

---

[9] This is true for the Wechsler intelligence scales (WAIS, WISC, and WPPSI). The Stanford-Binet test has a mean of 100 also, but the standard deviation is 16. Flynn (1987) suggests that the actual population mean IQ is well *above* 100 in many countries.

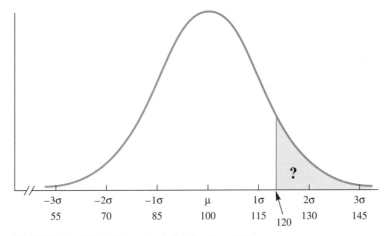

FIGURE 6.10 Theoretical distribution of IQ scores

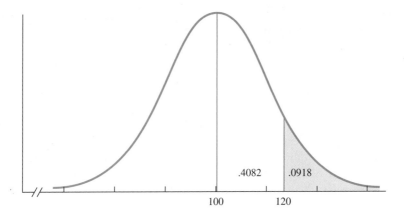

FIGURE 6.11 Proportion of the population with an IQ of 120 or higher

an IQ of 120 or higher tells you that .0918 has an IQ of 80 or lower. An IQ of 80 has a z score of −1.33.

You can answer questions of "how many" as well as questions of proportions using the normal distribution. Suppose 500 first-graders are entering school. How many would be expected to have IQs of 120 or higher? You just found that 9.18 percent of the population would have IQs of 120 or higher. If the population is 500, then calculating 9.18 percent of 500 gives you the number of children. Thus, (.0918)(500) = 45.9. So 46 of the 500 first-graders would be expected to have an IQ of 120 or higher.

There are 18 more normal curve problems for you to do in the rest of this chapter. Do you want to maximize your chances of working every one of them correctly the first time? Here's how. Start by sketching a normal curve. Read the problem and write the givens and the unknowns on your curve. Estimate the answer. Apply the z-score formula. Compare your answer with your estimate; if they don't agree, decide which is in error and make any changes that are appropriate. Confirm your answer by checking the answer in back of the book. If you decide to go for 18 out of 18, good luck!

> **error detection**
>
> Sketching a normal distribution is the best way to understand a problem and avoid errors. Draw vertical lines at the scores you are interested in. Write in proportions.

PROBLEMS

11. Many school districts place children with IQs of 70 or lower in special education classes. What proportion of the general population would be expected to be in these classes?

12. In a school district of 4000 students, how many would be expected to be in special education?

13. What proportion of the population would be expected to have IQs of 110 or higher?

14. Answer the following questions for 250 first-grade students.
   a. How many would you expect to have IQs of 110 or higher?
   b. How many would you expect to have IQs lower than 110?
   c. How many would you expect to have IQs lower than 100?

Finding the Score That Separates the Population into Two Proportions

Instead of starting with an IQ score and calculating proportions, you can work backward and answer questions about scores if you are given proportions. For example, what IQ score is required to be in the top 10 percent of the population?

My picture of this problem is shown as **Figure 6.12.** I began by sketching a more or less bell-shaped curve and writing in the mean (100). Next, I separated the "top 10 percent" portion with a vertical line. Because I need to find a score, I put a question mark on the score axis. The next step is to look in Table C under the column "area beyond z" for .1000. It is not there. You have a choice between .0985 and .1003. Be-

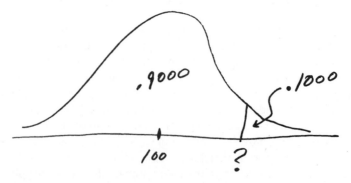

FIGURE 6.12 Sketch of a theoretical distribution of IQ scores divided into an upper 10 percent and a lower 90 percent

cause .1003 is closer to the desired .1000, use it.[10] The z score that corresponds to a proportion of .1003 is 1.28. Now you have all the information you need to solve for X.

To begin, solve the basic z-score formula for X:

$$z = \frac{X - \mu}{\sigma}$$

Multiplying both sides by σ produces

$$(z)(\sigma) = X - \mu$$

Adding μ to both sides isolates X. Thus, when you need to find a score (X) associated with a particular proportion of the normal curve, the formula is

$$X = \mu + (z)(\sigma)$$

Returning to the 10 percent problem and substituting numbers for the mean, the z score, and the standard deviation, you get

$$X = 100 + (1.28)(15)$$
$$= 100 + 19.20$$
$$= 119.2$$
$$= 119 \qquad \text{(IQs are usually expressed as whole numbers.)}$$

Therefore, the IQ score required to be in the top 10 percent of the population is 119.

Here is a similar problem. Suppose a mathematics department wants to restrict the remedial math course to those who really need it. The department has the scores on the math achievement exam taken by entering freshmen for the past 10 years. The scores on this exam are distributed in an approximately normal fashion, with $\mu = 58$ and $\sigma = 12$. The department wants to make the remedial course available to those students whose mathematical achievement places them in the bottom third of the freshman class. The question is: What score will divide the lower third from the upper two-thirds? **Figure 6.13** should help to clarify the question.

The first step is to look in column C of Table C to find .3333. Again, such a proportion is not listed. The nearest proportion is .3336, which has a z value of $-.43$. (This time you are dealing with a z score below the mean, where all z scores are negative.) Applying $z = -.43$, you get

$$X = \mu + (z)(\sigma)$$
$$= 58 + (-.43)(12)$$
$$= 58 - 5.16$$
$$= 52.84 = 53$$

Using the theoretical normal curve to establish a cutoff score is efficient. All you need are the mean, the standard deviation, and confidence in your assumption that the

---

[10] You might use **interpolation** to determine a more accurate z score for a proportion of .1000. This extra precision (and labor) is unnecessary because the final result is rounded to the nearest whole number. For IQ scores, the extra precision does not make any difference in the final answer.

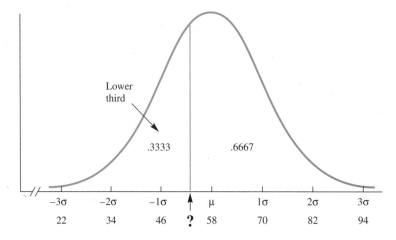

Lower
third

.3333   .6667

-3σ	-2σ	-1σ	↑	μ	1σ	2σ	3σ
22	34	46	?	58	70	82	94

FIGURE 6.13 **Distribution of scores on a math achievement exam**

scores are distributed normally. The empirical alternative for the mathematics department is to sort physically through all scores for the past 10 years, arrange them in a frequency distribution, and calculate the score that separates the bottom one-third.

PROBLEMS

15. Mensa is an organization of people who have high IQs. To be eligible for membership, a person must have an IQ "higher than 98 percent of the population." What IQ is required to qualify?

16. The mean height of American women aged 20–29 is 64.6 inches, with a standard deviation of 2.5 inches (*Statistical Abstract of the United States: 1999,* 1999).
   a. What height divides the tallest 5 percent of the population from the rest?
   b. The minimum height required for women to join the U.S. Army is 58 inches. What proportion of the population is excluded?

*17. The mean height of American men aged 20–29 is 69.8 inches, with a standard deviation of 3.0 inches (*Statistical Abstract of the United States: 1999,* 1999).
   a. The minimum height required for men to join the U.S. Army is 60 inches. What proportion of the population is excluded?
   b. What proportion of the population is taller than Napoleon Bonaparte, who was 5'2"?

18. The weight of many manufactured items is approximately normally distributed. For new U.S. pennies, the mean is 3.11 grams and the standard deviation is 0.05 gram (Youden, 1962).
   a. What proportion of all new pennies would you expect to weigh more than 3.20 grams?
   b. What weights separate the middle 80 percent of the pennies from the lightest 10 percent and the heaviest 10 percent?

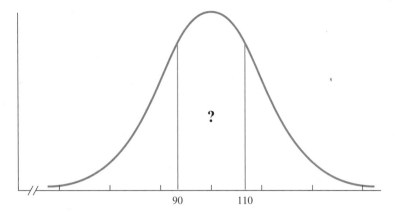

FIGURE 6.14 The normal distribution showing the IQ scores that define the "average" range

Finding the Proportion of the Population Between Two Scores

Table C in Appendix C can also be used to determine the proportion of the population between two scores. For example, IQ scores that fall in the range from 90 to 110 are often labeled "average." What proportion of the population falls in this range? **Figure 6.14** is a picture of the problem.

In this problem you must add an area on the left of the mean to an area on the right of the mean. First you need z scores that correspond to the IQ scores of 90 and 110:

$$z = \frac{90 - 100}{15} = \frac{-10}{15} = -.67$$

$$z = \frac{110 - 100}{15} = \frac{10}{15} = .67$$

The proportion of the distribution between the mean and $z = .67$ is .2486, and, of course, the same proportion is between the mean and $z = -.67$. Therefore, $(2)(.2486) = .4972$ or 49.72 percent. So approximately 50 percent of the population is classified as "average," using the "IQ = 90 to 110" definition.

What proportion of the population would be expected to have IQs between 70 and 90? **Figure 6.15** illustrates this question. There are two approaches to this problem. One is to find the area from 100 to 70 and then subtract the area from 90 to 100. The other way is to find the area beyond 90 and subtract from it the area beyond 70. I'll illustrate with the second approach. The corresponding z scores are

$$z = \frac{90 - 100}{15} = -.67 \quad \text{and} \quad z = \frac{70 - 100}{15} = -2.00$$

The area beyond $z = -.67$ is .2514, and the area beyond $z = -2.00$ is .0228. Subtracting the second proportion from the first, you find that .2286 of the population has an IQ in the range of 70 to 90.

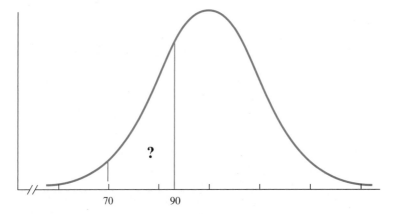

FIGURE 6.15 The normal distribution illustrating the area bounded by IQ scores of 90 and 70

PROBLEMS

*19. The distribution of 800 test scores in an introduction to psychology course was approximately normal, with $\mu = 35$ and $\sigma = 6$.
   a. What proportion of the students had scores between 30 and 40?
   b. What is the probability that a randomly selected student would score between 30 and 40?

20. Now that you know the proportion of students with scores between 30 and 40, would you expect to find the same proportion between scores of 20 and 30? If so, why? If not, why not?

21. Calculate the proportion of scores between 20 and 30. Be careful with this one; drawing a picture is especially advised.

22. How many of the 800 students would be expected to have scores between 20 and 30?

Finding the Extreme Scores in a Population

The extreme scores in a distribution are important in many statistical applications. Most often extreme scores in *either* direction are of interest. For example, many applications focus on the extreme 5 percent of the distribution. Thus, the upper $2\frac{1}{2}$ percent and the lower $2\frac{1}{2}$ percent receive attention. Turn to Table C and find the z score that separates the extreme $2\frac{1}{2}$ percent of the curve from the rest. (Of course, the z score associated with the lowest $2\frac{1}{2}$ percent of the curve will have a negative value.) Please memorize the z score you just looked up. This number will turn up many times in future chapters.

Here is an illustration. What two heart rates (beats per minute) separate the middle 95 percent of the population from the extreme 5 percent? According to studies summarized by Milnor (1990), the mean heart rate for humans is 71 beats per

FIGURE 6.16 Sketch showing the separation of the extreme 5 percent of the population from the rest

minute and the standard deviation is 9 beats per minute. To find the two scores, use the formula $X = \mu + (z)(\sigma)$. Substituting the values given, plus the z score you looked up, you get

Upper score	Lower score
$X = \mu + (z)(\sigma)$	$X = \mu - (z)(\sigma)$
$= 71 + (1.96)(9)$	$= 71 - (1.96)(9)$
$= 71 + 17.6$	$= 71 - 17.6$
$= 88.6$ or 87	$= 53.4$ or 53

Thus, 95 percent of the population is expected to have pulse rates between 53 and 87, which leaves 5 percent of the population above 87 or below 53. **Figure 6.16** is my sketch of this problem.

clue to the future

The idea of finding scores and proportions that are extreme in either direction will come up again after Chapter 7. In particular, the extreme 5 percent and the extreme 1 percent will be important.

PROBLEMS

23. What IQ scores are so extreme that they are achieved by only 1 percent of the population? Set this problem up using the "extreme 5 percent" example as a model.

24. What is the probability that a randomly selected person has an IQ higher than 139 *or* lower than 61?

25. Look at Figure 6.9 and suppose that the union leadership decided to ask for $0.85 per hour as a minimum wage. For those 185,822 workers, the mean was $0.99 with a standard deviation of $0.17. If $0.85 per hour was established as a minimum, how many workers would get raises?

26. Look at Figure 6.8 and suppose that a timber company decided to harvest all trees 8 inches DBH (diameter breast height) or larger from a 100-acre tract. On a 1-acre tract there were 199 trees with $\mu = 13.68$ and $\sigma = 4.83$. How many trees would be expected to be harvested from 100 acres?

COMPARISON OF THEORETICAL AND EMPIRICAL ANSWERS

You have been using the theoretical normal distribution to find probabilities and to calculate scores and proportions of IQs, diameters, wages, and other measures. Earlier in this chapter, the claim was made that *if* the empirical observations are distributed like a normal curve, accurate predictions can be made. A reasonable question is: How accurate are all these predictions I've just made? A reasonable answer can be fashioned from a comparison of the predicted proportions (from the theoretical curve) and the actual proportions (computed from empirical data). Figure 6.7 is based on 261 IQ scores of fifth-grade public school students. You worked through examples that produced proportions of people with IQs higher than 120, lower than 90, and between 90 and 110. These actual proportions can be compared with those predicted from the normal distribution. **Table 6.2** shows these comparisons.

As you can see by examining the Difference column of Table 6.2, the accuracy of the predictions ranges from excellent to not so good. Some of this variation can be explained by the fact that the mean IQ of the fifth-grade students was 101 and the standard deviation 13.4. Both the higher mean (101, compared with 100 for the normal curve) and the lower standard deviation (13.4, compared with 15) are due to the systematic exclusion of very-low-IQ children from regular public schools. Thus, the actual proportion of students with IQs lower than 90 is less than predicted, which is because our school sample is not representative of all 10- to 11-year-old children. This problem of sampling will be the major topic of the next chapter.

Although IQ scores *are* distributed approximately normally, many other scores are not. Karl Pearson recognized this, as have others. Theodore Micceri (1989) made this point again in an article titled, "The Unicorn, the Normal Curve, and Other Improbable Creatures." Caution is always in order when you are using theoretical distributions to make predictions about empirical events. However, don't let undue caution prevent you from getting the additional understanding that statistics offers.

T A B L E 6.2 Comparison of predicted and actual proportions

IQs	Predicted from normal curve	Calculated from actual data	Difference
Higher than 120	.0918	.0920	.0002
Lower than 90	.2514	.2069	.0445
Between 90 and 110	.4972	.5249	.0277

OTHER THEORETICAL DISTRIBUTIONS

In this chapter you have learned a little about rectangular distributions and binomial distributions and quite a bit about normal distributions. Later in this book you will encounter other distributions such as the *t* distribution, the *F* distribution, and the chi square distribution. In addition to the distributions in this book, mathematical statisticians have identified yet others, all of which are useful in particular circumstances. Some have interesting names such as the Poisson distribution; others have complicated names (hypergeometric distribution is an example). In every case, however, a distribution is used because it provides reasonably accurate probabilities about particular events.

PROBLEMS

27. For human infants born weighing 5.5 pounds or more, the mean gestation period is 268 days, which is just less than 9 months. The standard deviation is 14 days (McKeown and Gibson, 1951). What proportion of the gestation periods are expected to last 10 months or longer (300 days)?

28. The height of residential door openings in the United States is 6′8″. Use the information in problem 17 to determine the number of men among 10,000 who have to duck to enter a room.

29. An imaginative anthropologist measured the stature of 100 hobbits (using the proper English measure of inches) and found these values:

$$\Sigma X = 3600 \qquad \Sigma X^2 = 130{,}000$$

Assume that the heights of hobbits are normally distributed. Find μ and σ and answer the questions.

a. The Bilbo Baggins Award for Adventure is 32 inches tall. What proportion of the hobbit population is taller than the award?

b. Three hundred hobbits entered a cave that had an entrance 39 inches high. The orcs chased them out. How many hobbits could exit without ducking?

c. Gandalf is 46 inches tall. What is the probability that he is a hobbit?

30. Please review the objectives at the beginning of the chapter. Can you do what is asked?

 ## WEBSITE REFERENCES

http://psychology.wadsworth.com

Open "Choose Your Quiz" and select Spatz, 7e. Find the index of chapters, and click on Chapter 6, *Theoretical Distributions Including the Normal Distribution.*

Open "Choose Your Area," scroll down, and select "Research Methods and Statistics." Click on "Statistics" in the graph. Click on "Student Activities." Click on the Web Page button for "Z Scores."

chapter**seven**

SAMPLES, SAMPLING DISTRIBUTIONS, AND CONFIDENCE INTERVALS

OBJECTIVES FOR CHAPTER SEVEN

Study the text. Work the problems as you come to them. Afterward, you should be able to:

1. Define a random sample and obtain one if you are given a population of data
2. Define and identify biased sampling methods
3. Distinguish between random samples and the usual research samples
4. Define a sampling distribution and a sampling distribution of the mean
5. Describe the Central Limit Theorem
6. Calculate a standard error of the mean either from σ and N or from sample data
7. Describe the effect of N on the standard error of the mean
8. Use the z-score formula to find the probability that a sample mean or one more extreme was drawn from a population with a specified mean
9. Describe the t distribution
10. Calculate and interpret confidence intervals about sample means

P*lease be especially attentive to the concept of a sampling distribution. In inferential statistics, the idea of a **sampling distribution** is clearly the central concept. This book of "Tales of Distributions" is a book of tales of sampling distributions. Every chapter after this one is about some kind of sampling distribution; thus, you might consider this paragraph to be a superclue to the future.*

You will recall that *inferential statistics* is a method that takes chance factors into account when samples are used to reach conclusions about populations. The two central ideas in this definition are "takes chance factors into account" and "samples are used."

How can the effect of chance factors be determined? By using a sampling distribution, that's how. And what about getting a sample? I'll describe sampling, including some pitfalls. Thus, this chapter will cover topics that are central to inferential statistics.

Here is a story that brings up questions that can be answered using techniques presented in this chapter.

Late one afternoon, two students were discussing the average family income of students on their campus.

"Well, the average family income for college students is $57,000, nationwide," said the junior, while examining Sax (1999). "I'm sure the mean for this campus is at least that much."

"I don't think so," the sophomore replied. "I know lots of students who have only their own resources or come from pretty poor families. I'll bet you a dollar the mean for students here at State U. is below the national average."

"You're on," grinned the junior.

Together the two went out with a pencil and pad and asked ten students how much their family income was. The mean of these ten answers was $51,000.

Now the sophomore grinned. "I told you so; the mean here is $6000 less than the national average."

Disappointed, the junior immediately began to review their procedures. Rather quickly, a light went on. "Actually, this mean of $51,000 is meaningless — here's why. Those ten students aren't representative of the whole student body. They are late-afternoon students, and lots of them support themselves with temporary jobs while they go to school. Most students are supported from home by parents who have permanent and better-paying jobs. This sample is no good. We need results from the whole campus or at least from a representative sample."

To get the results from the whole student body, the two went the next day to the director of the financial aid office, who told them that family incomes for the student body are not public information. Sampling, therefore, was a necessity.

Though discouraged, the two students sought the advice of a friendly statistics professor. The professor explained how to obtain a random sample of students and suggested that 40 replies would be a practical sample size. Forty students, selected randomly, were identified. After three days of phone calls, visits, and call backs, the 40 responses produced a mean of $54,700, or $2300 less than the national average.

"Pay," demanded the sophomore.

"OK, OK, . . . here!"

Later, the junior began to think. What about another random sample and its mean? It would be different and it *could* be higher. Maybe the mean of $54,700 was just bad luck, just a chance event.

Shortly afterward, the junior confronted the sophomore with thoughts about repeated sampling. "How do we know that the random sample we got told us the truth about the population? Like, maybe the mean for the entire student body *is* $57,000. Maybe a sample with a mean of $54,700 would occur just by chance fairly often. I wonder what the chances are of getting a sample mean of $54,700 from a population with a mean of $57,000?"

"Well, that statistics professor told us that if we had any more questions to come back and get an explanation of sampling distributions. In the meantime, why don't I just apply my winnings toward a couple of ice cream cones."

The two statistical points of the story are that random samples are good (the junior paid off only after the results were based on a random sample) and that uncertainty about random samples can be reduced if you know about sampling distributions.

This is a chapter about getting a sample, drawing a conclusion about the population the sample came from, and knowing how much faith to put in your conclusion. Of course, there is some peril in this. Even the best methods of sampling produce variable results. How can you be sure that the sample you use will lead to a correct decision about its population? Unfortunately, you *cannot* be absolutely sure. **To draw a sample is to agree to accept some uncertainty about the results.** However, a sampling distribution allows you to measure the uncertainty. On the one hand, if a great deal of uncertainty exists, the sensible thing to do is to say you are uncertain. On the other hand, if there is very little uncertainty, the sensible thing to do is to reach a conclusion about the population, even though there *is* a small risk of being wrong. *Reread this paragraph; it is important.*

In this chapter, the first three sections are about samples. Following that, sampling distributions are explained and one method of drawing a conclusion about a population mean is presented. In the final sections, I will explain a second sampling distribution (the *t* distribution). The *t* distribution is used to calculate a **confidence interval,** an important statistic that provides additional information about a population mean.

RANDOM SAMPLES

When it comes to finding out about a population, the best sample is a random sample. In statistics, *random* refers to the method used to obtain the sample; it does not mean haphazard or unplanned. Any method that allows every possible sample of size N an equal chance to be selected produces a **random sample.** To obtain a random sample, you must do the following:

1. Define the population of numbers (scores).
2. Identify every member of the population.
3. Select numbers in such a way that every sample has an equal probability of being selected.

To illustrate, I will use the population of numbers in **Table 7.1,** for which $\mu = 9$ and $\sigma = 2$. Using these 20 scores allows us to satisfy requirements 1 and 2. As for requirement 3, here are two methods of selecting numbers so that all the possible samples have an equal probability of being selected. For this illustration, $N = 8$.

One method of getting a random sample is to write each number in the population on a slip of paper, put the 20 slips in a box, jumble them around, and draw out eight slips. The numbers on the slips are a random sample. This method works fine if the slips are all the same size, they are jumbled thoroughly, and the population has only a few members. If the population is large, this method is tedious.

TABLE 7.1 20 numbers used as a population; $\mu = 9$, $\sigma = 2$

9	7	11	13	10	8	10	6	8	8
10	12	9	10	5	8	9	6	10	11

TABLE 7.2 Assignment of identifying numbers to a population of numbers

ID number	Population number	ID number	Population number
01	9	11	10
02	7	12	12
03	11	13	9
04	13	14	10
05	10	15	5
06	8	16	8
07	10	17	9
08	6	18	6
09	8	19	10
10	8	20	11

A second (usually easier) method of selecting a random sample is to use a table of random numbers, such as Table B in Appendix C. To use the table, you must first assign an identifying number to each of the population numbers. **Table 7.2** is an example. The population numbers do not have to be arranged in any order.

Each population number is identified by a two-digit number. Now turn to **Table B** and pick an intersection of a row and a column. Any haphazard method will work; close your eyes and put your finger on a spot. Suppose you found yourself at row 80, columns 15–19 (page 364). Find that place. Reading horizontally, you find the digits 82279. You need only two digits to identify any member of your population, so you might as well use the first two (columns 15 and 16), which give you 8 and 2 or 82. Unfortunately, 82 is larger than any of the identifying numbers, so it doesn't match a number in the population, but at least you are started. From this point you can read two-digit numbers in any direction — up, down, or sideways — but the decision should be made before you look at the numbers. If you decide to read down, you find 04, which corresponds to population number 13 in Table 7.2, so 13 becomes the first number in the sample. The next identifying number is 34, which again does not correspond to a population number. Indeed, the next ten numbers are too large. The next usable ID number is 16, which places an 8 in the sample. Continuing the search, you reach the bottom of the table. At this point you can go in any direction; I moved to the right and started back up the two outside columns (18 and 19). The first number, 83, was too large, but the next identifying number 06 corresponded to an 8, which went into the sample. Next, a 15 and a 20 identified population numbers of 5 and 11 for the sample. The next number that is between 01 and 20 is 15, but it has already been used, so it should be ignored. Identifying numbers of 18, 02, and 14 produces population numbers of 6, 7, and 10, which completes the sampling task. Thus, the random sample of eight consists of these numbers: 13, 8, 8, 5, 11, 6, 7, and 10.

PROBLEM

*1. A random sample is supposed to yield a statistic similar to the population parameter. Find the mean of the random sample of eight numbers selected in the text.

What is this table of random numbers? In Table B (and in any set of random numbers), the probability of occurrence of any digit from 0 to 9 at any place in the table is the same: .10. Thus, you are just as likely to find 000 as 123 or 397. Incidentally, you cannot generate random numbers out of your head. Certain sequences begin to recur, and (unless warned) you will not include enough repetitions like 000 and 555. If warned, you will produce too many.

Here are some suggestions for using a table of random numbers efficiently.

1. Make a check beside the identifying number of a score when it is chosen for the sample. This helps to prevent duplications.
2. If the population is large (more than 100), it is more efficient to get all the identifying numbers from the table first. As you select them, put them in some rough order to help prevent duplications. After you have all the identifying numbers, go to the population to select the sample.
3. If the population has exactly 100 members, let 00 be the identifying number for 100. By doing this, you can use two-digit identifying numbers, each one of which matches a population score. This same technique works for populations of 10 or 1000 members.

Random sampling has two uses. (1) It is the best method of sampling if you want to generalize to a population. The two sections that follow, on biased samples and research samples, will both address the issue of generalizing from the sample to the population. (2) In an entirely separate use, random sampling is the mathematical basis for creating sampling distributions, which are central to inferential statistics. Researchers use sampling distributions independently of the method used to gather their sample of data.

PROBLEMS

2. Draw a random sample of 10 from the population in Table 7.1. Calculate $\overline{X}$.

3. Draw a random sample with $N = 12$ from the following scores:

76	47	81	70	67	80	64	57	76	81
68	76	79	50	89	42	67	77	80	71
91	72	64	59	76	83	72	63	69	
78	90	46	61	74	74	74	69	83	

BIASED SAMPLES

A **biased sample** is one obtained by a method that systematically underselects or overselects from certain groups in the population. Thus, with a biased sampling technique, every sample of a given size does *not* have an equal opportunity of being selected. With biased sampling techniques, you are much more likely to get an unrepresentative sample than you are with random sampling.

For example, it is reasonable that some conclusions based on mailed questionnaires are not valid because the sampling methods were biased. Usually an investigator defines the population, identifies each member, and mails the questionnaire to a randomly selected sample. Suppose that 60 percent of the recipients respond. Can valid results for the population be based on the questionnaires returned? Probably not. There is good reason to suspect that the 60 percent who responded are different from the 40 percent who did not. Thus, although the population is made up of both kinds of people, the sample reflects only one kind. Therefore, the sample is biased. The probability of bias is particularly high if the questionnaire elicits feelings of pride or despair or disgust or apathy in *some* of the recipients.

PROBLEMS

4. Suppose a large number of questionnaires about educational accomplishments are mailed out. Do you think that some recipients will be more likely to return the questionnaire than others? Which ones? If the sample is biased, will it overestimate or underestimate the educational accomplishments of the population?

5. Sometimes newspapers sample opinions of the local population by printing a "ballot" and asking readers to mark it and mail it in. Evaluate this sampling technique.

6. Consider as a population the students at State U. whose names are listed alphabetically in the student directory. Are the following samples biased or random?
  a. Every fifth name on the list
  b. Every member of the junior class
  c. 150 names drawn from a box that contains all the names in the directory
  d. One name chosen at random from the directory
  e. A random sample from those taking the required English course

A very famous case of a biased sample occurred in a poll that was to predict the results of the 1936 election for president of the United States. The *Literary Digest* (a popular magazine) mailed 10 million "ballots" to those on its master mailing list, a list of more than 10 million people compiled from "all telephone books in the U.S., rosters of clubs, lists of registered voters," and other sources. More than 2 million "ballots" were returned and the prediction was clear: Alf Landon by a landslide over Franklin Roosevelt. As you may have learned, the actual results were just the opposite; Roosevelt got 61 percent of the vote.

From the 10 million who had a chance to express a preference, 2 million very interested persons had selected themselves. This 2 million had more than its proportional share of those who were disgruntled with Roosevelt's depression-era programs. The 2 million ballots were a biased sample; the results were not representative of the population. In fairness, it should be noted that the *Literary Digest* used a similar master list in 1932 and predicted the popular vote within 1 percentage point.[1]

---

[1] See the *Literary Digest,* August 22, 1936, and November 14, 1936.

Obviously, researchers want to avoid biased samples; random sampling seems like the appropriate solution. The problem, however, is step 2 on page 136. For almost all research problems, it is impossible to identify every member of the population. Thus, researchers are confronted with a dilemma.

RESEARCH SAMPLES

Researchers want to avoid a biased sample, but they know that, for most research problems, it is *impossible* to obtain a random sample. Populations that researchers are interested in, such as college students, newborn infants, and people with passive-aggressive personality disorders, just cannot be sampled from randomly.[2] This means that a nonrandom sample will have to do if research is to be conducted.

Researchers have a number of ways of maintaining respect for their work even though they cannot use random samples. The best of these ways is to use *random assignment* of participants to groups. With random assignment, the available participants are partitioned into groups using a table of random numbers, which creates groups that are equivalent at the start of an experiment. Samples that have been randomly assigned have many of the same characteristics as random samples.

In addition, researchers who are aware of the problems of sampling may develop and then describe their efforts to avoid a biased sample. Also, one experiment does not stand alone. Other researchers are conducting similar investigations. Usually, the results based on research samples are true for other samples from the same population.

The fact that a conclusion is not based on a random sample should not alarm you. Nonrandom (though carefully controlled) samples are used by a very wide variety of people and organizations. The public opinion polls reported by Gallup, Lou Harris, and the Roper organization are based on carefully selected samples. ASCAP, the musicians' union, samples the broadcasts of more than 9000 U.S. radio stations and then collects royalties for members based on sample results. Inventory procedures in large retail organizations involve sampling. What you know about your blood is based on the analysis of a small sample. (Thank goodness!)

The reason for this widespread use of sampling is that it works; you can find out what you want to know by using a sample. This is true even though research samples are seldom random samples.

SAMPLING DISTRIBUTIONS

A sampling distribution is always a sampling distribution of a particular statistic. The statistic might be the mean, the variance, the correlation coefficient, or some other statistic. To conceptualize a sampling distribution, think about many equal-sized random samples being drawn from a population. The same statistic is calculated for each sample. Next, all these calculated statistics (one from each sample) are arranged into a frequency distribution. Now, imagine a frequency polygon of this frequency distri-

---

[2] Technically, the population consists of the measurements of the members and not the members themselves.

bution. Finally, if the frequency polygon is a normal curve, then you can use z scores to find probabilities.

What you just read describes the *empirical* construction of a sampling distribution. The sampling distributions you will use for your problems will be *theoretical* ones that have been derived from formulas. Both are the same, though, in what they can show. A sampling distribution will give you the probability associated with a particular portion of the sampling distribution. That is, a sampling distribution will give you the probability of the occurrence of statistics with particular values when random samples are drawn from a population.

Sampling distributions are so important that statisticians have special names for their mean and standard deviation. The mean of a sampling distribution is called the **expected value** and the standard deviation is called the **standard error.**[3] I will not have much more to say about expected value, but if you continue your study of statistics, you will encounter it. The standard error, however, will be used many times in this text.

THE SAMPLING DISTRIBUTION OF THE MEAN

Remember the population of 20 scores that you sampled from earlier in this chapter? For problem 2, you drew a sample of 10 and calculated the mean. For almost every sample in your class, the mean was different. Each, however, was an *estimate* of the population mean.

My example problem in the section on random sampling used $N = 8$; my one sample produced a mean of 8.50. Now, think about drawing 200 samples with $N = 8$ from the population of Table 7.1 scores, calculating the 200 means, and constructing a frequency polygon. You may already be thinking, "That's a job for a computer." Right. The result is Figure 7.1.

Look at **Figure 7.1.** Notice how nicely centered the sampling distribution is about the population mean of 9.00. Most of the sample means ($\overline{X}$'s) are fairly close to the population parameter, μ.

The characteristics of a sampling distribution of the mean are:

1. Every sample is drawn randomly from a specified population.
2. The sample size (N) is the same for all samples.
3. The number of samples is very large.
4. The mean ($\overline{X}$) is calculated for each sample.[4]
5. The sample means are arranged into a frequency distribution.

I hope that when you looked at Figure 7.1, you were at least suspicious that it might be the ubiquitous normal curve. It is. Now you are in a position that educated people often find themselves: What you learned in the past (how to use the normal curve for scores, X) can be used for a different problem — describing the relationship between $\overline{X}$ and μ.

---

[3] In this and in other statistical contexts, the term *error* means deviations or random variation. The word *error* is left over from the 19th century, when random variation was referred to as the "normal law of error." Of course, *error* sometimes means mistake, so you will have to be alert to the context when this word appears (or you may make an error).

[4] To create a sampling distribution of a statistic other than the mean, substitute that statistic at this step.

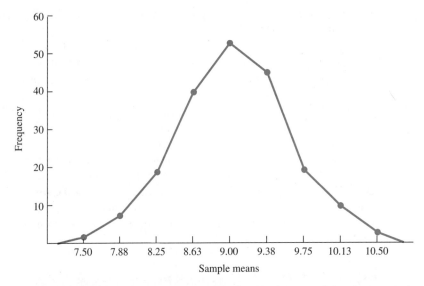

FIGURE 7.1 Empirical sampling distribution of the mean from the population in Table 7.1. For each sample mean, N = 8.

Of course, the normal curve is a theoretical curve, and I presented you with an empirical curve that only appears normal. I would like to let you prove for yourself that the form of a sampling distribution of the mean is a normal curve, but, unfortunately, that requires mathematical sophistication beyond that assumed for this course. So I will resort to a time-honored teaching technique — an appeal to authority.

CENTRAL LIMIT THEOREM

The authority I appeal to is mathematical statistics, which has proved a theorem called the **Central Limit Theorem:**

> For any population of scores, regardless of form, the sampling distribution of the mean will approach a normal distribution as N (sample size) gets larger. Furthermore, the sampling distribution of the mean will have a mean (the expected value) equal to μ and a standard deviation (the standard error) equal to $\sigma/\sqrt{N}$.

This appeal to authority resulted in a lot of information. Now you know not only that sampling distributions of the mean approach normal curves but also that, if you know the population parameters σ and μ, you can determine the parameters of the sampling distribution. The standard error of the sampling distribution is the standard deviation of the population (σ) divided by the square root of the sample size. The expected value of the sampling distribution is the same as the population mean, μ. One final point: The standard error of the mean is symbolized $\sigma_{\overline{X}}$, and the expected value of the mean is symbolized $E(\overline{X})$.

The most remarkable thing about the Central Limit Theorem is that it works *regardless* of the form of the original population. The left part of **Figure 7.2** shows two populations you are familiar with. One is the rectangular distribution of playing cards

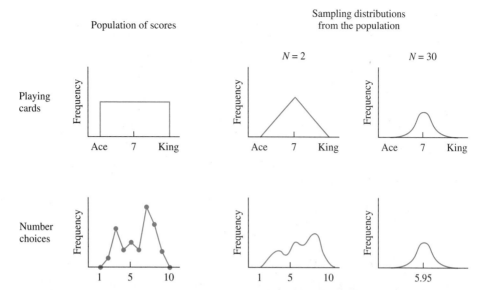

FIGURE 7.2 Populations of playing cards and number choices. Sampling distributions from each population with sample sizes of *N* = 2 and *N* = 30.

(Figure 6.1), and the other is the bimodal distribution of choices of numbers between 1 and 10 (Figure 6.5). The next two panels are sampling distributions for *N* = 2 and *N* = 30, which show the form of the distribution becoming more normal as *N* increases. Figure 7.2 illustrates that, regardless of the form of the population, the form of the sampling distribution of the mean approaches normal if *N* is large enough.

How large must *N* be for the sampling distribution of the mean to be normal? The traditional answer is 30 or more. However, if the population itself is symmetrical, then sampling distributions of the mean will be normal with *N*'s much smaller than 30. In contrast, if the population is severely skewed, *N*'s of more than 30 will be required.

Finally, the Central Limit Theorem does not apply to all sample statistics. Sampling distributions of the median, standard deviation, variance, and correlation coefficient are not normal distributions. The Central Limit Theorem does apply to the mean, which is a most important and popular statistic. (In a frequency count of all statistics, the mean is the mode.)

PROBLEMS

7. The standard deviation of a sampling distribution is called the _____ and the mean is called the _____.

8. a. Write the steps needed to construct an empirical sampling distribution of the range.
   b. What is the name of the standard deviation of this sampling distribution?
   c. Think about the expected value of the range compared to the population range. Write a statement about the relationship.

9. Describe the Central Limit Theorem in your own words.

10. In Chapter 5 you learned how to use a regression equation to predict a score ($\hat{Y}$) if you are given an X score. $\hat{Y}$ is a statistic, so naturally it has a sampling distribution with its own standard error. What can you conclude if the standard error is very small? Very large?

Calculating the Standard Error of the Mean

Calculating the standard error of the mean is fairly simple. I will illustrate with an example that you will use again. Recall that for the population in Table 7.1, σ is 2. For a sample size of eight, the standard error of the mean is

$$\sigma_{\bar{X}} = \frac{\sigma}{\sqrt{N}} = \frac{2}{\sqrt{8}} = \frac{2}{2.828} = 0.707$$

The Effect of Sample Size on the Standard Error of the Mean

As you can see by looking at the formula for the standard error of the mean, $\sigma_{\bar{X}}$ becomes smaller as N gets larger. **Figure 7.3** shows four sampling distributions of the mean, all based on the population of numbers in Table 7.1. The sample sizes are 2, 4, 8, and 16. A sample mean of 10 is included in all four figures as a reference point. Notice that, as N increases, a sample mean of 10 becomes less and less likely. The importance of sample size will become more apparent as this book progresses.

Determining Probabilities About Sample Means

To summarize where we are at this point: Mathematical statisticians have produced a mathematical invention, the sampling distribution. The sampling distribution of the mean, they tell us, is a normal curve. Fortunately, having worked problems in Chapter 6 about normally distributed *scores,* we are in a position to check this claim about normally distributed *means.*

One check is fairly straightforward, given that I already have the 200 sample means from the population in Table 7.1. To make this check, I will determine the proportion of sample means that are above a specified point, *using the theoretical normal curve* (Table C). I can then compare this theoretical proportion to the proportion of the 200 sample means that are *actually* above the specified point. If the two figures are similar, I have evidence that the normal curve can be used to answer questions about sample means.

The z score for a sample mean drawn from a sampling distribution with mean μ and standard error $\sigma/\sqrt{N}$ is

$$z = \frac{\bar{X} - \mu}{\sigma_{\bar{X}}}$$

Any sample mean will do for this comparison; I will use 10.0. The mean of the population is 9.0 and the standard error (for $N = 8$) is 0.707. Thus,

$$z = \frac{\bar{X} - \mu}{\sigma_{\bar{X}}} = \frac{10.0 - 9.0}{0.707} = 1.41$$

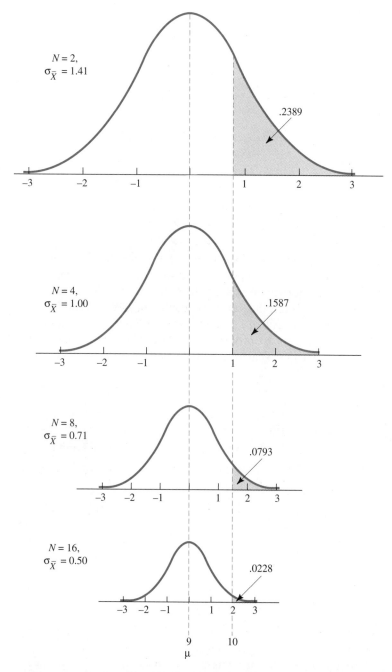

FIGURE 7.3 Sampling distributions of the mean for four different sample sizes. All samples are drawn from the population in Table 7.1. Note how a sample mean of 10 becomes rarer and rarer as $\sigma_{\bar{X}}$ becomes smaller.

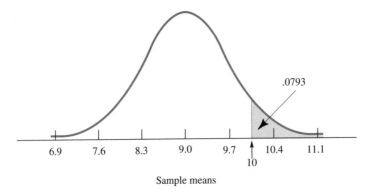

FIGURE 7.4 Theoretical sampling distribution of the mean from the population in Table 7.1. For each sample, $N = 8$.

By consulting Table C, I see that the proportion of the curve above a z value of 1.41 is .0793. **Figure 7.4** shows the sampling distribution of the mean for $N = 8$. Sample means are on the horizontal axis; the .0793 area above $\overline{X} = 10$ is shaded.

How accurate is this theoretical prediction of .0793? When I looked at the distribution of sample means that I used to construct Figure 7.1, I found that 13 of the 200 had means of 10 or more, a proportion of .0650. Thus, the theoretical prediction is off by less than $1\frac{1}{2}$ percent. That's not bad; the normal curve model passes the test.[5]

PROBLEMS

*11. When the population parameters are known, the standard error of the mean is $\sigma_{\bar{X}} = \sigma/\sqrt{N}$. The following table gives four σ values and four N values. For each combination, calculate $\sigma_{\bar{X}}$ and enter it in the table.

	σ			
N	1	2	4	8
1				
4				
16				
64				

12. On the basis of the table you constructed in problem 11, write a precise verbal statement about the relationship between $\sigma_{\bar{X}}$ and N.

13. To reduce $\sigma_{\bar{X}}$ to one-fourth its size, you must increase N by how much?

14. For the population in Table 7.1, and for samples with $N = 8$, what proportion of the sample means will be 8.5 or less?

[5] As you might suspect, the validity of the normal curve model for the sampling distribution of the mean has been established with a mathematical proof rather than an empirical check.

15. For the population in Table 7.1, and for samples with $N = 16$, what proportion of the sample means will be 8.5 or less? 10 or greater?

16. As you know from the previous chapter, for IQs: $\mu = 100$ and $\sigma = 15$. What is the probability that a first-grade classroom of 25 students who are chosen randomly from the population will have a mean IQ of 105 or greater? 90 or less?

17. Now you are in a position to return to the story of the two students at the beginning of this chapter. Find, for the junior, the probability that a sample of 40 from a population with $\mu = \$57,000$ and $\sigma = \$15,000$ would produce a mean of $54,700.

Using your answer to problem 17, the junior might say to the sophomore,

"... and so, the mean of $54,700 isn't very reliable. The standard deviation is large, and with $N = 40$, samples can bounce all over the place. Why, if the real mean for our campus is the same as the $57,000 national average, we would *expect* about one-sixth of all our random samples of 40 students to have means of $54,700 or less."

"Yeah," said the sophomore, "but that *if* is a big one. What if the campus mean is really $54,700? Then the sample we got was right on the nose. After all, a sample mean is an *unbiased estimator* of the population parameter."

"I see your point. And I see mine, too. It seems like either of us could be correct. That leaves me uncertain about the real campus parameter."

"Me, too."

Not all statistical stories end with so much uncertainty. However, I said that one of the advantages of random sampling is that you can measure the uncertainty. You measured it, and there was a lot. Remember, if you agree to draw a sample, you agree to accept some uncertainty about the results.

CONSTRUCTING A SAMPLING DISTRIBUTION WHEN σ IS NOT AVAILABLE

Let's review what you just did. You answered some questions about sample means by relying on a table of the normal curve. Your justification for saying that sample means are distributed normally was the Central Limit Theorem. The Central Limit Theorem always applies when the sample size is adequate and you know σ, both of which were true for the problems you worked. For those problems, you were given σ or calculated it from the population of data you had available.

In the world of empirical research, however, you often do not know σ and you don't have population data to calculate it. Because researchers are always inventing new dependent variables, an unknown σ is common. Without σ, the justification for using the normal curve evaporates. What to do if you don't have σ? Can you suggest something?

One solution is to use $\hat{s}$ as an estimate of σ. (Was that your suggestion?) This was the solution used by researchers about a century ago. They knew that $\hat{s}$ was only an estimate and that the larger the sample, the better the estimate. Thus, they chose problems for which they could gather huge samples. (Remember Karl Pearson and

Alice Lee's data on the 1376 pairs of father–daughter heights?) Very large samples produce an $\hat{s}$ that is identical to σ, for all practical purposes.

Other researchers, however, could not gather that much data. One of those we remember today is W. S. Gosset (1876–1937), who worked for Arthur Guinness, Son & Co., a brewery in Dublin, Ireland. Gosset had majored in chemistry and mathematics at Oxford, and his job at Guinness was to make scientifically based recommendations about brewing. Many of these recommendations were based on sample data.

Gosset was familiar with the normal curve and the strategy of using large samples to accurately estimate σ. Unfortunately, though, his samples were small. Gosset (and other statisticians) knew that such small-sample $\hat{s}$ values were not accurate estimators of σ and thus the normal curve could not be relied on for accurate probabilities.

Gosset's solution was to work out the mathematics of distributions based on $\hat{s}$. He found that the distribution depended on the sample size, with a different distribution for each N. These distributions make up a family of curves that have come to be called the **t distribution**.[6] The t distribution is an important tool for those who analyze data. I will use it for confidence interval problems in this chapter and for four other kinds of problems in later chapters.

THE t DISTRIBUTION

The different curves that make up the t distribution are distinguished from one another by their **degrees of freedom.** Degrees of freedom (abbreviated df) range from 1 to ∞. Knowing the degrees of freedom for your data tells you which t distribution to use.[7]

Determining the correct number of degrees of freedom for a particular problem can become fairly complex. For the problems in this chapter, however, the formula is simple: $df = N - 1$. Thus, if the sample consists of 12 members, $df = 11$. In later chapters I will give you a more thorough explanation of degrees of freedom (and additional formulas).

Figure 7.5 is a picture of three t distributions. Their degrees of freedom are 2, 9, and ∞. You can see that as the degrees of freedom increase, less and less of the curve is in the tails. Note that the t values on the horizontal axis are quite similar to the z scores used with the normal curve.

The t Distribution Table

You won't be surprised to learn that, in the same way that the normal curve table is used to obtain probabilities, a t distribution table is used to obtain probabilities. It is Table D in Appendix C. Table D, however, is arranged differently from the z-score table.

---

[6] Gosset spent several months in 1906–07 studying with Karl Pearson in London. It was during this period that the t distribution was developed.

[7] Traditionally, the t distribution is written with a lowercase t. A capital T is used for another distribution (covered in Chapter 14) and as a standardized score. However, because some computer programs do not print lowercase letters, t becomes T on some printouts (and often in text based on that printout). Be alert.

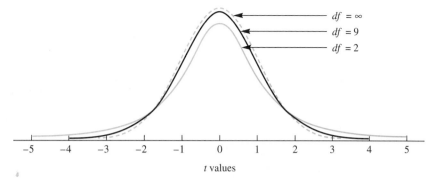

FIGURE 7.5 **Three different *t* distributions**

Look at **Table D** (page 368). The first column shows degrees of freedom, ranging from 1 to ∞. Across the top are three rows; the row you use depends on the kind of problem you are using the *t* distribution for. (In this chapter, you will use the top row for confidence intervals.) Each column is associated with a percent or probability. The body of the table contains *t* values.

The most common use of Table D is to find a *t* value. Based on an analysis of the problem, degrees of freedom are determined.[8] Based on considerations by the researcher, one of the top three rows and a probability figure are chosen. The intersection of row and column gives a *t* value, which is used in the analysis.

Dividing a Distribution into Portions

Look at **Figure 7.6,** which shows separate versions of the three distributions in Figure 7.5. Each is divided into a large middle portion and smaller tail portions. In all three curves in Figure 7.6, the middle portion is 95 percent and the other 5 percent is divided evenly between the two tails.

Notice that the *t* values that separate equal portions of the three curves are different. The fewer the *df*, the larger the *t* value that is required to separate out the extreme 5 percent of the curve.

The *t* values in Figure 7.6 came from Table D. Look at **Table D** for the 9 *df* row. Now look in the 95% confidence interval column. The *t* value at the intersection is 2.26, the number used in the middle curve of Figure 7.6.

Finally, look at the top curve in **Figure 7.6.** When *df* = ∞, the *t* scores that separate the middle 95 percent from the extreme 5 percent are −1.96 and +1.96. You have seen these numbers before. In fact, ±1.96 are the *z* scores you used in the preceding chapter to find heart rates that characterize the extreme 5 percent of the population. (See Figure 6.16.) Thus, a *t* distribution with *df* = ∞ is a normal curve. In summary, as *df* increases, the *t* distribution approaches the normal distribution.

Your introduction to the *t* distribution is in place. After working the problems that follow, you will be ready to calculate confidence intervals.

---

[8] If your data have degrees of freedom that are between two tabled values, be conservative and use the *t* value for the smaller *df*.

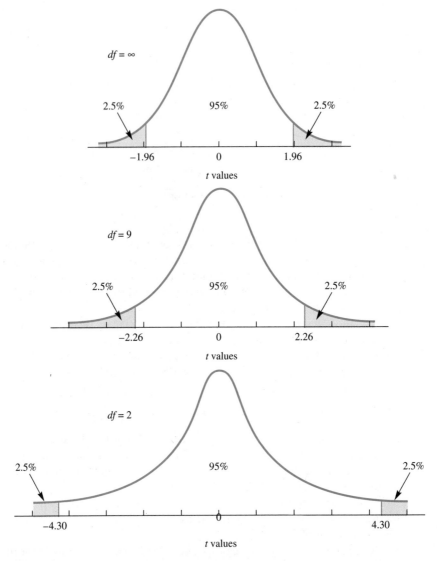

FIGURE 7.6 Three *t* distributions showing the *t* values that enclose 95 percent of the cases

PROBLEMS

18. Fill in the blanks in the statements with "*t*" or "normal."

a. There is just one _____ distribution, but there are many _____ distributions.

b. Given a value of +2.00 on the horizontal axis, a _____ distribution has a greater proportion to the right than does a _____ distribution.

c. A _____ distribution has a larger standard deviation than does a _____ distribution.

19. To know which t distribution to use, determine _____.

20. Who invented the t distribution? For what purpose?

CONFIDENCE INTERVAL ABOUT A SAMPLE MEAN

A confidence interval about a sample mean is an inferential statistic. Based on sample data from a population, a lower and an upper limit are calculated. The mean of the sample is in the middle of the interval. With a certain degree of confidence (usually 95 percent or 99 percent), you can state that the population mean is in the interval.

To find the lower and upper limits of a confidence interval about a population mean, use the formulas

$$LL = \overline{X} - t(s_{\overline{X}})$$
$$UL = \overline{X} + t(s_{\overline{X}})$$

where $\overline{X}$ is the mean of the sample from the population

t is a value from the t distribution table

$s_{\overline{X}}$ is the standard error of the mean, calculated from a sample

The formula for $s_{\overline{X}}$ is quite similar to the formula for $\sigma_{\overline{X}}$:

$$s_{\overline{X}} = \frac{\hat{s}}{\sqrt{N}}$$

where $\hat{s}$ is the standard deviation of the sample (using $N - 1$).

A confidence interval is useful for evaluating quantitative claims such as those made by manufacturers. For example, some light bulb manufacturers claim that their 60-watt bulbs will last for 1000 hours. On the average, is this true? One way to answer the question is to gather data on the number of hours 60-watt bulbs burn. The following are fictitious data:

$$N = 16$$
$$\Sigma X = 15{,}600$$
$$\Sigma X^2 = 15{,}247{,}500$$

PROBLEM

*21. For the data on the lives of 60-watt light bulbs, calculate the mean ($\overline{X}$), the standard deviation ($\hat{s}$), and the standard error of the mean ($s_{\overline{X}}$).

You can calculate a 95 percent confidence interval about the sample mean in problem 21 and use it to evaluate the claim that bulbs last 1000 hours. With $N = 16$, $df = N - 1 = 15$. The t value in Table D for a 95 percent confidence interval with $df = 15$ is 2.131. Using the answers you found in problem 21, you get

$$\text{LL} = \overline{X} - t(s_{\overline{X}}) = 975 - 2.131(12.5) = 948.36 \text{ hours}$$
$$\text{UL} = \overline{X} + t(s_{\overline{X}}) = 975 + 2.131(12.5) = 1001.64 \text{ hours}$$

Thus, you are 95 percent confident that the interval of 948.36 to 1001.64 hours contains the true mean lighting time of 60-watt bulbs. The advertised claim of 1000 hours *is* in the interval, but just barely. A reasonable interpretation is to recognize that samples are changeable, variable things and that you do not have good evidence showing that the lives of bulbs are shorter than claimed. But because the interval *almost* does not capture the advertised claim, you might have an incentive to gather more data.

To obtain a 99 percent or a 90 percent confidence interval, use the *t* values under those headings in Table D.

As another example, normal body temperature is commonly given as 98.6°F. In Chapters 2 and 3 you were given some data based on actual measurements (Mackowiak, Wasserman, and Levine, 1992) that showed a mean temperature of *less* than 98.6°F. Is it true that we have been wrong all these years?

Or, perhaps the study's sample mean is within the expected range of sample means taken from a population with a mean of 98.6°F. A 99 percent confidence interval will help in choosing between these two alternatives. The summary statistics are:

$$N = 40$$
$$\Sigma X = 3928.0$$
$$\Sigma X^2 = 385,749.24$$

These summary statistics produce the following values:

$$\overline{X} = 98.2$$
$$\hat{s} = 0.710$$
$$s_{\overline{X}} = 0.112$$
$$df = 39$$

The next task is to find the appropriate *t* value in Table D. Look under the heading for a 99 percent confidence interval. With 39 *df* be conservative and use 30 *df* so t_{99} (30 *df*) = 2.75. The 99 percent confidence limits about the sample mean are:

$$\text{LL} = \overline{X} - t(s_{\overline{X}}) = 98.2 - 2.704(0.112) = 97.9°\text{F}$$
$$\text{UL} = \overline{X} + t(s_{\overline{X}}) = 98.2 + 2.704(0.112) = 98.5°\text{F}$$

What conclusion do you reach about the 98.6°F standard? We *have* been wrong all these years. On the basis of actual measurements, you can be 99 percent confident that normal body temperature is between 97.9°F and 98.5°F.[9]

In summary, confidence intervals for population means produce an interval statistic (lower and upper limits) that is destined to contain μ 95 percent of the time (or 99 or 90). Whether or not a particular confidence interval contains the unknowable μ is uncertain, but you have control over the degree of that uncertainty.

---

[9] C. R. A. Wunderlich, a German investigator in the mid-19th century, was influential in establishing 98.6°F as the normal body temperature.

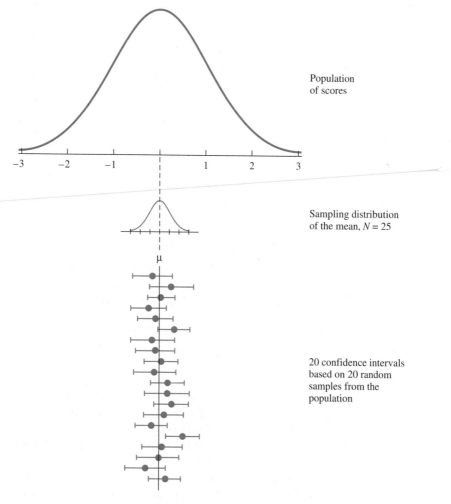

FIGURE 7.7 Twenty 95 percent confidence intervals, each calculated from a random sample of 25 from a normal population. Each sample mean is represented by a dot.

The concept of confidence intervals has been known to be troublesome. Here is another explanation of confidence intervals, this time with a picture.

Look at **Figure 7.7,** which has these features:

- A population of scores (top curve) [10]
- A sampling distribution of the mean when $N = 25$ (small curve)
- Twenty 95 percent confidence intervals based on random samples ($N = 25$) from the population

On each of the 20 horizontal lines, the endpoints represent the lower and upper limits, and the filled circle shows the mean. As you can see, nearly all the confidence in-

---

[10] It is not necessary that the population be normal.

tervals have "captured" μ. One has not. Find it. Does it make sense to you that 1 out of 20 of the 95 percent confidence intervals does not contain μ?

CATEGORIES OF INFERENTIAL STATISTICS

Two of the major categories of inferential statistics are confidence intervals and hypothesis testing. A confidence interval establishes an interval within which a population parameter is expected to lie (with a certain degree of confidence). You will see confidence interval problems again in Chapter 9.

The technique of hypothesis testing allows you to use sample data to test a hypothesis about a population parameter. The result is a yes/no decision. Hypothesis testing will be a topic in each of the remaining chapters in this book.

PROBLEMS

22. A social worker conducted an 8-week assertiveness training workshop. Afterward, the 14 clients took the Door Manifest Assertiveness Test, which has a national mean of 24.0. Use the data that follow to construct a 95 percent confidence interval about the sample mean. Write an interpretation about the effectiveness of the workshop.

$$\Sigma X = 378 \qquad \Sigma X^2 = 10,400$$

23. Airplane manufacturers use only slow-burning materials. For the plastic in coffee pots the maximum burn rate is 4 inches per minute. Raytheon Aircraft randomly sampled from a box of 120 Krups coffee pots and obtained burn rates of 1.20, 1.10, and 1.30 inches per minute. Calculate a 99.9 percent confidence interval and write an interpretation that tells about using Krups coffee makers in airplanes.

24. This problem will give you practice in constructing confidence intervals and in discovering how to reduce their size (without sacrificing confidence). Smaller confidence intervals tell you the value of μ more precisely.
  a. How wide is the 95 percent confidence interval about a sample mean when $\hat{s} = 4$ and $N = 4$?
  b. How much smaller is the 95 percent confidence interval about a sample mean when N is increased fourfold to 16 ($\hat{s} = 4$)?
  c. Compared to your answer in part a, how much smaller is the 95 percent confidence interval about a sample mean when N remains 4 and $\hat{s}$ is reduced to 2?

25. In Figure 7.7, the 20 lines that represent confidence intervals vary in length. What aspect of the sample causes this variation?

26. Using Figure 7.7 as a model of 95 percent confidence intervals, imagine a graphic with confidence intervals based on $N = 100$.
  a. How many of the 20 intervals are expected to capture μ?
  b. Will the confidence intervals be wider or narrower? By how much?

27. Using Figure 7.7 as a model, imagine that the lines are 90 percent confidence intervals rather than 95 percent confidence intervals.

a. How many of the 20 intervals are expected to capture μ?

b. Will the confidence intervals be wider or narrower?

28. In the accompanying figure, look at the curve of performance over trials, which is generally upward except for the dip at trial 2. Is the dip just a chance variation, or is it a reliable, nonchance drop? By indicating the lower and upper confidence limits for each mean, a graph can convey at a glance the reliability that goes with each point. Calculate a 95 percent confidence interval for each mean. Pencil in the lower and upper limits for each point on the graph. Write a sentence explaining the dip at trial 2.

	Trials			
	1	2	3	4
$\overline{X}$	8	6	14	16
$\hat{s}$	5	5	5	5
N	25	25	25	25

29. Here's a problem you probably were expecting. Review the objectives at the beginning of the chapter. You *are* using this valuable, memory-consolidating technique, aren't you?

 ## WEBSITE REFERENCES

http://psychology.wadsworth.com

Open "Choose Your Quiz" and select Spatz, 7e. Find the index of chapters, and click on Chapter 7, *Samples, Sampling Distributions, and Confidence Intervals.*

Open "Choose Your Area," scroll down, and select "Research Methods and Statistics." Click on "Statistics" in the graph. Click on "Student Activities." Click on the Web Page button for "Standard Error."

transition page

you are in the midst of material on inferential statistics, which, as you probably recall, is a technique that allows you to use sample data to help make decisions about populations. In the last chapter you studied and worked problems on *confidence intervals.* Those confidence intervals allowed you to decide something about the size of a population mean of an unmeasured population.

The next chapter, Chapter 8, will cover the basics of *hypothesis testing,* a more widely used inferential statistics technique. Hypothesis testing results in a yes/ no decision about a population parameter. The decision about the population parameter is based on a probability that is obtained from a sampling distribution.

In Chapter 8 you will test hypotheses about population means and population correlation coefficients. In Chapter 9 you will test hypotheses about the difference between the means of two populations. In addition you will calculate confidence intervals about a difference in sample means. In both of these two chapters you will determine probabilities with a *t* distribution. Also, both chapters will include the effect size index, the descriptive statistic you learned about in Chapter 4.

One of the most important uses of statistics is to analyze data from experiments. Chapter 9 will discuss the basics of simple experiments.

chapter **eight**

HYPOTHESIS TESTING AND EFFECT SIZE: ONE-SAMPLE DESIGNS

OBJECTIVES FOR CHAPTER EIGHT

Study the text. Work the problems as you come to them. Afterward, you should be able to:

1. Explain the procedure called hypothesis testing
2. Define the null hypothesis in words
3. Define three alternative hypotheses in words
4. Define α, *significance level, rejection region,* and *critical value*
5. Use a one-sample *t* test to decide if a sample mean came from a population with a hypothesized mean and write an interpretation
6. Decide if a sample *r* came from a population in which the correlation between the two variables is zero
7. Interpret the meaning of *p* in the phrase *p* < .05
8. Explain what rejecting the null hypothesis means and what retaining the null hypothesis means
9. Distinguish between Type I and Type II errors
10. Describe the difference between a one-tailed and a two-tailed statistical test
11. Calculate and interpret an effect size index

L et's start with a researcher who has a belief about how things are out there in the state of nature. This belief is often expressed as a comparison, such as: Group A is better than the national average or The effect of this treatment is to reduce time in therapy.

A researcher uses **hypothesis testing** to support beliefs about such comparisons. It is an objective method that allows strong claims of "more than," "faster," or "slower" *if the data come out according to the researcher's expectations.*

Let's move from a description of hypothesis testing to an explanation of its logic. For this example, let's say that the researcher believes that the mean of the population of interest is greater than some standard. The logic of hypothesis testing sets out to establish this relationship of *greater than* by showing that alternative relationships aren't very likely. The alternative relationships are *equal to* the standard and *less than* the standard.

To begin, data from the population of interest are gathered. Suppose the sample mean is greater than the standard. Samples fluctuate, however, and perhaps this example of *greater than* is just a chance fluctuation.

What hypothesis testing gives you is the probability of the data gathered, if the *equal to* relationship is true. If the probability is small, then the *equal to* relationship can be discarded. In addition, such data are even less probable if the relationship is *less than*. So, with *equal to* and *less than* eliminated, the relationship that remains is *greater than,* which is what the researcher believed from the beginning.

Reading through this logic, you can find places to say, "Yeah, but what if. . . ." The "what ifs" will be dealt with as the chapter progresses. But go ahead and write your "what ifs" now. Then, as you study the chapter, fill in the answers. When you finish, you will have an explanation of hypothesis testing in your own words.

In all hypothesis testing, the hypothesis being tested is a hypothesis about equality. The researcher thinks that the equality hypothesis is not true, and by showing that the data do not fit it, the equality hypothesis can be rejected.

The endpoint of hypothesis testing is a decision. Decision makers in fields that range from anthropology and business to zoology and zymurgy use hypothesis testing because it gives them an objective way to determine if a relationship is *greater than* or *less than.* Of course, hypothesis-testing decisions are based on data from samples, so some uncertainty always goes with the decision.

Every chapter from this point on will assume that you understand the concepts in hypothesis testing, so I will flesh out these ideas with care as the chapter progresses. In all, this book covers 14 different statistical tests that use hypothesis-testing logic. All of these statistical tests are used to reach conclusions about populations by using data from samples. The first of these tests allows you to test a hypothesis about a population mean μ_0 by finding the probability that sample data with a mean $\overline{X}$ came from the population with the mean μ_0.

PROBLEM

1. Hypothesis-testing techniques produce probability figures such as .20. What event has a probability of .20?

HYPOTHESIS TESTING—AN EXAMPLE PROBLEM

Determining whether a sample with a mean $\overline{X}$ came from a population with a mean μ_0 is useful for evaluating quantitative claims, such as that of the Frito-Lay® company, which claims that bags of Doritos® tortilla chips contain 269.3 grams. In this case, the equality hypothesis is that $\mu_0 = 269.3$ grams. This claim may or may not be true. When you buy a bag of chips, how much do you get?

Of course, you could weigh one bag and obtain an exact answer for that one bag. Suppose the bag weighs less than 269.3 grams. The conclusion that this observation supports, "*This bag* has less than claimed," is not very far-reaching. A conclusion about the *population* of actual weights would be much more interesting.

The population of weights of Doritos bags does have a mean, which I will symbolize as μ_1. Of course, it is impossible to know what this mean, μ_1, is. But, by using sampling and a hypothesis-testing technique called the *t* test, you can reach a conclusion about μ_1, the population of actual weights.

The result of hypothesis testing is not an exact value for μ_1 or even a confidence interval that captures μ_1. Rather, the result of hypothesis testing is a description of the *relationship* of μ_1 to μ_0. Here are the three possible relationships between μ_1 and μ_0, followed by what the relationships mean in the Doritos example.

$\mu_1 = \mu_0$: The contents of Doritos bags are about what is expected if the company is trying to put 269.3 grams into each package.

$\mu_1 < \mu_0$: The contents of Doritos bags weigh less than the company claims.

$\mu_1 > \mu_0$: The contents of Doritos bags weigh more than the company claims.

To reach a decision about μ_1 (and to have an example for this chapter), I weighed the contents of eight packages of Doritos tortillas (Nacho cheese flavor) on a scale in the lab. These are the summary statistics:

$N = 8$

$\Sigma X = 2165.4$ grams

$\Sigma X^2 = 586,121.56$ grams

The first thing to do with summary statistics is to calculate descriptive statistics to see what they tell you. Stop and decide which descriptive statistics will be helpful. What do you expect to find out from them?

As for the descriptive statistics:

$\overline{X} = 270.675$ grams

$\hat{s} = 0.523$ gram

Most important, the sample mean is *greater than* the company claim of 269.3 grams. Also, note that the standard deviation about the sample mean is quite small.

The next step for researchers is to use these sample statistics to reach a conclusion about the population of actual weights. For you to be in a position to take this step, you need a more thorough understanding of hypothesis testing, including some conventions that researchers use.

THE LOGIC OF HYPOTHESIS TESTING

This section presents terms and logic that will be important throughout the remainder of this book. Take notes here, and get answers to any questions you have. The presenting situation is that you want to know about a parameter of a population that cannot be measured.

The basic logic of hypothesis testing is to set up two hypotheses that cover all the possibilities about the parameter. The first is a hypothesis of equality that states that the population parameter is exactly a specified value. The second hypothesis is that the population parameter is *not* the specified value. You can see that logically one of these hypotheses *must* be true.

Next, obtain sample data from the population, and from the data calculate the statistic that corresponds to the parameter you are interested in. Now, select a sampling distribution (which shows the probability of the various values of the statistic). This sampling distribution is based on the equality hypothesis; that is, the probabilities show what happens when the equality hypothesis is true. Using the sampling distribution, find the probability associated with the statistic.

If the probability is very small, give up the belief that the equality hypothesis is correct. Giving up the equality hypothesis leaves only the other hypothesis, which you then conclude is correct. To restate this: If you can show that the data that you actually obtained are unlikely when the equality hypothesis is true, then you should reject the equality hypothesis.

On the other hand, if the probability you find is large, then the sample data are consistent with the equality hypothesis. Unfortunately, such data are *also* consistent with the other hypothesis. Given this state of affairs, the data are not helpful in eliminating one of the hypotheses.

Now I want to expand this explanation by giving you the statistical terms that are used in hypothesis testing. The application to the Doritos problem is presented in a different typeface.

1. Begin with a population mean you want to know about and data from a sample from the population. **For the Doritos example, the population mean is that of *all* bags; the sample mean ($N = 8$) is 270.675 grams.**

2. Recognize two hypotheses: A hypothesis of equality, called the **null hypothesis** and symbolized H_0 ("H sub oh" or "H sub zero"), and an **alternative hypothesis**, symbolized H_1 ("H sub one"):

   H_0: The mean weight of Doritos bags is 269.3 grams, the weight claimed by the Frito-Lay company.
   H_1: The mean weight of Doritos bags is not 269.3 grams.

3. Tentatively assume that H_0, the null hypothesis, is true. When the null hypothesis is true, whatever difference there is between the population mean and the sample mean is the result of chance variation. **For the Doritos example, a true null hypothesis means that the difference between 269.3 and 270.675, 1.375 grams, is just a chance fluctuation.**

4. Choose a sampling distribution that shows the probabilities of differences. These probabilities are correct when the null hypothesis is true and only chance is at work. **The *t* distribution is appropriate for the Doritos problem.**

5. Compare the difference observed in the data **(1.375 grams)** to the differences shown by the sampling distribution.

   a. If the observed difference is *unexpected* (the probability is low), conclude that the null hypothesis is false. Differences this large are unlikely when the null hypothesis is true, so the most reasonable conclusion is to reject the null hypothesis. Rejecting the null hypothesis leaves you with the alternative hypothesis, H_1. **The mean weight of Doritos bags is not 269.3 grams.**

b. If the observed difference is expected (the probability is high), the data are consistent with the null hypothesis. Unfortunately, such a difference is also consistent with the alternative hypothesis, so the data do not allow you to choose between the null and the alternative hypotheses. Thus, both the null hypothesis and the alternative hypothesis are retained. **The mean weight of the Doritos bag** *could* **be 269.3 grams.**

6. Write a conclusion about the population that was measured. Use language that is specific to the problem. **For the Doritos data, tell the story of bags, mean weight, and claims by the Frito-Lay company.**

Central to the logic of hypothesis testing is establishing a null hypothesis (step 2). It is usually a good idea to state the null hypothesis explicitly by using the formal language of mathematics. For the Doritos data:

H_0: $\mu_0 = 269.3$ grams

In words, this null hypothesis says that the mean of the population of Doritos weights is equal to the published claim, which is 269.3 grams. Step 3, tentatively assuming that the null hypothesis is true, is common to all hypothesis testing.

Of course, the researcher hopes to reject H_0. If the data allow rejection of H_0, then an alternative hypothesis, H_1, is accepted. Because the researcher has an expected conclusion in mind from the start, H_1 is selected so that it will encompass the researcher's belief. In practice, a researcher chooses one of the three H_1's, *before the data are collected.* The three alternative hypotheses for the Doritos data are:

H_1: $\mu_1 \neq 269.3$ grams. This alternative says simply that the mean of the population of Doritos weights is not 269.3 grams; it doesn't specify whether the actual mean is greater or less than 269.3 grams. If you reject H_0 and accept this H_1, you must examine the sample mean to determine whether its population mean is greater or less than the one specified by H_0.

H_1: $\mu_1 < 269.3$ grams. This alternative hypothesis says that the sample is from a population with a mean less than 269.3 grams.

H_1: $\mu_1 > 269.3$ grams. This alternative hypothesis says that the sample is from a population with a mean greater than 269.3 grams.

For the Doritos example, I chose the alternative hypothesis H_1: $\mu_1 \neq 269.3$ grams before I gathered the data. Indeed, as soon as I dreamed up this problem, I knew I wanted this alternative hypothesis because it would allow me to conclude that the Doritos bags contain more than *or* contain less than the company claims. The section on one-tailed and two-tailed tests, which comes later in this chapter, will give you more information on choosing an alternative hypothesis.

In summary, the null hypothesis meets with one of two fates at the hands of the data. It may be *rejected,* which allows you to accept an alternative hypothesis, or it may be *retained.* If it is retained, *it is not proved as correct;* it is simply retained as one among many possibilities. Hypothesis testing is a procedure in which a rejected null hypothesis leads to a strong conclusion involving *greater than* (or *less than*). A retained null hypothesis leaves you where you began — with both the null hypothesis and an alternative hypothesis. You do, however, have the data you gathered. They are informative, even if hypothesis testing does not lead to a strong conclusion.

clue to the future

You have just completed the section that explains the reasoning that is at the heart of inferential statistics. This reasoning is a part of every test in every chapter that follows in this book (and will also turn up in your future reading).

PROBLEMS

2. Is the null hypothesis a statement about a statistic or a parameter?

3. Distinguish between the null hypothesis and the alternative hypothesis, giving the symbol for each.

4. In your own words, outline the logic of hypothesis testing, using technical terms and symbols.

5. Agree or disagree: The text's sample of Doritos weights is a random sample.

USING THE *t* DISTRIBUTION FOR HYPOTHESIS TESTING

You will need some way to determine probabilities if you are going to test the hypothesis that a sample with a mean $\overline{X}$ came from a population with a mean μ_0. Fortunately, you already know about sampling distributions, a statistical device that was invented to reveal probabilities. Even more fortunately, the sampling distribution you need for the Doritos problem is a *t* distribution, which you learned about in the previous chapter. For the task at hand (testing the null hypothesis that $\mu_0 = 269.3$ grams using a sample of eight), a *t* distribution with 7 *df* will give you correct probabilities.

The *t* distribution for 7 *df* is found in Table D in Appendix C. The *t* distributions in Table D are theoretical distributions that come from a mathematical formula.

Before progressing with the Doritos example, I'd like you to think about constructing an empirical *t* distribution with 7 *df*. Imagine a set of scores that you designate as the population. Calculate the mean, which is a parameter. From the population draw a random sample of 8 and calculate a sample mean, which is a statistic. Subtract the parameter from the statistic, producing a difference. Draw another random sample ($N = 8$) and repeat the procedure. Do this many, many times and then arrange all the differences into a frequency distribution.

Think carefully for a moment about the mean (the expected value) of this distribution of differences. Stop reading and decide what its numerical value will be.

The mean of the sampling distribution of differences is zero because positive and negative differences occur equally often, giving you an algebraic sum close to zero. If you were successful on this problem, congratulations! Few students get it right when they first read the chapter.

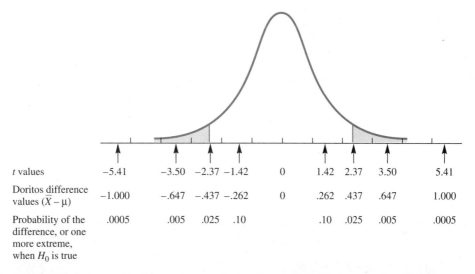

t values	−5.41		−3.50	−2.37	−1.42	0	1.42	2.37	3.50		5.41
Doritos difference values $(\bar{X} - \mu)$	−1.000		−.647	−.437	−.262	0	.262	.437	.647		1.000
Probability of the difference, or one more extreme, when H_0 is true	.0005		.005	.025	.10		.10	.025	.005		.0005

F I G U R E 8.1 A *t* distribution. The *t* values and probability figures are those that go with 7 *df*. The difference values apply only to the Doritos data.

Now, let's return to a theoretical *t* distribution with 7 *df*, **Figure 8.1.** It is based on the assumption that the sample of Doritos weights is from a population with a mean of 269.3 grams. That is, *the numbers in Figure 8.1 are based on the assumption that the null hypothesis is true.*

Examine Figure 8.1, paying careful attention to the three sets of values on the abscissa. The *t* values and the probability figures apply to any problem that uses a *t* distribution with 7 *df*. Sandwiched between are differences between sample means and the hypothesized population mean — differences that are specific to the Doritos data.

In your examination of Figure 8.1, note the following characteristics:

1. The mean is zero.
2. As the size of the difference between the sample mean and the population mean increases, *t* values increase.
3. As differences and *t* values become larger, probabilities become smaller.
4. As always, small probabilities go with small areas at each end of the curve.

Using Figure 8.1, you can make a decision about the null hypothesis for the Doritos data. (Perhaps you have already anticipated the conclusion?) The difference between the mean weight of the sample and the advertised weight is 1.375 grams. This difference is not even shown on the distribution, so it must be extremely rare — *if the null hypothesis is true.* How rare is it?

Let's take the rarest events shown and use them as a reference point. As seen in **Figure 8.1,** a difference as extreme as 1.000 or greater occurs with a probability of .0005. In addition, a difference as extreme as −1.000 or less occurs with the same probability, .0005. Adding these two probabilities together gives a probability of .001.[1]

---

[1] I'll explain this step in the section on one- and two-tailed tests.

Thus, the probability is .001 of getting a difference of ±1.000 or greater. The difference found in the data (1.375 grams) is even more rare, so the *p* value of the data actually obtained is <.001. Thus, if the null hypothesis is true (the population mean is 269.3 grams), the probability of getting the difference I got is less than 1 in 1000 ($p < .001$).

The reasonable thing to do is abandon (reject) the null hypothesis. With the null hypothesis gone, only the alternative hypothesis remains: H_1: $\mu_1 \neq 269.3$ grams. Knowing that the actual weight is *not* 269.3 grams, the next step is to ask whether it is greater than or less than 269.3 grams. The sample mean is greater, so a conclusion can be written: "The contents of Doritos bags weigh more than the company claims, $p < .001$."

To summarize, I determined the probability of the data when the null hypothesis is true. The probability was quite small. Rather than conclude that a very rare event had occurred, I concluded that the null hypothesis was false.

A PROBLEM AND THE ACCEPTED SOLUTION

The hypothesis that the sample of Doritos weights came from a population with a mean of 269.3 grams was so unlikely ($p < .001$) that it was easy to reject the hypothesis. But what if the probability had been .01, or .05, or .25, or .50? The problem that this section addresses is where on the probability continuum should you change the decision from "reject the null hypothesis" to "retain the null hypothesis."

It may already be clear to you that any solution will be an arbitrary one. Breaking a continuum into two parts often leaves you uncomfortable near the break. Nevertheless, those who use statistics have adopted a solution.

Setting Alpha (α) or Establishing a Significance Level

The widely accepted solution to the problem of where to break the continuum is to use the .05 mark. The choice of a probability value (whether at .05 or some other figure) is called *setting* α. It is also called establishing a **significance level.**

In an experiment, the researcher sets α, gathers data, and uses a sampling distribution to find the probability (p) of such data. This p value is correct only when the null hypothesis is true. If $p < \alpha$, reject H_0. Thus, if $\alpha = .05$ and $p \leq .05$, reject H_0. If $\alpha = .05$ and $p > .05$, retain H_0. Of course, if $p = .03$, or .01, or .001, reject H_0. If the probability is .051 or greater, retain H_0. When H_0 is rejected, the difference is described as **statistically significant.** Sometimes, when the context is clear, this is shortened to simply "significant." When H_0 is retained (the difference is not significant), the abbreviation **NS** is often used.

In some statistical reports, an α level is not specified; only data-based probability values are given. Thus, in the same report, some differences may be reported as significant at the .001 level, some at the .01 level, and some at the .05 level. Regardless of how the results are reported, however, researchers view .05 as an important cutoff (Nelson, Rosenthal, and Rosnow, 1986). When .10 or .20 is used as an α level, a justification should be given.

Rejection Region

The **rejection region** is the area of the sampling distribution that includes all the differences that have a probability *equal to* or *less than* α. Thus, any event in the rejection region leads to rejection of the null hypothesis.[2] In **Figure 8.1** the rejection region is shaded (for a significance level of .05). Thus, t values more extreme than ± 2.37 lead to rejecting H_0.

Critical Values

Most statistical tables do not provide you with probability figures for every outcome of a statistical test. What they provide are statistical test values that correspond to commonly chosen α values. These specific test values are called **critical values.**

I will illustrate critical values using the t distribution table (**Table D** in Appendix C). The t values in that table are called critical values when they are used in hypothesis testing. The α levels that researchers use are listed across the top in rows 2 and 3. Degrees of freedom are in column 1. Remember that if you have a df that is not in the table, it is conventional to use the next *smaller df* or to interpolate a t value that corresponds to the exact df.

The t values in the table separate the rejection region from the rest of the sampling distribution (sometimes referred to as the *acceptance region*). Thus, data-produced t values that are equal to or greater than the critical value fall in the rejection region, and the null hypothesis is rejected.

I will indicate critical values with an expression such as $t_{.05}$ (14 df) = 2.145. This expression indicates the sampling distribution that was used (t), α level (.05), degrees of freedom (14), and critical value from the table (2.145, for a two-tailed test).

PROBLEMS

6. True or false? Sampling distributions used for hypothesis testing are based on the assumption that H_0 is false.
7. What is a significance level? What is the largest significance level you can use without giving a justification?
8. Circle all the phrases that go with an event that is in the rejection region of a sampling distribution.

   p is small
   Retain the null hypothesis
   Accept H_1
   Middle section of the sampling distribution

9. Use Table D to find critical values for two-tailed tests for these values.
   a. α level of .01; $df = 17$
   b. α level of .001; $df = 46$
   c. α level of .05; $df = \infty$

---

[2] Some texts use the term **critical region** instead of rejection region.

THE ONE-SAMPLE *t* TEST

Some students suspect that it is not necessary to construct an entire sampling distribution such as the one in Figure 8.1 each time you have a hypothesis to test. You may be such a student and, if so, you are correct. The conventional practice is to analyze the data using the **one-sample *t* test**. The formula for the one-sample *t* test is

$$t = \frac{\overline{X} - \mu_0}{s_{\overline{X}}}; \quad df = N - 1$$

where $\overline{X}$ is the mean of the sample
μ_0 is the hypothesized mean of the population
$s_{\overline{X}}$ is the standard error of the mean

The question to be answered is whether the Frito-Lay company is justified in claiming that bags of Doritos tortilla chips weigh 269.3 grams. The null hypothesis is that the mean weight of the chips is 269.3 grams. The alternative hypothesis is that the mean weight is *not* 269.3 grams. A sample of eight bags produced a mean of 270.675 grams with a standard deviation (using $N - 1$) of 0.523 gram. To test the hypothesis that the sample weights are from a population with a mean of 269.3 grams, you can calculate a *t* value using the one-sample *t* test. Thus,

$$t = \frac{\overline{X} - \mu_0}{s_{\overline{X}}} = \frac{\overline{X} - \mu_0}{\hat{s}/\sqrt{N}} = \frac{270.675 - 269.3}{0.523/\sqrt{8}} = 7.44; \quad 7 \, df$$

To interpret a *t* value of 7.44 with 7 *df*, turn to Table D and find the appropriate critical value. For α levels for two-tailed tests, go down the .05 column until you reach 7 *df*. The *t* value, 7.44, is greater than the critical value 2.365, so the difference of 1.375 grams is significant at the .05 level. To find out if it is significant at an even lower α level, move to the right, looking for a number that is greater than 7.44. There isn't one. Because 7.44 is greater than 5.408, you can conclude that the difference is significant at the .001 level. Thus, you can reject the null hypothesis at the .001 level.

The last step is to write an informative conclusion that tells the story of the variable that was measured. For example, "The mean weight of the eight Doritos bags was 270.675 grams, which is significantly greater than the company's claim of 269.3 grams, $p < .001$."

A good interpretation is more informative than simply, "The null hypothesis was rejected, $p < .05$ (or retained, $p > .05$)." A good interpretation always (1) uses the terms of the experiment and (2) gives the direction of the difference.

The algebraic sign of the *t* in the *t* test is ignored when you are finding the critical value for a two-tailed test. Of course, the sign tells you whether the sample mean is greater or less than the null hypothesis mean — an important point in any statement of the results.

I want to address two additional issues about the Doritos conclusion; both have to do with generalizability. The first issue is the nature of the sample: It most certainly was not a random sample of the population in question. Because of this, my conclusion is not as secure as it would be for a random sample. How much less secure is it?

I don't really know, and inferential statistics provides me with no guidelines on this issue. So, I state both my conclusion (contents weigh more than company claims) and my methods (nonrandom sample, $N = 8$) and then see whether this conclusion will be challenged or supported by others with more or better data. (I'll just have to wait to see who responds to my claim.)

The other generalizability issue is whether the conclusion applies to other-sized bags of Doritos, to other Frito-Lay products, or to other manufacturers of tortilla chips. These are good questions but, again, inferential statistics does not give any guidelines. You'll have to use aids such as your experience, library research, correspondence with the company, or more data gathering.

Finally, I want to address the question of the usefulness of the one-sample t test. The t test is useful for checking manufacturers' claims (where an advertised claim provides the null hypothesis), comparing a sample to a known norm (such as an IQ score of 100), and comparing behavior against a "no error" standard (as used in studies of lying or visual illusions). In addition to its use in analyzing data, the one-sample t test is probably the simplest example of hypothesis testing (an important consideration when you are being introduced to the topic).

PROBLEM

10. What are the two characteristics of a good interpretation?

AN ANALYSIS OF POSSIBLE MISTAKES

To some forward-thinking students, the idea of adopting an α level of 5 percent seems preposterous. "You gather the data," they say, "and if the difference is large enough, you reject H_0 at the .05 level. You then state that the sample came from a population with a mean other than the hypothesized one. But in your heart you are uncertain. Perhaps a rare event happened."

This line of reasoning is fairly common. Many thoughtful students take the next step. "I don't have to use the .05 level. I'm going to use an α level of 1 in 1 million. That way I can reduce the uncertainty."

It is true that adopting a .05 α level leaves some room for mistaking a chance difference for a real difference. It is probably clear to you that lowering the α level (to a probability such as .01 or .001) *reduces* the probability of this kind of mistake. Unfortunately, lowering the α level *increases* the probability of a different kind of mistake.

Look at **Table 8.1,** which shows the two ways to make a mistake. Cell 1 shows the situation when the null hypothesis is true and you reject it. Your sample data and hypothesis testing have produced a mistake—a mistake called a **Type I error.** However, if H_0 is true and you retain it, you have made a correct decision (cell 2).

Now, suppose that the null hypothesis is actually false (the second column). If, on the basis of your sample data, you reject H_0, you have made a correct decision (cell 3).

TABLE 8.1 Type I and Type II errors*

		The true situation in the population	
		H_0 true	H_0 false
The decision made on the basis of sample data	Reject H_0	1. Type I error	3. Correct decision
	Retain H_0	2. Correct decision	4. Type II error

* In this case, *error* means mistake.

Finally, if the null hypothesis is false and you retain it (cell 4), you have made a mistake — a mistake called a **Type II error.**

Let's attend to a Type I error. The probability of a Type I error is α, and you control α when you adopt a significance level. If the data produce a p value that is less than α, you reject H_0 and conclude that the difference observed is statistically significant. If you reject H_0, you could be making a Type I error, but the probability of this error is controlled by your choice of α level. Note that rejecting H_0 is wrong only when the null hypothesis is true.

Now for a Type II error. A Type II error is possible only if you retain the null hypothesis. If the null hypothesis is false and you retain it, you have made a Type II error. The probability of a Type II error is symbolized by β.

Controlling α by setting a significance level is straightforward; determining the value of β is much more complicated. For one thing, a Type II error requires that the sample be from a population that is different from the null hypothesis population. Naturally, the more different the two populations are, the more likely you are to detect it and, thus, the lower β is.

In addition, the values of α and β are inversely related. As α is reduced (from, say, .05 to .01), β goes up. To explain, suppose you insist on a larger difference between means before you say the difference is "not chance." (You reduce α.) In this case, you will be less able to detect a small, real, nonchance difference. (β goes up.) The following description further illustrates the relationship between α and β.

Figure 8.2 shows two populations. The population on the left has a mean $\mu_0 = 10$; the one on the right has a mean $\mu_1 = 14$. An investigator draws a large sample from the population on the right and uses it to test the null hypothesis $H_0: \mu_0 = 10$. In the real world of data, every decision carries some uncertainty, but in this textbook example, the correct decision is clear: Reject the null hypothesis.

The actual decision, however, will be made by evaluating the difference between the sample mean $\overline{X}$ and the mean specified by the null hypothesis, $\mu_0 = 10$.

Now, let's suppose that the sample produced a mean of 14. That is, the sample tells the exact truth about its population. Under these circumstances, will the hypothesis-testing procedure produce a correct decision?

Figure 8.3 shows the sampling distribution for this problem. As you can see, if the 5 percent α level is used (*all* the blue areas), a difference of 4 falls in the rejection region. Thus, if $\alpha = .05$, you will correctly reject the null hypothesis. However, if a 1 per-

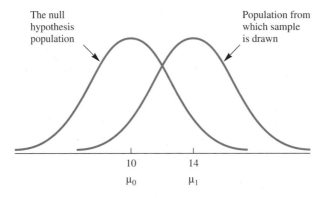

FIGURE 8.2 Frequency distribution of scores when H_0 is false

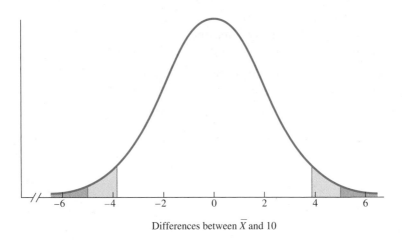

Differences between $\bar{X}$ and 10

FIGURE 8.3 A sampling distribution from the population on the left in Figure 8.2

cent α level is used (the *dark* blue areas only), a difference of 4 *does not* fall in the rejection region. Thus, if $\alpha = .01$, you will not reject H_0. This failure to reject the false null hypothesis is a Type II error.

At this point, I can return to the discussion of setting the α level. The suggestion was: Why not reduce the α level to 1 in 1 million? From the analysis of the potential mistakes, you can answer that when you lower the α level so you can reduce the chance of a Type I error, you increase β, the probability of a Type II error. More protection from one kind of error increases the liability to another kind of error. Most who use statistics adopt an α level (usually at .05) and let β fall where it may.

To summarize, there is always some uncertainty that goes with a conclusion based on statistical evidence. A famous statistics textbook put it this way, "If you agree to use a sample, you agree to accept some uncertainty about the results."

THE MEANING OF p IN $p < .05$

The symbol p (as in $p < .05$) is quite common in data-based investigations, regardless of the topic of investigation. p deserves a special section.

Every statistical test such as the t test has its own sampling distribution. A sampling distribution shows the probability of various sample outcomes when the null hypothesis is true. Thus, p is always the probability of the statistical test value when H_0 is true. If your t test gives you $p < .05$, it means that the sample results you actually got occur fewer than 5 times in 100 *when the null hypothesis is true.*

Although the meaning of p is straightforward, it is commonly misinterpreted. Everitt and Hay (1992) reported that among 70 academic psychologists, only 3 scored 100 percent on a six-item test on the meaning of p. Here are some cautions about the meaning of p:

- p is not the probability that H_0 is true.
- p is not the probability of a Type I error.
- p is not the probability of making a wrong decision.
- p is not the probability that the sample statistic is due to chance.

Perhaps the best thing for beginners to do is to memorize a definition of p and use it (or close variations) in the interpretations they write. In statistics, p is the probability of the data obtained (or the probability of the statistical test value), if the null hypothesis is true.

What about the situation when the null hypothesis is false? Statisticians address this situation under the topic of *power.* There will be a short discussion of power in the next chapter.

ONE- AND TWO-TAILED TESTS

As mentioned earlier, researchers choose, before the data are gathered, one of three possible alternative hypotheses. The choice of an alternative hypothesis determines the conclusions you will have available after the statistical analysis is finished. Here's the general expression of these alternative hypotheses and the conclusions they allow you:

Two-tailed test
$H_1: \mu_1 \neq \mu_0$. The population means differ, but no direction of the difference is specified. Conclusions that are possible: The sample is from a population with a mean *less than* that of the null hypothesis *or* the sample is from a population with a mean *greater than* that of the null hypothesis.

One-tailed tests
$H_1: \mu_1 < \mu_0$. Conclusion that is possible: The sample is from a population with a mean less than that of the null hypothesis.
$H_1: \mu_1 > \mu_0$. Conclusion that is possible: The sample is from a population with a mean greater than that of the null hypothesis.

If you want to have two different conclusions available to you — that the sample is from a population with a mean *less than* that of the null hypothesis *or* that the sample is from a population with a mean *greater than* that of the null hypothesis — you must have a rejection region in each tail of the sampling distribution. Such a test is called a **two-tailed test of significance** for reasons that should be obvious from **Figure 8.3** and **Figure 8.1.**

If your only interest is in showing (if the data will permit you) that the sample comes from a population with a mean greater than that of the null hypothesis or from a population with a mean less than that of the null hypothesis, you should run a **one-tailed test of significance,** which puts all of the rejection region into one tail of the sampling distribution. **Figure 8.4** illustrates this for a *t* distribution with 30 *df.*

You will find probability figures for two-tailed tests in the second row of **Table D;** probability figures for one-tailed tests are in row 3.

For most research situations, a two-tailed test is the appropriate one. The usual goal of a researcher is to discover the way things actually are, and a one-tailed test does not allow you to discover, for example, that the populations are exactly reversed from the way you expect them to be. Researchers always have expectations (but they sometimes get surprised).

In many applied situations, a one-tailed test is appropriate. Some new procedure, product, or person, for example, is being compared to an already existing standard. The only interest is in whether the new is better than the old; that is, only a difference in one direction will produce a decision to change. In such a situation, a one-tailed test seems appropriate.

I chose a two-tailed test for the Doritos data because I wanted to be able to conclude that the Frito-Lay company was giving more than they claimed *or* that they were giving less than they claimed. Of course, the data would have their say about my conclusion, but either outcome was interesting to me.

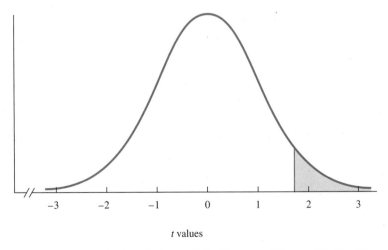

t values

FIGURE 8.4 A one-tailed test of significance, with $\alpha = .05$, $df = 30$

I'll end this section on one- and two-tailed tests by telling you that statistics instructors sometimes smile or even laugh at the subtitle of this book. The phrase "tales of distributions" brings to mind "tails of distributions."

PROBLEMS

11. Distinguish among α, Type I error, and p.

12. What is a Type II error?

13. Suppose the actual situation is that the sample comes from a population that is not the one hypothesized by the null hypothesis. If the significance level is .05, what is the probability of making a Type I error? (Careful on this one.)

14. Suppose a researcher chooses to use $\alpha = .01$ rather than .05. What is the effect on the probability of Type I and Type II errors?

15. The table has blanks for you to fill in. Your task is to look up critical values in Table D, decide if H_0 should be rejected or retained, and give the probability figure that characterizes your confidence in that decision. I designed this table so that some common misconceptions will be revealed — just in case you acquired one.

N	α level	Two-tailed or one-tailed test?	Critical value	t-test value	Reject H_0 or retain H_0?	p value
10	.05	Two	_____	2.25	_____	_____
20	.01	One	_____	2.57	_____	_____
35	.05	Two	_____	2.03	_____	_____
6	.001	Two	_____	6.72	_____	_____
24	.05	One	_____	1.72	_____	_____
56	.02	Two	_____	2.41	_____	_____

16. The Minnesota Multiphasic Personality Inventory (MMPI) has one scale that measures paranoia (a mistaken belief that others are out to get you). The population mean on the paranoia scale is 50. Suppose 24 police officers filled out the inventory, producing a mean of 54.3. Write interpretations of these t test values.
a. $t = 2.10$ **b.** $t = 2.05$

17. A large mail order company with ten distribution centers had an average return rate of 5.3 percent. In an effort to reduce this rate, a set of testimonials was included with every shipment. The return rate was reduced to 4.9 percent. Write interpretations of these t test values.
a. $t = 2.10$ **b.** $t = 2.50$

*18. Project Head Start is a program for preschool children from educationally disadvantaged environments. Head Start began over 30 years ago, and the data that follow are representative of early research on the effects of the program.

A group of children participated in the program for a year and then enrolled in first grade. They took an IQ test, the national norm of which was 100. Use a *t* test to test the hypothesis that the Head Start program had no effect on the children's IQ. Begin your work by stating the null hypothesis and end by writing a conclusion about the effects of the program.

$$\Sigma X = 5250 \qquad \Sigma X^2 = 560,854 \qquad N = 50$$

19. Odometers measure automobile mileage. How close to the truth is the number that is registered? Suppose 12 cars traveled exactly 10 miles (measured by surveying) and the following mileage figures were recorded by the odometers. State the null hypothesis, choose an alternative hypothesis, perform a *t* test, and write a conclusion.

9.8   10.1   10.3   10.2   9.9   10.4   10.0   9.9   10.3   10.0   10.1   10.2

*20. The NEO-PI-R is a personality test. The E stands for extroversion, one of the personality characteristics the test measures. The population mean for the extroversion test is 50. Suppose 13 people who were successful in used car sales took the test and produced a mean of 56.10 and a standard deviation of 10.00. Perform a *t* test and write a conclusion.

EFFECT SIZE INDEX

Hypothesis testing answers the question of whether the population a sample is drawn from is different from the population specified by the null hypothesis. If the answer is yes, another question a reasonable person would ask is, How *big* is the difference?

You probably recognize that this is a question of **effect size,** which you studied in Chapter 4. You may recall that in that chapter, you calculated an effect size index, *d*, to determine just how different *two* sample means were. The formula for *d* that corresponds to a one-sample *t* test is

$$d = \frac{|\overline{X} - \mu_0|}{\sigma}$$

where $\overline{X}$ is the mean of the sample
   μ_0 is the mean specified by the null hypothesis
   σ is the standard deviation of the null hypothesis population

Can you recall the guidelines that help you evaluate the absolute size of *d*? Cohen proposed these guidelines in 1969 and they have been widely adopted by researchers:

Small effect	$d = .20$
Medium effect	$d = .50$
Large effect	$d = .80$

In problem 18 you concluded that the Head Start program has a statistically significant effect on the IQs of preschool children. You found that the mean IQ of children in the program (105) was significantly higher than 100. But how big is this effect? The statistic d provides an answer. Knowing that $\sigma = 15$ for IQ tests, you get

$$d = \frac{|\overline{X} - \mu_0|}{\sigma} = \frac{105 - 100}{15} = .33$$

An effect size index of .33 indicates that the effect of Head Start is midway between a small effect and a medium effect.

Many research situations, unlike those that use IQ scores, do not provide σ. In these cases, researchers use $\hat{s}$, calculated from sample data, as an acceptable approximation of σ.

OTHER SAMPLING DISTRIBUTIONS

You have been learning about the sampling distribution of the mean. There are times, though, when the statistic necessary to answer a researcher's question is not the mean. For example, to find the degree of relationship between two variables, you need a correlation coefficient. To determine whether a treatment causes more variable responses, you need a standard deviation. And, as you know, proportions are commonly used statistics. In each of these cases (and indeed, for any statistic), researchers often use the basic hypothesis-testing procedure you have just learned. Hypothesis testing always involves a sampling distribution of the statistic.

Where do you find sampling distributions for statistics other than the mean? In the rest of this book you will encounter new statistics and new sampling distributions. Tables E–L in Appendix C represent sampling distributions from which probability figures can be obtained. In addition, some statistics have sampling distributions that are t distributions or normal curves. Still other statistics and their sampling distributions are covered in other books.

Along with every sampling distribution comes a standard error. Just as every statistic has its sampling distribution, every statistic has its standard error. For example, the standard error of the median is the standard deviation of the sampling distribution of the median. The standard error of the variance is the standard deviation of the sampling distribution of the variance. Worst of all, the standard error of the standard deviation is the standard deviation of the sampling distribution of the standard deviation. If you follow that sentence, you probably understand the definition of standard error quite well.

The next step in your exploration of sampling distributions will be brief, though useful and (probably) interesting. The t distribution will be used to answer a question about correlation coefficients.

PROBLEMS

21. Calculate the effect size index for the Doritos data on page 159 and interpret it using the conventions used by researchers.

22. In problem 20 you found that people who are successful in used car sales have significantly higher extroversion scores than those in the general population. What is the effect size index for those data and is the effect small, medium, or large?

23. In Chapters 2 and 7 you worked with data on normal body temperature. You found that the mean body temperature was 98.2°F rather than the conventional 98.6°F. The standard deviation for those data was 0.7°F. Calculate an effect size index and interpret it.

24. In addition to weighing Doritos chips, I weighed six Snickers candy bars manufactured by Mars, Inc. Test the company's advertised claim that the net weight of a bar is 58.7 grams by performing a t test on the following data, which are in grams. Find d. Write a conclusion about Snickers bars.

59.1   60.0   58.6   60.2   60.8   62.1

USING THE t DISTRIBUTION TO TEST THE SIGNIFICANCE OF A CORRELATION COEFFICIENT

In Chapter 5 you learned to calculate a Pearson product-moment correlation coefficient, a descriptive statistic that indicates the degree of relationship between two variables in a bivariate distribution. This section is on testing the statistical significance of correlation coefficients. The techniques of inferential statistics will be used to test the hypothesis that a sample-based r came from a population with a parameter coefficient of .00.

The present-day symbol for the Pearson product-moment correlation coefficient of a population is ρ (rho). The null hypothesis that is being tested is

$$H_0: \rho = .00$$

The alternative hypothesis is

$$H_1: \rho \neq .00 \quad \text{(a two-tailed test)}$$

For $\rho = .00$, the sampling distribution of r is a t distribution with a mean of .00 and a standard error of $s_r = \sqrt{(1 - r^2)/(N - 2)}$. The test statistic for a sample r is a t value calculated from the formula

$$t = \frac{r - \rho}{s_r}$$

Note that this formula has the same form as that of the t test you used to test the significance of a sample mean. This form, which you will see again, shows a difference between a statistic and a parameter divided by the standard error of the statistic.

An algebraic manipulation produces a formula that is easier to use:

$$t = (r)\sqrt{\frac{N - 2}{1 - r^2}}; \quad df = N - 2, \text{ where } N = \text{number of } pairs$$

This test is the first of several instances in which you will find a t value whose df is *not* one less than the number of observations. For testing the null hypothesis, H_0: $\rho =$ .00, $df = N - 2$. (Having fulfilled my promise to explain one- and two-tailed tests, I hope you will accept my promise to explain degrees of freedom more fully in the next chapter.)

Do you remember your work with correlation coefficients in Chapter 5?[3] Think about a sample $r = -.40$. How likely is a sample $r = -.40$ if the population $\rho = .00$? Does it help to know that the coefficient is based on a sample of 22 pairs? Let's answer these questions by testing the $r = -.40$ against the $\rho = .00$, using a t test. Applying the formula for t, you get

$$t = (r)\sqrt{\frac{N-2}{1-r^2}} = (-.40)\sqrt{\frac{22-2}{1-(-.40)^2}} = -1.95$$

$$df = N - 2 = 22 - 2 = 20$$

Table D shows that, for 20 df, a t value of ± 2.09 (two-tailed test) is required to reject the null hypothesis. The obtained t for $r = -.40$, where $N = 22$, is less than the tabled t, so the null hypothesis is retained. That is, a coefficient of $-.40$ would be expected by chance alone more than 5 times in 100 from a population in which the true correlation is zero.

In fact, for $N = 22$, $r = \pm.43$ is required for significance at the .05 level and $r = \pm.54$ for the .01 level. As you can see, even medium-sized correlations can be expected by chance alone for samples as small as 22 when the null hypothesis is true. Most researchers strive for N's of 30 or more for correlation problems.

The critical values under "α level for two-tailed test" in Table D are ones that allow you to reject the null hypothesis if the sample r is a large *positive* coefficient or a large *negative* one.

PROBLEM

25. Determine whether the Pearson product-moment correlation coefficients are significantly different from .00.

    a. $r = .62$; $N = 10$ **b.** $r = -.19$; $N = 122$

    c. $r = .50$; $N = 15$ **d.** $r = -.34$; $N = 64$

Here is a think problem. Suppose you have a summer job in an industrial research laboratory testing the statistical significance of hundreds of correlation coefficients based on varying sample sizes. An α level of .05 has been adopted by the management, and your task is to determine whether each coefficient is "significant" or "not significant." How can you construct a table of your own, using Table D and the t formula for testing the significance of a correlation coefficient, that will allow you to label each coefficient without working out a t value for each? Imagine or sketch out your answer.

The table you mentally designed already exists. One version is reproduced in Table A in Appendix C. The α values of .10, .05, .02, .01, and .001 are included there.

---

[3] If not, take a few minutes to review.

Use **Table A** in the future to determine whether a Pearson product-moment correlation coefficient is significantly different from .00, but remember that this table is based on the *t* distribution.

A word of caution is in order about testing ρ's other than .00. The sampling distributions for $\rho \neq .00$ are *not* *t* distributions. Also, if you want to know whether the *difference* between two sample-based coefficients is statistically significant, you should not use a *t* distribution because it will not give you correct probabilities. Fortunately, if you have questions that can be answered by testing these kinds of hypotheses, you won't find it difficult to understand the instructions given in intermediate-level texts. (See Ferguson and Takane, 1989, p. 205, or Howell, 1997, p. 251.)

This chapter has introduced you to two uses of the *t* distribution:

1. To test a sample mean against a hypothesized population mean
2. To test a sample correlation coefficient against a hypothesized population correlation coefficient of .00

The *t* distribution has other uses, too, and it has been important in the history of statistics.

Here is a little more background on W. S. Gosset, who invented the *t* distribution so he could assess probabilities for experiments conducted by the Guinness brewery. Gosset was one of a small group of scientifically trained employees who developed ways for the Guinness Company to improve its products.

Gosset wanted to publish his work on the *t* distribution in *Biometrika,* a journal founded in 1901 by Francis Galton, Karl Pearson, and W. R. F. Weldon. However, at the Guinness Company there was a rule that employees could not publish (the rule there, apparently, was publish *and* perish). Because the rule was designed to keep brewing secrets from escaping, there was no particular ferment within the company when Gosset, in 1908, published his new mathematical distribution under the pseudonym "Student." The distribution later came to be known as "Student's *t*." (No one seems to know why the letter *t* was chosen. E. S. Pearson surmises that *t* was simply a "free letter"; that is, no one had yet used *t* to designate a statistic.) Gosset worked for the Guinness Company all his life, so he continued to use the pseudonym "Student" for his publications in mathematical statistics. Gosset was devoted to his company, working hard and rising through the ranks. He was appointed head brewer a few months before his death in 1937.[4]

WHY .05?

The story of how science and other disciplines came to adopt .05 as the arbitrary point separating differences attributed to chance from those not attributed to chance is not usually covered in textbooks. The story takes place in England at the turn of the century.

It appears that the earliest explicit rule about α was in a 1910 article by Wood and Stratton, "The Interpretation of Experimental Results," which was published in *The Journal of Agricultural Science.* Thomas B. Wood, the principal editor of this journal,

---

[4] For more information, see "Gosset, W. S." in *Dictionary of National Biography. 1931–1940* (London: Oxford University Press, 1949), or see McMullen and Pearson (1939, pp. 205–253).

advised researchers to take "30:1 as the lowest odds which can be accepted as giving practical certainty that a difference . . . is significant" (Wood and Stratton, 1910). "Practical certainty" meant "enough certainty for a practical farmer." A consideration of the circumstances of Wood's journal provides a plausible explanation of his advice to researchers.

At the turn of the century, farmers in England were doing quite well. For example, Biffen (1905) reported that wheat production averaged 30 bushels per acre compared to 14 in the United States, 10 in Russia, and 7 in Argentina. Politically, of course, England was near the peak of its influence.

In addition, a science of agriculture was beginning to be established. Practical scientists, with an eye on increased production and reduced costs, conducted studies of weight gains in cattle, pigs, and sheep; amounts of dry matter in milk and mangels (a kind of beet); and yields of many grains. In 1905, *The Journal of Agricultural Science* was founded so these results could be shared around the world. Conclusions were reached after calculating probabilities, but the researchers avoided or ignored the problem created by probabilities between .01 and .25.

Wood's recommendation in his "how to interpret experiments" article was to adopt an α level that would provide a great deal of protection against a Type I error. (Odds of 30:1 against the null hypothesis convert to $p = .0323$.) A Type I error in this context would be a recommendation that, when implemented, did not produce improvement. I think that Wood and his colleagues wanted to be *sure* that when they made a recommendation, it would be correct.

A Type II error (failing to recommend an improvement) would not cause much damage; farmers would just continue to do well (though not as well as they might). A Type I error, however, would, at best, result in no agricultural improvement *and* in a loss of credibility and support for the fledgling institution of agriculture science.

To summarize my argument, it made sense for the agricultural scientists to protect themselves and their institution against Type I errors. They did this by not making a recommendation unless the odds were 30:1 against a Type I error.

When tables were published as an aid to researchers, probabilities such as .10, .05, .02, and .01 replaced odds (e.g., Fisher, 1925). So 30:1 ($p = .0323$) gave way to .05. Those tables were used by researchers in many areas besides agriculture, who appear to have adopted .05 as the most important significance level. (See Nelson, Rosenthal, and Rosnow, 1986, for a survey of psychologists.)

PROBLEMS

26. In Chapter 5 you encountered data for 11 countries on per capita cigarette consumption and the male death rate from lung cancer 20 years later. The correlation coefficient was .74. After consulting Table A, write a response to the statement, "The correlation is just a fluke; chance is the likely explanation for this correlation."

27. One of my colleagues gathered data on the time students spent on a midterm examination and their grade. The X variable is time and the Y variable is examination grade. Calculate r. Use Table A to determine if it is statistically significant.

$$\Sigma X = 903 \qquad \Sigma X^2 = 25{,}585 \qquad \Sigma XY = 46{,}885$$

$$\Sigma Y = 2079 \qquad \Sigma Y^2 = 107{,}707 \qquad N = 42$$

28. Please review the objectives at the beginning of the chapter. Can you do them?

 ## WEBSITE REFERENCES

http://psychology.wadsworth.com

Open "Choose Your Quiz" and select Spatz, 7e. Find the index of chapters, and click on Chapter 8, *Hypothesis Testing and Effect Size: One-Sample Designs.*

Open "Choose Your Area," scroll down, and select "Research Methods and Statistics." Click on "Statistics" in the graph. Click on "Student Activities." Click on the Web Page button for "Hypothesis Testing." After completing this workshop, return to the menu and click on "T Test for One Sample."

chapter **nine**

HYPOTHESIS TESTING, EFFECT SIZE, AND CONFIDENCE INTERVALS: TWO-SAMPLE DESIGNS

OBJECTIVES FOR CHAPTER NINE

Study the text. Work the problems as you come to them. Afterward, you should be able to:

1. Describe the logic of a simple experiment
2. Explain the testing of a null hypothesis when there are two samples
3. Explain some of the reasoning in finding degrees of freedom
4. Distinguish between independent-samples designs and correlated-samples designs
5. Calculate *t*-test values for both independent-samples designs and correlated-samples designs and write interpretations
6. Distinguish between statistically significant results and important results
7. Calculate and interpret an effect size index
8. Calculate confidence intervals for both independent-samples designs and correlated-samples designs and write interpretations
9. List and explain the assumptions for using the *t* distribution
10. Describe the factors that affect the rejection of the null hypothesis

In Chapter 8 you learned to use statistical hypothesis testing to answer questions about a population when data from one sample are available. In this chapter, the same hypothesis-testing reasoning will be used, but you will have data from two samples. Two-sample hypothesis testing is used frequently to analyze data from experiments.

One of the greatest benefits of studying statistics is that it helps you understand experiments and the experimental method. The experimental method is probably the most powerful method of finding out about natural phenomena. Besides being powerful, experiments can be interesting. They can answer such questions as these:

- Would you rather have a pigeon or a person searching for you if you were lost at sea?
- Does the removal of 20 percent of the cortex of the brain have an effect on the memory of tasks learned before the operation?
- Can you *reduce* people's ability to solve a problem by educating them?

In this chapter I will discuss the simplest kind of experiment and then show you how the logic of hypothesis testing, which you studied in the preceding chapter, can be expanded to answer questions like those above.

A SHORT LESSON ON HOW TO DESIGN AN EXPERIMENT

The basic ideas of a simple two-group experiment are not very complicated.

The logic of an experiment: Start with two equivalent groups. Treat them exactly alike except for one thing. Measure both groups and attribute any statistically significant difference between the two to the one way in which they were treated differently.

This summary of an experiment is described more fully in **Table 9.1.** The question that the experiment in the table sets out to answer is: What is the effect of Treatment A on a person's ability to perform task Q? In formal statistical terms, the question is: For task Q scores, is the mean of the population of those who have had Treatment A different from the mean of the population of those who have not had Treatment A?

To conduct this experiment, a group of participants is identified and two samples are assembled. These samples should be (approximately) equivalent. Treatment A is then administered to one group (generically called the **experimental group**) but not to the other group (generically called the **control group**). Except for Treatment A, both groups are treated in exactly the same way; that is, extraneous variables are held constant or balanced out for the two groups. Both groups perform task Q, and the mean score for each group is calculated.

The two sample means almost surely will differ. The question is whether the difference is due to Treatment A or is just the usual difference that would be expected of two samples from the same population.[1] This question can be answered by applying hypothesis-testing logic.

The first step in this logic is to assume there is no difference between two population means. Next, let the data tell you whether the assumption is reasonable. If the assumption is not reasonable, you are left with only one alternative: The populations have different means.

This generalized example has an independent variable with two levels (Treatment A and no Treatment A) and one dependent variable (scores on task Q). The word **treatment** is recognized by all experimentalists; it refers to different levels of the independent variable. The experiment in Table 9.1 has two treatments.

---

[1] Two samples from one population are the same statistically as two samples from two identical populations.

TABLE 9.1 Summary of a Simple Experiment

Key words	Tasks for the researcher	
Population	A population of participants	
	↙	↘
Assignment	Randomly assign one-half of *available* participants to the experimental group.	Randomly assign one-half of *available* participants to the control group.
	↓	↓
Independent variable	Give Treatment A to participants in experimental group.	Withhold Treatment A from participants in control group.
	↓	↓
Dependent variable	Measure participants' behavior on task Q.	Measure participants' behavior on task Q.
	↓	↓
Descriptive statistics	Calculate mean score on task Q, $\overline{X}_e$.	Calculate mean score on task Q, $\overline{X}_c$.
	↘	↙
Descriptive statistics	Calculate effect size index, d.	
	↓	
Inferential statistics	Compare $\overline{X}_e$ and $\overline{X}_c$ using a hypothesis-testing statistic. Reject or retain the null hypothesis.	
	↓	
Interpretation	Write a conclusion about the effect of Treatment A on task Q scores.	

The experimental procedure is a versatile one. Experiments have been used to decide a wide variety of issues such as how much sugar to use in a cake recipe, what kind of tea tastes the best, whether a drug is useful in treating cancer, and the effect of alcoholic parents on the personality of their children.

In many experiments, it is obvious that there are *two* populations of participants to begin with — for example, a population of men and a population of women. The question, however, is whether they are equal on the dependent variable.

In some experimental designs, participants are randomly assigned to treatments by the researcher; in others, the researcher uses a group of participants who have already been "treated" (for example, being males or being children of alcoholic parents). In either of these designs, the methods of inferential statistics are the same, although the interpretation of the first kind of experiment is usually less open to attack.[2]

---

[2] I am bringing up an issue that is beyond the scope of this statistics book. Courses with titles such as "Experimental Design" and "Research Methods" cover the intricacies of interpreting experiments.

The *raison d'être*[3] of experiments is to be able to tell a story about the universal effect of one variable on a second variable. All the work with samples is just a way to get evidence to support the story, but samples, of course, have chance built into them. Hypothesis testing takes chance factors into account when the story is told.

HYPOTHESIS TESTING: THE TWO-SAMPLE EXAMPLE

The researcher obtains data by following the procedures dictated by the design of the experiment. Once the data are gathered, some sort of hypothesis testing usually follows.

Hypothesis testing with two samples is similar to hypothesis testing with one sample; certainly, the logic is very similar. If you studied the section in Chapter 8 called "The Logic of Hypothesis Testing" until you understood it, it will be easier to add the new points here. If you aren't confident of your understanding, read, take notes, ask yourself questions, reread, ask questions of others, and so on until you do feel confident.

Here is the logic of hypothesis testing for a two-sample experiment. In a well-designed, well-executed experiment, all imaginable results are included in this statement: Either Treatment A has an effect or it does not have an effect. Begin by making a tentative assumption that Treatment A does *not* have an effect. Gather data. Using a sampling distribution based on the assumption that Treatment A has no effect, find the probability of the data obtained. If the probability is low, abandon your tentative assumption and conclude that Treatment A has an effect. If the probability is not low, you are back where you began: The data have not given you evidence against either of the two logical possibilities.

Now I will combine the logic of hypothesis testing with the language of an experiment:

1. Begin with two logical possibilities, the null hypothesis, H_0, and an alternative hypothesis, H_1.

   H_0: Treatment A *does not* have an effect; that is, the mean of the population of scores of those who receive Treatment A is equal to the mean of the population of scores of those who do not receive Treatment A. The difference between population means is zero. In statistical language:

   $$H_0: \mu_A - \mu_{no\ A} = 0 \qquad \text{or} \qquad H_0: \mu_A = \mu_{no\ A}$$

   H_1: Treatment A *does* have an effect; that is, the mean of the population of scores of those who receive Treatment A is *not* equal to the mean of the population of scores of those who do not receive Treatment A. The alternative hypothesis, which should be chosen before the data are gathered, can be two-tailed or one-tailed. An alternative hypothesis consists of *one* of the three H_1's that follow.

---

[3] *Raison d'être* means "reason for existence." (Ask someone who can pronounce French phrases to teach you this one. If you know how to pronounce the phrase, use it in conversation to identify yourself as someone to ask.)

Two-tailed alternative: $H_1: \mu_A \neq \mu_{no\ A}$

In the example of the simple experiment, this hypothesis says that Treatment A has an effect, but it does not indicate whether the treatment improves or disrupts performance on task Q. A two-tailed test allows either conclusion.

One-tailed alternatives: $H_1: \mu_A > \mu_{no\ A}$

$H_1: \mu_A < \mu_{no\ A}$

The first one-tailed alternative hypothesis allows you to conclude that Treatment A increases scores. However, no outcome of the experiment can lead to the conclusion that Treatment A produces lower scores than no treatment. The second alternative hypothesis permits a conclusion that Treatment A reduces scores, but not that it increases scores.

2. Tentatively assume that Treatment A has no effect (that is, assume H_0). If H_0 is true, the two samples will be alike except for the usual chance variations in samples.
3. Decide on an α level. (Usually, $\alpha = .05$.)
4. Choose an appropriate inferential statistical test. This test will have (a) a test statistic that can be calculated from the data, (b) a sampling distribution of the test statistic that shows its distribution when H_0 is true, and (c) a critical value for the α level. (For my two-group example, a t test and the t distribution with its critical values are appropriate.)
5. Calculate a test statistic using the sample data.
6. Compare the test statistic to the critical value from the sampling distribution. If the test statistic's probability is less than α, reject H_0. If larger, retain H_0. (For the t distribution, if the data-based t value is greater than the critical value in the table, reject H_0. If the data-based t value is less than the critical value, retain H_0.)
7. Write a conclusion that uses the terms of the experiment. Your conclusion should describe how the levels of the independent variable differ on the dependent variable in the populations the samples came from.

Researchers sometimes are interested in statistics other than the mean, and they often have more than two samples in an experiment. These situations call for sampling distributions that will be covered later in this book — sampling distributions such as F, chi square, U, and the normal distribution. I described the steps in this seven-item list somewhat broadly so that the list will make sense as an explanation not only for two-sample t tests but also for analyses you will be doing in the chapters that follow. I hope you will mark this section and return to it.

The sampling distribution that you will use to analyze the two-group experiments in this chapter is the t distribution. Not only are sample means distributed as t, but differences between sample means are also distributed as t.[4] To understand *which* t distribution you need, you will need to know more about degrees of freedom. But first, some problems.

●

[4] This statement is true *if* the assumptions discussed near the end of this chapter are true.

PROBLEMS

1. In your words, outline the logic of an experiment.

2. What procedure is used in the experiment described in Table 9.1 to ensure that the two samples are equivalent before treatment?

3. For each experiment that follows, identify the independent variable and its levels. Identify the dependent variable. Write a statement of the null hypothesis.

   a. A psychologist compared people with Type A personalities (high drive, time-oriented, and angry) to people with Type B personalities (do not exhibit these characteristics) on a test that measures satisfaction in life.

   b. A sociologist compared the numbers of years served in prison by those convicted of robbery and those convicted of embezzlement.

4. In your own words, outline the logic of hypothesis testing for a two-group experiment.

5. What is the purpose of an experiment, according to this textbook?

DEGREES OF FREEDOM

In your use of the t distribution, you have been determining degrees of freedom by rule-of-thumb techniques: $N - 1$ when you determine whether a sample mean $\overline{X}$ comes from a population with a mean μ_0, and $N - 2$ when you determine whether a correlation coefficient r is significantly different from .00. Now it is time to explain the concept more thoroughly.

The "freedom" in *degrees of freedom* refers to the freedom of a number to have any possible value. If you are asked to pick two numbers and there are no restrictions, both numbers are free to vary (take any value) and you have two degrees of freedom. If a restriction is imposed — say, $\Sigma X = 0$ — then one degree of freedom is lost because of that restriction; that is, when you now pick the two numbers, only one of them is free to vary. As an example, if you choose 3 for the first number, the second number *must be* -3. Because of the restriction that $\Sigma X = 0$, the second number is not free to vary. In a similar way, if you are to pick five numbers with a restriction that $\Sigma X = 0$, you have four degrees of freedom. Once four numbers are chosen (say, -5, 3, 16, and 8), the last number (-22) is determined.

The restriction that $\Sigma X = 0$ may seem to you to be an "out-of-the-blue" example and unrelated to your earlier work in statistics, but some of the statistics you have calculated have had such a restriction built in. For example, when you found $s_{\overline{X}}$, as required in the formula for the one-sample t test, you used some algebraic version of

$$s_{\overline{X}} = \frac{\hat{s}}{\sqrt{N}} = \frac{\sqrt{\dfrac{\Sigma(X - \overline{X})^2}{N - 1}}}{\sqrt{N}}$$

The restriction that is built in is that $\Sigma(X - \overline{X})$ is always zero and, in order to meet that requirement, one of the X's is determined. All X's are free to vary except one, and the degrees of freedom for $s_{\overline{X}}$ is $N - 1$. Thus, for the problem of using the t distribution to determine whether a sample came from a population with a mean μ_0,

$df = N - 1$. Walker (1940) summarized this reasoning by stating: "A universal rule holds: The number of degrees of freedom is always equal to the number of observations minus the number of necessary relations obtaining among these observations." A necessary relationship for $s_{\bar{X}}$ is that $\Sigma(X - \bar{X}) = 0$.

Another approach to explaining degrees of freedom is to emphasize the parameters that are being estimated by statistics. The rule of this approach is that *df* is equal to the number of observations minus the number of parameters that are estimated with a sample statistic. In the case of $s_{\bar{X}}$, 1 *df* is subtracted because $\bar{X}$ is used as an estimate of μ.

Now, let's turn to your use of the *t* distribution to determine whether a correlation coefficient *r* is significantly different from .00. The test statistic was calculated using the formula

$$t = (r)\sqrt{\frac{N - 2}{1 - r^2}}$$

The *r* in the formula is a *linear* correlation coefficient, which is based on a linear regression line. The formula for the regression line is

$$\hat{Y} = a + bX$$

There are two parameters in the regression formula, *a* and *b*. Each of these parameters costs one degree of freedom, so the *df* for testing the significance of correlation coefficients is $N - 2$.

These explanations of *df* for the one-sample *t* test and for testing the significance of *r* will prepare you for the reasoning I will give for determining *df* when you are analyzing two-group experiments.

CORRELATED-SAMPLES DESIGNS AND INDEPENDENT-SAMPLES DESIGNS

An experiment with two groups can be either a **correlated-samples design** or an **independent-samples design.**[5] These different designs require different *t*-test formulas, so you must decide what kind of design you have before you analyze the data.

You cannot tell the difference between the two designs by knowing the independent variable and the dependent variable. And, after the data are analyzed, you cannot tell the difference if you know the *t*-test value or if you have the interpretation of the experiment. To tell the difference, you must know whether scores in one group are paired with scores in a second group.

clue to the future

Most of the rest of this chapter will be organized around independent-samples and correlated-samples designs. In Chapters 10 and 11 the procedures you will learn are appropriate for only independent samples. Chapter 12 is a correlated-samples chapter. Three-fourths of Chapter 14 is also organized around these two designs.

---

[5] The terminology used in statistics and research design is confusing. Words such as *independent* and *correlated* are used to label several different concepts. My advice is to first attend to the concept itself and later memorize the label (or labels) that are used for the concept.

Correlated-Samples Design

The widely used correlated-samples design[6] may be created in three different ways. Fortunately, the actual arithmetic of calculating a *t*-test value is the same for all three. The three types of correlated-samples designs are natural pairs, matched pairs, and repeated measures.

Natural Pairs. In a **natural-pairs** investigation, the researcher does not assign the participants to one group or the other; the pairing occurs naturally, prior to the investigation. **Table 9.2** identifies one way in which natural pairs may occur in family relationships — in this case, fathers and sons. In such an investigation, you might ask whether fathers are shorter than their sons (or more religious, or more racially prejudiced, or whatever). Notice, though, that it is easy to decide that these are correlated-samples data: There is a logical pairing of the scores. That is, the two 74s belong together because they both came from the Smiths. Thus, pairing is based on membership in the same family.

Matched Pairs. In some situations, the researcher has control over the ways pairs are formed and a match can be arranged. One method is for two participants to be paired on the basis of similar scores on a pretest that is related to the dependent variable. This ensures that the two groups are fairly equivalent at the beginning of the experiment.

For example, in an experiment on the effect of hypnosis on problem solving, participants might initially be paired on the basis of IQ scores. IQ scores are known to be related to problem-solving ability. One member of each pair is randomly assigned to the hypnosis group (they solve problems while hypnotized) or the control group (they solve problems without hypnosis). If the two groups differ in problem-solving ability, you have some assurance that it wasn't due to IQ because the pairing produced groups equivalent in IQ.

Another variation of **matched pairs** is the split-litter technique used with animal subjects. A pair from a litter is split, and one is put into each group. In this way, the genetics and prenatal conditions are matched for the two groups. The same technique has been used in human experiments with twins or siblings. Gosset's barley experiments used this design; Gosset started with two similar subjects (adjacent plots of ground) and assigned them at random to one of two treatments.

TABLE 9.2 Illustration of a correlated-samples design

Father	Height (in.) X	Son	Height (in.) Y
Michael Smith	74	Mike, Jr.	74
Christopher Johnson	72	Chris, Jr.	72
Matthew Williams	70	Matt, Jr.	70
Joshua Brown	68	Josh, Jr.	68
Andrew Jones	66	Andy, Jr.	66
James Miller	64	Jim, Jr.	64

[6] Other terms used for this design are *within subjects, paired samples,* and *dependent samples.*

A third example of the matched-pairs technique occurs when a pair is formed *during* the experiment. Two subjects are "yoked" so that what happens to one, happens to the other (except for the independent and dependent variables). For example, if the procedure allows an animal in the experimental group to get a varying amount of food (or exercise or punishment), then that same amount of food (or exercise or punishment) is administered to the "yoked" control subject.

The difference between the matched-pairs design and a natural-pairs design is that, with the matched pairs, the investigator can randomly assign one member of the pair to a treatment. In the natural-pairs design, the investigator has no control over assignment. Although the statistics are the same, the natural-pairs design is usually open to more interpretations than the matched-pairs design.

Repeated Measures. A third kind of correlated-samples design is called a **repeated-measures** design because more than one measure is taken on each participant. This design may take the form of a before-and-after experiment. A pretest is given, some treatment is administered, and a post-test is given. The mean of the scores on the post-test is compared with the mean of the scores on the pretest to determine the effectiveness of the treatment. Clearly, two scores should be paired: the pretest and the post-test scores of each participant. In such an experiment, each person is said to serve as his or her own control.

For correlated-samples designs, researchers typically label one level of the independent variable X and the other Y. The characteristic that distinguishes correlated-samples designs from independent-samples designs is that there is a logical reason for each X score to be paired with its corresponding Y score. That is, the first score under treatment X is paired with the first score under treatment Y because the two scores are a natural pair, a matched pair, or repeated measures on the same subject.

Independent-Samples Design

The experiment outlined in Table 9.1 is an independent-samples design.[7] In this design, researchers typically label the two levels of the independent variable X_1 and X_2. The distinguishing characteristic of an independent-samples design is that there is no reason to pair a score in one treatment (X_1) with a particular score in the other treatment (X_2).

In creating an independent-samples design, researchers often begin with the pool of participants and then randomly assign individuals to the groups. Thus, there is no reason to suppose that the first score under X_1 should be paired with the first score under X_2. Random assignment can be used with correlated-samples designs, too. In a correlated-samples design, however, one member of a *pair* is assigned randomly.

The basic difference between the designs is that with a correlated-samples design there is a logical reason to pair up scores from the two groups; with the independent-samples design there is not. Thus, in a correlated-samples design, the two scores in the first row of numbers belong there because they come from two sources that are logically connected.

[7] Other terms used for this design are *between subjects, unpaired, randomized,* and *uncorrelated.*

PROBLEMS

6. Describe the two ways to state the rule governing the number of degrees of freedom.

7. Identify each of the following as an independent-samples design or a correlated-samples design. For each, identify the independent variable and its levels and the dependent variable. Work all six problems before checking your answers.

  a. An investigator gathered many case histories of situations in which identical twins were raised apart — one in a "good" environment and one in a "bad" environment. The group raised in the "good" environment was compared with that raised in the "bad" environment on attitude toward education.

  *b. A researcher counted the number of aggressive encounters among children who were playing with worn and broken toys. Next, the children watched other children playing with new, shiny toys. Finally, the first group of children resumed playing with the worn and broken toys, and the researcher again counted the number of aggressive encounters. (For a similar experiment, see Barker, Dembo, and Lewin, 1941.)

  c. Alaskans with seasonal affective disorder (depression) spent 2 hours a day under bright artificial light. Before treatment the mean depression score was 20.0, and at the end of one week of treatment the mean depression score was 6.4. (See Hellekson, Kline, and Rosenthal, 1986.)

  d. A researcher formed two groups. One was deprived of REM (rapid eye movement) sleep and the other was not. At the sleep lab, participants were paired and randomly assigned to one of the two groups. When those in the deprivation group began to show REM, they were awakened, thus depriving them of REM sleep. The other member of the pair was then awakened for an equal length of time during non-REM sleep. The next day all participants filled out a mood questionnaire.

  e. A researcher formed two groups. One was deprived of REM sleep and the other was not. At the sleep lab, participants were randomly assigned to one of the two groups. When those in the deprivation group began to show REM, they were awakened, thus depriving them of REM sleep. The other participants were awakened during non-REM sleep. The next day all participants filled out a mood questionnaire.

  f. Thirty-two freshmen applied for a sophomore honors course. Only 16 could be accepted, so the dean flipped a coin for each applicant, with the result that 16 were selected. At graduation, the dean compared the mean grade point average of those who had taken the course with the GPA of those who had not to see if the sophomore honors course had an effect on GPA.

THE *t* TEST FOR INDEPENDENT-SAMPLES DESIGNS

When a *t* test is used to decide whether two populations have the same mean, the null hypothesis is

$$H_0: \mu_1 = \mu_2$$

where the subscripts 1 and 2 are assigned arbitrarily to the two populations. If this null hypothesis is true, any difference between the two sample means is due to chance. The task is to establish an α level, calculate a t-test value, and compare that value with a critical value of t in Table D. If the t value calculated from the data is greater than the critical value (that is, less probable than α), reject H_0 and conclude that the two samples came from populations with different means. If the data-based t value is not as large as the critical value, retain H_0. I expect that this sounds familiar to you. For an independent-samples design, the formula for the **t test** is

$$t = \frac{\overline{X}_1 - \overline{X}_2}{s_{\overline{X}_1 - \overline{X}_2}}$$

The term $s_{\overline{X}_1 - \overline{X}_2}$ is the **standard error of a difference,** and **Table 9.3** shows formulas for calculating it. Use the formula at the top of the table when the two samples have an unequal number of scores. In the situation $N_1 = N_2$, the formula simplifies to those shown on the bottom of Table 9.3.

The previous t-test formula is the "working formula" for the more general case in which the numerator is $(\overline{X}_1 - \overline{X}_2) - (\mu_1 - \mu_2)$. For the cases in this book, the hypothesized value of $\mu_1 - \mu_2$ is zero, which reduces the general case to the working formula shown above. Thus, the t test, like many other statistical tests, consists of a difference between a statistic and a parameter divided by the standard error of the statistic.

The formula for degrees of freedom for independent samples is $df = N_1 + N_2 - 2$. Here is the reasoning. For each sample, the number of degrees of freedom is $N - 1$ because, for each sample, a mean has been calculated with the restriction that $\Sigma(X - \overline{X}) = 0$. Thus, the total degrees of freedom is $(N_1 - 1) + (N_2 - 1) = N_1 + N_2 - 2$.

TABLE 9.3 Formulas for $s_{\overline{X}_1 - \overline{X}_2}$, the standard error of a difference, for independent-samples t tests

If $N_1 \neq N_2$:

$$s_{\overline{X}_1 - \overline{X}_2} = \sqrt{\left(\frac{\Sigma X_1^2 - \dfrac{(\Sigma X_1)^2}{N_1} + \Sigma X_2^2 - \dfrac{(\Sigma X_2)^2}{N_2}}{N_1 + N_2 - 2}\right)\left(\frac{1}{N_1} + \frac{1}{N_2}\right)}$$

If $N_1 = N_2$:

$$s_{\overline{X}_1 - \overline{X}_2} = \sqrt{s_{\overline{X}_1}^2 + s_{\overline{X}_2}^2}$$

$$= \sqrt{\left(\frac{\hat{s}_1}{\sqrt{N_1}}\right)^2 + \left(\frac{\hat{s}_2}{\sqrt{N_2}}\right)^2}$$

$$= \sqrt{\frac{\Sigma X_1^2 - \dfrac{(\Sigma X_1)^2}{N_1} + \Sigma X_2^2 - \dfrac{(\Sigma X_2)^2}{N_2}}{N_1(N_2 - 1)}}$$

The following example shows the use of an independent-samples design to evaluate the effect of a drug on learning a complex problem-solving task. The drug group (seven monkeys) received pills, while the other group (six monkeys) was given an inert substance (a placebo). Pills and training on the task continued for six days. The total number of errors that each monkey made was the dependent variable.

The null hypothesis is that the drug has no effect on errors. A two-tailed test is called for because the investigator is interested in any effect the drug has, either to enhance or to inhibit performance.

The number of errors each animal made and the t test are shown in **Table 9.4.** Because the N's are unequal for the two samples, the longer formula for the standard error must be used. The t value for these data is -2.99, a negative value. For a two-tailed test, always use the absolute value of t. Thus, $|-2.99| = 2.99$.

TABLE 9.4 Number of errors for each monkey during six days of training

Drug group, X_1	Placebo group, X_2
34	39
52	57
26	68
47	74
42	49
37	57
40	
ΣX 278	344
ΣX^2 11,478	20,520
N 7	6
$\overline{X}$ 39.71	57.33

Applying the t test gives

$$t = \frac{\overline{X}_1 - \overline{X}_2}{s_{\overline{X}_1 - \overline{X}_2}} = \frac{\overline{X}_1 - \overline{X}_2}{\sqrt{\left(\dfrac{\Sigma X_1{}^2 - \dfrac{(\Sigma X_1)^2}{N_1} + \Sigma X_2{}^2 - \dfrac{(\Sigma X_2)^2}{N_2}}{N_1 + N_2 - 2}\right)\left(\dfrac{1}{N_1} + \dfrac{1}{N_2}\right)}}$$

$$= \frac{39.71 - 57.33}{\sqrt{\left(\dfrac{11,478 - \dfrac{(278)^2}{7} + 20,520 - \dfrac{(344)^2}{6}}{7 + 6 - 2}\right)\left(\dfrac{1}{7} + \dfrac{1}{6}\right)}}$$

$$= \frac{-17.62}{\sqrt{\left(\dfrac{437.43 + 797.33}{11}\right)(0.31)}} = \frac{-17.62}{5.90} = -2.99$$

$$df = N_1 + N_2 - 2 = 7 + 6 - 2 = 11$$

To evaluate a data-based *t* value of 2.99 with 11 *df,* you need a critical value from **Table D.** Begin by finding the row with 11 *df.* The critical value in the column for a two-tailed test with $\alpha = .05$ is 2.201; that is, $t_{.05}(11) = 2.201$. Thus, the null hypothesis can be rejected. Because 2.99 is also greater than the critical value at the .02 level (2.718), the results would usually be reported as "significant at the .02 level."

The final step is to interpret the results. Stop for a moment and compose your interpretation of what this experiment shows.

"The drug group averaged fewer errors than the placebo group (37.71 vs. 57.33). Thus, the drug *facilitated* learning ($p < .02$)."

PROBLEMS

*8. Imagine being lost at sea, bobbing around in an orange life jacket. The Coast Guard sends out two kinds of observers to spot you, fellow humans and pigeons. All observers are successful; the search time for each is given in the table in minutes. Decide on a one- or two-tailed test, analyze the data, and write a conclusion.

Fellow humans	Pigeons
45	31
63	24
39	20

*9. Karl Lashley (1890–1958) studied how the brain stores memories. In one experiment, rats learned a simple maze. Afterward, they were anesthetized and operated on. For half the rats, 20 percent of the cortex of the brain was removed. For the other half, the same operation was performed, except that no brain tissue was removed (a sham operation). After the rats recovered, they were given 20 trials in the maze, and the number of errors was recorded. Identify the independent variable, its levels, and the dependent variable. Analyze the data in the table. Write a conclusion about the effect that a 20 percent loss of cortex has on retention of a simple maze.

	Percent of cortex removed	
	0	20
ΣX	208	252
ΣX^2	1706	2212
N	40	40

10. A mail order firm is considering a software package that provides information on products (availability, history, cost, and so forth). To decide whether to purchase the new package, the next 17 new employees were randomly assigned to training on the old or the new package. After training, each employee completed a search report on 40 products. Analyze the data, which are in minutes, and write a conclusion about the new software package.

New package	Old package
4.3	6.3
4.7	5.8
6.4	7.2
5.2	8.1
3.9	6.3
5.8	7.5
5.2	6.0
5.5	5.6
4.9	

*11. Two sisters attend different universities. Each is sure that she has to compete with brighter students than her sister does. Although both are majoring in literature, each has an appreciation for quantitative thinking, so they decide to settle their discussion with ACT admission scores of freshmen at the two schools. Suppose they bring you the following data and ask for an analysis. The N's represent all freshmen for one year.

	The U.	State U.
$\overline{X}$	23.4	23.5
$\hat{s}$	3	3
N	8000	8000

The sisters agree to a significance level of .05. Begin by deciding whether to use a one-tailed or a two-tailed test. Calculate a t test and interpret your results. Be sure to carry several decimal places in your work on this problem.

THE t TEST FOR CORRELATED-SAMPLES DESIGNS

The formula for a t test for data from a correlated-samples design has a familiar theme: a difference between means divided by the standard error of a difference. The standard error of a difference between means of correlated samples is symbolized $s_{\overline{D}}$.

One formula for a t test between correlated samples is

$$t = \frac{\overline{X} - \overline{Y}}{s_{\overline{D}}}$$

where $\quad s_{\bar{D}} = \sqrt{s_{\bar{X}}^2 + s_{\bar{Y}}^2 - 2r_{XY}(s_{\bar{X}})(s_{\bar{Y}})}$

$df = N - 1$, where N = number of pairs

The number of degrees of freedom in the correlated-samples case is the number of *pairs* minus one. Although each pair has two values, once one value is determined, the other is restricted to a similar value. (After all, they are called *correlated* samples.) In addition, another degree of freedom is subtracted when $s_{\bar{D}}$ is calculated. This loss is similar to the loss of 1 *df* when $s_{\bar{X}}$ is calculated.

Compare the denominator of the correlated-samples *t* test with the denominator of the independent-samples *t* test (Table 9.3, for $N_1 = N_2$). The difference in the two procedures is the term $2r_{XY}(s_{\bar{X}})(s_{\bar{Y}})$. Of course, when $r_{XY} = 0$, this term drops out of the formula, and the standard error is the same as for independent samples.

Now, notice what happens to the standard error term in the correlated-samples design when $r > 0$: The standard error is *reduced*. Such a reduction will increase the size of the *t*-test value. Whether this reduction increases the likelihood of rejecting the null hypothesis depends on *how much t* is increased because there are fewer degrees of freedom in a correlated-samples design than in an independent-samples design.

The formula $s_{\bar{D}} = \sqrt{s_{\bar{X}}^2 + s_{\bar{Y}}^2 - 2r_{XY}(s_{\bar{X}})(s_{\bar{Y}})}$ is used only for illustration. The algebraically equivalent but arithmetically easier *direct-difference method* does not require you to calculate *r*. To find $s_{\bar{D}}$ by the direct-difference method, find the difference between each pair of scores, calculate the standard deviation of these difference scores, and divide the standard deviation by the square root of the number of pairs. Thus,

$$t = \frac{\bar{X} - \bar{Y}}{s_{\bar{D}}} = \frac{\bar{X} - \bar{Y}}{\hat{s}_D/\sqrt{N}}$$

where $\quad \hat{s}_D = \sqrt{\dfrac{\Sigma D^2 - \dfrac{(\Sigma D)^2}{N}}{N - 1}}$

$D = X - Y$

N = number of *pairs* of scores

Here is an example of a correlated-samples design and a *t*-test analysis. Suppose you are interested in the effects of interracial contact on racial attitudes. You have a fairly reliable test of racial attitudes in which high scores indicate more positive attitudes. You administer the test one Monday morning to a biracial group of fourteen 12-year-old girls who do not know one another but who have signed up for a week-long community day camp. The campers then spend the next week taking nature walks, playing ball, eating lunch, swimming, making things, and doing the kinds of things that camp directors dream up to keep 12-year-old girls busy. On Saturday morning the girls are again given the racial attitude test. Thus, the data consist of 14 pairs of before-and-after scores. The null hypothesis is that the mean of the population of "after" scores is equal to the mean of the population of "before" scores or, in terms of the specific experiment, that a week of interracial contact has no effect on racial attitudes.

TABLE 9.5 Hypothetical data from a racial attitudes study

Name*	Racial attitude scores		D	D^2
	Before day camp X	After day camp Y		
Brittany	34	38	−4	16
Ashley	22	19	3	9
Jessica	25	36	−11	121
Amanda	31	40	−9	81
Sarah	27	36	−9	81
Megan	32	31	1	1
Caitlin	38	43	−5	25
Samantha	37	36	1	1
Stephanie	30	30	0	0
Katherine	26	31	−5	25
Emily	16	34	−18	324
Lauren	24	31	−7	49
Kayla	26	36	−10	100
Rachel	29	37	−8	64
Sum	397	478	−81	897
Mean	28.36	34.14		

* The girls' names are the 14 most common in the United States, in order (Dunkling, 1993).

Suppose the data in **Table 9.5** are obtained. Using the sums of the D and D^2 columns in Table 9.5, you can find $\hat{s}_D$:

$$\hat{s}_D = \sqrt{\frac{\Sigma D^2 - \dfrac{(\Sigma D)^2}{N}}{N-1}} = \sqrt{\frac{897 - \dfrac{(-81)^2}{14}}{13}} = \sqrt{32.95} = 5.74$$

$$s_{\bar{D}} = \frac{\hat{s}_D}{\sqrt{N}} = \frac{5.74}{\sqrt{14}} = 1.53$$

Thus,

$$t = \frac{\bar{X} - \bar{Y}}{s_{\bar{D}}} = \frac{28.36 - 34.14}{1.53} = \frac{-5.78}{1.53} = -3.78$$

$$df = N - 1 = 14 - 1 = 13$$

Because $t_{.01}(13\ df) = 3.012$, a t value of 3.78 is significant beyond the .01 level; that is, $p < .01$. Because the "after" mean is greater than the "before" mean, conclude that racial attitudes were significantly more positive after camp than before.

Note that $\bar{X} - \bar{Y} = \bar{D}$, the mean of the difference scores. In this problem, $\Sigma D = -81$ and $N = 14$, so $\bar{D} = \Sigma D/N = -81/14 = -5.78$.

PROBLEMS

12. Give a formula and definition for each symbol.

a. $\hat{s}_D$

b. D

c. $s_{\bar{D}}$

d. $\bar{Y}$

13. The table shows data based on the aggression/toys experiment described in problem 7b. Reread the problem, analyze the scores, and write a conclusion.

Before	After
16	18
10	11
17	19
4	6
9	10
12	14

*14. Here are data from a consulting job I did some time ago. A lawyer asked me to analyze salary data for employees at a large rehabilitation center to determine whether there was evidence of sex discrimination. Among all employees, men's salaries were about 25 percent higher than women's. To rule out explanations like "more educated" and "more experienced," a subsample of employees with bachelor's degrees was separated out. From this group, men and women with equal experience were paired, resulting in an N of 16 pairs. Analyze the data. In your conclusion (which you can direct to the judge), explain whether education, experience, or chance could account for the observed difference. *Note:* You will have to think to set up this problem. Start by looking at the data you have.

	Women, X	Men, Y
ΣX or ΣY	\$204,516	\$251,732
ΣX^2 or ΣY^2	2,697,647,000	4,194,750,000
ΣXY	3,314,659,000	

*15. Which do you respond faster to, a visual signal or an auditory signal? The reaction times (RT) of seven students were measured for both signals. Decide whether the study is an independent- or correlated-samples design, analyze the data, and write a conclusion.

Auditory RT (seconds)	Visual RT (seconds)
0.16	0.18
0.19	0.20
0.14	0.20
0.14	0.17
0.13	0.14
0.18	0.17
0.20	0.21

16. To study the effects of primacy versus recency two groups were chosen randomly from a large sociology class. Both groups were given a two-page description of a person at work, including a paragraph telling how the person had been particularly helpful to a new employee. For half of the descriptions the "helping" paragraph was near the beginning (primacy), and for the other half the "helping" paragraph was near the end (recency). Afterward, each participant wrote a paragraph on how the worker might spend leisure time. The dependent variable was the number of positive adjectives (words like *good, exciting, cheerful*) in the paragraph. Based on these data, write a conclusion about primacy and recency.

Recency	Primacy
1	0
6	6
8	10
4	8
4	5

17. Asthma occurs in as much as 5 percent of the population. In one experiment, a group whose attacks began before age 3 was compared with a group whose attacks began after age 6. The dependent variable was the number of months the asthma persisted. Analyze the difference between means and write a conclusion.

	Onset of asthma	
	Before age 3	After age 6
$\overline{X}$	60	36
$\hat{s}$	12	6
N	45	45

SIGNIFICANT RESULTS AND IMPORTANT RESULTS

When I was a rookie instructor, I did some research with an undergraduate student, David Cervone, who was interested in hypnosis. We used two rooms in the library to conduct the experiment. When the results were analyzed, two of the groups were significantly different. One day, a librarian asked how the experiment had come out.

"Oh, we got some significant results," I said.

"I imagine David thought they were more significant than you did," was the reply.

At first I was confused. Why would David think they were more significant than I would? Point oh-five was point oh-five. Then I realized that the librarian and I were using the word *significant* in two quite different ways. I was using *significant* in the statistical sense: The difference was a *reliable* one that would be expected to occur again if the study was run again. The librarian meant *important*. Of course, *important* is important so let's pursue the librarian's meaning and ask about the importance of a difference.

First of all, if a difference is not statistically significant, then it is not important. "Not significant" means that the difference may be unreliable. If a difference is NS, you have no assurance that a second experiment will produce a similar difference.

If a difference is statistically significant, you have to go beyond hypothesis testing to decide if the difference is important. Arthur Irion (1976) captured this limitation of hypothesis testing (and the two meanings of the word *significant*) when he reported that his statistical test "reveals that there is significance among the differences although, of course, it doesn't reveal what significance the differences have."

In two problems that you worked, you found differences that were statistically significant. In problem 11 the difference in college ACT admissions scores between two schools was one-tenth of a point, which doesn't qualify as important. In contrast (problem 17), the difference in how long children's asthma lasts (60 months or 36 months) seems like an important difference.

Thus, hypothesis testing tests a difference only for statistical significance. The next two sections will describe other statistics that help you decide about importance.

EFFECT SIZE INDEX

The effect size index, *d*, may help you decide whether a significant difference is important. You probably recall from Chapter 4 that the mathematical formula for *d* is

$$d = \frac{|\mu_1 - \mu_2|}{\sigma}$$

You may also recall that for problems in both Chapter 4 and Chapter 8 you were given a value for σ. Now that you know about degrees of freedom, you will be able to estimate σ and thus calculate a value for *d* from the raw data of an experiment.

Effect Size Index for Independent Samples

The working formula for d, the effect size index in the independent-samples case, is

$$d = \frac{|\overline{X}_1 - \overline{X}_2|}{\hat{s}}$$

Calculating $\overline{X}_1$ and $\overline{X}_2$ requires no explanation. The calculation of $\hat{s}$ depends on whether or not the sample sizes are equal. When $N_1 = N_2$,

$$\hat{s} = \sqrt{N_1}(s_{\overline{X}_1 - \overline{X}_2})$$

where N_1 is the sample size for *one group*. By giving a little thought, you can see that this formula follows from your work in Chapter 8, where the standard error of the mean was found by dividing $\hat{s}$ by $\sqrt{N}$.

In the case $N_1 \neq N_2$, you cannot find $\hat{s}$ directly from $s_{\overline{X}_1 - \overline{X}_2}$. To find $\hat{s}$ when $N_1 \neq N_2$, use the formula

$$\hat{s} = \sqrt{\frac{\hat{s}_1^2(df_1) + \hat{s}_2^2(df_2)}{df_1 + df_2}}$$

where df_1 and df_2 are the degrees of freedom $(N - 1)$ for each of the two samples
$\hat{s}_1$ and $\hat{s}_2$ are the sample standard deviations
The formula for $s_{\overline{X}_1 - \overline{X}_2}$ for $N_1 \neq N_2$ (Table 9.3) does not include separate elements for $\hat{s}_1$ and $\hat{s}_2$. Thus, you have to calculate $\hat{s}_1$ and $\hat{s}_2$ from the raw data for each sample.

Effect Size Index for Correlated Samples

The working formula for d for correlated samples is

$$d = \frac{|\overline{X} - \overline{Y}|}{\hat{s}}$$

The formula for $\hat{s}$ for correlated samples is

$$\hat{s} = \sqrt{N}(s_{\overline{D}})$$

where N is the number of pairs of participants.

Interpretation of d

The conventions proposed by Cohen (1969) are used to interpret d:

Small effect	$d = .20$
Medium effect	$d = .50$
Large effect	$d = .80$

To illustrate the interpretation of d, I will calculate an effect size index for two problems that you have already worked. For both of these, the t test revealed a significant difference.

When you compared the ACT scores at two state universities (problem 11), you found that the mean for State U. freshmen (23.5) was significantly higher than the mean for freshmen at The U. (23.4). The standard error of the difference for this $N_1 = N_2$ study was 0.0474. Thus,

$$d = \frac{|\overline{X}_1 - \overline{X}_2|}{\hat{s}} = \frac{|23.5 - 23.4|}{\sqrt{8000(0.0474)}} = \frac{0.1}{4.240} = .02$$

An effect size index of .02 is very, very small. Thus, although the difference between the two schools is statistically significant, the size of this difference seems to be of no consequence.

In problem 14 you found that there was a significant difference between the salaries of men and women. An effect size index will reveal whether this difference can be considered small, medium, or large.

$$d = \frac{|\overline{X} - \overline{Y}|}{\hat{s}} = \frac{12{,}782.25 - 15{,}733.25}{\sqrt{16(2872)}} = \frac{2951}{718} = 1.02$$

An effect size index of 1.02 qualifies as large. Thus, the statistical analysis of these data, which includes a t test and an effect size index, supports the conclusion that sex discrimination was taking place and that the effect was large.

ESTABLISHING A CONFIDENCE INTERVAL ABOUT A MEAN DIFFERENCE

Another way to measure the size of the effect that an independent variable has is to establish a confidence interval. You probably recall from Chapter 7 that a confidence interval establishes a range of values that captures a parameter with a certain degree of confidence. In this section the parameter that is being captured is the *difference* between population means, $\mu_1 - \mu_2$. Researchers usually choose 95 percent or 99 percent as the amount of confidence they want.

Confidence Intervals for Independent Samples

To find the lower and upper limits of the confidence interval about a mean difference for data from an independent-samples design, use these formulas:

$$\text{LL} = (\overline{X}_1 - \overline{X}_2) - t(s_{\overline{X}_1 - \overline{X}_2})$$
$$\text{UL} = (\overline{X}_1 - \overline{X}_2) + t(s_{\overline{X}_1 - \overline{X}_2})$$

Of course, $\overline{X}_1$ and $\overline{X}_2$ come from the two samples. The standard error of the difference, $s_{\overline{X}_1 - \overline{X}_2}$, is calculated using the appropriate formula from Table 9.3. The value for t comes from Table D. Look again at **Table D.** The top row of the table gives you commonly used confidence interval percents.

As an example, let's return to problem 8, the time required by humans and pigeons to find a person lost at sea. The data produce a mean of 49 minutes for the three humans and 25 minutes for the three pigeons. The standard error of the difference, $s_{\bar{X}_1-\bar{X}_2}$, is 7.895. For this example, let's find the lower and upper limits of a 95 percent confidence interval about the true difference between humans and pigeons. The t (4 df) value from Table D for 95 percent confidence is 2.776. You get

$$\text{LL} = (\bar{X}_1 - \bar{X}_2) - t(s_{\bar{X}_1-\bar{X}_2}) = (49 - 25) - 2.776(7.895) = 2.08 \text{ minutes}$$
$$\text{UL} = (\bar{X}_1 - \bar{X}_2) + t(s_{\bar{X}_1-\bar{X}_2}) = (49 - 25) + 2.776(7.895) = 45.92 \text{ minutes}$$

The interpretation of this confidence interval is that we can expect, with 95 percent confidence, that the true difference between the people and the pigeons in the time required to find a person lost at sea is between 2.08 to 45.92 minutes.

Not only does a confidence interval about a mean difference tell you the size of the difference, but *it also tests the null hypothesis* in the process. The null hypothesis is $H_0: \mu_1 - \mu_2 = 0$. A 95 percent confidence interval gives you a lower limit for $\mu_1 - \mu_2$ and an upper limit for $\mu_1 - \mu_2$. If zero is not between the two limits, you can be 95 percent confident that the difference between μ_1 and μ_2 is not zero. This is equivalent to rejecting the null hypothesis at the .05 level. Because the interval for search time, 2.08 to 45.92 minutes, does not contain zero, you can reject the hypothesis that humans and pigeons take equal amounts of time to spot a person lost at sea. This is the conclusion you reached when you first worked problem 8.

Thus, one of the arguments for confidence intervals is that they do everything that hypothesis testing does, plus they give you an estimate of just how much difference there is between the two groups.

Confidence Interval for Correlated Samples

To find the confidence interval about a mean difference for correlated-samples data, use these formulas:

$$\text{LL} = (\bar{X} - \bar{Y}) - t(s_{\bar{D}})$$
$$\text{UL} = (\bar{X} - \bar{Y}) + t(s_{\bar{D}})$$

Remember that with correlated samples data, df is $N - 1$, where N is the number of pairs of data.

If you are a little hazy in your understanding of confidence intervals about a mean difference, it might help to reread the material in Chapter 7 that introduced confidence intervals.

Both the effect size index and confidence intervals about a mean difference are statistics that may help you decide about the importance of a difference that is statistically significant.

error detection

For confidence intervals for either independent or correlated samples, use a t value from Table D, not a t-test value calculated from the data.

PROBLEMS

18. In an experiment in teaching German, the holistic method produced higher scores on a test than the lecture–discussion method. A report of the experiment included the phrase "$p < .01$." What event does the p refer to?

19. Write sentences that distinguish between a statistically significant difference and an important difference.

20. Problem 9 revealed that there was no significant loss in the memory of rats that had 20 percent of their cortex removed. What is the effect size index for those data? My interpretation will note that the nonsignificant t test was based on 80 rats, which is quite a large sample. Use this hint to write your own interpretation based on both the effect size index and the t test.

21. In the example problem for the correlated-samples t test (page 195), you found that a week of camp activities significantly improved racial attitudes. Calculate and interpret an effect size index for those data.

22. How does sleep affect memory? Eight volunteers learned a list of 10 nonsense words. Half the group then slept for 4 hours and half engaged in daytime activities. Each person then recalled the list. The next day everyone learned a new list. Those who had slept stayed awake and those who were awake the day before slept for 4 hours. Each recalled the new list. The table shows the number of words each person recalled under the two conditions. Establish a 95 percent confidence interval about the difference between the two means. Write an interpretation that includes a statement about the fate of the null hypothesis when $\alpha = .05$. (Based on a 1924 study by Jenkins and Dallenbach)

Asleep	Awake
4	2
3	2
5	4
3	0
2	4
4	3
3	1
6	4

23. Twenty depressed patients were randomly assigned to two groups of ten. Each group spent 2 hours each day working and visiting at the clinic. One group worked in bright artificial light; the other did not. At the end of the study, the mean depression score for those who received light was 6; for those who did not, the mean score was 11. (High scores mean more depression.) The standard error of the difference was 1.20. Establish a 99 percent confidence interval about the mean difference and write an interpretation.

REACHING CORRECT CONCLUSIONS

In the section on the logic of hypothesis testing (page 184), you learned that the last step is to write a conclusion. Of course, a conclusion is called for when you calculate confidence intervals. How can you be sure that your conclusions are correct? You can't be completely sure, of course, but here are some considerations that can affect your confidence.

Nature of the Populations the Samples Come From

For an independent-samples *t* test, the two populations the samples come from should (1) be normally distributed and (2) have variances that are equal. The reason the populations should have these characteristics is that the probability figures in Table D were derived mathematically from populations that were normally distributed and had equal variances. What if you are uncertain about the characteristics of the populations in your own study? Are you likely to reach the wrong conclusion by using Table D?

There is no simple answer to the preceding question. Several studies suggest that the answer is no because the *t* test is a robust test. (*Robust* means that a test gives fairly accurate probabilities even when the populations do not have the assumed characteristics.) Robustness, however, has not been defined quantitatively (Bradley, 1978), and several studies show sizable departures from accuracy in a variety of situations. (See, for example, Blair and Higgins, 1985.) In actual practice, researchers routinely use *t* tests unless the data are clearly not normally distributed or the variances are not equal.

Control of Extraneous Variables

Extraneous variables are bad news when it comes time to draw conclusions. If a study has an uncontrolled extraneous variable, you cannot write a simple cause-and-effect conclusion relating the independent variable to the dependent variable.

The most common way for researchers to control for extraneous variables is to randomly assign the subjects that are available to treatment groups. Using random assignment creates two groups that are approximately equal on all extraneous variables.

Unfortunately, for many interesting investigations, random assignment is just not possible. Studies of gender, socioeconomic status, and age are examples. When random assignment isn't possible, researchers typically use additional control groups and logic to reduce the number of extraneous variables.

This section introduced three considerations that affect your confidence when you draw conclusions from an independent-samples *t* test. These three considerations are sometimes referred to as "assumptions." The general topic "assumptions of the test" will be addressed again in later chapters.

To prepare you for the next section, remember that the probabilities the t distribution gives are for differences that occur when the null hypothesis is true. When those probabilities are correct and you use an α level of .05, you are assured that if you reject the null hypothesis, the probability that you've made a Type I error is .05 or less. Thus, you have good protection against making a mistake when the null hypothesis is true. However, what about when the null hypothesis is false?

REJECTING THE NULL HYPOTHESIS—THE TOPIC OF POWER

When the null hypothesis is false, you should reject it. But will you? Factors that affect rejecting a false null hypothesis come under the topic of **power.** Power is a broad topic, but the statistical definition of power is simple: Power $= 1 - \beta$.

With a little thought, you can make sense out of the expression $1 - \beta$. As you know from my discussion of **Table 8.1,** β is the probability of a Type II error. A Type II error occurs when you *fail* to reject a false null hypothesis. So, if H_0 is false, the probability that you *will* reject it is $1 - \beta$. Thus, the more powerful the experiment, the more likely you are to detect a false null hypothesis.

Of course, rejecting H_0 is what a researcher wants. When H_0 is rejected, a clear-cut conclusion is reached; the two populations are different. If you can't reject H_0, you are left with uncertainty about the relationship between the two populations. Rejecting H_0 depends on power, and there are three factors that affect the power of an experiment: effect size, standard error, and α.

1. *Effect size.* The larger the effect size, the more likely that you will reject H_0. For example, the greater the difference between the mean of the experimental group population and the mean of the control group population, the more likely it is that the samples will lead you to correctly conclude that the populations are different. Of course, determining effect size *before* the data are gathered is difficult, and experience helps.

2. *The standard error of a difference.* Look at the formulas for t on pages 190 and 193. You can see that as $s_{\bar{X}_1 - \bar{X}_2}$ and $s_{\bar{D}}$ get smaller, t gets larger and you are more likely to reject H_0. Here are two ways you can reduce the size of the standard error:

a. *Sample size.* The larger the sample, the smaller the standard error of a difference. **Figure 7.3** shows this to be true for the standard error of the mean; it is also true for the standard error of a difference. How big should a sample be? Cohen (1992) provides a helpful table (as well as a discussion of the other factors that affect power). Many times, of course, sample size is dictated by practical considerations — time, money, or availability of participants.

b. *Sample variability.* Reducing the variability in the sample data will produce a smaller standard error. You can reduce variability by using reliable measuring instruments, recording data correctly, being consistent, and, in short, reducing the "noise" or random error in your experiment.

3. α. The larger α is, the more likely you are to reject H_0. As you know, the conventional limit is .05. Values above .05 begin to raise doubts in readers' minds, although there has been a healthy recognition in recent years of the arbitrary nature

of the .05 rule. Researchers whose results come out with $p = .08$ (or near there) are very likely to remain convinced that a real difference exists and proceed to gather more data, expecting that the additional data will establish a statistically significant difference.

Actually incorporating these three factors into a power analysis is a topic covered by intermediate textbooks.[8] Researchers typically use a power analysis before gathering data to help decide how large N should be. A power analysis after the data are gathered and analyzed is used to help determine the value of an experiment in which the null hypothesis was retained.

I will close this section on power by asking you to imagine that you are a researcher directing a project that could make a Big Difference. (Because you are imagining this, the difference can be in anything you would like to imagine — the health of millions, the destiny of nations, your bank account, whatever.) Now suppose that the success or failure of the project hinges on one final statistical test. One of your assistants comes to you with the question, "How much power do you want for this last test?"

"All I can get," you answer.

If you examine the list of factors that influence power, you will find that there is only one item that you have some control over and that is the standard error of a difference. Of the factors that affect the size of the standard error of a difference, the one that most researchers can best control is N. So, allocate plenty of power to important statistical tests — use large N's.

PROBLEMS

24. Your confidence that the probabilities are accurate when you run an independent-samples t test depends on certain assumptions. List them.

25. Using words, define statistical power.

26. List the four factors that determine the power of an experiment. [If you can recall them without looking at the text, you know that you have been reading actively (and effectively).]

27. What is the effect of set (previous experience) on problem-solving? In this experiment (based on Birch and Rabinowitz, 1951), the problem was to tie together two strings (A and B) that were suspended from the ceiling 12 feet apart. Unfortunately, a participant could not reach both strings at one time. The plight is shown in the illustration. The solution to the problem is to tie a weight to string A, swing it out, and then catch it on the return. The researcher supplied the weight (an electrical light switch) under one of two conditions. In one condition, participants had previously wired the switch into an electrical circuit and used it to turn on a light. In the other condition, participants had no experience with the switch until they were confronted with the two-string problem. The question was whether experience with the

[8] See Howell (1997, Chap. 8) or Aron and Aron (1999, Chap. 8).

switch has any effect on the time required to solve the problem. State the null hypothesis, analyze the data with a *t* test and an effect size index, and write a conclusion.

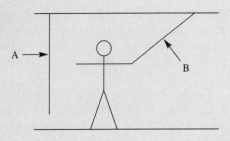

	Wired the switch into circuit	No previous experience with switch
$\overline{X}$	7.40 minutes	5.05 minutes
$\hat{s}$	2.13 minutes	2.31 minutes
N	20	20

28. A French teacher wanted to test the claim that "concrete, immediate experience enhances vocabulary learning." Driving from the college to the city (about 30 miles), he recorded an audiotape describing in French the terrain, signs, distances, and so forth. ("L'auto est sur le pont. Carrefour prochain est dangereux.") For the next part of his investigation, his French II students were ranked from 1 to 10 in competence. The ten were divided in such a way that numbers 1, 4, 5, 8, and 9 listened to the tape while riding to the city (concrete, immediate experience) and numbers 2, 3, 6, 7, and 10 listened to the tape in the language laboratory. The next day a vocabulary test was given and the number of errors for each student was recorded. Think about how to set up these data, analyze them, and write a conclusion about immediate, concrete experience.

| #1 7; | #2 7; | #3 15; | #4 7; | #5 12; |
| #6 22; | #7 24; | #8 12; | #9 21; | #10 32 |

29. Skim over the objectives at the beginning of the chapter as a way to consolidate what you have learned.

WEBSITE REFERENCES

http://psychology.wadsworth.com

Open "Choose Your Quiz" and select Spatz, 7e. Find the index of chapters, and click on Chapter 9, *Hypothesis Testing, Effect Size, and Confidence Intervals: Two-Sample Designs.*

Open "Choose Your Area," scroll down, and select "Research Methods and Statistics." Click on "Statistics" in the graph. Click on "Student Activities." Click on the Web Page button for "T Test for Between Groups and Related Groups."

WHAT WOULD YOU RECOMMEND? chapters six–nine

It is time again for you to do a set of *What would you recommend?* problems. Your task is to choose an appropriate statistical technique from among the several you have learned in the previous four chapters. No calculations are necessary. Reviewing is allowed. For each problem that follows, (1) recommend a statistic that will either answer the question or make the comparison, and (2) explain why you recommend that statistic.

a. The numbers of smiles by 10-month-old infants were recorded for 12 minutes. There was an audience during half the time and no audience during the other half. The researchers — Jones, Collins, and Hong (1991) — wanted to find out whether smiles are for communication (require an audience) or whether they express emotion (don't require an audience).

b. McCrae and Costa (1990, p. 105) describe an adult social worker who has a score of 73 on a personality inventory that measures gregariousness (the revised NEO Personality Inventory). The scores on this inventory are distributed normally (population mean = 50 and standard deviation = 10). How can you find the percent of the population that is less gregarious than this social worker?

c. Suppose a psychologist conducted a workshop to help people improve their social skills. At the end of the workshop, all the participants fill out the revised NEO Personality Inventory. Suppose the mean gregariousness score of the participants was 55. What statistical test can help determine whether the workshop produced a group whose mean gregariousness score was greater than the population mean?

d. A sampling of high school students takes the SAT test each year. The mean score for the students in a particular state is readily available. What statistical technique will give you a set of brackets that "captures" a population mean?

e. One interpretation of the statistical technique in problem D is that the brackets capture the *state's* population mean. This interpretation is not correct. Why not?

f. In a sample of 100 people, a correlation coefficient of .20 was found between the number of illnesses in the past year and scores on a measure of stress. How can you determine the likelihood of such a correlation coefficient if there were actually no relationship between the two variables?

g. Colleges and universities in the United States are sometimes divided into six categories: public universities, four-year public institutions, two-year public institutions, private universities, four-year private institutions, and two-year private institutions. In any given year, each category has a certain number of students. Which of the six categories does your institution belong to? How can you find the probability that a student, chosen at random, comes from that category?

h. Children (mean age = 6 years) who ate breakfast at school participated in an experiment to determine the effect of sugar on activity. On some days a particular child received sugar with breakfast and on other days the child did not. This was true for all the children in the study. Later in the day each child was observed and activity recorded. [See Rosen et al. (1988), who found that the effect of sugar was quite small.]

transition page

*t*he *t* test, at which you are now skilled, is the traditional way to analyze the results from an experiment that is designed to compare two treatments. The next step is to learn a technique for analyzing the results of an experiment that has more than two treatments.

This technique is called the **analysis of variance** (ANOVA for short, pronounced uh-ṅove-uh). ANOVA is based on the very versatile concept of assigning the variability in the data to various sources. (This often means that ANOVA identifies the variables that produce changes in the scores.) ANOVA was invented by Sir Ronald A. Fisher, an English biologist and statistician. It is used more frequently than any other inferential statistical technique by social and behavioral scientists.

So the transition this time is from a *t* test and its sampling distribution, the *t* distribution, to ANOVA and its sampling distribution, the *F* distribution. In Chapter 10 you will use ANOVA to compare means from an independent-samples design that involves more than two treatments. Chapter 11 will show you that ANOVA can be used to analyze experiments in which there are two *independent variables*, each of which may have two or more levels of treatment. In Chapter 12 you will use ANOVA to compare means from a correlated-samples design that has more than two treatments (but just one independent variable).

The analysis of variance is a widely used statistical technique, and Chapters 10–12 will be devoted to an introduction to its elementary forms. Intermediate and advanced books explain more sophisticated (and complicated) analysis-of-variance designs. [Howell (1997) is particularly accessible.]

chapter**ten**

ANALYSIS OF VARIANCE: ONE-WAY CLASSIFICATION

OBJECTIVES FOR CHAPTER TEN

Study the text. Work the problems as you come to them. Afterward, you should be able to:

1. Identify the independent and dependent variables in a one-way ANOVA
2. Explain the rationale of ANOVA
3. Define *F* and explain its relationship to *t*
4. Compute sums of squares, degrees of freedom, mean squares, and *F* for an ANOVA
5. Construct a summary table of ANOVA results
6. Interpret the *F* value for a particular experiment and explain what the experiment shows
7. Distinguish between *a priori* and *post hoc* tests
8. Use the Tukey Honestly Significant Difference (HSD) test to make all pairwise comparisons
9. List and explain the assumptions of ANOVA
10. Calculate and interpret *f,* an effect size index

$\mathbf{I}$n this chapter you will work with the simplest analysis-of-variance design, a design called **one-way ANOVA.** (ANOVA stands for analysis of variance.) The *one* in one-way means that one independent variable is being analyzed. One-way ANOVA is used to find out if there are any differences among three or more population means.[1] Experiments with more than two treatment levels are common in all disciplines that use statistics. Here are examples, some of which will be covered as you progress through this chapter:

[1] As will become apparent, ANOVA can be used also when there are just two groups.

1. Samples of lower-, middle-, and upper-class persons were compared on their attitudes toward religion.
2. Four groups learned a task. A different schedule of reinforcement was used for each group. Afterward, response persistence was measured.
3. Suicide rates were compared for countries that represented low, medium, and high degrees of modernization.

Except for the fact that these experiments have more than two groups, they are like those you analyzed with an *independent-samples t test* in Chapter 9. There is one independent and one dependent variable. The null hypothesis is that the population means are the same for all groups. The subjects in each group are independent of the subjects in the other groups. The only difference is that the independent variable has more than two levels. To analyze such designs, use a one-way ANOVA.[2]

In example 1, the independent variable is social class, and it has three levels. The dependent variable is attitude toward religion. The null hypothesis is that the religious attitudes are the same in all three populations of social classes — that is, H_0: $\mu_{lower} = \mu_{middle} = \mu_{upper}$.

PROBLEM

*1. For examples 2 and 3, identify the independent variable, the number of levels of the independent variable, the dependent variable, and the null hypothesis.

Although this chapter does not have "hypothesis testing" or "effect size" in the title (as did the two previous chapters on *t* tests), ANOVA is a hypothesis-testing technique, and a new effect size index will be explained.

A common first reaction to the task of determining if there is a difference among three or more population means is to run *t* tests on all possible pairs of sample means. For three populations, three *t* tests would be required ($\overline{X}_1$ vs. $\overline{X}_2$, $\overline{X}_1$ vs. $\overline{X}_3$, and $\overline{X}_2$ vs. $\overline{X}_3$). For four populations, six tests are needed.[3] This multiple *t*-test approach *will not work* (and not just because doing lots of *t* tests is tedious).

Here is the reasoning: Suppose you have 15 samples all drawn from the same population. (Because there is just one population, the null hypothesis is clearly true.) These 15 sample means will vary from one another as a result of chance factors. Now suppose you ran every possible *t* test (all 105 of them), retaining or rejecting each null hypothesis at the .05 level. How many times would you reject the null hypothesis?

The answer is about five. When the null hypothesis is true (as in this example) and α is .05, 100 *t* tests will produce about five Type I errors.

Now, let's move from this theoretical analysis to the reporting of an experiment. Suppose you conducted a 15-group experiment, ran 105 *t* tests, and found five significant differences. If you then pulled out those five and said they were reliable differences (that is, differences that are not due to chance), people who understand the

[2] The ANOVA technique for analyzing a correlated-samples design will be covered in Chapter 12.
[3] The formula for the number of combinations of *n* things taken two at a time is $[n(n - 1)]/2$.

preceding two paragraphs would realize that you don't understand statistics (and would recommend that your manuscript not be published). You can protect yourself from such a disaster if you use a statistical technique that keeps the overall risk of a Type I error at an acceptable level (such as .05 or .01).

Sir Ronald A. Fisher (1890–1962) developed such a technique. Fisher did brilliant work in genetics, but it has been overshadowed by his fundamental work in statistics. In genetics, it was Fisher who showed that Mendelian genetics is compatible with Darwinian evolution. (For a while after 1900, genetics and evolution were in opposing camps.) And, among other contributions, Fisher showed how a recessive gene can become established in a population.

In statistics, Fisher invented ANOVA, the topic of this chapter and the next two. He also discovered the exact sampling distribution of r (1915), developed a general theory of estimation, and wrote *the* book on statistics. (*Statistical Methods for Research Workers* was first published in 1925 and went through 14 editions and several translations by 1973.)

Before getting into biology and statistics in such a big way, Fisher worked for an investment company for 2 years and taught in a public school for 4 years. [See Greene (1966), Yates (1981), or Hald (1998, pp. 734–739).]

RATIONALE OF ANOVA

The question that ANOVA addresses is whether the populations that the samples come from have the same μ. If the answer is no, then at least one of the populations has a different μ. I've grouped the rationale of ANOVA into subsections on (1) hypothesis testing and (2) the F distribution.

Hypothesis Testing

1. The populations you are comparing (the different treatments) may be exactly the same *or* one or more of them may have a different mean. These two descriptions cover all the logical possibilities and are, of course, the null hypothesis (H_0) and the alternative hypothesis (H_1).
2. Tentatively assume that the null hypothesis is correct. If the populations are all the same, any differences among sample means will be the result of chance.
3. Choose a sampling distribution that will show the probability of various differences among sample means when H_0 is true. For more than two sample means, use the one that Fisher discovered, which is now called the F distribution.
4. Obtain data from the populations you are interested in. (This usually means you must conduct an experiment.) Perform calculations on the data using the procedures of ANOVA until you have an F value.
5. Compare your calculated F value to the critical value of F in the sampling distribution (F distribution). From this you can determine the probability of obtaining the data you did, *if the null hypothesis is true.*

6. Come to a conclusion about H_0. If the probability is .05 or less, reject H_0. With H_0 eliminated, you are left with H_1. If the probability is greater than .05, retain H_0, leaving you with both H_0 and H_1 as possibilities.

7. Tell the story of what the data show. If you reject H_0 when there are three or more treatments, a comparison between specific groups will require further data analysis. If you retain H_0, then your data leave you with both H_0 and H_1.

I'm sure that the ideas in this list are more than somewhat familiar to you. In fact, I suspect that your only uncertainties are about step 3, the F distribution, and step 4, calculating an F value. The next several pages will explain the F distribution and then what to do with the data to get an F value.

The F Distribution

The **F distribution,** like the t distribution, is a sampling distribution. As such, it gives the probabilities of various outcomes when the null hypothesis is true (all samples come from identical populations).

The F value that is calculated from the data is a ratio; one estimate of the population variance is divided by a second estimate. Thus,

$$F = \frac{\text{estimate of } \sigma^2}{\text{estimate of } \sigma^2}$$

These two estimates of σ^2 are obtained by different methods. The numerator of the F ratio is obtained by a method that does a good job of estimating σ^2 *only* when H_0 is true. If H_0 is false, the estimate of σ^2 in the numerator will be too large. The denominator of the F ratio is obtained by a method that does a good job of estimating σ^2 *regardless* of whether the null hypothesis is true or false.

Thus, when the null hypothesis is true, the expected value of F is about 1.00 because both numerator and denominator are estimators of σ^2, and $\sigma^2/\sigma^2 \cong 1.00$. If an F value is much larger than 1.00, there is cause to suspect that H_0 is false. Of course, when the null hypothesis is true, values somewhat larger or smaller than 1.00 are to be expected because of sampling fluctuation.

Another way to describe the F ratio is that

$$F = \frac{\text{variation between treatment means}}{\text{variation within treatments}}$$

Both of the explanations that follow illustrate this version of the F ratio.

An algebraic explanation of the F ratio. Assume that the null hypothesis is true and that several treatments have produced a set of *treatment* (group) *means*.

1. Using the treatment means as data, calculate a standard deviation. This standard deviation of treatment means is an acquaintance of yours, the standard error of the mean, $s_{\bar{X}}$.

2. Remember that $s_{\bar{X}} = \hat{s}/\sqrt{N}$. Squaring both sides gives $s_{\bar{X}}^2 = \hat{s}^2/N$. Multiplying both sides by N and rearranging give $\hat{s}^2 = N s_{\bar{X}}^2$.

3. $\hat{s}^2$ is, of course, an estimator of σ^2.

　　　　Thus, starting with a null hypothesis that is true and a set of treatment means, you can obtain an estimate of the variance of the populations that the samples come from. This estimate of σ^2 is called the *between-treatments estimate* (also called the between-means estimate and the between-groups estimate).

　　　　Notice that the validity of the between-treatments estimate of σ^2 depends on the assumption that the samples are all drawn from the same population. If one or more samples come from a population with a larger or smaller mean, the treatment means will vary more, and the between-treatments estimate will be larger.

　　　　The other estimate of σ^2 (the denominator) is obtained from the variability within each of the samples. Because each sample variance is an independent estimator of σ^2, averaging them produces a more reliable estimate. This within-treatments estimate is called the **error estimate** (also called the *within-groups estimate*). The error estimate is an unbiased estimate even if the null hypothesis is false. Finally, the two estimates can be put into a ratio. If the between-treatments estimate is much larger than the error estimate, the null hypothesis is rejected.

A graphic explanation of the F ratio.　　Here is an illustrated explanation of the *F* ratio. **Figure 10.1** shows the situation when the null hypothesis is true ($\mu_A = \mu_B = \mu_C$). Three samples have been drawn from identical populations and a mean calculated for each sample. (See where the means are projected onto the dependent-variable line.) The means are fairly close together, so the variability among these three means (the between-treatments estimate) is small. Examine **Figure 10.1.**

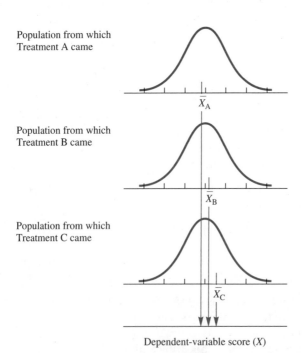

Population from which
Treatment A came

$\overline{X}_A$

Population from which
Treatment B came

$\overline{X}_B$

Population from which
Treatment C came

$\overline{X}_C$

Dependent-variable score (X)

FIGURE　10.1　　H_0 is true. All three treatment groups are drawn from identical populations. Not surprisingly, samples from those populations produce means that are similar. (Note that the points are close together on the X axis.)

Population from which
Treatment A came

Population from which
Treatment B came

Population from which
Treatment C came

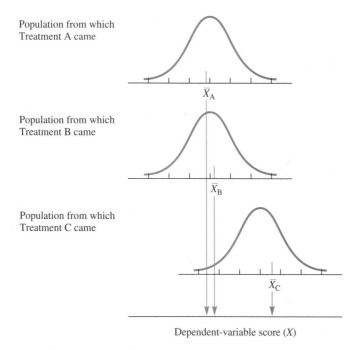

Dependent-variable score (X)

FIGURE 10.2 H_0 is false. Two treatment groups come from identical populations. A third group comes from a population with a larger mean. The sample means show more variability than those in Figure 10.1.

In **Figure 10.2,** the null hypothesis is false (one group comes from a population with a larger μ). The projection of the means this time shows that population C produces a sample mean that increases the variability among the three sample means. Examine **Figure 10.2.**

By studying Figures 10.1 and 10.2, you can convince yourself that the between-treatments estimate is larger when the null hypothesis is false. So, if you have a small amount of variability between treatment means, retain H_0. If you have a large amount of variability between treatment means, reject H_0. *Small* and *large,* however, are relative terms and, in this case, they are relative to the population variance.

Figure 10.3 shows the same situation as Figure 10.1 except that the populations have larger variances. Because of this, the variation among the sample means projected onto the dependent-variable line is greater than in Figure 10.1. This shows that the amount of between-treatments variability is influenced by the population variance.

So, a large between-treatments estimate may be due to large population variances (Figure 10.3) or to a false null hypothesis (Figure 10.2). Dividing the between-treatment estimate by an estimate of the population variance (the error estimate) gives the F ratio, which is large when H_0 is false and about 1.00 when H_0 is true.

The F distribution is a sampling distribution that shows the probability of various ratios of variances when the null hypothesis is true. Each variance has a number of degrees of freedom associated with it. **Table F** in Appendix C gives you critical values of F for α levels of .05 and .01. The first version of this table was compiled by George W.

Population from which
Treatment A came

Population from which
Treatment B came

Population from which
Treatment C came

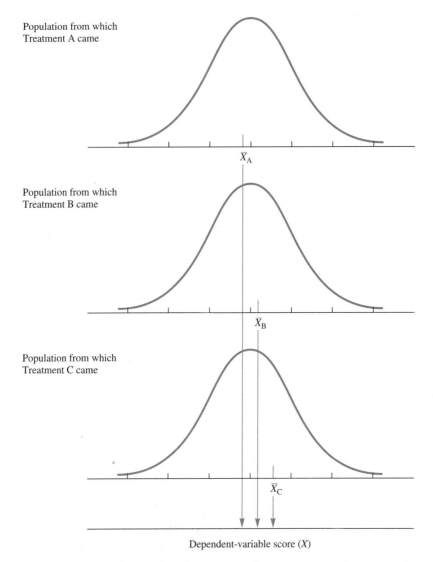

Dependent-variable score (X)

FIGURE 10.3 H_0 is true. All three treatment groups are drawn from identical populations. Because the populations have larger variances than those in Figure 10.1, the sample
means show more variability; that is, the arrow points are spaced farther apart.

Snedecor of Iowa State University in 1934. At that time, Snedecor named the variance
ratio F in honor of Fisher. **Figure 10.4** is a picture of two F distributions.

There are a number of similarities between the F distribution and the t distribution. Like t, the F distribution is also a family of curves, with each member of the family characterized by different degrees of freedom. A particular F distribution depends
on the degrees of freedom in both the numerator estimate of σ^2 and the denominator
estimate of σ^2. As the number of degrees of freedom approaches infinity, the F curve
approaches a normal distribution, which is true for t as well.

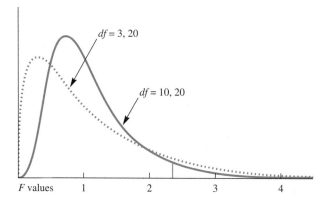

FIGURE 10.4 Two *F* distributions, showing their positively skewed nature.

The mathematical relationship between t and F is that $t^2 = F$ (for an F with 1 *df* in the numerator). For example, the square of a t value with 15 *df* is equal to F with 1, 15 *df*.

Finally, Fisher and Gosset were friends for years, each respectful of the other's work [Fisher's daughter reported their correspondence (Box, 1981).] Fisher recognized that the Student's t test was really a ratio of two measures of variability and that such a concept was applicable to experiments with more than two groups.[4]

PROBLEMS

2. Suppose three means came from a population with $\mu = 100$ and a fourth from a population with a smaller mean, $\mu = 50$. Will the between-treatments estimate of variability be larger or smaller than the between-treatments estimate of variability of four means drawn from a population with $\mu = 100$?

3. How can two estimates of a population variance be obtained?

4. Define F.

5. Interpret the meaning of the F values below for the experiments described in examples 1 and 3 on the first pages of this chapter.
   a. A very small F value for example 1
   b. A very large F value for example 3

6. In your own words, explain the rationale of ANOVA. Identify the two components that combine to produce an F value. Explain what each of these components measures, both when the null hypothesis is true and when it is false.

---

[4] That is,

$$t = \frac{\overline{X}_1 - \overline{X}_2}{s_{\overline{X}_1 - \overline{X}_2}}$$

$$= \frac{\text{a range}}{\text{standard error of a difference between means}} = \frac{\text{a measure of variability between treatments}}{\text{a measure of variability within treatments}}$$

MORE NEW TERMS

Sum of squares   In the "Clue to the Future" on page 63 you read that parts of the formula for the standard deviation would be important in future chapters. That future is now. The numerator of the basic formula for the standard deviation $(X - \overline{X})^2$ is the **sum of squares** (abbreviated *SS*). So, $SS = \Sigma(X - \overline{X})^2$. A more descriptive name for sum of squares is "sum of the squared deviations."

Mean square   **Mean square** (*MS*) is the ANOVA term for a variance $\hat{s}^2$. The mean square is a sum of squares divided by its degrees of freedom.

Grand mean   The **grand mean** is the mean of all scores; it is computed without regard for the fact that the scores come from different groups (samples).

tot   A subscript *tot* (total) after a symbol means that the symbol stands for all such numbers in the experiment; ΣX_{tot} is the sum of all X scores.

t   The subscript *t* after a symbol means that the symbol applies to a treatment group; for example, $\Sigma(\Sigma X_t)^2$ tells you to sum the scores in each group, square each sum, and then sum these squared values. N_t is the number of scores in one treatment group.

K K is the number of treatments in the experiment. This is the same as the number of levels of the independent variable.

F test   An **F test** is the result of an analysis of variance.

clue to the future

In this chapter and the next two you will work problems with several sets of numbers. Unless you are using a computer, you will need the sum, the sum of squares, and other values for each set. Most calculators will produce both the sum and the sum of squares with just one entry of each number. If you know how (or take the time to learn) to exploit your calculator's capabilities, you can spend less time on statistics problems and make fewer errors as well.

SUMS OF SQUARES

Analysis of variance is called that because the variability of all the scores in an experiment is assigned to two or more sources. In the case of simple analysis of variance, just two sources contribute all the variability to the scores. One source is the variability between treatments, and the other source is the variability within the scores of each treatment. The sum of these two sources is equal to the total variability.

Now that you have studied the rationale of ANOVA, it is time to learn to do the calculations. **Table 10.1** shows fictional data from an experiment in which there were three treatments for patients with psychological disorders. The independent variable was drug therapy. Three levels of the independent variable were used — Drug A, Drug B, and Drug C. The dependent variable was the number of psychotic episodes each patient had during the therapy period.

TABLE 10.1 Computation of sums of squares of fictional data from a drug study

	Drug A		Drug B		Drug C	
	X_A	X_A^2	X_B	X_B^2	X_C	X_C^2
	9	81	9	81	4	16
	8	64	7	49	3	9
	7	49	6	36	1	1
	5	25	5	25	1	1
Σ	29	219	27	191	9	27
$\overline{X}$	7.25		6.75		2.25	

$$\Sigma X_{tot} = 29 + 27 + 9 = 65$$

$$\Sigma X_{tot}^2 = 219 + 191 + 27 = 437$$

$$SS_{tot} = \Sigma X_{tot}^2 - \frac{(\Sigma X_{tot})^2}{N_{tot}} = 437 - \frac{65^2}{12} = 84.917$$

$$SS_{drugs} = \Sigma\left[\frac{(\Sigma X_t)^2}{N_t}\right] - \frac{(\Sigma X_{tot})^2}{N_{tot}}$$

$$= \frac{29^2}{4} + \frac{27^2}{4} + \frac{9^2}{4} - \frac{65^2}{12}$$

$$= 210.25 + 182.25 + 20.25 - 352.083 = 60.667$$

$$SS_{error} = \Sigma\left[\Sigma X_t^2 - \frac{(\Sigma X_t)^2}{N_t}\right] = \left(219 - \frac{29^2}{4}\right) + \left(191 - \frac{27^2}{4}\right) + \left(27 - \frac{9^2}{4}\right)$$

$$= 8.750 + 8.750 + 6.750 = 24.250$$

Check: $SS_{tot} = SS_{treat} + SS_{error} = 84.917 = 60.667 + 24.250$

In this experiment (and in all one-way ANOVAs), the null hypothesis is that the populations are identical. In terms of treatment means in the drug experiment, H_0: $\mu_A = \mu_B = \mu_C$. If the null hypothesis is true, the differences among the numbers of psychotic episodes for the three treatment groups is due to sampling fluctuation and not to the drugs administered.

In this particular experiment, the researcher believed that Drug C would be better than Drug A or B, which were in common use. In this case, *better* means fewer psychotic episodes. The researcher's belief is an alternative hypothesis (sometimes called a private hypothesis or experimenter's hypothesis). This belief, however, is not ANOVA's alternative hypothesis, H_1. The alternative hypothesis in ANOVA is that one or more populations are different from the others. No *greater than* or *less than* is specified.

After you have read the description of an experiment and understand what it is about, the next step in an ANOVA is to calculate the *sum of squares,* which is one measure of variability. I will illustrate with calculations for the data in Table 10.1.

First, please focus on the *total* variability as measured by the total sum of squares (SS_{tot}). As you will see, you are already familiar with SS_{tot}. To find SS_{tot}, subtract the grand mean from each score, square these deviation scores, and sum them:

$$SS_{tot} = \Sigma(X - \overline{X}_{tot})^2$$

To compute SS_{tot}, use the algebraically equivalent raw-score formula:

$$SS_{tot} = \Sigma X_{tot}^2 - \frac{(\Sigma X_{tot})^2}{N_{tot}}$$

which you may recognize as the numerator of the raw-score formula for $\hat{s}$. To calculate SS_{tot}, square each score and sum the squared values to obtain ΣX_{tot}^2. Next, the scores are summed and the sum squared. That squared value is divided by the total number of scores to obtain $(\Sigma X_{tot})^2/N_{tot}$. Subtraction of $(\Sigma X_{tot})^2/N_{tot}$ from ΣX_{tot}^2 yields the total sum of squares. For the data in Table 10.1,

$$SS_{tot} = \Sigma X_{tot}^2 - \frac{(\Sigma X_{tot})^2}{N_{tot}} = 437 - \frac{65^2}{12} = 84.917$$

Thus, the total variability of all the scores in Table 10.1 is 84.917 when measured by the sum of squares. This total comes from two sources: the between-treatments sum of squares and the error sum of squares. Each of these can be computed separately.

The *between-treatments sum of squares* (SS_{treat}) is the variability of the group means from the grand mean of the experiment, weighted by the size of the group:

$$SS_{treat} = \Sigma[N_t(\overline{X}_t - \overline{X}_{tot})^2]$$

SS_{treat} is more easily computed by the raw-score formula:

$$SS_{treat} = \Sigma\left[\frac{(\Sigma X_t)^2}{N_t}\right] - \frac{(\Sigma X_{tot})^2}{N_{tot}}$$

This formula tells you to sum the scores for each treatment group and then square the sum. Each squared sum is then divided by the number of scores in its group. These values (one for each group) are then summed, giving you $\Sigma[(\Sigma X_t)^2/N_t]$. From this sum is subtracted the value $(\Sigma X_{tot})^2/N_{tot}$, a value that was obtained previously in the computation of SS_{tot}.

Although the general case of an experiment is described in general terms, *treatment* and SS_{treat}, a specific case is described in words specific to that experiment. For the experiment in Table 10.1, that word is *drugs*. Thus,

$$SS_{drugs} = \Sigma\left[\frac{(\Sigma X_t)^2}{N_t}\right] - \frac{(\Sigma X_{tot})^2}{N_{tot}}$$

$$= \frac{29^2}{4} + \frac{27^2}{4} + \frac{9^2}{4} - \frac{65^2}{12} = 60.667$$

The other source of variability is the *error sum of squares* (SS_{error}), which is the sum of the variability within each of the groups. SS_{error} is defined as

$$SS_{error} = \Sigma(X_1 - \overline{X}_1)^2 + \Sigma(X_2 - \overline{X}_2)^2 + \cdots + \Sigma(X_K - \overline{X}_K)^2$$

or the sum of the squared deviations of each score from the mean of its treatment group added to the sum of the squared deviations from all other groups for the experiment. As with the other SS values, one arrangement of the arithmetic is easiest. For SS_{error}, it is

$$SS_{error} = \Sigma\left[\Sigma X_t^2 - \frac{(\Sigma X_t)^2}{N_t}\right]$$

This formula tells you to square each score in a treatment group and sum them (ΣX_t^2). Subtract from this a value that you obtain by summing the scores, squaring the sum, and dividing by the number of scores in the group: $(\Sigma X_t)^2/N_t$. For each group, a value is calculated, and these values are summed to get SS_{error}. For the data in Table 10.1,

$$SS_{error} = \Sigma\left[\Sigma X_t^2 - \frac{(\Sigma X_t)^2}{N_t}\right]$$
$$= \left(219 - \frac{29^2}{4}\right) + \left(191 - \frac{27^2}{4}\right) + \left(27 - \frac{9^2}{4}\right) = 24.250$$

Please pause and look at what is happening in calculating the sums of squares, which is the first step in any ANOVA problem. The total variation (SS_{tot}) is found by squaring *every* number (score) and adding them up. From this total, you subtract $(\Sigma X_{tot})^2/N_{tot}$. In a similar way, the variation that is due to treatments is found by squaring the sum of each treatment and dividing by N_t, adding these up, and then subtracting the same component you subtracted before, $(\Sigma X_{tot})^2/N_{tot}$. This factor,

$$\frac{(\Sigma X_{tot})^2}{N_{tot}}$$

is sometimes referred to as the *correction factor* and is found in all ANOVA analyses. Finally, the error sum of squares is found by squaring each number *within* a treatment group and subtracting that group's correction factor. Summing the differences from all the groups gives you SS_{error}.

Another way of expressing these ideas is to say that SS_{error} is the variation *within* the groups and SS_{treat} is the variation *between* the groups. In a one-way ANOVA, the total variability (as measured by SS_{tot}) consists of these two components, SS_{treat} and SS_{error}. So,

$$SS_{tot} = SS_{treat} + SS_{error}$$

Thus, $84.917 = 60.667 + 24.250$ for the drug experiment. This relationship is shown graphically in **Figure 10.5.**

error detection

$SS_{tot} = SS_{treat} + SS_{error}$. All sums of squares are always zero or positive, never negative. This check will not catch errors made in summing the scores (ΣX) or in summing squared scores (ΣX^2).

FIGURE 10.5 $SS_{tot} = SS_{treat} + SS_{error}$

PROBLEMS

*7. Here are three groups of numbers. This problem and the ones that use this data set are designed to illustrate some relationships and to give you practice calculating ANOVA values. To begin, show that $SS_{tot} = SS_{treat} + SS_{error}$.

X_1	X_2	X_3
5	6	12
4	4	8
3	2	4

*8. Emile Durkheim (1858–1917), a founder of sociology, believed that modernization produced social problems such as suicide. He compiled data from several European countries and published *Suicide* (1897), a book that helped establish quantification in social science disciplines. The data in this problem were created so that the conclusion you reach will mimic that of Durkheim (and present-day sociologists).

On variables such as electricity consumption, newspaper circulation, and gross national product, 13 countries were classified as to their degree of modernization (low, medium, or high). The numbers in the table are suicide rates per 100,000 persons. For the data in the table, name the independent variable, the number of levels it has, and the dependent variable. Compute SS_{tot}, SS_{mod}, and SS_{error}.

Degree of modernization		
Low	Medium	High
4	17	20
8	10	22
7	9	19
5	12	9
		14

*9. The best way to cure phobias is known. Bandura, Blanchard, and Ritter (1969) conducted an experiment that provided individuals with one of four treatments for their intense fear of snakes. One group worked with a model — a person who handled a 4-foot king snake and encouraged others to imitate her. One group watched a film of adults and children who enjoyed progressively closer contact with a king snake. A third group received desensitization therapy, and a fourth group served as a control, receiving no treatment. The numbers of snake-approach responses after treatment are listed in the table. Name the independent variable and its levels. Name the dependent variable. Compute $SS_{tot} = SS_{treat} + SS_{error}$.

Model	Film	Desensitization	Control
29	22	21	13
27	18	17	12
27	17	16	9
21	15	14	6

MEAN SQUARES AND DEGREES OF FREEDOM

After you find SS, the next step in an ANOVA is to find the mean squares. A mean square is simply a sum of squares divided by its degrees of freedom. It is an estimate of the population variance, σ^2, when the null hypothesis is true.

Each sum of squares has a particular number of degrees of freedom associated with it. In a one-way classification, the df are df_{tot}, df_{treat}, and df_{error}. The relationship among degrees of freedom is the same as that among sums of squares:

$$df_{tot} = df_{treat} + df_{error}$$

The formula for df_{tot} is $N_{tot} - 1$. The df_{treat} is the number of treatment groups minus one $(K - 1)$. The df_{error} is the sum of the degrees of freedom for each group $[(N_1 - 1) + (N_2 - 1) + \cdots + (N_K - 1)]$. If there are equal numbers of scores in the K groups, the formula for df_{error} reduces to $K(N_t - 1)$. A little algebra will reduce this still further:

$$df_{error} = K(N_t - 1) = KN_t - K$$

However, $KN_t = N_{tot}$, so $df_{error} = N_{tot} - K$. This formula for df_{error} works whether or not the numbers in all groups are the same. In summary,

$$df_{tot} = N_{tot} - 1$$
$$df_{treat} = K - 1$$
$$df_{error} = N_{tot} - K$$

error detection

$df_{tot} = df_{treat} + df_{error}$. Degrees of freedom are always positive.

Mean squares, then, can be found using the following formulas:[5]

$$MS_{treat} = \frac{SS_{treat}}{df_{treat}} \qquad \text{where} \quad df_{treat} = K - 1$$

$$MS_{error} = \frac{SS_{error}}{df_{error}} \qquad \text{where} \quad df_{error} = N_{tot} - K$$

For the data in Table 10.1,

$$df_{drugs} = K - 1 = 3 - 1 = 2$$
$$df_{error} = N_{tot} - K = 12 - 3 = 9$$

Check: $df_{tot} = df_{treat} + df_{error} = 2 + 9 = 11$

$$MS_{drugs} = \frac{SS_{drugs}}{df_{drugs}} = \frac{60.667}{2} = 30.333$$

$$MS_{error} = \frac{SS_{error}}{df_{error}} = \frac{24.250}{9} = 2.694$$

Notice that, although $SS_{tot} = SS_{treat} + SS_{error}$ and $df_{tot} = df_{treat} + df_{error}$, mean squares are *not* additive; that is, $MS_{tot} \neq MS_{treat} + MS_{error}$.

CALCULATION AND INTERPRETATION OF F VALUES USING THE F DISTRIBUTION

The next two steps in an ANOVA are to calculate an F value and to interpret it using the F distribution.

The F Test: Calculation

You learned earlier that F is a ratio of two estimates of the population variance. MS_{treat} is an estimate based on the variability between groups. MS_{error} is an estimate based on the sample variances. An *F* test consists of dividing MS_{treat} by MS_{error} to obtain an F value:

$$F = \frac{MS_{treat}}{MS_{error}}$$

Every F value has two degrees of freedom associated with it. The first is the df associated with MS_{treat} (the numerator) and the second is the df associated with MS_{error} (the denominator). For the data in Table 10.1,

$$F = \frac{MS_{drugs}}{MS_{error}} = \frac{30.333}{2.694} = 11.26; \quad df = 2, 9$$

---

[5] MS_{tot} is not used in ANOVA; only MS_{treat} and MS_{error} are calculated.

TABLE 10.2 Summary table of ANOVA for data in Table 10.1

Source	SS	df	MS	F	p
Drugs	60.667	2	30.333	11.26	<.01
Error	24.250	9	2.694		
Total	84.917	11			

$F_{.01}(2, 9\ df) = 8.02$

F Value: Interpretation

The next question is: What is the probability of obtaining $F = 11.26$ if all three samples come from populations with the same mean? If that probability is less than α, then the null hypothesis should be rejected.

Turn now to **Table F** in Appendix C, which gives the critical values of F when $\alpha = .05$ and when $\alpha = .01$. Across the top of the table are degrees of freedom associated with the numerator (MS_{treat}). For Table 10.1, $df_{drugs} = 2$, so 2 is the column you want. Along the left side of the table are degrees of freedom associated with the denominator (MS_{error}). In this case, $df_{error} = 9$, so look for 9 along the side. The tabled value for 2 and 9 df is 4.26 at the .05 level (lightface type) and 8.02 at the .01 level (boldface type). If the F value from the data is as great as or greater than the tabled value, reject the null hypothesis. If the F value from the data is not as great as the tabled value, the null hypothesis must be retained.

Because $11.26 > 8.02$, you can reject the null hypothesis at the .01 level. The three samples do not have a common population mean. At least one of the drugs produced scores (psychotic episodes) that were different from the rest.

At this point, your interpretation must stop. An ANOVA does not tell you which of the population means is different from the others. Such an interpretation requires additional statistical analysis, which will be covered later in this chapter.

It is customary to present all of the calculated ANOVA statistics in a *summary table*. **Table 10.2** is an example. Look at the right side of the table under p (for probability). The notation "$p < .01$" is shorthand for "the probability is less than 1 in 100 of obtaining treatment means as different as the ones actually obtained, if, in fact, the samples all came from identical populations."

Sometimes Table F does not contain an F value for the df in your problem. For example, an F with 2, 35 df or 4, 90 df is not tabled. When this happens, be conservative; use the F value that is given for *fewer df* than you have. Thus, the proper F values for those two examples would be based on 2, 34 df and 4, 80 df, respectively.[6]

---

[6] The F distribution may be used to test hypotheses about variances as well as hypotheses about means. To determine the probability that two sample variances came from the same population (or from populations with equal variances), form a ratio with the larger sample variance in the numerator. The resulting F value can be interpreted with Table F. The proper df are $N_1 - 1$ and $N_2 - 1$ for the numerator and denominator, respectively. For more information, see Kirk (1999, p. 430) or Ferguson and Takane (1989, p. 202).

SCHEDULES OF REINFORCEMENT—A LESSON IN PERSISTENCE

Persistence is shown when you keep on trying even though the rewards are scarce or nonexistent. It is something that seems to vary a great deal from person to person, from task to task, and from time to time. What causes this variation? What has happened to lead to such differences in people, tasks, and times?

Persistence can often be explained if you know how frequently reinforcement (rewards) occurred in the past. I will illustrate with data produced by pigeons, but the principles illustrated hold for other forms of life, including students and professors.

The data in **Table 10.3** are typical results from a study of *schedules of reinforcement*. A hungry pigeon is taught to peck at a disk on the wall. A peck produces food according to a schedule the experimenter has set up. For some pigeons every peck produces food — a continuous reinforcement schedule (crf). For some birds every

T A B L E   10.3   Partial analysis of the data on the number of minutes to extinction after four schedules of reinforcement during learning

	Schedule of reinforcement during learning			
	crf	FR2	FR4	FR8
	3	5	8	10
	5	7	12	14
	6	9	13	15
	2	8	11	13
	5	11	10	11
		10		
		6		
ΣX	21	56	54	63
ΣX^2	99	476	598	811
$\overline{X}$	4.2	8.0	10.8	12.6

$$\Sigma X_{\text{tot}} = 21 + 56 + 54 + 63 = 194$$

$$\Sigma X^2_{\text{tot}} = 99 + 476 + 598 + 811 = 1984$$

$$SS_{\text{tot}} = \Sigma X^2_{\text{tot}} - \frac{(\Sigma X_{\text{tot}})^2}{N_{\text{tot}}} = 1984 - \frac{(194)^2}{22} = 1984 - 1710.727 = 273.273$$

$$SS_{\text{schedules}} = \Sigma\left[\frac{(\Sigma X_t)^2}{N_t}\right] - \frac{(\Sigma X_{\text{tot}})^2}{N_{\text{tot}}} = \frac{(21)^2}{5} + \frac{(56)^2}{7} + \frac{(54)^2}{5} + \frac{(63)^2}{5} - \frac{(194)^2}{22} = 202.473$$

$$SS_{\text{error}} = \Sigma\left[\Sigma X_t^2 - \frac{(\Sigma X_t)^2}{N_t}\right]$$

$$= \left(99 - \frac{(21)^2}{5}\right) + \left(476 - \frac{(56)^2}{7}\right) + \left(598 - \frac{(54)^2}{5}\right) + \left(811 - \frac{(63)^2}{5}\right)$$

$$= 70.800$$

$$df_{\text{tot}} = N_{\text{tot}} - 1 = 22 - 1 = 21$$

$$df_{\text{schedules}} = K - 1 = 4 - 1 = 3$$

$$df_{\text{error}} = N_{\text{tot}} - K = 22 - 4 = 18$$

TABLE 10.4 Summary table of ANOVA analysis of the schedules of reinforcement study

Source	SS	df	MS	F	p
Schedules	202.473	3	67.491	17.16	<.01
Error	70.800	18	3.933		
Total	273.273	21			

$F_{.05}(3, 18) = 3.16$ $F_{.01}(3, 18) = 5.09$

other peck produces food — a fixed-ratio schedule of 2:1 (FR2). A third group of pigeons get food after every fourth peck — an FR4 schedule. Finally, for a fourth group, eight pecks are required to produce food — an FR8 schedule. After all groups receive 100 reinforcements, no more food is given (extinction begins). Under such conditions pigeons will continue to peck for a while and then stop. The dependent variable is the number of minutes a bird continues to peck (persist) after the food stops. As you can see in Table 10.3, three groups had five birds and one had seven.

Table 10.3 shows the raw data and the steps required to get the three SS figures and the three df figures. Work through **Table 10.3** now. When you finish, examine the summary table (**Table 10.4**), which continues the analysis by giving the mean squares, F test, and the probability of such an F.

When an experiment produces an F with a probability value less than .05, a conclusion is called for. Here it is: The schedule of reinforcement used during learning has a significant effect on the persistence of responding during extinction. It is unlikely that the four samples have a common population mean.

You might feel unsatisfied by that conclusion. You might ask which groups differ from the others, or which groups are causing that "significant difference." Good questions. You will find answers in the next section, but first, here are a few problems to reinforce what you learned in this section.

PROBLEMS

10. Suppose you obtained $F = 2.56$ with 5, 75 df. How many treatment groups are in the experiment? What are the critical values for the .05 and .01 levels? What conclusion should you reach?

11. Here are two more questions about those raw numbers you found sums of squares for in problem 7.
  a. Perform an F test and compose a summary table. Look up the critical value of F and make a decision about the null hypothesis. Write a sentence explaining what the analysis has shown.
  b. Perform an F test on the data from group 2 and group 3. Perform a t test on these same two groups.

*12. Perform an F test on the modernization and suicide data in problem 8 and write a conclusion.

*13. Perform an F test on the data from Bandura and colleagues' phobia treatment study (problem 9). Write a conclusion.

COMPARISONS AMONG MEANS

A significant F by itself tells you only that it is not likely that the samples have a common population mean. Usually you want to know more, but exactly what you want to know often depends on the particular problem you are working on. For example, in the schedules of reinforcement study, many would ask which schedules are significantly different from others (all pairwise comparisons). Another question might be whether there are sets of schedules that do not differ among themselves but do differ from other sets. In the drug study, the focus of interest is the new experimental drug. Is it better than either of the standard drugs or is it better than the average of the standards? This is just a *sampling* of the kinds of questions that can be asked.

Questions such as these are answered by analyzing the data using specialized statistical tests. As a group, these tests are referred to as **multiple-comparisons tests.** Unfortunately, no one multiple-comparisons method is satisfactory for all questions. In fact, not even three or four methods will handle all questions satisfactorily.

The reason there are so many methods is that specific questions are best answered by tailor-made tests. Over the years, statisticians have developed many tests for specific situations, and each test has something to recommend it. For an excellent chapter on the tests and their rationale, see Howell (1997, "Multiple Comparisons Among Treatment Means").

With this background in place, here is the plan for this elementary statistics book. I will discuss the two major *categories* of tests subsequent to ANOVA (*a priori* and *post hoc*) and then cover one *post hoc* test (the Tukey HSD test).

A Priori and *Post Hoc* Tests

A priori **tests** require that you *plan* a limited number of tests before gathering the data. *A priori* means "based on reasoning, not on immediate observation." Thus, for *a priori* tests, you must choose the groups that will be compared when you design the experiment.

Post hoc **tests** can be chosen after the data are gathered and you have examined the means. Any and all differences for which a story can be told are fair game for *post hoc* tests. *Post hoc* means "after this" or "after the fact."

The necessity for two kinds of tests is evident if you think again about the 15 samples that were drawn from the same population. Before the samples are drawn, what is your expectation that two randomly chosen means will be significantly different if tested with an independent-measures t test? About .05 is a good guess. Now, what if the data are drawn, the means compiled, and you then pick out the largest and the smallest means and test them with a t test? What is your expectation that these two will be significantly different? Much higher than .05, I would suppose. Thus, if you want to make comparisons after the data are gathered (and keep α at .05), then larger differences should be required before a difference is declared "not due to chance."

The preceding paragraphs should give you a start on understanding that there is a good reason for having more than one statistical test to answer questions subsequent to ANOVA. For this text, however, I have confined us to just one test, a *post hoc* test named the Tukey (Too′ Key) Honestly Significant Difference test.

The Tukey Honestly Significant Difference Test

For many research problems, the principal question is: Are any of the treatments different from the others? John W. Tukey designed a test to answer this question when N is the same for all samples. A **Tukey Honestly Significant Difference** (HSD) test pairs each sample mean with every other mean, a procedure called *pairwise comparisons*. For each pair, the Tukey HSD test tells you if one sample mean is significantly larger than the other.

When N_t is the same for all samples, the formula for HSD is

$$\text{HSD} = \frac{\overline{X}_1 - \overline{X}_2}{s_{\overline{X}}}$$

where $s_{\overline{X}} = \sqrt{\dfrac{MS_{\text{error}}}{N_t}}$

N_t = the number in each treatment; the number that an $\overline{X}$ is based on

The critical value against which HSD is compared is found in Table G in Appendix C. Turn to **Table G.** Critical values for $HSD_{.05}$ are on the first page; those for $HSD_{.01}$ are on the second. For either page, enter the row that gives the *df* for the MS_{error} from the ANOVA. Go over to the column that gives K for the experiment.

If the HSD you calculated from the data is equal to or greater than the number at the intersection in Table G, reject the null hypothesis and conclude that the two means are significantly different. Finally, describe that difference using the names of the independent and dependent variables.

I will illustrate the Tukey HSD test by doing pairwise comparisons on the three groups in the drug experiment (Tables 10.1 and 10.2). The three separate pairwise comparisons can be done in any order, but there is a strategy that may save some arithmetic. The trick is to pick a difference between means of intermediate size and test that difference first. If this difference proves to be significant, then all larger differences also will be significant. If this difference fails to reach significance, then all smaller differences also will be not significant. This trick always works if the sample sizes are equal, but it *may not* work if the sample sizes are not equal, an exception pointed out to me by an undergraduate user of an earlier edition of this book.

In the drug study, there are three means and so there will be three comparisons. The "mean difference of intermediate size" is the difference between the experimental drug (C) and a standard drug (B):

$$\text{HSD} = \frac{\overline{X}_B - \overline{X}_C}{s_{\overline{X}}} = \frac{6.75 - 2.25}{\sqrt{2.694/4}} = \frac{4.50}{0.821} = 5.48$$

Because 5.48 is greater than the tabled value for HSD at the .01 level ($HSD_{.01} = 5.43$), you can conclude that the experimental drug (C) is significantly better than Drug B at reducing psychotic episodes ($p < .01$). Because the difference between Drug C and Drug A is even greater than the difference between Drugs C and B, you can also conclude that the experimental drug is significantly better than Drug A in reducing psychotic episodes ($p < .01$).

One comparison remains, the comparison between the two standard drugs, A and B:

$$\text{HSD} = \frac{7.25 - 6.75}{\sqrt{2.694/4}} = 0.61$$

Because $\text{HSD}_{.05} = 3.95$, you can conclude that this study does not provide evidence of the superiority of Drug B over Drug A.

Tukey HSD when N_t's are not equal.   The Tukey HSD test was developed for studies in which the N_t's are equal for all groups. Sometimes, despite the best of intentions, unequal N_t's occur. A modification in the denominator of HSD will produce a statistic that can be tested with the values in Table G. The modification, recommended by Kramer (1956) and shown to be accurate (Smith, 1971), is

$$s_{\bar{X}} = \sqrt{\frac{MS_{\text{error}}}{2}\left(\frac{1}{N_1} + \frac{1}{N_2}\right)}$$

Tukey HSD and ANOVA

In the analysis of an experiment, it has been common practice to calculate an ANOVA and then, if the F value is significant, conduct Tukey HSD tests. Mathematical statisticians have shown, however, that this practice is overly conservative (Zwick, 1993). That is, if the *only* thing you want to know from your data is whether any *pairs* of treatments show a significant difference, then it is legitimate to run Tukey HSD tests without first showing that a one-way ANOVA produced a significant F. For the beginning statistics student, however, I recommend that you learn both procedures and apply them in the order of ANOVA and then HSD.

PROBLEMS

14. Distinguish between the two categories of tests subsequent to ANOVA: *a priori* and *post hoc.*
15. For Durkheim's modernization and suicide data, make all possible pairwise comparisons and write a conclusion telling what the data show.
16. For the schedules of reinforcement data, make all possible pairwise comparisons and write a conclusion telling what the data show.
*17. For the data on treatment of phobias, compare the best treatment with the second best, and the poorest treatment with the control group. Write a conclusion about the study.

ASSUMPTIONS OF THE ANALYSIS OF VARIANCE

Reaching conclusions by analyzing data with an ANOVA is fairly straightforward. But are there any hidden requirements? Of course.

Two of the requirements are about the populations the samples come from. If these "requirements" (assumptions) are met, you are assured that the probability figures in Table F (.05 and .01) are correct for the *F* test values in the table. If the third "requirement" is met, cause-and-effect conclusions are supported.

1. *Normality.* The dependent variable is assumed to be normally distributed in the populations from which samples are drawn. In some areas of research populations are known to be skewed, and researchers in those fields may decide that ANOVA is not appropriate for their data analysis. However, unless there is a reason to suspect that populations depart severely from normality, the inferences made from the *F* test will probably not be affected. ANOVA is robust. (It gives correct probabilities even when the populations are not exactly normal.) Where there is suspicion of a severe departure from normality, however, use the nonparametric method explained in Chapter 14.

2. *Homogeneity of variance.* The variances of the dependent-variable scores for each of the populations are assumed to be equal. Figures 10.1, 10.2, and 10.3, which were used to illustrate the rationale of ANOVA, show populations with equal variances. Several methods of testing this assumption are presented in advanced texts such as Winer, Brown, and Michels (1991), Kirk (1995), and Keppel (1982). When variances are grossly unequal, the nonparametric method discussed in Chapter 14 may be called for. [See Wilcox (1987) for a discussion of this issue.]

3. *Random assignment.* Random assignment distributes the effect of extraneous variables equally over all levels of the independent variable. Using random assignment allows you to conclude (if the *F* test is significant) that a difference between means is *caused* by a difference in levels of the independent variable.

I hope these assumptions seem familiar to you. They are the same as those you learned for the *t* distribution in Chapter 9. This makes sense; *t* is the special case of *F* when there are two samples.

EFFECT SIZE INDEX

The important question Are the populations different? can be answered with the hypothesis-testing *F* test. The equally important question, Are the differences large, medium, or small? requires an *effect size index.*

I'm sure you recall calculating an effect size index, *d,* in Chapters 4, 8, and 9. For ANOVA, the effect size index is symbolized by *f* and there is an entirely new formula:

$$f = \frac{\sigma_{\text{treat}}}{\sigma_{\text{error}}}$$

To find *f, estimators* of σ_{treat} and σ_{error} are required. The estimator of σ_{treat} is symbolized by $\hat{s}_{\text{treat}}$ and is found using the formula[7]

$$\hat{s}_{\text{treat}} = \sqrt{\frac{K-1}{N_{\text{tot}}}(MS_{\text{treat}} - MS_{\text{error}})}$$

Likewise, the estimator of σ_{error} is symbolized by $\hat{s}_{\text{error}}$, which is simply the square root of the error variance:

$$\hat{s}_{\text{error}} = \sqrt{MS_{\text{error}}}$$

---

[7] This formula is appropriate for an ANOVA with equal numbers of participants in each treatment group (Kirk, 1995, p. 181). For unequal *N*'s, see Cohen (1988, pp. 359–362).

Here's an example of calculating an effect size index for an ANOVA using the drug data in Table 10.1:

$$f = \frac{\sqrt{\dfrac{K-1}{N_{tot}}(MS_{treat} - MS_{error})}}{\sqrt{MS_{error}}} = \frac{\sqrt{\dfrac{3-1}{12}(30.333 - 2.694)}}{\sqrt{2.694}} = 1.31$$

The next step is to interpret an f value of 1.31. Jacob Cohen, who promoted the awareness of effect size and also contributed formulas, provided guidance for judging f values as small, medium, and large (Cohen, 1969, 1992):

Small	$f = .10$
Medium	$f = .25$
Large	$f = .40$

Thus, an f value of 1.31 indicates that the effect size in the drug study was just huge. (This is not surprising because textbook authors often manufacture fictional data so that an example will produce a large effect.)

Here is the final set of problems on one-way ANOVA. All that you learn here will be of use to you in Chapter 11, which will explain how to use ANOVA to analyze a more complex experiment.

PROBLEMS

18. List the three assumptions of the analysis of variance.

19. For the data from Bandura and colleagues' phobia treatment experiment (problems 9, 13, and 17), find f, the effect size index, and write a sentence of interpretation.

20. Hermann Rorschach, a Swiss psychiatrist, began to get an inkling of an idea in 1911. By 1921 he had developed, tested, and published his test, which measures motivation and personality. One of the several ways that the test is scored is to count the number of responses the test-taker gives (R scores). The data in the table are representative of those given by normals, schizophrenics, and depressives. Analyze the data with an F test and a set of Tukey HSDs. Write a conclusion about Dr. Rorschach's test.

Normals	Schizophrenics	Depressives
23	17	16
18	20	11
31	22	13
14	19	10
22	21	12
28	14	10
20	25	
22		

21. Almost all health professionals recommend that women practice monthly breast self-examination (BSE) as a way to detect breast cancer in its early stages (when treatment is much more effective). Several methods have been used to teach BSE. T. S. Spatz (1991) compared the method recommended by the American Cancer Society (ACS) to the 4MAT method, a method based on different learning styles that different people have. She also included an untrained control group. The dependent variable was the amount of information retained three months after training. Analyze the data in the table with an ANOVA, HSD tests, and an effect size index. Write a conclusion.

	ACS	4MAT	Control
ΣX	178	362	62
ΣX^2	2619	7652	1192
N	20	20	20

22. Common sense says that a valuable goal is worth suffering for. Is that the way it works? In a classic experiment in social psychology (Aronson and Mills, 1959), college women had to "qualify" to be in an experiment. The qualification activity caused severe embarrassment, mild embarrassment, or no embarrassment. After qualifying, the women listened to a recorded discussion, which was, according to the experimenters, "one of the most worthless and uninteresting discussions imaginable." The dependent variable was the women's rating of the discussion. Analyze the ratings using the techniques in this chapter (high scores = favorable ratings). Write an interpretation.

Degree of Embarrassment

Severe	Mild	None
18	18	17
23	12	15
14	14	9
20	15	12
25	10	13

23. Look over the objectives at the beginning of the chapter. Can you do them?

WEBSITE REFERENCES

http://psychology.wadsworth.com

Open "Choose Your Quiz" and select Spatz, 7e. Find the index of chapters, and click on Chapter 10, *Analysis of Variance: One-Way Classification.*

Open "Choose Your Area," scroll down, and select "Research Methods and Statistics." Click on "Statistics" in the graph. Click on "Student Activities." Click on the Web Page button for "One Way Anova."

chapter **eleven**

ANALYSIS OF VARIANCE: FACTORIAL DESIGN

OBJECTIVES FOR CHAPTER ELEVEN

Study the text. Work the problems as you come to them. Afterward, you should be able to:

1. Define the terms *factorial design, factor, level,* and *cell*
2. Name the sources of variance in a factorial design
3. Compute sums of squares, degrees of freedom, mean squares, and *F* values in a factorial design
4. Determine whether or not *F* values are significant
5. Describe the concept of interaction
6. Interpret interactions and main effects
7. Make pairwise comparisons between means with Tukey HSD tests
8. List the assumptions required for a factorial design ANOVA

In Chapter 10 you learned the basic concepts and procedures for conducting a simple one-way ANOVA. This chapter will expand those concepts and procedures to cover a popular and more complex design: the **factorial ANOVA.**

In ANOVA terminology, **factor** means independent variable. A factorial ANOVA is an ANOVA with two or more factors (two or more independent variables). A big advantage of a factorial ANOVA is that it tells you not only about the factors but also about the **interaction** between the factors.

A review of the designs you have studied in previous chapters, using factorial terminology, will help prepare you to understand factorial ANOVA and its bonus information, the interaction.

In Chapter 9 you learned to analyze data from a two-group experiment, schematically shown in **Table 11.1.** There is *one independent variable,* Factor A, with data from two levels, A_1 and A_2. For the t test, the question is whether $\overline{X}_{A_1}$ is significantly different from $\overline{X}_{A_2}$. You learned to determine this for both independent and correlated-samples designs.

TABLE 11.1 Illustration of a two-treatment design that can be analyzed with a *t* test

Factor *A*

A_1	A_2
Scores on the dependent variable	Scores on the dependent variable
$\overline{X}_{A_1}$	$\overline{X}_{A_2}$

TABLE 11.2 Illustration of a four-treatment design that can be analyzed with an *F* test

Factor *A*

A_1	A_2	A_3	A_4
Scores on the dependent variable	Scores on the dependent variable	Scores on the dependent variable	Scores on the dependent variable
$\overline{X}_{A_1}$	$\overline{X}_{A_2}$	$\overline{X}_{A_3}$	$\overline{X}_{A_4}$

In Chapter 10 you learned to analyze data from a two-or-more-treatment experiment, schematically shown in **Table 11.2** for a four-treatment design. There is *one independent variable,* Factor *A,* with data from four levels, A_1, A_2, A_3, and A_4. In the one-way ANOVA of Chapter 10, the question is whether the four treatment means could have come from populations with a common mean. You learned to determine this for independent samples. (One-way ANOVA for correlated samples will be covered in Chapter 12.)

FACTORIAL DESIGN

In this chapter you will learn to analyze data from a design in which there are *two independent variables* (factors), each of which may have two or more levels. **Table 11.3** illustrates this design. One factor (Factor *A*) has three levels (A_1, A_2, and A_3) and another factor (Factor *B*) has two levels (B_1 and B_2).

A factorial design has *two or more independent variables.* Each level of a factor occurs with all levels of the other factor(s). This chapter will cover only two-factor designs for independent samples.[1]

---

[1] Advanced textbooks discuss the analysis of three-or-more-factor designs and correlated factorial designs. See Howell (1997), Kirk (1995), or Winer, Brown, and Michels (1991).

TABLE 11.3 Illustration of a 2 × 3 factorial design that can be analyzed with *F* tests

Factor *A*

		A_1	A_2	A_3	
Factor *B*	B_1	Cell A_1B_1 Scores on the dependent variable	Cell A_2B_1 Scores on the dependent variable	Cell A_3B_1 Scores on the dependent variable	$\overline{X}_{B_1}$
	B_2	Cell A_1B_2 Scores on the dependent variable	Cell A_2B_2 Scores on the dependent variable	Cell A_3B_2 Scores on the dependent variable	$\overline{X}_{B_2}$
		$\overline{X}_{A_1}$	$\overline{X}_{A_2}$	$\overline{X}_{A_3}$	

Factorial Design Terms

Factorial designs are identified with a shorthand notation such as "2 × 3" (two by three) or "3 × 5." The general term is $R \times C$ (Rows × Columns). The first number is the number of levels of one factor; the second number is the number of levels of the other factor. The design in Table 11.3 is a 2 × 3 design. Assignment of a factor to a row or column is arbitrary; Table 11.3 could be rearranged into a 3 × 2 table.

In Table 11.3 there are six cells. The participants in each **cell** are treated differently. Participants in the upper left cell are given treatment A_1 *and* treatment B_1; that cell is identified as Cell A_1B_1. Participants in the lower right cell are given treatment A_3 and treatment B_2, and that cell is called Cell A_3B_2.

Factorial Design Results[2]

To begin, a factorial ANOVA gives you the same information that you get from two separate one-way ANOVAs. Look again at **Table 11.3.** A factorial ANOVA will determine the probability of obtaining means such as $\overline{X}_{A_1}$, $\overline{X}_{A_2}$, and $\overline{X}_{A_3}$ from identical populations (a one-way ANOVA with three groups). The same factorial ANOVA will determine the probability of $\overline{X}_{B_1}$ and $\overline{X}_{B_2}$ coming from the same population (a second one-way ANOVA).

In addition, a factorial ANOVA helps you decide whether the two factors *interact* with each other to affect scores on the dependent variable. An interaction between two factors means that *the effect that one factor has on the dependent variable depends on which level of the other factor is being administered.*

---

[2] See Kirk (1995, p. 431) for both the advantages and the disadvantages of factorial ANOVA.

Table 11.3 would have an interaction if the difference between Level B_1 and Level B_2 depended on whether you looked at the A_1 column, the A_2 column, or the A_3 column. To expand on this, each of the three columns represents an experiment that compares B_1 to B_2. If there is *no* interaction (that is, Factor A does not interact with Factor B), then the difference between B_1 and B_2 will be the same for all three experiments. If there *is* an interaction, however, the difference between B_1 and B_2 will *depend* on whether you are examining the experiment at A_1, at A_2, or at A_3.

Factorial Design Examples

Perhaps a couple of factorial design examples will help. The first has an interaction; the second doesn't. Suppose several students are sitting in a dormitory lounge one Monday discussing the big party they all went to last Saturday. There is lots of disagreement; some loved the party, while others just shrug. Now, imagine that everyone gets quantitative and rates the party on a scale from 1 to 10. The resulting numbers are quite variable. Why are they so variable?

Well, for one thing there were two activities at the party, talking and playing games. For another thing, some people went to the party with one special person and some went with several friends. With these two variables identified, plus the numerical ratings of the party, you have the makings of a factorial ANOVA.

Look at **Table 11.4.** Locate the two factors, activity and companions. Note that each factor has two levels, which creates four groups of people. Scores on the dependent variable are commonly presented in the cells of the table. Shown in Table 11.4 are the cell means.

In Table 11.4 there is an interaction. A general explanation is that the effect of the number of companions depended on whether the principal activity was talk or games. Or, to say the same thing, the effect of the activity depended on the number of companions a person went to the party with.

A more specific explanation is that those who went with one companion and spent the time talking rated the party high. However, those who went with one companion and played games didn't have such a good time. Going with several companions and talking produced poor ratings; going with several companions and playing games resulted in a high mean rating.

T A B L E 11.4 Illustration of a factorial design with an interaction (The numbers are mean ratings of a party.)

		Activity	
		Talk	Games
Companions	One	8	3
	Several	3	8

TABLE 11.5 Illustration of a factorial design with no interaction (The numbers are mean ratings of a party.)

		Activity	
		Talk	Games
Personality	Shy	3	3
	Outgoing	8	8

As a second example, let's take a different Monday group who attended the same party where the activities were talking and playing games. In this group, there were two personality types, shy and outgoing.

Table 11.5 shows the outcome with this second group. All the shy people gave the party a low rating and all the outgoing people gave it a high rating. In Table 11.5 there is no interaction. The effect of personality type does not depend on the activity. Shy people rated the party the same regardless of whether they talked or played games, and outgoing people rated the party the same regardless of whether they talked or played games. Or, to say the same thing, talking produced low ratings from shy people and high ratings from the outgoing; games produced the same story, low ratings from shy people and high ratings from the outgoing.

PROBLEMS

1. Identify each factorial design using $R \times C$ notation.
   *a. Men and women worked complex problems at either 7:00 A.M. or 7:00 P.M.
   b. Three methods of teaching Spanish were used with 4-year-olds and 8-year-olds.
   c. Four strains of mice were infected with three types of virus.
   d. (This one is a little different.) Four varieties of peas were grown with three amounts of fertilizer in soil of low, medium, or high acidity.
2. Group together the designs that have the same number of factors.
   a. Correlated t test
   b. Independent t test
   c. One-way ANOVA with four treatments
   d. A 2×4 factorial ANOVA
3. Group together the designs that have an independent variable with two levels.
   a. Correlated t test
   b. Independent t test
   c. One-way ANOVA with four treatments
   d. A 2×4 factorial ANOVA
4. In your own words define an interaction.
5. Make up an outcome for problem 1a that has an interaction.

MAIN EFFECTS AND INTERACTION

A two-way factorial ANOVA produces three statistical tests, two **main effects** and one **interaction.** A main effect is like the between-treatments test in a one-way ANOVA. Thus, a factorial ANOVA gives you a test for Factor A and a separate test for Factor B. The test for the interaction (symbolized AB) tells you whether the scores for the different levels of Factor A depend on the levels of Factor B.

An Example That Does Not Have an Interaction

Table 11.6 shows a 2×3 factorial ANOVA. The numbers in the cells are cell means. I'll use Table 11.6 to illustrate how to examine a table of cell means and to look for the two main effects and the interaction.

Main effects. Look at the margin means for Factor B ($B_1 = 30$, $B_2 = 70$). The B main effect will give the probability of obtaining means of 30 and 70 if both samples came from identical populations. Thus, with this probability you can decide whether the null hypothesis H_0: $\mu_{B_1} = \mu_{B_2}$ should be rejected or not.

The A main effect will give you the probability of getting means of A_1, A_2, and A_3 (30, 40, and 80) if H_0: $\mu_{A_1} = \mu_{A_2} = \mu_{A_3}$. Again, this probability can result in a rejection or a retention of the null hypothesis.

Notice that a comparison of B_1 to B_2 satisfies the requirements for an experiment (page 181) in that, except for getting either B_1 or B_2, the groups were treated alike. That is, Group B_1 and Group B_2 are alike in that both groups have equal numbers of participants who received A_1. Whatever the effect of receiving A_1, it will occur as much in the B_1 group as it does in the B_2 group; B_1 and B_2 differ only in their levels of Factor B. It may be worthwhile for you to apply this same line of reasoning to the question of whether comparisons of A_1, A_2, and A_3 satisfy the requirement for an experiment.

Interaction. To check for an interaction, look at the two cell means in the A_1 column of **Table 11.6.** Changing from B_1 to B_2 increases the mean score by 40 points. This increase of 40 points is seen at every level of A. (Find it for A_2 and then for A_3.) Thus,

TABLE 11.6 Illustration of a 2 × 3 factorial design (The number in each cell is the mean for all the scores in the cell. *N* is the same for each cell.)

		Factor A			
		A_1	A_2	A_3	Factor B means
Factor B	B_1	10	20	60	30
	B_2	50	60	100	70
	Factor A means	30	40	80	Grand mean 50

the effect of changing levels of Factor B *does not* depend on Factor A; the 40-point effect is the same, regardless of the level of A.

In a similar way, if you start with the B_1 row, you see that the means increase 10 points and then another 40 points (A_1 to A_2 to A_3). This same change is found in the B_2 row (an increase of 10 points and then an increase of 40 points). Thus, the effect of changing levels of Factor A *does not* depend on Factor B; the change is the same regardless of which level of B you look at.

Graphing the interaction.   Data from factorial ANOVAs are often presented graphically because graphs are such a good aid to understanding. **Figure 11.1** is a line graph of the data in Table 11.6. Note that the two curves in Figure 11.1 are parallel, which means that there is no interaction between the two factors.

In Figure 11.1 there is a curve for B_1 and a second curve for B_2. The data of Table 11.6 could, however, be graphed with three curves, one each for A_1, A_2, and A_3. The two levels of B (B_1 and B_2) would then be points on the X axis. The result would be three parallel curves. Again, the parallel curves indicate that there is no interaction between the two factors.

If the curves in a graph of a factorial design are not parallel, it *may* indicate that there is an interaction between the two factors. However, all data are subject to sampling error. Thus, curves may not be parallel as a result of an interaction or as a result of sampling error. A statistical test of the interaction is used to help you decide between these two explanations.

An Example That Has an Interaction

Table 11.7 shows a 2 × 3 factorial design in which there *is* an interaction between the two independent variables. The main effect of Factor A is seen in the margin means

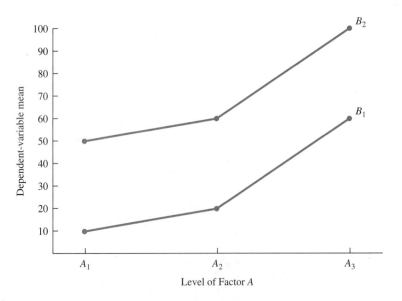

FIGURE 11.1 Graphic representation of data in Table 11.6; no interaction exists.

TABLE 11.7 Illustration of a 2 × 3 factorial design with an interaction between factors (The numbers represent the mean of all scores within that cell.)

Factor A

	A_1	A_2	A_3	Factor B means
B_1	10	20	30	20
B_2	100	70	40	70
Factor A means	55	45	35	Grand mean 45

Factor B (labels B_1 and B_2 at left)

along the bottom. The average effect of a change from A_1 to A_2 to A_3 is to *reduce* the mean score by 10 (55 to 45 to 35). But look at the cells. For B_1, the effect of changing from A_1 to A_2 to A_3 is to *increase* the mean score by 10 points. For B_2, the effect is to *decrease* the score by 30 points. These data have an interaction because the effect of one factor *depends* on which level of the other factor you administer.

Let me describe this interaction in Table 11.7 another way. Look at the difference between Cell A_1B_1 and Cell A_2B_1 — a difference of -10 ($10 - 20$). If there were no interaction, this same difference (-10) should be found between Cells A_1B_2 and A_2B_2. But this latter difference is $+30$ ($100 - 70$); it is in the opposite direction. Something about B_2 reverses the effect of changing from A_1 to A_2 that was found under condition B_1.

Figure 11.2 is a line graph of the data in Table 11.7. You can see that B_1 increases across the levels of Factor A. B_2, however, decreases as you go from A_1 to A_2 to A_3. The lines for the two levels of B are not parallel.[3]

Every factorial ANOVA, with or without an interaction, has main effects. When there is an interaction, however, a straightforward interpretation of the main effects can be misleading. Indeed, for the data in Table 11.7, the kind of interpretation I did for Table 11.6 would produce erroneous statements about one of the two factors. I will have more to say about this problem after you have worked some problems.

Bar Graphs of Factorial ANOVA Data

Besides line graphs such as Figures 11.1 and 11.2, factorial ANOVA data are often displayed as bar graphs. **Figure 11.3** shows the cell means of Table 11.4, the first party-rating example (in which there was an interaction). As with line graphs, the dependent-variable scores are on the Y axis. One of the factors is on the X axis (number of

---

[3] Students often have trouble with the concept of interaction. Usually, having the same idea presented in different words facilitates understanding. Two good references are Howell (1999, pp. 343–345) and Minium, Clarke, and Coladarci (1999, pp. 358–362).

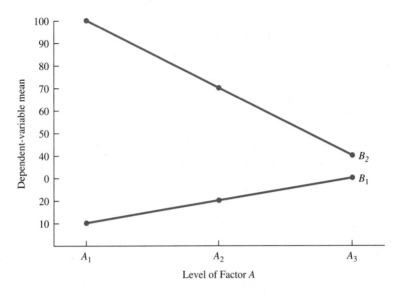

FIGURE 11.2 Graphic representation of the interaction of Factors *A* and *B* from Table 11.7

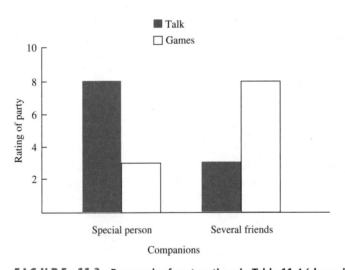

FIGURE 11.3 Bar graph of party ratings in Table 11.4 (shows interaction)

companions). The second factor (activity at the party) is indicated by shaded bars. Study **Figure 11.3.**

An interaction is suggested by a bar graph if the relationship of the shaded bar to the unshaded bar is reversed for the two levels of the factor on the *X* axis. Thus, in Figure 11.3 talking is rated higher than games by those who came with one companion but lower by those who came with several companions.

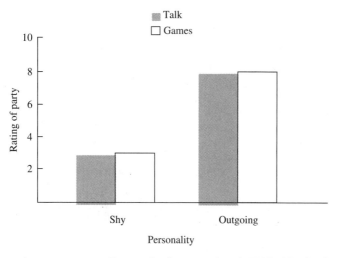

FIGURE 11.4 **Bar graph of party ratings in Table 11.5 (no interaction)**

As a second example of using bar graphs to display factorial design data, look at **Figure 11.4,** which shows the data from Table 11.5, the second party-rating example. There was no interaction in those data. Figure 11.4 shows that the relationship of the shaded bars to the unshaded bars is the same for both personality types. (It isn't necessary for the shaded bar to be the same height as the unshaded bar. If the *difference* in the shaded and unshaded bars is the same for both levels of the factor on the *X* axis, then there is no interaction.)

PROBLEMS

6. What is a main effect?

7. A description of the dependent variable and a table with cell means are given for the five problems below. High scores indicate better or positive. For each problem, compose five answers, a–e.

   a. Graph the cell means. Use a line graph or bar graph as indicated.

   b. Decide whether an interaction appears to be present. In these problems you may state that there appears to be *no* interaction or that there appears to be one. Interactions, like main effects, are subject to sampling variations.

   c. Write a statement explaining how the two variables interact or do not interact.

   d. For problems with no interaction, decide whether either main effect appears to be significant.

   e. Explain the main effects in those problems that do not show an interaction.

   i. The dependent variable for this ANOVA is attitude scores. High scores mean more positive attitudes. Attitudes were expressed toward card games and computer games by the young and the old. (Use a bar graph.)

Factor A (age)

		A_1 (young)	A_2 (old)
Factor B (games)	B_1 (cards)	50	75
	B_2 (computer)	75	50

ii. This is a study of the effect of the dose of a new drug on three types of schizophrenia. The dependent variable is scores on a competency test. (Use a line graph.)

A (diagnosis)

		A_1 (disorganized)	A_2 (catatonic)	A_3 (paranoid)
B (dose)	B_1 (small)	5	10	70
	B_2 (medium)	20	25	85
	B_3 (large)	35	40	100

iii. The dependent variable is an improvement score measured after a 90-day period of therapy. (Use a bar graph.)

A (therapy)

		A_1 (psychoanalytic)	A_2 (cognitive)	A_3 (behavior modification)
B (verbal ability)	B_1 (low)	15	60	30
	B_2 (high)	75	120	90

iv. The dependent variable is the score on a test that measures self-consciousness. (Use a line graph.)

A (grade)

		A_1 (5th grade)	A_2 (9th grade)	A_3 (12th grade)
B (height)	B_1 (short)	10	30	20
	B_2 (tall)	5	65	35

v. The dependent variable is attitude toward lowering the taxes on profits from investments. (Use a line graph.)

		A_1 (low)	A_2 (middle)	A_3 (high)
		A (socioeconomic status)		
B (gender)	B_1 (men)	10	35	60
	B_2 (women)	15	35	55

A SIMPLE EXAMPLE OF A FACTORIAL DESIGN

As you read the following story, identify the dependent variable and the two independent variables (factors).

The term was over. A dozen students who had all been in the same class were talking about what a bear of a test the comprehensive final exam had been. Competition surfaced.

"Of course, I was ready. Those of us who major in natural science disciplines just study more than you humanities types."

"Poot, poot, and balderdash," exclaimed one of the humanities types, who was a sophomore.

A third student said, "Well, there are some of both types in this group; let's just gather some data. How many hours did each of you study for that final?"

"Wait a minute," exclaimed one of the younger students. "I'm a natural science type, but this is my first term in college. I didn't realize how much time I would need to cover the readings. I don't want my score to pull down my group's tally."

"Hmmm," mused the data-oriented student. "Let's see. Look, some of you are in your first term and the rest of us have had some college experience. We'll just take experience into account. Maybe the dozen of us will divide up evenly."

This story establishes the conditions for a 2 × 2 factorial ANOVA. One of the factors is area of interest; the two levels are natural science and humanities. The other factor is previous experience in college; the two levels are none and some. The dependent variable is hours of study for a final examination.

It probably won't surprise you to find out that when I created a data set to go with this story, the dozen did divide up evenly — there were three students in each of the four cells. The data, cell means, and ΣX and ΣX^2 are shown in **Table 11.8.**

Look at the margin means in Table 11.8. Compare the mean number of hours of study for natural science and humanities. What about the effect of experience in college (the other margin means)? Is the difference as large or larger there? Now, look at the four cell means. Does there appear to be an interaction? This kind of preliminary inspection of the data will improve your understanding of a study and will help alert you to gross computational errors if they occur later.

TABLE 11.8 Scores, cell means, and summary statistics for the study time data

Previous college experience (Factor A)

		Some (A_1)		None (A_2)			
		X	X^2	X	X^2		
	Humanities (B_1)	12	144	8	64		
		11	121	7	49		
		10	100	7	49		
		$\Sigma X_{cell} = 33$		$\Sigma X_{cell} = 22$		$\Sigma X_{B_1} = 55$	
Major (Factor B)		$\Sigma X_{cell}^2 = 365$		$\Sigma X_{cell}^2 = 162$		$\Sigma X_{B_1}^2 = 527$	
		$\overline{X}_{cell} = 11.00$		$\overline{X}_{cell} = 7.3333$		$\overline{X}_{B_1} = 9.1667$	
	Natural science (B_2)	12	144	10	100		
		10	100	9	81		
		9	81	8	64		
		$\Sigma X_{cell} = 31$		$\Sigma X_{cell} = 27$		$\Sigma X_{B_2} = 58$	
		$\Sigma X_{cell}^2 = 325$		$\Sigma X_{cell}^2 = 245$		$\Sigma X_{B_2}^2 = 570$	
		$\overline{X}_{cell} = 10.3333$		$\overline{X}_{cell} = 9.00$		$\overline{X}_{B_2} = 9.6667$	
		$\Sigma X_{A_1} = 64$		$\Sigma X_{A_2} = 49$		$\Sigma X_{tot} = 113$	
		$\Sigma X_{A_1}^2 = 690$		$\Sigma X_{A_2}^2 = 407$		$\Sigma X_{tot}^2 = 1097$	
		$\overline{X}_{A_1} = 10.6667$		$\overline{X}_{A_2} = 8.1667$		$\overline{X}_{tot} = 9.4167$	

Sources of Variance and Sums of Squares

Remember that in Chapter 10 you identified three sources of variance in the one-way analysis of variance: (1) the total variance, (2) the between-treatments variance, and (3) the error variance. In a factorial design with two factors, the same sources of variance can be identified. However, the between-treatments variance, which is called the between-cells variance in factorial ANOVA, is further partitioned into *three* components. These are the two *main effects* and the *interaction*. Thus, of the variability among the means of the four cells in Table 11.8, some can be attributed to the A main effect, some to the B main effect, and the rest to the AB interaction. Calculation of the sums of squares is shown in **Table 11.9**.

Total sum of squares. Calculating the total sum of squares will be easy for you because it is the same as SS_{tot} in the one-way analysis. Defined as $\Sigma(X - \overline{X}_{tot})^2$, SS_{tot} is the sum of the squared deviations of each score in the experiment from the grand mean of the experiment. To compute SS_{tot}, use the formula

$$SS_{tot} = \Sigma X_{tot}^2 - \frac{(\Sigma X_{tot})^2}{N_{tot}}$$

TABLE 11.9 Sums of squares and computation check for the study time data

$$SS_{tot} = \Sigma X^2_{tot} - \frac{(\Sigma X_{tot})^2}{N_{tot}} = 1097 - \frac{(113)^2}{12} = 1097 - 1064.0833 = 32.9167$$

$$SS_{cells} = \Sigma \left[\frac{(\Sigma X_{cell})^2}{N_{cell}} \right] - \frac{(\Sigma X_{tot})^2}{N_{tot}} = \frac{(33)^2}{3} + \frac{(22)^2}{3} + \frac{(31)^2}{3} + \frac{(27)^2}{3} - \frac{(113)^2}{12}$$

$$= 363 + 161.3333 + 320.3333 + 243 - 1064.0833 = 23.5833$$

$$SS_{experience} = \frac{(\Sigma X_{A_1})^2}{N_{A_1}} + \frac{(\Sigma X_{A_2})^2}{N_{A_2}} - \frac{(\Sigma X_{tot})^2}{N_{tot}} = \frac{(64)^2}{6} + \frac{(49)^2}{6} - \frac{(113)^2}{12}$$

$$= 682.6667 + 400.1667 - 1064.0833 = 1082.8334 - 1064.0833$$

$$= 18.7501$$

$$SS_{major} = \frac{(\Sigma X_{B_1})^2}{N_{B_1}} + \frac{(\Sigma X_{B_2})^2}{N_{B_2}} - \frac{(\Sigma X_{tot})^2}{N_{tot}} = \frac{(55)^2}{6} + \frac{(58)^2}{6} - \frac{(113)^2}{12}$$

$$= 504.1667 + 560.6667 - 1064.0833 = 1064.8334 - 1064.0833 = 0.7501$$

$$SS_{AB} = N_{cell}[(\overline{X}_{A_1B_1} - \overline{X}_{A_1} - \overline{X}_{B_1} + \overline{X}_{tot})^2 + (\overline{X}_{A_2B_1} - \overline{X}_{A_2} - \overline{X}_{B_1} + \overline{X}_{tot})^2$$

$$+ (\overline{X}_{A_1B_2} - \overline{X}_{A_1} - \overline{X}_{B_2} + \overline{X}_{tot})^2 + (\overline{X}_{A_2B_2} - \overline{X}_{A_2} - \overline{X}_{B_2} + \overline{X}_{tot})^2]$$

$$= 3[(11.00 - 10.6667 - 9.1667 + 9.4167)^2 + (7.3333 - 8.1667 - 9.1667 + 9.4167)^2$$

$$+ (10.3333 - 10.6667 - 9.6667 + 9.4167)^2 + (9.00 - 8.1667 - 9.6667 + 9.4167)^2]$$

$$= 4.0836$$

Check: $SS_{AB} = SS_{cells} - SS_A - SS_B = 23.5833 - 18.7501 - 0.7501 = 4.0831$

$$SS_{error} = \Sigma \left[\Sigma X^2_{cell} - \frac{(\Sigma X_{cell})^2}{N_{cell}} \right]$$

$$= \left(365 - \frac{(33)^2}{3} \right) + \left(162 - \frac{(22)^2}{3} \right) + \left(325 - \frac{(31)^2}{3} \right) + \left(245 - \frac{(27)^2}{3} \right)$$

$$= (365 - 363) + (162 - 161.3333) + (325 - 320.3333) + (245 - 243)$$

$$= 2 + 0.6667 + 4.6667 + 2 = 9.3333$$

Check: $SS_{tot} = SS_{cells} + SS_{error}$

$$32.9167 = 23.5833 + 9.3333$$

For the study time data (Table 11.9),

$$SS_{tot} = 1097 - \frac{(113)^2}{12} = 32.9167$$

Between-cells sums of squares. To find the main effects and interaction, first find the between-cells variability and then partition it into its component parts. SS_{cells} is defined $\Sigma[N_{cell}(\overline{X}_{cell} - \overline{X}_{tot})^2]$. A "cell" in a factorial ANOVA is a group of participants treated alike; for example, humanities majors with no previous college experience constitute a cell.

The computational formula for SS_{cells} is similar to that for a one-way analysis:

$$SS_{cells} = \Sigma \left[\frac{(\Sigma X_{cell})^2}{N_{cell}} \right] - \frac{(\Sigma X_{tot})^2}{N_{tot}}$$

For the study time data (Table 11.9),

$$SS_{cells} = \frac{(33)^2}{3} + \frac{(22)^2}{3} + \frac{(31)^2}{3} + \frac{(27)^2}{3} - \frac{(113)^2}{12} = 23.5833$$

After SS_{cells} is obtained, it is partitioned into its three components: the A main effect, the B main effect, and the AB interaction.

Main effects sum of squares. The sum of squares for each main effect is somewhat like a one-way ANOVA. The sum of squares for Factor A ignores the existence of Factor B and considers the deviations of the Factor A means from the grand mean. Thus,

$$SS_A = N_{A_1}(\overline{X}_{A_1} - \overline{X}_{tot})^2 + N_{A_2}(\overline{X}_{A_2} - \overline{X}_{tot})^2$$

where N_{A_1} is the total number of scores in the A_1 cells. For Factor B,

$$SS_B = N_{B_1}(\overline{X}_{B_1} - \overline{X}_{tot})^2 + N_{B_2}(\overline{X}_{B_2} - \overline{X}_{tot})^2$$

Computational formulas for the main effects are like formulas for SS_{treat} in a one-way design:

$$SS_A = \frac{(\Sigma X_{A_1})^2}{N_{A_1}} + \frac{(\Sigma X_{A_2})^2}{N_{A_2}} - \frac{(\Sigma X_{tot})^2}{N_{tot}}$$

And for the study time data (Table 11.9),

$$SS_{experience} = \frac{(64)^2}{6} + \frac{(49)^2}{6} - \frac{(113)^2}{12} = 18.7501$$

The computational formula for the B main effect simply substitutes B for A in the previous formula. Thus,

$$SS_B = \frac{(\Sigma X_{B_1})^2}{N_{B_1}} + \frac{(\Sigma X_{B_2})^2}{N_{B_2}} - \frac{(\Sigma X_{tot})^2}{N_{tot}}$$

For the study time data (Table 11.9),

$$SS_{major} = \frac{(55)^2}{6} + \frac{(58)^2}{6} - \frac{(113)^2}{12} = 0.7501$$

Interaction sum of squares. To find the sum of squares for the interaction, use this formula:

$$\begin{aligned} SS_{AB} = N_{cell}[(\overline{X}_{A_1B_1} - \overline{X}_{A_1} - \overline{X}_{B_1} + \overline{X}_{tot})^2 \\ + (\overline{X}_{A_2B_1} - \overline{X}_{A_2} - \overline{X}_{B_1} + \overline{X}_{tot})^2 \\ + (\overline{X}_{A_1B_2} - \overline{X}_{A_1} - \overline{X}_{B_2} + \overline{X}_{tot})^2 \\ + (\overline{X}_{A_2B_2} - \overline{X}_{A_2} - \overline{X}_{B_2} + \overline{X}_{tot})^2] \end{aligned}$$

For the study time data (Table 11.9),

$$SS_{AB} = 3[(11.00 - 10.6667 - 9.1667 + 9.4167)^2$$
$$+ (7.3333 - 8.1667 - 9.1667 + 9.4167)^2$$
$$+ (10.3333 - 10.6667 - 9.6667 + 9.4167)^2$$
$$+ (9.00 - 8.1667 - 9.6667 + 9.4167)^2]$$
$$= 4.0836$$

Because SS_{cells} contains only the components SS_A, SS_B, and the interaction SS_{AB}, you can also obtain SS_{AB} by subtraction:

$$SS_{AB} = SS_{cells} - SS_A - SS_B$$

For the study time data (Table 11.9),

$$SS_{AB} = 23.5833 - 18.7501 - 0.7501 = 4.0831$$

Error sum of squares. The error variability is due to the fact that participants treated alike differ from one another on the dependent variable. Because all were treated the same, this difference must be due to uncontrolled variables or random variation. SS_{error} for a 2 × 2 design is defined as

$$SS_{error} = \Sigma(X_{A_1B_1} - \overline{X}_{A_1B_1})^2 + \Sigma(X_{A_2B_1} - \overline{X}_{A_2B_1})^2 + \Sigma(X_{A_1B_2} - \overline{X}_{A_1B_2})^2$$
$$+ \Sigma(X_{A_2B_2} - \overline{X}_{A_2B_2})^2$$

In words, SS_{error} is the sum of the sums of squares for each *cell* in the experiment. The computational formula is

$$SS_{error} = \Sigma\left[\Sigma X_{cell}^2 - \frac{(\Sigma X_{cell})^2}{N_{cell}}\right]$$

For the study time data (Table 11.9),

$$SS_{error} = \left(365 - \frac{(33)^2}{3}\right) + \left(162 - \frac{(22)^2}{3}\right)$$
$$+ \left(325 - \frac{(31)^2}{3}\right) + \left(245 - \frac{(27)^2}{3}\right) = 9.3333$$

As in a one-way ANOVA, the total variability in a factorial ANOVA is composed of two components: SS_{cells} and SS_{error}. Thus,

$$32.9167 = 23.5833 + 9.3333$$

As you read earlier, SS_{cells} can be partitioned among SS_A, SS_B, and SS_{AB} in a factorial ANOVA. Thus,

$$23.5833 = 18.7501 + 0.7501 + 4.0831$$

These relationships among sums of squares in a factorial ANOVA are shown graphically in **Figure 11.5.**

Did you notice that the direct calculation of SS_{AB} produced 4.0836 but that the value obtained by subtraction was 4.0831? The difference is due to rounding. By the

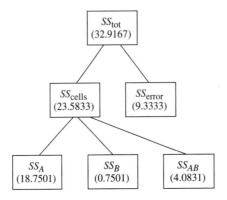

FIGURE 11.5 The partition of sums of squares in a factorial ANOVA problem

time mean squares and an F test are calculated (in the next section), the difference disappears.

I will interrupt the analysis of the study time data so that you may practice what you have learned about the sums of squares in a factorial ANOVA.

error detection

One computational check for a factorial ANOVA is the same as that for the one-way classification: $SS_{tot} = SS_{treat} + SS_{error}$. A more complete version is $SS_{tot} = SS_A + SS_B + SS_{AB} + SS_{error}$. As before, this check will not catch errors in calculating ΣX or ΣX^2.

PROBLEMS

*8. You know that women live longer than men. What about right-handed and left-handed people? Coren and Halpern's (1991) review included data on both gender and handedness, as well as age at death. An analysis of the age-at-death numbers in this problem will produce conclusions like those reached by Coren and Halpern.

		Gender	
		Women	Men
Handedness	Left-handed	76 74 69	67 61 58
	Right-handed	82 78 74	76 72 68

a. Identify the design using $R \times C$ notation.

b. Name the independent variables and the dependent variable.

c. Calculate SS_{tot}, SS_{cells}, SS_{gender}, $SS_{handedness}$, SS_{AB}, and SS_{error}.

*9. An educational psychologist was interested in whether different teacher responses affected boys and girls differently. Three classrooms with both boys and girls were used in the experiment. In one classroom, the teacher's response was one of "neglect." The children were not even observed during the time they were working arithmetic problems. In a second classroom, the teacher's response was "reproof." The children were observed, and errors were corrected and criticized. In the third classroom, children were "praised" for correct answers, and incorrect answers were ignored. The summary numbers in the table are the numbers of errors the children made on an arithmetic test at the end of the experiment. ($N = 10$ per cell.)

a. Identify the design using $R \times C$ notation.

b. Name the independent variables.

c. Name the dependent variable.

d. Calculate SS_{tot}, SS_{cells}, $SS_{response}$, SS_{gender}, SS_{AB}, and SS_{error}.

Teacher response

		Neglect (A_1)	Reproof (A_2)	Praise (A_3)
Gender	Boys (B_1)	$\Sigma X = 169$ $\Sigma X^2 = 3129$ $\overline{X} = 16.9$	$\Sigma X = 136$ $\Sigma X^2 = 2042$ $\overline{X} = 13.6$	$\Sigma X = 118$ $\Sigma X^2 = 1620$ $\overline{X} = 11.8$
	Girls (B_2)	$\Sigma X = 146$ $\Sigma X^2 = 2244$ $\overline{X} = 14.6$	$\Sigma X = 167$ $\Sigma X^2 = 2955$ $\overline{X} = 16.7$	$\Sigma = 109$ $\Sigma X^2 = 1291$ $\overline{X} = 10.9$

*10. Many experiments have investigated the concept of *state-dependent memory*. Participants learn a task in a particular state — say, under the influence of a drug or not under the influence of the drug. They recall what they have learned in the same state or in the other state. The dependent variable is recall score (memory). The data in the table are designed to illustrate the phenomenon of state-dependent memory. Calculate SS_{tot}, SS_{cells}, SS_{learn}, SS_{recall}, SS_{AB}, and SS_{error}.

Learn with

		Drug	No drug
Recall with	Drug	$\Sigma X = 43$ $\Sigma X^2 = 400$ $N = 5$	$\Sigma X = 20$ $\Sigma X^2 = 100$ $N = 5$
	No drug	$\Sigma X = 25$ $\Sigma X^2 = 155$ $N = 5$	$\Sigma X = 57$ $\Sigma X^2 = 750$ $N = 5$

Degrees of Freedom, Mean Squares, and F Tests

Now that you are skilled at calculating sums of squares, you can proceed with the analysis of the study time data. Mean squares, as before, are found by dividing the sums of squares by their appropriate degrees of freedom. Degrees of freedom for the sources of variance are listed next.

In general:	*For the study time data:*
$df_{tot} = N_{tot} - 1$	$df_{tot} = 12 - 1 = 11$
$df_A = A - 1$	$df_{experience} = 2 - 1 = 1$
$df_B = B - 1$	$df_{major} = 2 - 1 = 1$
$df_{AB} = (A - 1)(B - 1)$	$df_{AB} = (1)(1) = 1$
$df_{error} = N_{tot} - (A)(B)$	$df_{error} = 12 - (2)(2) = 8$

In these equations, A and B stand for the number of levels of Factor A and Factor B, respectively.

error detection

$$df_{tot} = df_A + df_B + df_{AB} + df_{error}$$
$$df_{cells} = df_A + df_B + df_{AB}$$

Mean squares for the study time data are

$$MS_{experience} = \frac{SS_{experience}}{df_{experience}} = \frac{18.7501}{1} = 18.7501$$

$$MS_{major} = \frac{SS_{major}}{df_{major}} = \frac{0.7501}{1} = 0.7501$$

$$MS_{AB} = \frac{SS_{AB}}{df_{AB}} = \frac{4.0831}{1} = 4.0831$$

$$MS_{error} = \frac{SS_{error}}{df_{error}} = \frac{9.3333}{8} = 1.1667$$

The F values are computed by dividing main effect mean squares and the interaction mean square by the error mean square:

$$F_{experience} = \frac{MS_{experience}}{MS_{error}} = \frac{18.7501}{1.1667} = 16.07$$

$$F_{major} = \frac{MS_{major}}{MS_{error}} = \frac{0.7501}{1.1667} = 0.64$$

$$F_{AB} = \frac{MS_{AB}}{MS_{error}} = \frac{4.0831}{1.1667} = 3.50$$

TABLE 11.10 ANOVA summary table for the study time data

Source	SS	df	MS	F	p
A (experience)	18.7501	1	18.7501	16.07	<.01
B (major)	0.7501	1	0.7501	0.64	>.05
AB	4.0831	1	4.0831	3.50	>.05
Error	9.3333	8	1.1667		
Total	32.9167	11			

Results of a factorial ANOVA are usually presented in a summary table. Examine **Table 11.10,** the summary table for the study time data.

You now have three F values from the data. Each has 1 degree of freedom in the numerator and 8 degrees of freedom in the denominator. To find the probabilities of obtaining such F values if the null hypothesis is true, use Table F. Table F yields the following critical values: $F_{.05}(1, 8) = 5.32$ and $F_{.01}(1, 8) = 11.26$.

The next step is to compare the obtained F's to the critical value F's and then tell what your analysis reveals about the relationships among the variables. With factorial ANOVAs, you should interpret the interaction F first and then proceed to the F's for the main effects. (As you continue this chapter, be alert for the explanation of *why* you should use this order.)

The interaction F_{AB} (3.50) is less than 5.32, so the interaction between experience and major is not significant. Although the effect of experience was to increase study time by 50 percent for humanities majors and only about 15 percent for natural science majors, this interaction difference was not statistically significant. When *the interaction is not significant,* interpret each main effect as if it came from a one-way ANOVA.

The main effect for experience ($F_{experience}$) is significant beyond the .01 level (16.07 > 11.26). The null hypothesis $\mu_{some} = \mu_{none}$ can be rejected. By examining the margin means (8.17 and 10.67), you can conclude that experienced students study significantly more than those without experience. F_{major} was less than 1, and F values less than 1 are never significant. Thus, these data do not provide evidence that students in either major study more than the other.

PROBLEMS

11. For the longevity data in problem 8, compute *df, MS,* and *F* values. Arrange these in a summary table that footnotes appropriate critical values. Plot the cell means with a bar graph. Tell what the analysis shows.

*12. For the data on educational practices in problem 9, compute *df, MS,* and *F* values. Arrange these in a summary table and note the appropriate critical values. Plot the cell means using a line graph. Tell what the analysis shows.

*13. For the data on state-dependent memory in problem 10, compute *df, MS,* and *F* values. Arrange these in a summary table and note the appropriate critical values. Plot the cell means using a bar graph.

ANALYSIS OF A 2 × 3 DESIGN

Here's a question for you. Read the question, take 10 seconds, and compose an answer. *Who is taller, boys or girls?*

A halfway good answer is "It depends." A completely good answer gives the other variable that the answer depends on; for example, "It depends on the *age* of the boys and girls."

Factorial ANOVA designs are often used when a researcher thinks that the answer to a question about the effect of variable *A* is "It depends on variable *B*." As you may have already anticipated, the way to express this idea statistically is to talk about significant interactions.

The analysis that follows is for a 2 × 3 factorial design. The factor with two levels is gender: males and females. The second factor is age: 6, 12, and 18 years old. The dependent variable is height in inches.

I've constructed data that mirror fairly closely the actual situation for Americans (Berk, 1998). The data and the summary statistics are given in **Table 11.11.** Orient yourself to this table by noting the column headings and the row headings. Examine the six cell means and the five margin means. Note that $N = 4$ for each cell.

Table 11.12 shows the calculation of the components of a factorial ANOVA. The headings in this table are familiar to you: sums of squares, degrees of freedom, mean squares, and *F* values. Work your way through each line in **Table 11.12** now.

The results of a factorial ANOVA are presented as a summary table. **Table 11.13** is the summary table for the height data. First, check the interaction *F* value. It is statistically significant.

Interpreting a Significant Interaction

A significant interaction always calls for a graph of the cell means. A graph will help you understand and it is even more important for the people you are explaining an interaction to. Look at **Figure 11.6.** Boys and girls are about the same height at age 6, but during the next 6 years the girls grow faster, and they are taller at age 12. In the following 6 years, however, the boys grow faster, and at age 18 the boys are taller than the girls are. The question of whether boys or girls are taller *depends* on what age is being considered.

So, one way to interpret a significant interaction is to examine a graph of the cell means and describe the changes in the variables. Though a good start, this technique is just a beginning. A complete interpretation of interactions requires techniques that are beyond the scope of a textbook for elementary statistics. (To begin a quest for more understanding, see Rosnow and Rosenthal, 1999, for an explanation of residuals.)

Interpreting Main Effects When the Interaction Is Significant

When the interaction is *not* significant, the interpretation of a main effect is the same as the interpretation of a one-way ANOVA. When the interaction *is* significant, however, the interpretation of main effects becomes complicated.

TABLE 11.11 Data and summary statistics of heights

Gender (Factor A)

		Male (A_1)		Female (A_2)		Summary values
		X_{A_1}	$X^2_{A_1}$	X_{A_2}	$X^2_{A_2}$	
Age (Factor B)	Age 6 (B_1)	44	1936	44	1936	$\Sigma X_{B_1} = 362$
		45	2025	44	1936	$\Sigma X^2_{B_1} = 16{,}390$
		46	2116	46	2116	$\overline{X}_{B_1} = 45.25$
		47	2209	46	2116	
	Sum	182	8286	180	8104	
	Mean	45.50		45.00		
	Age 12 (B_2)	56	3136	58	3364	$\Sigma X_{B_2} = 474$
		57	3249	60	3600	$\Sigma X^2_{B_2} = 28{,}120$
		59	3481	61	3721	$\overline{X}_{B_2} = 59.25$
		60	3600	63	3969	
	Sum	232	13,466	242	14,654	
	Mean	58.00		60.50		
	Age 18 (B_3)	66	4356	63	3969	$\Sigma X_{B_3} = 544$
		69	4761	65	4225	$\Sigma X^2_{B_3} = 37{,}074$
		72	5184	67	4489	$\overline{X}_{B_3} = 68.00$
		73	5329	69	4761	
	Sum	280	19,630	264	17,444	
	Mean	70.00		66.00		
	Summary values	$\Sigma X_{A_1} = 694$ $\Sigma X^2_{A_1} = 41{,}382$ $\overline{X}_{A_1} = 57.83$		$\Sigma X_{A_2} = 686$ $\Sigma X^2_{A_2} = 40{,}202$ $\overline{X}_{A_2} = 57.17$		$\Sigma X_{tot} = 1380$ $\Sigma X^2_{tot} = 81{,}584$ $\overline{X}_{tot} = 57.50$

To illustrate, consider the main effect of gender in **Table 11.13.** The simple interpretation of $F = .59$ is that gender does not have a significant effect on height. But clearly, that isn't right! Gender *is* important, as the interpretation of the interaction revealed.

In a factorial ANOVA, the analysis of a main effect *ignores* the other independent variable. This causes no problems if there is no interaction, but it can lead to a faulty interpretation when the interaction is significant.

So, what should you do about interpreting a factorial ANOVA that has a significant interaction? First, interpret the interaction based on a graph of the cell means. As for the main effects, you might (1) seek the advice of your instructor, (2) acknowledge that you don't yet know the techniques needed for a thorough interpretation, or (3) study more advanced statistics textbooks (Kirk, 1995, pp. 383–389 or Howell, 1997, pp. 409–411).

TABLE 11.12 Calculation of the components of the 2 × 3 factorial ANOVA of heights

Sums of squares

$$SS_{tot} = \Sigma X_{tot}^2 - \frac{(\Sigma X_{tot})^2}{N_{tot}} = 81,584 - \frac{(1380)^2}{24} = 2234.000$$

$$SS_{cells} = \Sigma\left[\frac{(\Sigma X_{cell})^2}{N_{cell}}\right] - \frac{(\Sigma X_{tot})^2}{N_{tot}}$$

$$= \frac{(182)^2}{4} + \frac{(180)^2}{4} + \frac{(232)^2}{4} + \frac{(242)^2}{4} + \frac{(280)^2}{4} + \frac{(264)^2}{4} - \frac{(1380)^2}{24} = 2152.000$$

$$SS_{A(gender)} = \frac{(\Sigma X_{A_1})^2}{N_{A_1}} + \frac{(\Sigma X_{A_2})^2}{N_{A_2}} - \frac{(\Sigma X_{tot})^2}{N_{tot}} = \frac{(694)^2}{12} + \frac{(686)^2}{12} - \frac{(1380)^2}{24} = 2.667$$

$$SS_{B(age)} = \frac{(\Sigma X_{B_1})^2}{N_{B_1}} + \frac{(\Sigma X_{B_2})^2}{N_{B_2}} + \frac{(\Sigma X_{B_3})^2}{N_{B_3}} - \frac{(\Sigma X_{tot})^2}{N_{tot}} = \frac{(362)^2}{8} + \frac{(474)^2}{8} + \frac{(544)^2}{8} - \frac{(1380)^2}{24} = 2107.000$$

$$SS_{AB} = N_{cell}[(\bar{X}_{A_1B_1} - \bar{X}_{A_1} - \bar{X}_{B_1} + \bar{X}_{tot})^2 + (\bar{X}_{A_2B_1} - \bar{X}_{A_2} - \bar{X}_{B_1} + \bar{X}_{tot})^2$$
$$+ (\bar{X}_{A_1B_2} - \bar{X}_{A_1} - \bar{X}_{B_2} + \bar{X}_{tot})^2 + \cdots + (\bar{X}_{A_2B_3} - \bar{X}_{A_2} - \bar{X}_{B_3} + \bar{X}_{tot})^2]$$
$$= 4[(45.5 - 57.83 - 45.25 + 57.50)^2 + (45.00 - 57.17 - 45.25 + 57.50)^2$$
$$+ (58.00 - 57.83 - 59.25 + 57.50)^2 + (60.50 - 57.17 - 59.25 + 57.50)^2$$
$$+ (70.00 - 57.83 - 68.00 + 57.50)^2 + (66.00 - 57.17 - 68.00 + 57.50)^2]$$
$$= 4[10.583] = 42.333$$

Check: $SS_{AB} = SS_{cells} - SS_A - SS_B = 2152.000 - 2.667 - 2107.000 = 42.333$

$$SS_{error} = \Sigma \left[\Sigma X^2_{cell} - \frac{(\Sigma X_{cell})^2}{N_{cell}} \right]$$

$$= \left(8286 - \frac{(182)^2}{4} \right) + \left(8104 - \frac{(180)^2}{4} \right) + \left(13,466 - \frac{(232)^2}{4} \right) + \left(14,654 - \frac{(242)^2}{4} \right) + \left(19,630 - \frac{(280)^2}{4} \right) + \left(17,444 - \frac{(264)^2}{4} \right)$$

$$= 82.000$$

Check: $2234.000 = 2152.000 + 82.000$

Mean squares

$$MS_A = \frac{SS_A}{df_A} = \frac{2.667}{1} = 2.667$$

$$MS_B = \frac{SS_B}{df_B} = \frac{2107.000}{2} = 1053.500$$

$$MS_{AB} = \frac{SS_{AB}}{df_{AB}} = \frac{42.333}{2} = 21.167$$

$$MS_{error} = \frac{SS_{error}}{df_{error}} = \frac{82.000}{18} = 4.556$$

Degrees of freedom

$$df_A = A - 1 = 2 - 1 = 1$$

$$df_B = B - 1 = 3 - 1 = 2$$

$$df_{AB} = (A - 1)(B - 1) = (1)(2) = 2$$

$$df_{error} = N_{tot} - (A)(B) = 24 - (2)(3) = 18$$

$$df_{tot} = N_{tot} - 1 = 24 - 1 = 23$$

F values

$$F_A = \frac{MS_A}{MS_{error}} = \frac{2.667}{4.556} = .59$$

$$F_B = \frac{MS_B}{MS_{error}} = \frac{1053.500}{4.556} = 231.23$$

$$F_{AB} = \frac{MS_{AB}}{MS_{error}} = \frac{21.167}{4.556} = 4.65$$

TABLE 11.13 ANOVA summary table for the height data

Source	SS	df	MS	F	p
A (gender)	2.667	1	2.667	.59	>.05
B (age)	2107.000	2	1053.500	231.23	<.01
AB	42.333	2	21.167	4.65	<.05
Error	82.000	18	4.556		
Total	2234.000	23			

$F_{.05}(2, 18 \ df) = 3.55$ $\qquad F_{.01}(2, 18 \ df) = 6.01$

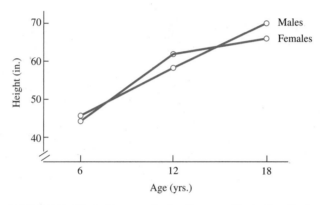

FIGURE 11.6 Line graph of cell means, illustrating the interaction in the height data

Sometimes, however, a simple interpretation of a main effect is not misleading. The height data provide an example. The interpretation of the significant age effect is that age has an effect on height. In Figure 11.6 this effect is seen in an increase from age 6 to age 12 and another increase from age 12 to age 18. In situations like this, a conclusion such as "The main effect of age was significant, $p < .01$" is appropriate. For more explanation of this approach to interpreting main effects even though the interaction is significant, see Howell (1999, pp. 343–345).

PROBLEMS

14. These problems are designed to help you learn to *interpret* the results of a factorial experiment. For each problem (a) identify the independent and dependent variables, (b) fill in the rest of the summary table, and (c) interpret the results.

   i. A clinical psychologist has been investigating the relationship between humor and aggression. The participant's task in these experiments is to think up as many captions as possible for four cartoons. The test lasts 8 minutes, and the psychologist simply counts the number of captions produced. In the psychologist's latest experiment, a participant is either insulted or treated in a neutral way by an experimenter. Next, the participant responds to the cartoons at the request of either the same experimenter or a different experimenter. The cell means and the summary table are given.

Participant was

Experimenter was		Insulted	Treated neutrally
	Same	31	18
	Different	19	17

Source	df	MS	F	p
A (between treatments)	1	675.00		
B (between experimenters)	1	507.00		
AB	1	363.00		
Error	44	74.31		

ii. An educational psychologist is interested in response biases. For example, a response bias occurs if the grade an English teacher puts on a theme written by a tenth grader is influenced by the student's name or by the occupation of his father. For this particular experiment, the psychologist used a "high-prestige" name (David) and a "low-prestige" name (Elmer). In addition, she made up two biographical sketches that differed only in the occupation of the student's father (research chemist or unemployed). The participants in this experiment read a biographical sketch and then graded a theme by that person using a scale of 50 to 100. The same theme was given to each participant.

Name

Occupation of father		David	Elmer
	Research chemist	86	81
	Unemployed	80	88

Source	df	MS	F	p
A (between names)	1	22.50		
B (between occupations)	1	2.50		
AB	1	422.50		
Error	36	51.52		

iii. What styles of presentations help statistics students learn? A statistics teacher gathered data on two variables. One variable, questions by the instructor, had two levels, present and absent. The other variable was demonstrations; the two levels were present and absent. The dependent variable was the scores on a comprehension test given after the presentation.

Questions

		Present	Absent
Demonstrations	Present	89	81
	Absent	78	73

Source	df	MS	F	p
A (questions)	1	211.25		
B (demonstrations)	1	451.25		
AB	1	11.25		
Error	76	46.03		

15. Write an interpretation of your analysis of the data from the state-dependent memory experiment (problems 10 and 13).

COMPARING LEVELS WITHIN A FACTOR

If the interaction in a factorial ANOVA is not significant, you can compare levels within a factor. Each main effect in a factorial ANOVA is like a one-way ANOVA; the question that remains after testing a main effect is: Are any of the levels of this factor significantly different from any others? You will be pleased to learn that the solution you learned for one-way ANOVA also applies to factorial ANOVA.

Of course, for some main effects you do not need a test subsequent to ANOVA; the *F* test on that main effect gives you the answer. This is true for any main effect with only *two levels*. Just like a *t* test, a significant *F* tells you that the two means are from different populations. Thus, if you are analyzing a 2×2 factorial ANOVA, you do not need a test subsequent to ANOVA for *either* factor.

When there are more than two levels of one factor in a factorial design, a Tukey HSD test can be used to test for significant differences between pairs of levels. There is one restriction, however, which is that the Tukey HSD test is appropriate *only if the interaction is **not** significant.*[4]

The HSD formula for a factorial ANOVA is the same as that for a one-way ANOVA:

$$\text{HSD} = \frac{\overline{X}_1 - \overline{X}_2}{s_{\overline{X}}}$$

---

[4] A nonsignificant interaction is a condition required by many *a priori* and *post hoc* tests.

where $s_{\bar{X}} = \sqrt{\dfrac{MS_{error}}{N_t}}$

Let me explain the $\bar{X}$'s and the N_t for the HSD formula. The $\bar{X}$'s are margin means and N_t is the number of scores used to calculate the $\bar{X}$ (a margin N). I'll illustrate using the data for the arithmetic errors of boys and girls who were taught by different methods. In that study (problems 9 and 12), the interaction was not significant, so Tukey HSD tests are appropriate. The actual data are shown in problem 9 (page 251). Look at that data table now. To compare the effect of *reproof* to that of *praise* with a Tukey HSD test, N is 20 (10 from boys and 10 from girls). The margin $\bar{X}$'s are 15.15 and 11.35 (303/20 and 227/20).

error detection

The N_t in $\sqrt{MS_{error}/N_t}$ is the N_t that each $\bar{X}$ in the numerator is based on. This rule applies to one-way ANOVA, to factorial ANOVA, and to one-factor correlated ANOVA (see Chapter 12).

To begin the analysis, I used the cell data to produce the following margin means and differences:

Neglect and Reproof	$15.75 - 15.15 = 0.60$ error
Neglect and Praise	$15.75 - 11.35 = 4.40$ errors
Reproof and Praise	$15.15 - 11.35 = 3.80$ errors

The intermediate difference is 3.80 (reproof minus praise), so it should be tested first:

$$\text{HSD} = \frac{\bar{X}_R - \bar{X}_P}{s_{\bar{X}}} = \frac{15.15 - 11.35}{\sqrt{19.89/20}} = \frac{3.80}{1.00} = 3.80$$

From Table G, HSD$_{.05}$ for $K = 3$ and $df_{error} = 40$ is 3.44. Because $3.80 > 3.44$, you can conclude that teaching by praising correct answers and ignoring mistakes results in significantly fewer errors on a comprehensive test than teaching by criticizing mistakes.

Because the difference between neglect and praise is even greater than the difference between reproof and praise, you can conclude that praise results in significantly fewer errors than neglect.

The other difference is between neglect and reproof:

$$\text{HSD} = \frac{\bar{X}_N - \bar{X}_R}{s_{\bar{X}}} = \frac{15.75 - 15.15}{\sqrt{19.89/20}} = \frac{0.60}{1.00} = 0.60$$

The HSD comparing neglect and reproof is not significant; there is no evidence that one is better than the other.

Note that the value of $s_{\bar{X}}$, 1.00, is the same in both HSD problems. For factorial ANOVA problems such as those in this text (cells all have the same N), $s_{\bar{X}}$ always has the same value for every comparison.

RESTRICTIONS AND LIMITATIONS

I have emphasized throughout this book the limitations that go with each statistical test you learn. For the factorial analysis of variance presented in this chapter, the restrictions include the three that you learned for one-way ANOVA. If you cannot recall those three, refresh your memory by rereading the section near the end of Chapter 10, "Assumptions of the Analysis of Variance." In addition to the basic assumptions of ANOVA, the factorial ANOVA presented in this chapter requires the following:

1. The number of scores in each cell must be equal. For techniques dealing with unequal N's, see Kirk (1995), Howell (1997), or Ferguson and Takane (1989).
2. The cells must be independent. This may be accomplished by randomly assigning a participant to only one of the cells. This restriction means that these formulas should not be used with any type of correlated-samples design.
3. The levels of both factors are chosen by the experimenter. The alternative is that the levels of one or both factors are chosen at random from several possible levels of the factor. The techniques of this chapter are used when the levels are *fixed* by the experimenter and not chosen randomly. For a discussion of fixed and random models of ANOVA, see Howell (1997); Winer, Brown, and Michels (1991); or Ferguson and Takane (1989).

PROBLEMS

16. List the six requirements that must be met if data are to be analyzed by a factorial ANOVA using techniques outlined in this chapter.
17. Two social psychologists asked freshman and senior college students to write an essay supporting recent police action on campus. (The students were known to be against the police action.) The students were given 50 cents, one dollar, five dollars, or ten dollars for their essay. Later, each student's attitude toward the police was measured on a scale of 1 to 20. Perform an ANOVA, fill out a summary table, and write an interpretation of the analysis at this point. (See Brehm and Cohen, 1962, for a similar study of Yale students.) If appropriate, calculate HSDs for all pairwise comparisons. Write an overall interpretation of the results of the study.

			Reward			
			$10	$5	$1	50¢
Students	Freshmen	ΣX	73	83	99	110
		ΣX^2	750	940	1290	1520
		N	8	8	8	8
	Seniors	ΣX	70	81	95	107
		ΣX^2	820	910	1370	1610
		N	8	8	8	8

18. The conditions that make for happy people have been researched by psychologists for a long time. The data that follow are based on a review of the literature by Diener and colleagues (1999). Participants were asked their marital status and how often they engaged in religious behavior. They also indicated how happy they were on a scale of 1 to 10. Analyze the data with a factorial ANOVA and, if appropriate, Tukey HSD tests.

Frequency of religious behavior

		Never	Occasionally	Often
Marital status	Married	6 2 4	3 7 5	7 8 9
	Unmarried	4 2 3	3 1 5	3 7 5

19. Review the objectives at the beginning of the chapter.

WEBSITE REFERENCES

http://psychology.wadsworth.com

Open "Choose Your Quiz" and select Spatz, 7e. Find the index of chapters, and click on Chapter 11, *Analysis of Variance: Factorial Design.*

Open "Choose Your Area," scroll down, and select "Research Methods and Statistics." Click on "Statistics" in the graph. Click on "Student Activities." Click on the Web Page button for "Two Way Anova."

chapter **twelve**

ANALYSIS OF VARIANCE: ONE-FACTOR CORRELATED MEASURES

OBJECTIVES FOR CHAPTER TWELVE

Study the text. Work the problems as you come to them. Afterward, you should be able to:

1. Distinguish between correlated-samples and independent-samples designs and identify the statistical tests that go with each design

2. Describe the characteristics of a one-factor correlated-measures analysis of variance (ANOVA)

3. For a one-factor correlated-measures ANOVA, identify the sources of variance and calculate their sums of squares, degrees of freedom, mean squares, and the *F* value

4. Interpret the *F* value from a one-factor correlated-measures ANOVA

5. Use a Tukey Honestly Significant Difference test to make pairwise comparisons

6. Distinguish between Type I and Type II errors

7. List and explain the advantages and limitations of a one-factor correlated-measures ANOVA

8. List and explain the assumptions of a one-factor correlated-measures ANOVA

Three chapters in this book are on analysis of variance (ANOVA). No other inferential statistical test gets as much space from me or as much time from you. This emphasis reflects its importance and widespread use.

This heavy emphasis on ANOVA in an introductory statistics book does make sense; among behavioral scientists, ANOVA is the most widely used inferential statistics technique. When I surveyed 20 empirical articles in *Psychological Science*, I found that 60 percent of the articles used some form of analysis of variance. Of those articles that used analysis of variance, two-thirds of them used a correlated-measures

TABLE 12.1 Identification of statistical tests based on kind of design

Independent variables and number of levels	Design of experiment	
	Correlated samples	Independent samples
One independent variable, two levels	Correlated-samples *t* test	Independent-samples *t* test
One independent variable, two or more levels	Correlated-measures ANOVA	One-way ANOVA
Two independent variables, each with two or more levels		Factorial ANOVA

ANOVA at least once in analyzing the data. It seems likely (if my sample is a clue to the future) that you will have opportunities to use what you are learning in this chapter as you continue your education.

You have already covered material that prepares you to answer the question What makes an analysis of variance a **correlated-measures ANOVA?** That material was in Chapter 9, where you learned to distinguish between an independent-samples *t* test and a correlated-samples *t* test. You probably recall reading the procedure and asking if there was some reason why the top two scores in each group were paired. If there was a reason [such as natural pairs, matched pairs, or same subjects (repeated measures)], then the design required a correlated-samples *t* test. If there was no reason for two scores to be paired, an independent-samples *t* test was called for.

The experimental designs in Chapter 10 (one-way ANOVA) and Chapter 11 (factorial ANOVA) both require that the analyzed scores be independent. This requirement is often satisfied by randomly assigning different participants to the different treatments.[1]

All the statistical tests covered in Chapters 9, 10, 11, and 12 are given in **Table 12.1.** Table 12.1 shows which tests are appropriate for correlated-samples designs and which are appropriate for independent-samples designs. In this chapter you will return to the analysis of data from a correlated-samples design. (Now is the time for you to decide for yourself whether you need to review the distinction between correlated-samples and independent-samples designs. If the answer is yes, reread pages 186–188.)

The designs that can be analyzed using correlated-measures ANOVA are the same designs that use a correlated *t* test: natural pairs, matched pairs, and repeated measures.[2]

One other reminder may be helpful: The "one-factor" in the chapter title means that just one independent variable is involved. The number of levels of this variable can be two or more. For this chapter, all the problems you do will have three or more levels of the independent variable. (When you have two levels, use a correlated *t* test.)

---

[1] To understand this chapter, you must have completed Chapter 10, Analysis of Variance: One-Way Classification. It isn't necessary (though it might be helpful) for you to have also studied Chapter 11, Analysis of Variance: Factorial Design.

[2] Some writers use the name *randomized blocks* for this design. Depending on the context, it is also referred to as *subjects × treatments, subjects × trials,* and *split plot.*

TABLE 12.2 A data set

| Subjects | Levels of independent variable, X | | |
	X_1	X_2	X_3
S_1	57	60	64
S_2	71	72	74
S_3	75	76	78
S_4	93	92	96
$\overline{X}$	74	75	78

EXAMINING THE NUMBERS

Look at the data in **Table 12.2.** There are three levels of the one independent variable (named simply X at this point). The mean score for each level is shown at the bottom of its column. Using the skills you learned in Chapter 10, estimate whether an F ratio for these data will be significant. Remember that $F = MS_{\text{treat}}/MS_{\text{error}}$. Go ahead, stop reading and make your estimate. (If you also write the steps in your thinking, you can compare them to the analysis that follows.)

Here's one line of thinking that might be expected from a person who remembers the lessons of Chapter 10:

> Well, OK, let's start with the means of the three treatments. They are pretty close together; 74, 75, and 78. Not much variability there. How about *within* each treatment? Uh-huh, scores range from 57 to 93 in the first column, so there is a good bit of variability there. Same story in the second and the third columns, too. There's lots of variability within the groups, so MS_{error} will be large. Nope, I don't think the F ratio will be very big. My guess is that F won't be significant — that the means are not significantly different.

Most of us who teach statistics would say that this reasoning shows a good understanding of what goes into the F value in a one-way ANOVA. With a correlated-measures ANOVA, however, there is an *additional* consideration.

The additional consideration for the data in Table 12.2 is that *each subject*[3] contributed a score to *each level* of the independent variable. To explain, look at the second column, X_2, which has scores of 60, 72, 76, and 92. The numbers are quite variable. But notice that some of this variability is *predictable* from the X_1 scores. That is, low scores (such as 60) are associated with S_1 (who had a low score of 57 on X_1). Notice, too, that in column X_3, S_1 had a score of 64, the lowest in the X_3 column. Thus, knowing that a score was made by S_1, you can predict that the score will be low. Please do a similar analysis on the scores for the person labeled S_4.

---

[3] "Participant" is usually preferable to "subject" when referring to people whose scores are analyzed in a behavioral science study. Statistical methods, however, are used by all manner of disciplines; dependent variable scores come from agricultural plots, production records, and *Rattus norvegicus* as well as from people. Thus, statistics books often use the more general term, *subjects.*

In a correlated-measures ANOVA, then, some of the variability of the scores within a level of the independent variable is predictable if you know which subject contributed the score. If you could *remove* the variability that goes with the differences between the subjects, you could *reduce* the variability within a level of the independent variable.

One way to reduce the variability between subjects is to use subjects who are alike. Although this would work well, finding people who are all alike is almost impossible. Fortunately, a correlated-measures ANOVA provides a statistical alternative that accomplishes the same goal.

ONE-FACTOR CORRELATED-MEASURES ANOVA: THE RATIONALE

Here's the paragraph that gives the overview of the topic of this chapter. A correlated-measures ANOVA allows you to calculate and then discard the variability that comes from the differences between the subjects. The remaining variability in the data set is then partitioned into two components: one that is due to the differences between treatments (between-treatment variance) and another that is due to the inherent variability in the measurements (the error variance). These two variances are put into a ratio; the between-treatments variance is divided by the error variance. This ratio is an F value. The F ratio is a sensitive measure of any differences that may exist among the treatments because the variability produced by the differences between individuals has been removed.

AN EXAMPLE PROBLEM

The numbers in Table 12.2 can be used to illustrate the procedures for calculating a one-factor correlated-measures ANOVA. To include the important step of telling what the analysis shows, however, you must know what the independent and dependent variables are. The numbers in the table (the dependent variable) are scores on multiple-choice tests. The independent variable is the instruction given to the test-takers. The three levels of instruction are:

1. Always return to your first choice if you become doubtful about an item.
2. Write notes and comments about the items on the test itself.
3. No instructions

Thus, the design compares two strategies and a control group.[4]

Do you use either of these strategies? Do you have an opinion about the effect of using either of these strategies? Perhaps you would be willing to express your opinion by deciding which of the columns in Table 12.2 represent which of the two strategies

---

[4] By the way, should you be one of those people who find yourself taking multiple-choice tests from time to time, understanding this problem will reveal a strategy for improving your score on such exams.

and which column represents the control condition (taking the test without instructions). (If you want to guess, do it now, because I'm going to label the columns in the next paragraph.)

The numbers in the X_2 column represent typical multiple-choice test performances (control condition). The mean is 75. The X_1 column shows scores when test-takers follow strategy 1 (go with your initial choice). The scores in column X_3 represent performance when test-takers make notes on the test itself (though these notes are not graded in any way). Does either strategy have a significant effect on scores? A correlated-measures ANOVA will give you an answer.

The calculations for this analysis are shown in Table 12.3 (which starts with the same data as Table 12.2, except that labels and sums have been added). Three subscripts (explained below) will be new to you. Examine **Table 12.3** now.

TABLE 12.3 A correlated-measures ANOVA comparing two test-taking strategies with a control condition (The data set is the same as Table 12.2.)

	Strategies			
Subjects	Use first choice	Control	Write notes	ΣX_s
S_1	57	60	64	181
S_2	71	72	74	217
S_3	75	76	78	229
S_4	93	92	96	281
ΣX_t	296	300	312	908
$\overline{X}$	74	75	78	

$\Sigma X_{tot} = 296 + 300 + 312 = 908$

$\Sigma X_{tot}^2 = 57^2 + 60^2 + \cdots + 92^2 + 96^2 = 70{,}460$

$SS_{tot} = \Sigma X_{tot}^2 - \dfrac{(\Sigma X_{tot})^2}{N_{tot}} = 70{,}460 - 68{,}705.333 = 1754.667$

$SS_{subjects} = \dfrac{181^2}{3} + \dfrac{217^2}{3} + \dfrac{229^2}{3} + \dfrac{281^2}{3} - \dfrac{908^2}{12}$

$\qquad = 70{,}417.333 - 68{,}705.333 = 1712.000$

$SS_{strategies} = \dfrac{296^2}{4} + \dfrac{300^2}{4} + \dfrac{312^2}{4} - \dfrac{908^2}{12}$

$\qquad = 68{,}740.000 - 68{,}705.333 = 34.667$

$SS_{error} = SS_{tot} - SS_{subjects} - SS_{strategies}$

$\qquad = 1754.667 - 1712.000 - 34.667 = 8.000$

$df_{tot} = N_{tot} - 1 = 12 - 1 = 11$

$df_{subjects} = N_s - 1 = 4 - 1 = 3$

$df_{strategies} = N_t - 1 = 3 - 1 = 2$

$df_{error} = (N_s - 1)(N_t - 1) = (3)(2) = 6$

Sums of Squares

In the calculations that follow the data set in Table 12.3, note the familiar symbols, ΣX_{tot} (the sum of all the scores) and ΣX_{tot}^2. I have illustrated the calculation of ΣX_{tot}^2 (the sum of all the squared scores) without using a column for X^2, as I did in the two previous chapters. The formula for SS_{tot} is the same as that for all ANOVA problems:

$$SS_{tot} = \Sigma X_{tot}^2 - \frac{(\Sigma X_{tot})^2}{N_{tot}}$$

For the data in Table 12.3, $SS_{tot} = 1754.667$.

$SS_{subjects}$ is the variability due to differences between subjects. As you know, people differ in their abilities when it comes to multiple-choice exams. These differences are reflected in the different row totals. The sum of one subject's scores (a row total) is symbolized ΣX_s. This sum depends on the number of treatments, which is symbolized N_t. The general formula for $SS_{subjects}$ is

$$SS_{subjects} = \Sigma \left[\frac{(\Sigma X_s)^2}{N_t} \right] - \frac{(\Sigma X_{tot})^2}{N_{tot}}$$

Work through the calculations in Table 12.3 that lead to the conclusion $SS_{subjects} = 1712.000$. Perhaps you have already noted that 1712 is a big portion of the total sum of squares.

SS_{treat} is the variability that is due to differences among the strategies the subjects used; that is, SS_{treat} is the variability due to the independent variable. The sum of the scores for one treatment level is ΣX_t. This sum depends on the number of subjects in the experiment, N_s. The general formula for SS_{treat} is

$$SS_{treat} = \Sigma \left[\frac{(\Sigma X_t)^2}{N_s} \right] - \frac{(\Sigma X_{tot})^2}{N_{tot}}$$

For the analysis in Table 12.3, the generic term *treatments* has been replaced with the term specific to these data: *strategies*. As you can see, $SS_{strategies} = 34.667$.

SS_{error} is the variability that remains in the data when the effects of the other identified sources have been removed. To find SS_{error}, calculate

$$SS_{error} = SS_{tot} - SS_{subjects} - SS_{treat}$$

For these data, $SS_{error} = 8.000$.

PROBLEMS

1. You have just worked through three SS values for a correlated-samples ANOVA. But is it a correlated-samples design? No description of the procedures was given. For this problem, describe a way to form the groups that would create a correlated-samples design.

2. In Chapter 10, page 222 (and in Chapter 11, page 250), the text included a graphic showing how SS_{tot} is partitioned into its components. Construct a graphic that shows the partition of the components of sums of squares in a one-factor correlated-measures ANOVA.

Degrees of Freedom, Mean Squares, and F

After you calculate sums of squares, the next step is to determine their degrees of freedom. Degrees of freedom for one-way correlated-measures ANOVA follow the general rule: number of observations minus one.

In general:	*For the strategy study:*
$df_{tot} = N_{tot} - 1$	$df_{tot} = 12 - 1 = 11$
$df_{subjects} = N_s - 1$	$df_{subjects} = 4 - 1 = 3$
$df_{treat} = N_t - 1$	$df_{strategies} = 3 - 1 = 2$
$df_{error} = (N_s - 1)(N_t - 1)$	$df_{error} = (3)(2) = 6$

As is always the case for ANOVAs, mean squares are found by dividing a sum of squares by its *df*. With mean squares in hand, you can form any appropriate *F* ratios. For a reason to be explained later, the only *F* ratio you will calculate is one that determines whether there are significant differences among the *treatments*. Because you will not test the significance of the factor *subjects*, there is no need to calculate $MS_{subjects}$.

$$MS_{strategies} = \frac{SS_{strategies}}{df_{strategies}} = \frac{34.667}{2} = 17.333$$

$$MS_{error} = \frac{SS_{error}}{df_{error}} = \frac{8.000}{6} = 1.333$$

To find *F*, divide $MS_{strategies}$ by the error term:

$$F = \frac{MS_{strategies}}{MS_{error}} = \frac{17.333}{1.333} = 13.003$$

The now familiar ANOVA summary table is the conventional way to present the results of an analysis of variance. **Table 12.4** shows the analysis that assessed strategies for taking multiple-choice exams.

Interpretation of F

The *F* from the data (13.00) is greater than F_{01} from the table (10.92), which allows you to conclude that the three strategies do not have a common population mean.

TABLE 12.4 ANOVA summary table
for the exam strategy study

Source	SS	df	MS	F
Subjects	1712.000	3		
Strategies	34.667	2	17.333	13.00
Error	8.000	6	1.333	
Total	1754.667	11		
	$F_{.05}(2, 6) = 5.14$		$F_{.01}(2, 6) = 10.92$	

Such a statement, of course, doesn't answer the obvious question: Should I use or avoid either of these strategies?

With this question in mind, look at the means of each of the strategies in **Table 12.3.** What pairwise comparison is the most interesting to you?

TUKEY HSD TESTS

The one pairwise comparison that I'll do for you is the comparison between the Write notes strategy and the Control group. The formula for a Tukey HSD is the same as the one you used previously:

$$\text{HSD} = \frac{\overline{X}_1 - \overline{X}_2}{s_{\overline{X}}}$$

where $s_{\overline{X}} = \sqrt{\dfrac{MS_{\text{error}}}{N_t}}$

N_t = the number of scores in each treatment; the number that an $\overline{X}$ is based on.

Applying the formula to the data for the Write notes strategy and the Control group, you get

$$\text{HSD}_{\text{write v. cont}} = \frac{78 - 75}{\sqrt{1.333/4}} = \frac{3}{0.577} = 5.20; \quad p < .05$$

The critical value of HSD at the .05 level for a study with three groups and with a df_{error} of 6 is 4.34. Because the data-produced HSD is greater than the critical value of HSD, you can conclude that making notes on a multiple-choice exam will produce a statistically significantly better score.[5]

OK, now it is time for you to put into practice what you have been learning.

PROBLEMS

3. Many studies have shown that interspersing rest between practice sessions (spaced practice) produces better performance than a no-rest condition (massed practice). Most of these studies used short rests, such as minutes or hours. Bahrick et al. (1993), however, studied the effect of weeks of rest. Thirteen practice sessions of learning the English equivalents of foreign words were interspersed with rests of 2 weeks, 4 weeks, or 8 weeks. The scores (percent of recall) for all four participants in this study are shown. Analyze the data with a one-factor correlated-measures ANOVA, construct a summary table, perform Tukey HSD tests, and write an interpretation.

---

[5] For confirmation, see McKeachie, Pollie, and Speisman (1955).

Subject	Rest interval (weeks)		
	2	4	8
1	40	50	58
2	58	56	65
3	44	70	69
4	57	61	74

4. Perform a Tukey HSD test using the data in Table 12.3. Compare the strategy of using the first choice to the control group.[6]

TYPE I AND TYPE II ERRORS

In problem 3 you reached the conclusion that an 8-week rest between practice sessions is better than a 2-week rest. Could this conclusion be in error? Of course it could be, although it isn't very likely. But suppose the conclusion is erroneous. What kind of a statistical error is it: a Type I error or a Type II error?

In problem 4 you compared the strategy of *return to your initial choice* to the *control* condition. You did not reject the null hypothesis. If this conclusion is wrong, then you've made an error. Is it a Type I error or a Type II error? (A Type I error is rejecting the null hypothesis when it is true. A Type II error is failing to reject a false null hypothesis. For a review of these concepts, see pages 167–169 and 204–205.)

SOME BEHIND-THE-SCENES INFORMATION ABOUT CORRELATED-MEASURES ANOVA

Now that you can analyze data from a one-factor correlated-measures ANOVA, you are ready to appreciate something about what is really going on when you perform this statistical test. I'll begin by addressing *the fundamental issue* in any ANOVA: *partitioning the variance.*

Think about SS_{tot}. You may recall that the definitional formula for SS_{tot} is $\Sigma(X - \overline{X}_{tot})^2$. When you calculate SS_{tot}, you get a number that measures how variable the scores are. If the scores are all close together, SS_{tot} is small. (Think this through. Do you agree?) If the scores are quite different, SS_{tot} is large. (Agree?)

Regardless of the amount of total variance, an ANOVA divides it into separate, independent parts. Each part is associated with a different source.

Look again at the raw scores in **Table 12.3.** Some of the variability in those scores is associated with different subjects. You can see this in the *row totals*—there is a lot of variability from S_1 to S_2 to S_3 to S_4; that is, there is variability *between the subjects.*

---

[6] This comparison is based on data presented by Benjamin, Cavell, and Shallenberger (1984).

Now, focus on the variability *within the subjects*. That is, look at the variability within S_1. S_1's scores vary from 57 to 60 to 64. S_2's scores vary from 71 to 72 to 74. Where does this variability within a subject come from? Clearly, *treatments* are implicated because, as treatments vary from the first to the second to the third column, scores increase.

But the change in treatments doesn't account for all the variation. That is, for S_1 the change in treatments produced a 3-point or 4-point increase in the scores, but for S_2 the change in treatments produced only a 1-point or 2-point increase, going from the first to the third treatment. Thus, the change from the first treatment to the third is not consistent; although an increase usually comes with the next treatment, the change isn't the same for each subject. This means that some of the variability within a subject is accounted for by the treatments, but not all. The variability that remains in the data in Table 12.3 is used to calculate an error term. It is sometimes called the **residual variance.**

The analysis presented in this chapter for correlated-measures ANOVA is appropriate if (1) the levels of the treatments were chosen by the researcher (a fixed-effects model) rather than being selected randomly, and (2) the subjects are randomly assigned to rows. These conditions are common in behavioral science experiments. For other situations, see Howell (1997, Chap. 14) or Kirk (1995, Chap. 7).

Advantages of Correlated-Measures ANOVA

With correlated-measures ANOVA there is a reason the scores in each row belong together. That reason might be that the subjects were matched before the experiment began, that the subjects were "natural triplets" or "natural quadruplets," or that the same subject contributed all the scores in one row.

The case where the same subjects are used for all levels of the independent variable has the advantage of saving time and effort. To illustrate, it takes time to recruit participants, give explanations, and provide debriefing. So, if your study has three levels of the independent variable, and you can run each participant in all three conditions, you can probably save two-thirds of the time it takes to recruit, explain, and debrief. Clearly, a correlated-measures design can be more *efficient*.

What about statistical *power?*[7] Any procedure that reduces the size of the error term (MS_{error}) will produce a larger F ratio. When the between-subjects variability is removed from the analysis, the error term is reduced, making a correlated-measures design more powerful. To summarize, two of the advantages of a correlated-measures ANOVA are *efficiency* and *power*.

Cautions About Correlated-Measures ANOVA

The previous section, titled "Advantages of . . . ," leads good readers to respond internally with "OK, *disadvantages* is next." Well, this section will give you two *cautions* that go with correlated-measures ANOVAs. To address the first, work the problem that follows.

---

[7] You may recall that the more powerful the statistical test, the more likely it will reject a false null hypothesis; see pages 204–205.

PROBLEM

5. One of the following experiments does not lend itself to a correlated-measures design using the same participants in every treatment. Figure out which one and explain why.

a. You want to design an experiment to find out the physiological effects of meditation. Your plan is to measure oxygen consumption before, during, and after a meditation session.

b. A friend, interested in the treatment of depression, wants to compare insight therapy, drug therapy, and a combination of the two, using Beck Depression Inventory scores as a dependent variable.

The first caution about correlated-measures ANOVA is that it may not be useful if one level of the independent variable continues to affect the participant's response in the next treatment condition (a *carryover* effect). Now, there is no problem with the experiment in problem 5a because there is no reason to expect that measuring oxygen consumption before meditation will carry over and affect consumption at a later time. However, in problem 5b, the effect of therapy is long lasting and will affect depression scores at a later time. (To use a correlated-measures ANOVA on problem 5b, some method of matching patients could be used.)

Having cautioned you about carryover effects, I must mention that *sometimes* you want and expect carryover effects. For example, to evaluate a workshop, you might give the participants a pretest, a test after the workshop, and a follow-up test 6 months later. The whole issue is carryover effects, and a correlated-measures ANOVA is the statistical test to use to measure these effects.

The second caution about one-factor correlated-measures ANOVA is that it comes in more than one variety. The variety described in this chapter is appropriate when the levels of the independent variable are chosen by the researcher and the subjects are assigned randomly to treatments. To understand more about the different varieties of correlated-measures ANOVA, you need to understand the *statistical model* that ANOVA is based on. However, this introductory book doesn't cover models.[8]

You can learn more about statistical models by taking an intermediate course in statistics. In the meantime, if you are faced with analyzing data not in this textbook, do what many who analyze data do: Ask for advice from someone knowledgeable about statistics.

ASSUMPTIONS OF CORRELATED-MEASURES ANOVA

If the probability figure you get from Table F is to be accurate, then the data used to calculate the F value must have certain characteristics. Mathematical statisticians re-

---

[8] Writers of introductory textbooks must choose a framework that beginners can use. Unfortunately, in statistics, a simplified framework that works fine for elementary statistics is not adequate for more advanced concepts. Fortunately, learning a more complex framework when the time comes doesn't lead to too much confusion.

fer to these characteristics as the "assumptions of the statistical test." Here are two of the assumptions that go with correlated-measures ANOVA. If these assumptions are met, the probability figures in Table F (.05 and .01) will be accurate.

The first assumption is familiar to you: The populations the scores are drawn from are distributed as a normal curve. The second assumption is that the populations show covariance matrix sphericity, a topic that is often covered in advanced statistics courses (see Kirk, 1995).

PROBLEMS

6. What are two cautions about using a correlated-measures ANOVA?

7. List the assumptions required for assurance that the *F* distribution will give accurate probabilities for a set of data.

*8. Many college students are familiar with meditational exercises. One question about meditation that has been answered is: Are there physiological effects? The data that follow are representative of answers to this question. These data on oxygen consumption (in cubic centimeters per minute) are based on Wallace and Benson's article in *Scientific American* (1972). Analyze the data. Include HSD tests and an interpretation that describes the effect of meditation on oxygen consumption.

	Oxygen consumption (cc/min)		
Subjects	Before meditation	During meditation	After meditation
S_1	175	125	180
S_2	320	290	315
S_3	250	210	250
S_4	270	215	260
S_5	220	190	240

9. In problem 8 you drew some conclusions, each of which could be wrong. For each conclusion, identify what type of error is possible.

10. Most of the problems you have worked in this book were already "set up" (as was problem 8). Researchers, however, usually have to begin by arranging the data in a usable form. So, your first step in this problem will be to arrange the data.

About 10–20 percent of the population experience a period of major depression at some point in their lives. Several kinds of therapy are available. Many studies have tried to answer questions of whether one kind is better than others and whether an untreated control group improves as much as people who receive therapy. Robinson, Berman, and Neimeyer (1990) combined the results from 73 studies to reach the conclusions that are built into the data that follow. For these data, I have given you characteristics that allow you to match up a group of four subjects for each row. The dependent variable is Beck Depression Inventory scores (BDI) recorded at the end of treatment. Before you begin arranging, identify the independent variable. Analyze the data and write a conclusion.

Female, age 20, BDI = 10, drug therapy
Male, age 30, BDI = 9, behavior therapy
Female, age 45, BDI = 16, drug therapy
Female, age 20, BDI = 11, behavior therapy
Female, age 20, BDI = 8, cognitive therapy
Female, age 20, BDI = 18, no therapy
Female, age 45, BDI = 10, cognitive therapy
Male, age 30, BDI = 10, drug therapy
Male, age 30, BDI = 15, cognitive therapy
Female, age 45, BDI = 24, no therapy
Female, age 45, BDI = 16, behavior therapy
Male, age 30, BDI = 21, no therapy

11. Look over the objectives at the beginning of the chapter. Can you do them?

 WEBSITE REFERENCES

http://psychology.wadsworth.com

Open "Choose Your Quiz" and select Spatz, 7e. Find the index of chapters, and click on Chapter 12, *Analysis of Variance: One-Factor Correlated Measures.*

transition page

So far in your practice of hypothesis testing, you have used three different sampling distributions. The normal curve is appropriate when you know σ (Chapter 7). When you don't know σ, you can use the t distribution or the F distribution (Chapters 7–12) if the populations you are sampling from have certain characteristics (such as being normally distributed).

In the next two chapters, you will learn about statistical tests that require neither knowledge of σ nor that the data have the characteristics needed for t and F tests. These tests have sampling distributions that will be new to you. They do, however, have the same purpose as those you have been working with— providing you with the probability of obtaining the observed sample results, *if the null hypothesis is true.*

In Chapter 13, The Chi Square Distribution, you will learn to analyze frequency count data. These data result when observations are classified into categories and the frequencies in each category are counted. In Chapter 14, Nonparametric Statistics, you will learn four techniques for analyzing scores that are ranks or can be reduced to ranks.

The techniques in Chapters 13 and 14 are often described as "less powerful." This means that *if* the populations you are sampling from have the characteristics required for t and F tests, then a t or F test is more likely than a nonparametric test to reject H_0 if it should be rejected. To put this same idea another way, t and F tests have a smaller probability of a Type II error if the population scores have the characteristics the tests require.

chapter**thirteen**

THE CHI SQUARE DISTRIBUTION

OBJECTIVES FOR CHAPTER THIRTEEN

Study the text. Work the problems as you come to them. Afterward, you should be able to:

1. Identify the kind of data that require a chi square test for hypothesis testing
2. Distinguish between problems that require tests of independence and those that require goodness-of-fit tests
3. For tests of independence: State the null hypothesis and calculate and interpret chi square values
4. For goodness-of-fit tests: State the null hypothesis and calculate and interpret chi square values
5. Calculate a chi square value from a 2 × 2 table using the shortcut method
6. Calculate an effect size index for 2 × 2 chi square problems
7. Describe the issues associated with chi square tests based on small N's

 sociology class was deep into a discussion of methods of population control. As the class ended, the topic was abortion; it was clear that there was strong disagreement about using abortion as a method of population control. Both sides in the argument had declared that "educated people" supported their side. Two empirically minded students left the class determined to get actual data on attitudes toward abortion. They solicited the help of a friendly psychology professor, who encouraged them to distribute a questionnaire to his general psychology class. The questionnaire asked, "Do you consider abortion to be an acceptable or unacceptable method of population control?" The questionnaire also asked the respondent's gender. The results were presented to the sociology class in tabular form:

	Abortion is	
	Acceptable	Unacceptable
Women	59	29
Men	15	37
Σ	74	66
	53%	47%

The general conclusion of the class was that college students were pretty evenly divided on this issue, 53 percent to 47 percent.

The attitude data, then, did not seem to indicate that "educated people" were clearly on one side or the other on this issue. The class discussion resumed, with personal opinions dominating, until an observant student said, "Look! Look at that table! The women are for abortion and the guys are against it. Look, 59 of the 88 women say 'acceptable'— that's 67 percent. But only 15 of the 52 guys say 'acceptable'— that's just 29 percent. There is a big gender difference here!"

"Maybe," said another student, "and maybe not. That looks like a big difference, but maybe both groups really have the same attitude and the difference we see here is just the usual variation you find in samples."

Do the two positions in the story sound familiar? Indeed, the issue seems like one to resolve by comparing the sample results to a sampling distribution that shows what happens when the null hypothesis is true. In this case, the null hypothesis is that women and men do not differ in their attitudes. If this is true, how likely is it that two samples produce proportions of .67 and .29?[1] A chi square analysis will give you a probability figure to help you decide. Thus, chi square is another inferential statistics technique — one that leads you to reject or retain a null hypothesis.

Chi square (pronounced "ki," as in *kind,* "square," and symbolized χ^2) is a sampling distribution that gives probabilities about frequencies. Frequencies like those in the attitude–gender table are distributed approximately as chi square, so the chi square distribution will provide the probabilities needed for decision making about the difference between the attitudes of women and men.

The characteristic that distinguishes chi square from the techniques in previous chapters is the *kind of data* you get when you make an observation. For the *t* test and ANOVA, the data are *scores* on a quantitative variable. That is, the subject of the experiment is measured on a quantitative variable and scores, such as IQs, attitudes, time, or errors, result.

With chi square, the data are *frequency counts* of categories. Each subject in the study is observed and placed in one category. The frequency of observations in a category is counted, and the analysis is performed on the frequency counts. What a chi square analysis does is compare the observed frequencies of a category to frequencies that would be expected if the null hypothesis is true.

Karl Pearson (1857–1936), of Pearson product-moment correlation coefficient fame, published the first article on chi square in 1900. Pearson wanted a way to measure the "fit" between data generated by a theory and data obtained from observations. Until chi square was dreamed up, theory testers presented theoretical predictions and empirical data side by side, followed by a declaration such as "good fit" or "poor fit."

The most popular theories at the end of the 19th century predicted that data would be distributed as a normal curve. Many data gatherers had adopted Quételet's position that measurements of almost any social phenomenon would be normally distributed if the number of cases was large enough. Pearson (and others) thought this was not true, and they proposed other curves. By inventing chi square, Pearson provided everyone with a quantitative method of *choosing* the curve of best fit.

---

[1] Casting the chi square problems in this chapter into proportions will help you in *interpreting* the results. However, this book does not cover chi square tests of proportions (see Kirk, 1999, pp. 568–573).

Chi square turned out to be very versatile, being applicable to many problems besides curve-fitting ones. As a test statistic, it is used by psychologists, sociologists, health professionals, biologists, educators, political scientists, economists, foresters, and others. In addition to its use as a test statistic, the chi square *distribution* has come to occupy an important place in theoretical statistics. As further evidence of the importance of Pearson's chi square, the editors of *Science 84* chose it as one of the 20 most important discoveries of the 20th century (Hacking, 1984).

Pearson, too, turned out to be very versatile, contributing data and theory to both biology and statistics. He is credited with naming the standard deviation, giving it the symbol σ, and coining the word *biometry*. He (and Galton and Weldon) founded the journal *Biometrika* to publicize and promote the marriage of biology and mathematics.[2] In addition to his scientific efforts, Pearson was an advocate of women's rights and the father of an eminent statistician, Egon Pearson.

As a final note, Pearson's overall goal was not to be a biologist or a statistician. His goal was a better life for the human race. An important step in accomplishing this was to "develop a methodology for the exploration of life" (Walker, 1968, pp. 499–500).[3]

THE CHI SQUARE DISTRIBUTION

The **chi square distribution** is a theoretical distribution, just as t and F are. Like them, as the number of degrees of freedom increases, the shape of the distribution changes. **Figure 13.1** shows how the shape of the chi square distribution changes as degrees of freedom change from 1 to 5 to 10. Note that χ^2, like the F distribution, is a positively skewed curve.

Critical values for χ^2 are given in Table E in Appendix C for α levels of .10, .05, .02, .01, and .001. Look at **Table E.**

The design of the χ^2 table is similar to that of the t table: α levels are listed across the top and each row shows a different df value. Notice, however, that with chi square, as degrees of freedom increase (looking down columns), larger and larger values of χ^2 are required to reject the null hypothesis. *This is just the opposite of what occurs in the t and F distributions.* This will make sense if you examine the three curves in **Figure 13.1.**

Once again, I will symbolize a critical value by giving the statistic (χ^2), the α level, df, and the critical value. Here's an example (that you will see again): $\chi^2_{.05}(1\ df) = 3.84$.

To calculate a χ^2 to compare to a critical value from Table E, use the formula

$$\chi^2 = \Sigma\left[\frac{(O - E)^2}{E}\right]$$

where O = observed frequency
            E = expected frequency

Obtaining the **observed frequency** is simple enough — count the events in each category. Finding the **expected frequency** is a bit more complex. Two methods can be

[2] In 1901 when *Biometrika* began, The Royal Society, the principal scientific society in Great Britain, accepted articles on biology and articles on mathematics, but they would not permit papers that combined the two. (See Galton, 1901, for a thinly disguised complaint against the establishment.)
[3] For a short biography of Pearson, see Helen M. Walker (1968).

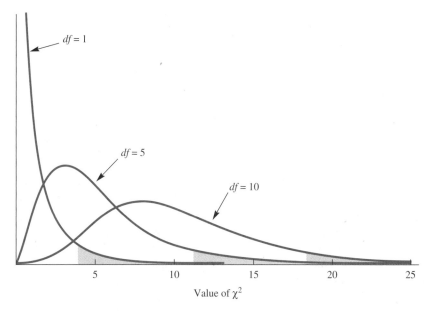

F I G U R E 13.1 Chi square distribution for three different degrees of freedom. Rejection regions at the .05 level are shaded.

used to determine the expected frequency. One method is used if the problem is a "test of independence" and another method is used if the problem is one of "goodness of fit."

CHI SQUARE AS A TEST OF INDEPENDENCE

Probably the most common use of chi square is to test the independence of two variables. The null hypothesis for a chi square test of independence is that the two variables are *independent* — that there is no relationship between the two. If the null hypothesis is rejected, you can conclude that the two variables are *related* and then tell how they are related.

The data gathered by the empirically minded students had two variables, gender and attitude.[4] The null hypothesis is that attitude and gender are independent — that knowing a person's gender gives you no clue to his or her attitude about abortion (and that knowing a person's attitude gives no clue about gender). Rejecting the null hypothesis supports the alternative hypothesis, which is that the two variables are related — that knowing a person's gender helps predict the person's attitude.

A chi square test of independence requires observed values and expected values. **Table 13.1** shows the observed values that you saw before. In addition, it includes the expected values (in parentheses). Expected frequencies are those that are expected if the null hypothesis is true. Please pay careful attention to the logic behind the calculation of expected frequencies.

[4] A table arranged in this form is often called a *contingency* table.

T A B L E 13.1 Hypothetical data on attitudes toward abortion (Expected frequencies are in parentheses.)

	Abortion is		
	Acceptable	Unacceptable	Σ
Women	59 (46.51)	29 (41.49)	88
Men	15 (27.49)	37 (24.51)	52
Σ	74	66	140

Let's start with an explanation of the expected frequency in the upper left corner, 46.51, the expected number of women who think abortion is acceptable. Of all the 140 subjects, 88 (row total) were women, so if you chose a subject at random, the probability that the person would be a woman is $88/140 = .6286$.

In a similar way, of the 140 subjects, 74 (column total) thought that abortion was acceptable. Thus, the probability that a randomly chosen person would think that abortion is acceptable is $74/140 = .5286$.

If you now ask what the probability is that a person chosen at random is a woman who thinks that abortion is acceptable, you can find the answer by multiplying together the probability of the two separate events.[5] Thus, $(.6286)(.5286) = .3323$.

Finally, you can find the expected frequency of such people by multiplying the probability of such a person by the total number of people. Thus, $(.3323)(140) = 46.52$. Notice what happens to the arithmetic when these steps are combined:

$$\left(\frac{88}{140}\right)\left(\frac{74}{140}\right)(140) = \frac{(88)(74)}{140} = 46.51$$

The slight difference in the two answers (46.52 vs. 46.51) is due to rounding. The second answer is the correct one even though I carried four decimal places in the first answer. (This serves as a lesson for those of you who value precision.)

Thus, the formula for the expected value of a cell is its row total, multiplied by its column total, divided by N.

In a similar way, the expected frequency of men with the attitude that abortion is unacceptable is the probability of a man times the probability of an attitude of unacceptable times the number of subjects. For each of the four cells, here are the calculations of the expected values:

$$\frac{(88)(74)}{140} = 46.51 \qquad \frac{(88)(66)}{140} = 41.49$$

$$\frac{(52)(74)}{140} = 27.49 \qquad \frac{(52)(66)}{140} = 24.51$$

Once the expected values are calculated, you can construct a table like **Table 13.2**, which leads to a χ^2 value of 19.14.

---

[5] This is like determining the chances of obtaining two heads in two tosses of a coin. For each toss, the probability of a head is $\frac{1}{2}$. The probability of two heads in two tosses is found by multiplying the two probabilities: $(\frac{1}{2})(\frac{1}{2}) = \frac{1}{4} = .25$. See the section "A Binomial Distribution" in Chapter 6 for a review of this topic.

TABLE 13.2 Calculation of χ^2 for the data in Table 13.1

O	E	$O - E$	$(O - E)^2$	$\dfrac{(O - E)^2}{E}$
59	46.51	12.49	156.00	3.35
15	27.49	−12.49	156.00	5.67
29	41.49	−12.49	156.00	3.76
37	24.51	12.49	156.00	6.36
				$\chi^2 = 19.14$

Every χ^2 value is accompanied by its *df*. To determine the *df* for any $R \times C$ (rows by columns) table such as Table 13.1, use the formula $(R - 1)(C - 1)$. In this case, $(2 - 1)(2 - 1) = 1$. To help understand this formula, remember the "freedom to vary" explanation (pp. 185–186).

For any $R \times C$ table, the *marginal* totals of the expected frequencies are fixed so that they equal the observed frequency.[6] In a 2×2 table, this leaves only one cell free to vary. Once a value is fixed for one cell, the other cell values are determined if you are to maintain the marginal totals. The general rule again is $df = (R - 1)(C - 1)$, where R and C refer to the number of rows and columns.

To determine the significance of $\chi^2 = 19.14$ with 1 *df*, look at Table E in Appendix C. In the first row, you will find that, if $\chi^2 = 10.83$, the null hypothesis may be rejected at the .001 level of significance. Because the χ^2 exceeds this, you can conclude that the attitudes toward abortion are influenced by gender; that is, gender and attitudes toward abortion in the population are not independent but related. By examining the proportions (.67 and .29), you can conclude that women are significantly more likely to consider abortions to be acceptable.

SHORTCUT FOR ANY 2 × 2 TABLE

If you have a calculator and are analyzing a 2×2 table, the following shortcut will save you time. With this shortcut, you do not have to calculate the expected frequencies, which reduces calculating time by as much as one-half. Here is the general case of a 2×2 table:

			Row totals
	A	B	$A + B$
	C	D	$C + D$
Column totals	$A + C$	$B + D$	N

---

[6] In Table 13.1, the marginal total of 88 must be the sum of the two expected frequencies, 46.51 and 41.49, as well as the sum of the obtained frequencies, 59 and 29.

To calculate χ^2 from a 2 × 2 table, use the formula:

$$\chi^2 = \frac{N(AD - BC)^2}{(A + B)(C + D)(A + C)(B + D)}$$

The term in the numerator, $AD - BC$, is the difference between the cross-products. The denominator is the product of the four marginal totals.

To illustrate the equivalence of the shortcut method and the method explained in the previous section, I will calculate χ^2 for the data in Table 13.1. Translating cell letters into cell totals produces $A = 59$, $B = 29$, $C = 15$, $D = 37$, and $N = 140$. Thus,

$$\chi^2 = \frac{(140)[(59)(37) - (29)(15)]^2}{(88)(52)(74)(66)} = 19.14$$

Both methods yield $\chi^2 = 19.14$.

A common error in interpreting a χ^2 problem is to tell less than you might. This is especially the case when the shortcut method is used. For example, the statement "There is a relationship between gender and abortion, $\chi^2(1) = 19.14$; $p < .001$" leaves unsaid the *direction* of the relationship. In contrast, "The attitude of women toward abortion is more favorable than the attitude of men, $\chi^2(1) = 19.14$; $p < .001$" tells the reader not only that there is a relationship but also what the relationship is. So, don't stop with a vague "There is a relationship." Tell what the relationship is.

EFFECT SIZE INDEX FOR 2 × 2 CHI SQUARE DATA

Although a χ^2 test allows you to conclude that two variables are related, it doesn't tell you the degree of relationship. To know the degree of relationship, you need an effect size index.

ϕ (written **phi** and pronounced "fee" by statisticians) is an effect size index for chi square that works for 2 × 2 tables. ϕ is calculated with the formula

$$\phi = \sqrt{\frac{\chi^2}{N}}$$

The interpretation of ϕ is much like that of a correlation coefficient. ϕ gives you the degree of relationship between the two variables in a chi square analysis. Thus, a value near 0 means that there is no relationship and a value near 1 means that there is an almost perfect relationship between the two variables.

Here are the guidelines for evaluating ϕ coefficients:

Small effect	$\phi = .10$
Medium effect	$\phi = .30$
Large effect	$\phi = .50$

Perhaps these values look familiar; they are the same as those you learned in Chapter 5 for evaluating the correlation coefficient r.

For the abortion debate data,

$$\phi = \sqrt{\frac{\chi^2}{N}} = \sqrt{\frac{19.14}{140}} = \sqrt{.1367} = .37$$

Thus, for these manufactured data, there is a medium-sized relationship between gender and attitude.

PROBLEMS

1. In the late 1930s and early 1940s an important sociology experiment took place in the Boston area. At that time 650 boys (median age = $10\frac{1}{2}$ years) participated in the Cambridge–Somerville Youth Study (named for the two economically depressed communities where the boys lived). The participants were *randomly* assigned to either a delinquency-prevention program or a control group. Boys in the delinquency-prevention program had a counselor and experienced several years of opportunities for enrichment. At the end of the study, police records were examined for evidence of delinquency among all 650 boys. Analyze the data in the table and write a conclusion.

	Received program	Control
Police record	114	101
No police record	211	224

2. A psychology student was interested in the effect of group size on the likelihood of joining an informal group. On one part of the campus, he had a group of *two* people looking intently up into a tree. On a distant part of the campus, a group of *five* stood looking up into a tree. Single passersby were classified as joiners or nonjoiners depending on whether they looked up for 5 seconds or longer or made some comment to the group. The data in the table were obtained. Use chi square techniques to determine whether group size had an effect on joining. (See Milgram, 1969.) Use the calculation method you did not choose for problem 1. Write a conclusion.

	Group size	
	2	5
Joiners	9	26
Nonjoiners	31	34

3. You have probably heard that salmon return to the stream in which they hatched. According to the story, after years of maturing in the ocean, the salmon arrive at the mouth of a river and swim upstream, choosing at each fork the stream that leads to the pool in which they hatched. Arthur D. Hasler did a number of interesting experiments on this homing instinct [reported in Hasler (1966) and Hasler and Scholz (1983)]. Here are two sets of data that he gathered.
   a. These data help answer the question of whether salmon really do make consistent choices at the forks of a stream in their return upstream. Forty-

six salmon were captured from the Issaquah Creek (just east of Seattle, Washington) and another 27 from its East Fork. All salmon were marked and released below the confluence of these two streams. All of the 46 captured in the Issaquah were recaptured there. Of the 27 originally captured in the East Fork, 19 were recaptured from the East Fork and 8 from the Issaquah. Use χ^2 to determine if salmon make consistent choices at the confluence of two streams. Calculate ϕ and write an explanation of the results.

b. Hasler believed that the salmon were making the choice at each fork on the basis of olfactory (smell) cues. He thought that young salmon become imprinted on the particular mix of dissolved mineral and vegetable molecules in their home streams. As adults, they simply make choices at forks on the basis of where the smell of home is coming from. To test this, he captured 70 salmon from the two streams, plugged their nasal openings, and released the salmon below the confluence of the two streams. The fish were recaptured above the fork in one stream or the other. Analyze using chi square techniques and write a sentence about Hasler's hypothesis.

Recapture site

		Issaquah	East Fork
	Issaquah	39	12
Capture site	East Fork	16	3

4. Here are some data that I analyzed for an attorney. Of the 4200 white applicants at a large manufacturing facility, 390 were hired. Of the 850 black applicants, 18 were hired. Analyze the data and write a conclusion. *Note:* You will have to work on the data some before you set up the table.

CHI SQUARE AS A TEST FOR GOODNESS OF FIT

In a test of independence, the expected frequencies come from calculations on the data at hand. In a **goodness-of-fit** test, the expected frequencies come from a hypothesis, theory, or model. With a goodness-of-fit test, you can determine whether or not there is a good fit between the theory and the data.

In a chi square goodness-of-fit test, the *null hypothesis is that the actual data fit the expected data.* A rejected H_0 means that the data do *not* fit the model — that is, the model is inadequate. A retained H_0 means that *the data are not at odds with the model.* A retained H_0 does not prove that the model, theory, or hypothesis is true because other models may also predict such results. A retained H_0 does, however, lend support for the model.

Where do hypotheses, theories, and models come from? At a very basic level, they come from our nature as human beings. Humans are always trying to understand things, and this often results in a guess at how things are. When these guesses are developed, supported, and discussed (and perhaps, published), they earn the more sophisticated label of hypothesis, theory, or model. Some hypotheses, theories, and models make quantitative predictions. Such predictions are expected frequencies. By comparing these predicted frequencies to observed frequencies, a chi square analysis provides a test of the theory.

The chi square goodness-of-fit test is used frequently in population genetics, where Mendelian laws predict ratios such as $3:1$, $1:2:1$, and $9:3:3:1$. For example, the law (model) might predict that crossing two pea plants will result in three times as many seeds in the smooth category as in the wrinkled category ($3:1$ ratio). If you perform the crosses, you might get 316 smooth seeds and 84 wrinkled seeds, for a total of 400. How well do these actual frequencies fit the expected frequencies of 300 and 100? Chi square will give you a probability figure to help you decide if the data you have are consistent with the theory.

Although the mechanics of χ^2 require you to manipulate raw frequency counts (such as 316 and 84), you can best understand this test by thinking of proportions. Thus, 316 out of 400 and 84 out of 400 represent proportions of .79 and .21. How likely are such proportions if the population proportions are .75 and .25? Chi square provides you with the probability of obtaining the observed proportions if the theory is true. As a result, you will have a quantitative way to decide "good fit" or "poor fit."

In the genetics example, the expected frequencies were predicted by a theory. Sometimes, in a goodness-of-fit test, the expected frequencies are predicted by chance. In such a case, a rejected H_0 means that something besides chance is at work. When you interpret the analysis, you should identify what that "something" is. Here is an example.

Suppose you were so interested in sex stereotypes and job discrimination that you conducted the following experiment, which is modeled after that of Mischel (1974). You made up four one-page resumes and four fictitious names — two female, two male. The names and resumes were randomly combined, and each participant was asked to read the resumes and "hire" one of the four "applicants" for a management trainee position. The null hypothesis tested is a "no discrimination" hypothesis. The hypothesis says that gender is *not* being used as a basis for hiring and therefore equal numbers of men and women will be hired. Thus, if the data cause the null hypothesis to be rejected, the hypothesis of "no discrimination" may be rejected.

Suppose you had 120 participants in your study, and they "hired" 75 men and 45 women. Is there statistical evidence that sex stereotypes are leading to discrimination? That is, given the hypothesized result of 60 men and 60 women, how good a fit are the observed data of 75 men and 45 women?

Applying the χ^2 formula, you get

$$\chi^2 = \Sigma\left[\frac{(O - E)^2}{E}\right] = \frac{(75 - 60)^2}{60} + \frac{(45 - 60)^2}{60} = 3.75 + 3.75$$
$$= 7.50$$

The number of degrees of freedom for this problem is the number of categories minus one. There are two categories here — hired men and hired women. Thus, $2 - 1 = 1$ *df*. Looking in Table E, you find in row 1 that, if $\chi^2 \geq 6.64$, the null hypothesis may be rejected at the .01 level. Thus, the model may be rejected.

The last step in data analysis (and perhaps the most important one) is to write a conclusion. Begin by returning to the descriptive statistics of the original data. Because 62.5 percent of the hires were men and 37.5 percent were women, you can conclude that discrimination in favor of men was demonstrated.

The term *degrees of freedom* implies that there are some restrictions. For both tests of independence and goodness of fit, one restriction is always that the sum of the expected events must be equal to the sum of the observed events; that is, $\Sigma E = \Sigma O$. If you manufacture a set of expected frequencies from a model, their sum must be equal to the sum of the observed frequencies. In our sex discrimination example, $\Sigma O = 75 + 45 = 120$ and $\Sigma E = 60 + 60 = 120$.

error detection

The sum of the expected frequencies must be equal to the sum of the observed frequencies: $\Sigma E = \Sigma O$.

PROBLEMS

5. Using the data on wrinkled and smooth pea seeds on page 287, test how well the data fit a 3:1 hypothesis.

6. This problem illustrates work of behaviorist John B. Watson (1879–1958) on emotions. Watson (1924) thought there were three basic, inherited emotions: fear, rage, and love. He suspected that the wide variety of emotions experienced by adults had been learned. Proving his suspicion would be difficult because "unfortunately there are no facilities in maternity wards for keeping mother and child under close observation for years." So Watson attacked the apparently simpler problem of showing that the emotions of fear, rage, and love could be distinguished in infants. (He recognized the difficulty of proving that these were the *only* basic emotions, and he made no such claim.) Fear could be elicited by dropping the child onto a soft feather pillow (but not by the dark, dogs, white rats, or a pigeon fluttering its wings in the baby's face). Rage could be elicited by holding the child's arms tightly at its sides, and love by "tickling, shaking, gentle rocking, and patting" among other things.

Here is an experiment reconstructed from Watson's conclusions. A child was stimulated in a way designed to elicit fear, rage, or love. An observer then looked at the child and judged the emotion the child was experiencing. Each judgment was scored as correct or incorrect — correct meaning that the judgment (say, fear) corresponded to the stimulus (say, dropping). Sixty observers made one judgment each, with the results shown in the table. To find the expected frequencies, think about the chance of being correct or incorrect when there are three possible outcomes. Analyze the data with χ^2 and write your conclusion in a sentence.

Correct	Incorrect
32	28

CHI SQUARE WITH MORE THAN ONE DEGREE OF FREEDOM

The χ^2 values you have found so far have been evaluated by comparing them to a chi square distribution with 1 *df*. In this section the problems will require chi square distributions with more than 1 *df*. Some of these will be tests of independence and some will be goodness-of-fit problems.

Here's a question for you. Suppose that for six days next summer you get a free vacation, all expenses paid. Which location would you choose for your vacation?

___ City
___ Mountains
___ Seashore

I asked 78 students that question and found, not surprisingly, that different people chose different locations.

A chi square analysis can test the hypothesis that all three of these locations are preferred equally. In this case, the null hypothesis is that, among those in the population, all three locations are equally likely as a first choice. Using sample data and a chi square test, you can reject or retain this hypothesis.

Think for a moment about the expected value for each cell in this problem. If each of the three locations is equally likely, then the expected value for a location is $\frac{1}{3}$ times the number of respondents; that is, $(\frac{1}{3})(78) = 26$.

The results I got from my sample are as follows:

	Choice		
City	Mountains	Seashore	Total
9	27	42	78

The arrangement of the arithmetic for this χ^2 problem is a simple extension of what you have already learned. The analysis is shown in **Table 13.3.**

What is the *df* for the χ^2 value in Table 13.3? The marginal total of 78 is fixed, and, because ΣE must equal ΣO, only two of the expected cell frequencies are free to vary. Once two are determined, the third is restricted to whatever value will make $\Sigma E = \Sigma O$. Thus, there are $3 - 1 = 2$ *df*. For this design, the number of *df* is the number of categories minus 1. From Table E, for $\alpha = .001$, χ^2 with 2 *df* = 13.82. Therefore, reject

T A B L E 13.3 Calculation of χ^2 for the vacation location preference data

Locations	O	E	$O - E$	$(O - E)^2$	$\dfrac{(O - E)^2}{E}$
City	9	26	−17	289	11.115
Mountains	27	26	1	1	0.038
Seashore	42	26	16	256	9.846
					$\chi^2 = 20.999$

the hypothesis of independence and conclude that the students were responding in a nonchance manner to the questionnaire.

One advantage of χ^2 is its *additive* nature. Each $(O - E)^2/E$ value is a measure of deviation of the data from the model, and the final χ^2 is simply the sum of these measures. Because of this, you can examine the $(O - E)^2/E$ values and see which deviations are contributing the most to χ^2. For the vacation preference data, it is a greater number of seashore choices and fewer city choices that make the final χ^2 significant. The mountain choices are about what would be predicted by the "equal likelihood" model.

Suppose you showed the vacation preference data to a few of your friends, and one of them (a senior psychology major) said, "That's typical of data in psychology — it's based only on college students! If you do any studies of your own, why don't you find out what the noncollege population thinks? Try some high school kids; try some business types downtown."

With this admonition in mind, you embark on a study of family planning. Your question reads, "A couple just starting a family this year should plan to have how many children?"

___ 0 or 1
___ 2 or 3
___ 4 or more

As respondents answer your question, you classify them as high school, college, or business types. Suppose you obtained the 275 responses that are shown in **Table 13.4.**

For this problem, you do not have a theory or model to tell you what the theoretical frequencies should be. The question is whether there is a relationship between attitudes toward family size and group affiliation. A chi square test of independence may answer this question. The null hypothesis is that there is *no* relationship — that is, that recommended family size and group affiliation are independent.

To get the expected frequency for each cell, assume independence (H_0) and apply the reasoning explained earlier about multiplying probabilities:

$$\frac{(92)(104)}{275} = 34.793 \qquad \frac{(92)(130)}{275} = 43.491 \qquad \frac{(92)(41)}{275} = 13.716$$

T A B L E 13.4 Recommended family size by those in high school, college, and business

Subjects	Number of children			
	0–1	2–3	4 or more	Σ
High school	26	57	9	92
College	61	38	18	117
Business	17	35	14	66
Σ	104	130	41	275

TABLE 13.5 Calculation of χ^2 for the family planning data

O	E	$O - E$	$(O - E)^2$	$\dfrac{(O - E)^2}{E}$
26	34.793	−8.793	77.317	2.222
57	43.491	13.509	182.493	4.196
9	13.716	−4.716	22.241	1.622
61	44.247	16.753	280.663	6.343
38	55.309	−17.309	299.602	5.417
18	17.444	0.556	0.309	0.018
17	24.960	−7.960	63.362	2.539
35	31.200	3.800	14.440	0.463
14	9.840	4.160	17.306	1.759
				$\chi^2 = 24.579$

$$\frac{(117)(104)}{275} = 44.247 \qquad \frac{(117)(130)}{275} = 55.309 \qquad \frac{(117)(41)}{275} = 17.444$$

$$\frac{(66)(104)}{275} = 24.960 \qquad \frac{(66)(130)}{275} = 31.200 \qquad \frac{(66)(41)}{275} = 9.840$$

These expected frequencies are incorporated into **Table 13.5,** which shows the calculation of the χ^2 value.

The *df* for a contingency table such as this is obtained from the formula

$$df = (R - 1)(C - 1)$$

Thus, $df = (R - 1)(C - 1) = (3 - 1)(3 - 1) = 4$.

For $df = 4$, $\chi^2 = 18.46$ is required to reject H_0 at the .001 level. The obtained χ^2 exceeds this value, so reject H_0 and conclude that attitudes toward family size are related to group affiliation.[7]

By examining the right-hand column in **Table 13.5,** you can see that 6.343 and 5.417 constitute a large portion of the final χ^2 value. By working backward on those rows, you will discover that college students chose the 0–1 category more often than expected and the 2–3 category less often than expected. The interpretation then is that college students think that families should be smaller than high school students and businesspeople do.

There is an effect size index for chi square problems that have more than one degree of freedom. It was proposed by Harald Cramér and has been variously referred to as Cramer's ϕ, ϕ_c, and Cramer's *V*. The formula and explanations are available in Aron and Aron (1999, pp. 448–449) and Howell (1997, pp. 158–159).

---

[7] The statistical analysis of these data is straightforward; the null hypothesis is rejected. However, a flaw in the experimental design has two variables confounded, leaving the interpretation clouded. Because the three groups differ in both age and education, we are unsure whether the differences in attitude are related to only age, to only education, or to both of these variables. A better design would have groups that differed only on the age or only on the education variable.

PROBLEMS

*7. Professor Stickler always grades "on the curve." "Yes, I see to it that my grades are distributed as 7 percent A's, 24 percent B's, 38 percent C's, 24 percent D's, and 7 percent flunks," he explained to a young colleague. The colleague, convinced that the professor is really a softer touch than he sounds, looked at Professor Stickler's grade distribution for the past year. He found frequencies of 20, 74, 120, 88, and 38, respectively, for the five categories, A through F. Run a χ^2 test for the colleague. What do *you* conclude about Professor Stickler?

8. Is problem 7 a problem of goodness of fit or of independence?

*9. "Snake eyes," a local gambler, let a group of sophomores know they would be welcome in a dice game. After 3 hours the sophomores went home broke. However, one sharp-minded, sober lad had recorded the results on each throw. He decided to see if the results of the evening fit an "unbiased dice" model. Conduct the test and write a sentence summary of your conclusions.

Number of spots	Frequency
6	195
5	200
4	220
3	215
2	190
1	180

10. Identify problem 9 as a test of independence or of goodness of fit.

*11. Remember the controversy, described in Chapter 1, over the authorship of 12 of *The Federalist* papers? The question was whether they were written by Alexander Hamilton or James Madison. Mosteller and Wallace (1989) selected 48 1000-word passages known to have been written by Hamilton and 50 1000-word passages known to have been written by Madison. In each passage they counted the frequency of certain words. The results for the word *by* are shown in the table. Is *by* a word that the two writers used with significantly different frequency? Explain how these results help in deciding about the 12 disputed papers.

Rate per 1000 words	Hamilton	Madison
0–6	21	5
7–12	27	31
13–18	0	14

12. Is problem 11 one of independence or of goodness of fit?

SMALL EXPECTED FREQUENCIES

The theoretical chi square distribution, which comes from a mathematical formula, is a continuous function that can have any positive numerical value. Chi square test statistics calculated from frequencies do not work this way. They change in discrete steps and, when the expected frequencies are very small, the discrete steps are quite large. It is clear to mathematical statisticians that, as the expected frequencies approach zero, the theoretical chi square distribution becomes a less and less reliable way to estimate probabilities. In particular, the fear is that such chi square analyses will reject the null hypothesis more often than warranted.

The question for researchers who hope to analyze their data with chi square has always been: Those expected frequencies — how small is too small? For years the usual recommendation was that if an expected frequency was less than 5, a chi square analysis was suspect. A number of studies, however, have led to a revision of the answer.

When *df* = 1

Several studies used a computer to draw thousands of random samples from a known population for which the null hypothesis was true. Each sample was analyzed with a chi square, and the proportion of rejections of the null hypothesis (Type I errors) was calculated. (If this proportion were approximately equal to α, the chi square analysis would be clearly appropriate.) The general trend of these studies has been to show that the theoretical chi square distribution gives accurate conclusions even when the expected frequencies are considerably less than 5.[8] (See Camilli and Hopkins, 1978, for an analysis of a 2 × 2 test of independence.)

When *df* > 1

When $df > 1$, the same uncertainty exists if one or more expected frequencies is small. Is the chi square test still advisable? Bradley and colleagues (1979) addressed some of the cases in which $df > 1$. Again, they used a computer to draw thousands of samples and analyzed each sample with a chi square test. The proportion of cases in which the null hypothesis was mistakenly rejected was calculated. Nearly all were in the .03 to .06 range but a few fell outside this range, especially when the sample sizes were small ($N = 20$). Once again, the conclusion was that the chi square test gives fairly accurate probabilities, even with sample sizes that are considered small.

---

[8] **Yates' correction** used to be recommended for 2 × 2 tables that had one expected frequency less than 5. The effect of this was to reduce the size of the obtained chi square. According to current thinking, Yates' correction overcorrected, resulting in α levels that were much smaller than the investigator claimed to be using. Therefore, I do not recommend Yates' correction for the kinds of problems found in most research. (See Camilli and Hopkins, 1978.)

An Important Consideration

Now, let's back away from these trees we've been examining and visualize the forest that you are trying to understand. The big question, as always, is: What is the nature of the population? The question is answered by studying a sample, which may lead you to reject or to retain the null hypothesis. Either decision could be in error. The concern of this section so far has been whether the chi square test is keeping the percentage of mistaken rejections (Type I errors) near an acceptable level (.05).

Because the big question is What is the nature of the population? you should keep in mind the other error you could make: retaining the null hypothesis when it should be rejected (a Type II error). How likely is a Type II error when expected frequencies are small? Overall (1980) has addressed this question and his answer is "very."

Overall's warning is that if the effect of the variable being studied is not just huge, then small expected frequencies have a high probability of dooming you to make the mistake of failing to discover a real difference. You may recall that this issue was discussed in Chapter 9 under the topic of power.

How can this mistake be avoided? Use large N's. The larger the N, the more power you have to detect any real differences. And, in addition, the larger the N, the more confident you are that the actual probability of a Type I error is your adopted α level.

Combining Categories

Combining categories is a technique that allows you to ensure that your chi square analysis will be more accurate. (It reduces the probability of a Type II error and keeps the probability of a Type I error close to α.) I will illustrate this technique with some data from Chapter 6 on the normal distribution.

In Chapter 6 I claimed that IQ scores are distributed normally and offered as evidence a frequency polygon, Figure 6.7. I asked you simply to look at it and note that this empirical distribution appeared to fit the normal curve. (I was using the pre-1900 method of presenting evidence.) Now you are in a position to use a Pearson goodness-of-fit test to determine whether the scores are distributed as a normal curve. For that set of 261 IQ scores, the mean was 101 and the standard deviation was 13.4. The question this goodness-of-fit test will answer is: Will a theoretical normal curve with a mean of 101 and a standard deviation of 13.4 fit the data graphed in Figure 6.7?

To find the expected frequency of each class interval, I constructed **Table 13.6.** The first three columns show the observed scores arranged into a grouped frequency distribution and the lower limit of each class interval. The z-score column shows the z score for the lower limit of the class interval, based on a mean of 101 and a standard deviation of 13.4. The next column was obtained by entering the normal curve table at the z score and finding the proportion of the curve above the z score. From this proportion I subtracted the proportion of the curve associated with class intervals with higher scores. (For example, from the proportion above $z = 1.75$, .0401, I subtracted .0166 to get .0235.) Finally, the proportion in the interval was multiplied by 261, the number of students, to get the expected frequency.

TABLE 13.6   IQ scores and expected frequencies for 261 fifth-grade students

Class interval	Observed frequency	Lower limit of class interval	z Score	Proportion of normal curve within class interval	Expected frequency (261 × proportion)
130 and up	4	129.5	2.13	.0166	4.3
125–129	5	124.5	1.75	.0235	6.1
120–124	18	119.5	1.38	.0437	11.4
115–119	19	114.5	1.01	.0724	18.9
110–114	25	109.5	0.63	.1081	28.2
105–109	28	104.5	0.26	.1331	34.7
100–104	36	99.5	−0.11	.1464	38.2
95–99	41	94.5	−0.49	.1441	37.6
90–94	31	89.5	−0.86	.1172	30.6
85–89	28	84.5	−1.23	.0856	22.3
80–84	14	79.5	−1.60	.0545	14.2
75–79	7	74.5	−1.98	.0309	8.1
70–74	4	69.5	−2.35	.0145	3.8
65–69	1	64.5	−2.72	.0061	1.6
64 and below	0		Below −2.72	.0033	0.9
Sum	261			1.0000	260.9

Now you have the ingredients necessary for a χ^2 test. The observed frequencies are in the second column. The frequencies expected if the data are distributed as a normal curve with a mean of 101 and a standard deviation of 13.4 are in the last column. I could combine these observed and expected frequencies into the now familiar $(O - E)^2/E$ formula, except that several categories have suspiciously small expected frequencies.

The solution to this problem is to combine some adjacent categories. If the top category is 125 and up, the expected frequency becomes 10.4 (4.3 + 6.1). In a similar fashion, the bottom category becomes 74 and below and has an expected frequency of 6.3 (0.9 + 1.6 + 3.8).[9] These combinations are shown in **Table 13.7**, a familiar χ^2 table with 12 categories. The χ^2 value is 7.98.

What is the *df* for this problem? This is a case in which *df* is not equal to the number of categories minus 1. The reason is that there is more than the usual one restriction on the expected frequencies. Usually, the requirement that $\Sigma E = \Sigma O$ accounts for the loss of 1 *df*. For this problem, however, I also required that the mean and standard deviation of the expected frequency be equal to the observed mean and observed standard deviation. Thus, *df* in this case is the number of categories minus 3, and 12 − 3 = 9 *df*.

---

[9] I have to confess that the guideline I used for how much combining to do was the old notion about not having expected frequencies of less than 5.

TABLE 13.7 Testing the goodness of fit of IQ data and a normal curve

Class interval	O	E	$O - E$	$(O - E)^2$	$\dfrac{(O - E)^2}{E}$
125 and up	9	10.4	−1.4	1.96	0.189
120–124	18	11.4	6.6	43.56	3.821
115–119	19	18.9	0.1	0.01	0.001
110–114	25	28.2	−3.2	10.24	0.363
105–109	28	34.7	−6.7	44.89	1.294
100–104	36	38.2	−2.2	4.84	0.127
95–99	41	37.6	3.4	11.56	0.307
90–94	31	30.6	0.4	0.16	0.005
85–89	28	22.3	5.7	32.49	1.457
80–84	14	14.2	−0.2	0.04	0.003
75–79	7	8.1	−1.1	1.21	0.149
74 and below	5	6.3	−1.3	1.69	0.268
Σ	261	260.9			$\chi^2 = 7.984$

By consulting Table E in Appendix C, you can see that a χ^2 of 16.92 is required to reject the null hypothesis at the .05 level. The null hypothesis for a goodness-of-fit test is that the data *do* fit the model. Thus, to reject H_0 in this case is to show that the data are not normally distributed. However, because the obtained $\chi^2 = 7.98$ is less than the critical value of 16.92, the conclusion is that these data do not differ significantly from a normal curve model in which $\mu = 101$ and $\sigma = 13.4$.

There are other situations in which small expected frequencies may be combined. For example, in an opinion survey, the response categories might be "strongly agree," "agree," "no opinion," "disagree," and "strongly disagree." If there are few respondents who check the "strongly" categories, those who do might simply be combined with their "agree" or "disagree" neighbors — reducing the table to three cells instead of five. The basic rule on combinations is that they must "make sense." It wouldn't make sense to combine the "strongly agree" and the "no opinion" categories in this example.

WHEN YOU MAY USE CHI SQUARE

1. Use chi square when categories are established and the measurement of subjects or events is that they are placed into their appropriate category. Thus, with chi square, the data are frequency counts.
2. The procedures presented in this chapter require that each measurement be **independent** of other measurements. This doesn't refer to the independence of the variables but to the independence of the measurements. This is the same independence required by scores for an independent *t* test or a one-way ANOVA. Thus, before-and-after frequency counts of the same people cannot be analyzed with chi square.

3. Like all the methods of hypothesis testing you have studied, chi square conclusions apply to the population if the samples are representative of the population. Random sampling assures representativeness.

PROBLEMS

13. An early hypothesis about schizophrenia (a form of severe mental illness) was that it has a simple genetic cause. In accordance with the theory, one-fourth (a 1:3 ratio) of the offspring of a certain selected group of parents would be expected to be diagnosed as schizophrenic. Suppose that of 140 offspring, 19.3 percent were schizophrenic (and 80.7 percent were not schizophrenic). Use this information to test the goodness of fit of a 1:3 model. Remember that a χ^2 analysis is performed on frequency counts.

14. What is the difference between the goodness-of-fit and the test of independence?

15. You may recall from Chapter 8 that I bought Doritos tortilla chips and Snickers candy bars to have data for problems. For this chapter I bought M&Ms. In a "fun size" package I got 4 browns, 6 reds, 8 yellows, 2 greens, 3 oranges, and 0 blues. Mars, Inc., the manufacturer, claims that the population breakdown is 30 percent browns, 20 percent reds, 20 percent yellows, 10 percent greens, 10 percent oranges, and 10 percent blues (M&M/Mars, 1993). Use chi square to test the fit of the data to the expected values.

16. Suppose a friend of yours decides to gather his own data on sex stereotypes and job discrimination. He makes up ten resumes and ten fictitious names, five male and five female. He asks each of his 24 subjects to "hire" five of the candidates rather than just one. The data show that 65 men and 55 women are hired. He asks you for advice on χ^2. What would you say to your friend?

17. For a political science class, students gathered data for a supporters profile of the two candidates for senator, Hill and Dale. One student stationed herself at a busy intersection and categorized the cars turning left according to whether or not they signaled the turn and whose bumper sticker was displayed. After 2 hours, she left with the frequencies shown in the table. Test these data with χ^2 to see whether there is a story to be told about the backers of the two candidates. Carefully phrase the student's contribution to the supporters profile.

Signaled turn	Sticker	Frequency
No	Hill	11
No	Dale	2
Yes	Hill	57
Yes	Dale	31

18. Another friend, who knows that you are almost finished with this book, comes to you for advice on the statistical analysis of the following data. He wants to know whether there is a significant difference between men and women. What do you tell your friend?

	Men	Women
Mean score	123	206
N	18	23

19. Our political science student from problem 17 has launched a new experiment — to determine whether affluence is related to voter choice. This time she looks for yard signs for the two candidates and then classifies the houses as brick or one-story frame. Her reasoning is that the one-story frame homes represent less affluent voters in her city. Houses that do not fit the categories are ignored. After 3 hours of driving and three-fourths of a tank of gas, the frequency counts are as shown in the table. Analyze the data using your chi square techniques. Carefully write a statement of the results that can be used by the political science class in the voter profile.

Type of house	Yard signs	Frequency
Brick	Hill	17
Brick	Dale	88
Frame	Hill	59
Frame	Dale	51

20. When using χ^2, what is the proper *df* for each of the following tables?
  a. 1×4
  b. 4×5
  c. 2×4
  d. 6×3

21. Please reread the chapter objectives. Can you do each one?

WEBSITE REFERENCES

http://psychology.wadsworth.com

Open "Choose Your Quiz" and select Spatz, 7e. Find the index of chapters, and click on Chapter 13, *The Chi Square Distribution.*

Open "Choose Your Area," scroll down, and select "Research Methods and Statistics." Click on "Statistics" in the graph. Click on "Student Activities." Click on the Web Page button for "Chi Square."

chapter**fourteen**

NONPARAMETRIC STATISTICS

OBJECTIVES FOR CHAPTER FOURTEEN

Study the text. Work the problems as you come to them. Afterward, you should be able to:

1. Describe the rationale of nonparametric statistical tests
2. Determine when a Mann–Whitney *U* test is appropriate, perform the test, and interpret the results
3. Determine when a Wilcoxon matched-pairs signed-ranks *T* test is appropriate, perform the test, and interpret the results
4. Determine when a Wilcoxon–Wilcox multiple-comparisons test is appropriate, perform the test, and interpret the results
5. Calculate a Spearman r_s correlation coefficient and test the hypothesis that the coefficient came from a population with a correlation of zero

As you read the story, try to identify the independent and dependent variables before I give them away.

Two child psychologists were talking shop over coffee one morning. (Much research begins with just such sessions.) The topic was intensive early training in athletics. Both psychologists were convinced that such training made the child less sociable as an adult, but one psychologist went even further. "I think that really intensive training of young kids is ultimately detrimental to their performance in the sport. Why, I'll bet that, among the top ten men's singles tennis players, those with intensive early training are not in the highest ranks."

"Well, I certainly wouldn't go that far," said the second psychologist. "I think all that early intensive training is quite helpful."

"Good. In fact, great. We disagree and we may be able to decide who is right. Let's get the ground rules straight. For tennis players, how early is early and what is intensive?"

"Oh, I'd say early is starting by age 7 and intensive is playing every day for 2 or more hours."[1]

---

[1] Because the phrase *intensive early training* can mean different things to different people, the first psychologist has provided the second with an operational definition. An **operational definition** is a definition that specifies how the concept can be measured.

"That seems reasonable. Now, let's see, our population is 'excellent tennis players' and these top ten will serve as our representative sample."

Can you identify the independent and dependent variables?

"Yes, indeed. What we would have among the top ten players would be two groups to compare. One had intensive early training, and the other didn't. The dependent variable is the player's rank. What we need is some statistical test that will tell us whether the difference in average ranks of the two groups is statistically significant."[2]

"Right. Now, a *t* test won't give us an accurate probability figure because *t* tests assume that the population of dependent variable scores is normally distributed. A distribution of ranks is rectangular with each score having a frequency of 1."

"I think there is a nonparametric test that is the proper one for ranks."

Here is a new category of tests that can be used to analyze experiments in which the dependent variable is ranks. This category of tests is called **nonparametric statistics.** As you will see, there are data other than ranks for which nonparametric tests are appropriate. To begin, study the rationale of these tests until you understand the ideas they are based on.

THE RATIONALE OF NONPARAMETRIC TESTS

Many of the reasoning steps you go through in doing nonparametric statistics are already familiar to you — they are the hypothesis-testing steps. By this time in your study of statistics, you should be able to write a coherent paragraph about the null hypothesis, the alternative hypothesis, gathering data, using a sampling distribution to find the probability of such results when the null hypothesis is true, making a decision about the null hypothesis, and telling a story that the data support. Can you do that? Right now?

The steps in the hypothesis-testing rationale that are unique to this chapter are those on sampling distributions of ranks. Here's an explanation of how a sampling distribution based on ranks might be constructed.

Suppose you drew two samples of equal size (for example, $N_1 = N_2 = 10$) from the same population.[3] You then arranged all the scores from both samples into one overall ranking, from 1 to 20. Because the samples are from the same population, the sum of the ranks of one group should be equal to the sum of the ranks of the second group. In this case, the expected sum for each group is 105. (With a little figuring, you can prove this for yourself now, or you can wait for an explanation later in the chapter.)

Although the expected sum is 105, actual sampling from the population would also produce sums greater than 105 and less than 105. After repeated sampling, all the sums could be arranged into a sampling distribution, which would allow you to determine the likelihood of any sum (if both samples come from the same population).

---

[2] This is how the experts convert vague questionings into comprehensible ideas that can be communicated to others — they identify the independent and the dependent variables.

[3] As always, drawing two samples from one population is statistically the same as starting with two identical populations and drawing a random sample from each.

TABLE 14.1 The function of some nonparametric and parametric tests

Nonparametric test	Function	Parametric test
Mann–Whitney U	Tests for a significant difference between two independent samples	Independent-samples t test
Wilcoxon matched-pairs signed-ranks T	Tests for a significant difference between two correlated samples	Correlated-samples t test
Wilcoxon–Wilcox multiple-comparisons	Tests for significant differences among all possible pairs of independent samples	One-way ANOVA and Tukey HSD tests
Spearman r_s	Describes the degree of correlation between two variables	Pearson product-moment correlation coefficient, r

Even if the two sample sizes are unequal, the same logic will work. A sampling distribution can be constructed that will show the expected variation in sums of ranks for one of the two groups.

With this rationale in mind, you are ready to learn four new techniques. The first three are hypothesis-testing methods that determine whether samples came from the same population. The fourth technique involves a correlation coefficient for ranked data, symbolized r_s. You will learn to calculate r_s and then test the hypothesis that a sample r_s came from a population with a correlation coefficient of .00.

The four nonparametric techniques in this chapter and their functions are listed in Table 14.1. In earlier chapters you studied parametric tests that have similar functions, which are listed on the right side of the table. Study **Table 14.1** carefully now.

There are many other nonparametric statistical methods beside the four that you will study in this chapter. Peter Sprent (1989) has written a superb handbook, *Applied Nonparametric Statistical Methods,* that covers many of these methods. Sprent's book describes applications in fields such as road safety, space research, trade, and medicine, as well as in traditional academic disciplines.

COMPARISON OF NONPARAMETRIC TO PARAMETRIC TESTS

In what ways are the nonparametric tests in this chapter similar to and in what ways are they different from parametric tests (such as t tests and ANOVA)? Here are the similarities:

- The hypothesis-testing logic is the same for both nonparametric and parametric tests. Both tests yield the probability of the observed data, *when the null hypothesis is true.* As you will see, though, the null hypotheses of the two kinds of tests are different.

■ Both nonparametric and parametric tests require you to assign participants randomly to subgroups (or to sample randomly from the populations).

As for the differences:

■ Nonparametric tests do not require the assumptions about the populations that parametric tests require. For example, parametric tests such as *t* tests and ANOVA produce accurate probabilities when the populations are normally distributed and have equal variances. Nonparametric tests do not assume the populations have these characteristics.

■ The null hypothesis for nonparametric tests is that the population distributions are the same. For parametric tests, the null hypothesis is usually that the population means are the same (H_0: $\mu_1 = \mu_2$). Because distributions can differ in form, variability, central tendency, or all three, the interpretation of a rejection of the null hypothesis may not be quite so clear-cut after a nonparametric test.

Recommendations on how to choose between a parametric and a nonparametric test have varied over the years. Two of the issues involved in the debate are the scale of measurement and the power of the tests.

In the 1950s and after, some texts recommended that nonparametric tests be used if the scale of measurement was nominal or ordinal. After a period of controversy this consideration was dropped. Later, it resurfaced. Currently, most researchers do not use a "scale of measurement" criterion in deciding between nonparametric and parametric tests.

The other issue, power, is also complicated. Power, you may recall, comes up when the null hypothesis *should* be rejected. A test's power is a measure of how likely the test is to reject a false null hypothesis. It is clear to mathematical statisticians that if the populations being sampled from are normally distributed and have equal variances, then parametric tests are more powerful than nonparametric ones.

If the populations are not normal or do not have equal variances, then it is less clear what to recommend. Early work on this question showed that parametric tests were robust, meaning that they gave approximately correct probabilities even though populations were not normal or did not have equal variances. However, this robustness has been questioned. For example, Blair, Higgins, and Smitley (1980) showed that a nonparametric test (Mann–Whitney *U*) is generally more powerful than its parametric counterpart (*t* test) for the nonnormal distribution they tested. Blair and Higgins (1985) arrived at a similar conclusion when they compared the Wilcoxon matched-pairs signed-ranks *T* test to the *t* test. I am sorry to leave these issues without giving you specific advice, but no simple rule of thumb about superiority is possible except for one: If the data are ranks, use a nonparametric test.

Finally, there is not even a satisfactory name for these tests. Besides nonparametric, they are also referred to as **distribution-free statistics.** Although *nonparametric* and *distribution-free* mean different things to statisticians, the two words are used almost interchangeably by research workers. I will use the term *nonparametric tests* and illustrate them in the order given in Table 14.1.

clue to the future

For each of the four tests in this chapter, you will have to *understand* the steps. Formulas aren't available that allow a mechanical "plug in the raw numbers and grind out an answer." To work problems correctly, think your way through them.

PROBLEMS

1. What is the name of the test that is the nonparametric relative of an independent-samples *t* test? Give the name for the relative of a correlated-samples *t* test.
2. How are nonparametric tests different from parametric tests?
3. Sketch out in your own words the rationale of nonparametric tests.
4. Two issues have dominated the discussion of how to choose between nonparametric and parametric tests. What are they?

THE MANN–WHITNEY *U* TEST

The **Mann–Whitney *U* test** is a nonparametric test for data from an independent-samples design.[4] Thus, it is the appropriate test for the child psychologists to use to test the difference in ranks of tennis players.

The Mann–Whitney *U* test produces a statistic, *U*, that is evaluated by consulting the sampling distribution of *U*. Like all the distributions you have encountered that can be used in the analysis of small samples, the sampling distribution of *U* depends on the sample size.

When sample sizes are relatively small, critical values of *U* can be found in **Table H** in Appendix C. Use Table H if $N_1 = 20$ or less *and* $N_2 = 20$ or less. If the number of scores in one of the samples is greater than 20, the statistic *U* is distributed approximately as a normal curve. In this case, a *z* score is calculated, and the values of ± 1.96 and ± 2.58 are used as critical values for $\alpha = .05$ and $\alpha = .01$.

Mann–Whitney *U* Test for Small Samples

To provide data to illustrate the Mann–Whitney *U* test, I made up the numbers in Table 14.2 about the intensive early training of the top ten male singles tennis players.

In **Table 14.2** the players are listed by initials in the left column; the right column indicates that they had intensive early training ($N_{yes} = N_1 = 4$) or that they did not

---

[4] The Mann–Whitney *U* test is identical to the **Wilcoxon rank-sum test.** Wilcoxon published his test first (1945), but when Mann and Whitney independently published a test based on the same logic (1947), they provided tables and a *name* for their statistic. Today researchers use the Mann–Whitney *U* test. The lesson: name your invention.

TABLE 14.2 Early training of top ten male singles tennis players

Players	Rank	Intensive early training?
Y.O.	1	No
U.E.	2	No
X.P.	3	No
E.C.	4	Yes
T.E.	5	No
D.W.	6	Yes
O.R.	7	No
D.S.	8	Yes
H.E.	9	Yes
R.E.	10	No

$\Sigma R_{yes} = 4 + 6 + 8 + 9 = 27$

$\Sigma R_{no} = 1 + 2 + 3 + 5 + 7 + 10 = 28$

($N_{no} = N_2 = 6$). Each player's rank is shown in the middle column. At the bottom of the table, the ranks of the *yes* players are summed (27), as are the ranks of the *no* players (28).

The sums of the ranks are used to calculate two **U values.** The smaller of the two U values is used to enter Table H, which yields a probability figure. For the *yes* group, the U value is

$$U = (N_1)(N_2) + \frac{N_1(N_1 + 1)}{2} - \Sigma R_1$$

$$= (4)(6) + \frac{(4)(5)}{2} - 27 = 7$$

For the *no* group, the U value is

$$U = (N_1)(N_2) + \frac{N_2(N_2 + 1)}{2} - \Sigma R_2$$

$$= (4)(6) + \frac{(6)(7)}{2} - 28 = 17$$

error detection

The sum of the two U values in a Mann–Whitney U test is equal to the product of $(N_1)(N_2)$.

Applying the error detection hint to the tennis ranking calculations, you get $7 + 17 = 24 = (4)(6)$. The calculations check.

Now, please examine **Table H.** It appears on two pages. On the first page of the table, the *lightface* type gives the critical values for α levels of .01 for a one-tailed test and .02 for a two-tailed test. The numbers in *boldface* type are critical values for $\alpha =$ .005 for a one-tailed test and $\alpha = .01$ for a two-tailed test. In a similar way, the second page gives larger α values for both one- and two-tailed tests. The commonly used two-tailed test with $\alpha = .05$ is on this second page (boldface type). Having chosen the page you want, you determine the column by N_1 and the row by N_2.

Now, you can test the $U = 7$ value from the tennis data. From the conversation of the two child psychologists, it is clear that a two-tailed test is appropriate; they are interested in knowing whether intensive early training *helps* or *hinders* players. Because an α level was not discussed, you should do what they would do — see if the difference is significant at the .05 level and, if it is, see if it is also significant at some smaller α level. Thus, in Table H begin by looking for the critical value of U for a two-tailed test with $\alpha = .05$. This number is on the second page in boldface type at the intersection of $N_1 = 4$, $N_2 = 6$. The critical value is 2.

Because the obtained value of U is 7, you must *retain* the null hypothesis and conclude that there is no evidence from the sample that the distribution of players trained early and intensively is significantly different from the distribution of those without such training.

Although you can easily find a U value using the preceding method and quickly go to Table H and reject or retain the null hypothesis, it would help your understanding of this test to think about small values of U. Under what conditions would you get a small U value? What kinds of samples would give you a U value of zero?

By examining the formula for U, you can see that $U = 0$ when the members of one sample all rank lower than every member of the other sample. Under such conditions, rejecting the null hypothesis seems reasonable. By playing with numbers in this manner, you can move from the rote memory level to the understanding level.

Note that the *smaller* the U value, the more different the two samples are. This is just the opposite of the situation with t tests, ANOVA, and χ^2.

Assigning Ranks and Tied Scores

Sometimes you may choose a nonparametric test for data that are not already in ranks.[5] In such cases you have to rank the scores. Two questions often arise. Is the largest or the smallest score ranked 1, and what should I do about the ranks for scores that are tied?

You will find the answer to the first question to be very satisfactory. For the Mann–Whitney test, it doesn't make any difference whether you call the largest or the smallest score 1. (This is not true for the test described next, the Wilcoxon T test.)

Ties are handled by giving all tied scores the same rank. This rank is the mean of the ranks the tied scores would have if no ties had occurred. For example, if a distribution of scores is 12, 13, 13, 15, and 18, the corresponding ranks are 1, 2.5, 2.5, 4, 5. The two scores of 13 would have been 2 and 3 if they had not been tied, and 2.5 is the

---

[5] Severe skewness or populations with very unequal variances are often reasons for such a choice.

mean of 2 and 3. As a slightly more complex example, the scores 23, 25, 26, 26, 26, 29 have ranks of 1, 2, 4, 4, 4, 6. Ranks of 3, 4, and 5 average out to be 4.

Ties do not affect the value of U if they are in the same group. If several ties involve both groups, a correction factor may be advisable.[6]

error detection

Assigning ranks is tedious. Practice (with feedback) leads to fewer errors. Dealing with ties is tedious *and* troublesome. Pay careful attention to the examples.

Mann–Whitney U Test for Larger Samples

If one sample (or both samples) has 21 scores or more, the normal curve should be used to assess probability. A **z test** value is obtained by the formula

$$z = \frac{(U + c) - \mu_U}{\sigma_U}$$

where $c = 0.5$, a correction factor explained later

$$\mu_U = \frac{(N_1)(N_2)}{2}$$

$$\sigma_U = \sqrt{\frac{(N_1)(N_2)(N_1 + N_2 + 1)}{12}}$$

Here c is a correction for continuity. It is used because the normal curve is a continuous function but the values of z obtained in this test are discrete. U, as before, is the smaller of the two possible U values.

Once again, the formula is familiar: the difference between a statistic based on data $(U + c)$ and the expected value of a parameter (μ_U) divided by a measure of variability, σ_U. After you have obtained z, critical values can be obtained from Table C, the normal curve table. For a two-tailed test, reject H_0 if $z \geq |1.96|$ ($\alpha = .05$). For a one-tailed test, reject H_0 if $z \geq 1.65$ ($\alpha = .05$). The corresponding values for $\alpha = .01$ are $z \geq |2.58|$ and $z \geq 2.33$.

Here is a problem for which the normal curve is necessary. An undergraduate psychology major was devoting a year to the study of memory. The principal independent variable was gender. Among her several experiments was one in which she asked the students in a general psychology class to write down everything they remembered that was unique to the previous day's class, during which a guest had lectured. Students were encouraged to write down every detail they remembered. This class was routinely videotaped so it was easy to check each recollection for accuracy and uniqueness. Because the samples indicated that the population data were severely skewed, the psychology major chose a nonparametric test.

The scores, their ranks, and the statistical analysis are presented in **Table 14.3**. (A plot of the scores in Table 14.3 shows skew.)

---

[6] See Kirk (1999, p. 594) for the correction factor.

TABLE 14.3 Numbers of items recalled by men and women, ranks, and a Mann–Whitney U test

Men ($N = 24$)		Women ($N = 17$)	
Items recalled	Rank	Items recalled	Rank
70	3	85	1
51	6	72	2
40	9	65	4
29	13	52	5
24	15	50	7
21	16.5	43	8
20	18.5	37	10
20	18.5	31	11
17	21	30	12
16	22	27	14
15	23	21	16.5
14	24.5	19	20
13	26.5	14	24.5
13	26.5	12	28.5
11	30.5	12	28.5
11	30.5	10	33
10	33	10	33
9	35.5		$\Sigma R_2 = 258$
9	35.5		
8	37.5		
8	37.5		
7	39		
6	40		
3	41		
	$\Sigma R_1 = 603$		

$$U = (N_1)(N_2) + \frac{N_1(N_1 + 1)}{2} - \Sigma R_1 = (24)(17) + \frac{(24)(25)}{2} - 603 = 105$$

$$\mu_U = \frac{(N_1)(N_2)}{2} = \frac{(24)(17)}{2} = 204$$

$$\sigma_U = \sqrt{\frac{(N_1)(N_2)(N_1 + N_2 + 1)}{12}} = \sqrt{\frac{(24)(17)(42)}{12}} = 37.79$$

$$z = \frac{(U + c) - \mu_U}{\sigma_U} = \frac{105 + 0.5 - 204}{37.79} = -2.61$$

The z score of -2.61 led to rejection of the null hypothesis, so the psychology major returned to the original data and their means to interpret the results. Because the mean rank of the women, 15 ($258 \div 17$), is greater than that of the men, 25 ($603 \div 24$), and because higher ranks (those closer to 1) mean more recollections, she concluded that women recalled significantly more items than men did.

Her conclusion singles out central tendency for emphasis. On the average, women did better than men. The Mann–Whitney test, however, compares distributions. What our undergraduate has done is what most researchers who use the Mann–Whitney do: Assume that the two populations have the same form but differ in central tendency. Thus, when a significant U value is found, it is common to attribute it to a difference in central tendency.

error detection

Here are two checks you can easily make. First, the last rank will be the sum of the two N's. In Table 14.3, $N_1 + N_2 = 41$, which is the lowest rank.

Second, when ΣR_1 and ΣR_2 are added, the sum will equal $N(N + 1)/2$, where N is the total number of scores. In Table 14.3, $603 + 258 = 861 = (41)(42)/2$.

From the information in the error detection box, you can see how I found the expected sum of 105 in the section on the rationale of nonparametric tests. There are 20 scores, so the overall sum of the ranks is

$$\frac{(20)(21)}{2} = 210$$

Half of this total should be found in each group, so the *expected* sum of ranks of each group, both of which come from the same population, is 105.

PROBLEMS

5. To determine the effects of estrogen on dominance, seven rats were given injections of that female hormone. The control group ($N = 8$) was injected with sesame oil, which is inert. Each rat engaged in a series of contests in a narrow runway. The rat that pushed the other one out was considered the more dominant of the two (see Work and Rogers, 1972). The number of bouts each won is given. Those marked with asterisks are the estrogen-injected rats. Analyze the results with a Mann–Whitney test and write a conclusion about the effects of estrogen.

14, 13, 12, 12, *11, 9, 8, *7, *6, 4, 3, *2, *2, *1, *0

*6. A friend of yours is trying to convince a mutual friend that the ride in an automobile built by (Toyota, Honda [you choose]) is quieter than the ride in an automobile built by (Honda, Toyota [no choice this time; you had just 1 degree of freedom — once the first choice was made, the second was determined]). This friend has arranged to borrow six fairly new cars — three made by each company — and to drive your blindfolded mutual friend around in the six cars (labeled A–F) until a stable ranking for quietness is achieved. So convinced is your friend that he insists on adopting $\alpha = .01$, "so that only the recalcitrant will not be convinced."

As the statistician in the group, you decide that a Mann–Whitney U test is appropriate. However, being the type who thinks through the statistics *before* gathering experimental data, you look at the appropriate subtable in Table H and find that the experiment, as designed, is doomed to retain the null hypothesis. Write an explanation, explaining why the experiment must be redesigned.

7. Suppose your friend in problem 6 talked three more persons out of their cars for an afternoon, labeled the cars A–I, changed α to .05, and conducted the "quiet" test. The results are shown in the table. Perform a Mann–Whitney U test.

Company Y		Company Z	
Car	ranks	Car	ranks
B	1	H	3
I	2	A	5
F	4	G	7
C	6	E	8
		D	9

8. Grackles, commonly referred to as blackbirds, are hosts for a parasitic worm that lives in the tissue around the brain. To see if the incidence of this parasite was changing, 24 blackbirds were captured and the number of brain parasites in each bird was recorded. These data were compared with the infestation of 16 birds captured from the same place 10 years earlier. Analyze the data with a Mann–Whitney test and write a conclusion. Be careful and systematic in assigning ranks. Errors here are quite frustrating.

Present day

20	16	12	11	51	8	23	68	23	44	0	78
0	28	53	20	44	20	36	32	64	16	101	0

Ten years earlier

16	19	43	90	16	72	29	62
103	39	70	29	110	32	87	57

THE WILCOXON MATCHED-PAIRS SIGNED-RANKS T TEST

The **Wilcoxon matched-pairs signed-ranks T test** (1945) is appropriate for testing the difference between two correlated samples. In Chapter 9 you learned of three kinds of correlated-samples designs: natural pairs, matched pairs, and repeated measures (before and after). In each of these designs, a score in one group is logically paired with a score in the other group. If you are not sure of your understanding of the difference between a correlated-samples and an independent-samples design, you should review the explanations and problems in Chapter 9. Knowing the difference is necessary when you decide between a Mann–Whitney U test and a Wilcoxon matched-pairs signed-ranks T test.

The result of a Wilcoxon matched-pairs signed-ranks test is a **T value,**[7] which is interpreted using **Table J.** Finding T involves some steps that are different from finding U, so be alert.

Table 14.4 provides you with a few numbers from a correlated-samples design. Using **Table 14.4,** work through the following steps, which lead to a T value for a Wilcoxon matched-pairs signed-ranks T:

1. Find a difference, D, for every pair of scores. The order of subtraction doesn't matter, but it *must* be the same for every pair.
2. Using the *absolute value* for each difference, rank the differences. The rank of 1 is given to the *smallest* difference, 2 goes to the next smallest, and so on.
3. Attach to each rank the sign of its difference. Thus, if a difference produces a negative value, the rank for that pair is negative.
4. Sum the positive ranks and sum the negative ranks.
5. T is the *smaller of the absolute values* of the two sums.[8]

Two cautions: For the Wilcoxon matched-pairs signed-ranks T test, it is the *differences* that are ranked, and *not the scores* themselves. For this test, a rank of 1 always goes to the smallest difference.

The rationale of the Wilcoxon matched-pairs signed-ranks T test is that, *if* there is no true difference between the two populations, the absolute value of the negative sum should be equal to the positive sum, with any deviations being due to sampling fluctuations.

Table J shows the critical values for both one- and two-tailed tests for several α levels. To enter the table, use N, the number of *pairs* of subjects. Reject H_0 when the obtained T is *equal to* or *less than* the critical value in the table. Like the Mann–Whitney test, your obtained statistic (T) must be equal to or less than the tabled critical value if you are to reject H_0.

To illustrate the calculation and interpretation of a Wilcoxon matched-pairs signed-ranks T test, here is an experiment based on some early work of Muzafer Sherif (1935). Sherif was interested in whether a person's basic perception could be influenced by others. The basic perception he used was a judgment of the size of the *autokinetic effect.* The autokinetic effect is obtained when a person views a stationary point of light in an otherwise dark room. After a few moments, the light appears to move erratically. Sherif asked his participants to judge how many inches the light moved. Under such conditions, judgments differ widely among individuals but they are fairly consistent for each individual.

---

[7] Be alert when you see a capital T in your outside readings; it has uses other than to symbolize the Wilcoxon matched-pairs signed-ranks test. Also note that this T is capitalized, whereas the t in the t test and t distribution is not capitalized except by some computer printers. Some writers avoid these problems by designating the Wilcoxon matched-pairs signed-ranks test with a W or a W_s.

[8] The Wilcoxon test is like the Mann–Whitney test in that you have a choice of two values for your test statistic. For both tests, choose the smaller value.

TABLE 14.4 Illustration of how to calculate *T*

Pair	Variable 1	Variable 2	D	Rank	Signed rank
A	16	24	−8	3	−3
B	14	17	−3	1	−1
C	23	18	5	2	2
D	23	9	14	4	4

$$\Sigma(\text{positive ranks}) = 6$$
$$\Sigma(\text{negative ranks}) = -4$$
$$T = 4$$

In the first phase of Sherif's experiment, participants worked alone. They estimated the distance the light moved and, after a few judgments, their estimates stabilized. These are the before measurements. Next, additional observers were brought into the dark room and everyone announced aloud their perceived movement. These additional observers were confederates of Sherif, and they always judged the movement to be somewhat less than that of the participant. Finally, the confederates left and the participant made another series of judgments (the after measurements). The perceived movements (in inches) of the light are shown in Table 14.5 for 12 participants.

Table 14.5 also shows the calculation of a Wilcoxon matched-pairs signed-ranks *T* test. The *D* column is simply the before measurements minus the after measurements. These *D* scores are then ranked by absolute size and the sign of the difference attached in the Signed ranks column. Notice that when $D = 0$, that pair of scores is dropped from further analysis and *N* is reduced by 1. The negative ranks have the smaller sum, so $T = 4$.

This *T* is less than the *T* value of 5 shown in Table J under $\alpha = .01$ (two-tailed test) for $N = 11$. Thus, the null hypothesis is rejected. The after scores represent a distribution that is different from the before scores. Now let's interpret this in terms of the experiment.

By examining the *D* column, you can see that all scores except two are positive. This means that, after hearing others give judgments smaller than their own, the participants saw the amount of movement as less. Thus, you may conclude (as did Sherif) that even basic perceptions tend to conform to perceptions expressed by others.

Tied Scores and *D* = 0

Ties among the *D* scores are handled in the usual way — that is, each tied score is assigned the mean of the ranks that would have been assigned if there had been no ties. Ties do not affect the probability of the rank sum unless they are numerous (10 percent or more of the ranks are tied). When there are numerous ties, the probabilities in Table J associated with a given critical *T* value may be too large. In this case, the test

TABLE 14.5 Wilcoxon matched-pairs signed-ranks analysis of the effect of peers on perception

Participant	Mean movement (in.) Before	Mean movement (in.) After	D	Signed ranks
1	3.7	2.1	1.6	4.5
2	12.0	7.3	4.7	10
3	6.9	5.0	1.9	6
4	2.0	2.6	−0.6	−3
5	17.6	16.0	1.6	4.5
6	9.4	6.3	3.1	8
7	1.1	1.1	0.0	Eliminated
8	15.5	11.4	4.1	9
9	9.7	9.3	0.4	2
10	20.3	11.2	9.1	11
11	7.1	5.0	2.1	7
12	2.2	2.3	−0.1	−1

$$\Sigma(\text{positive}) = 62$$
$$\Sigma(\text{negative}) = -4$$
$$T = 4$$
$$N = 11$$

Check: $62 + 4 = 66$ and $\dfrac{11(12)}{2} = 66$

is described as too conservative because it may fail to ascribe significance to differences that are in fact significant (Wilcoxon and Wilcox, 1964).

As you already know from Table 14.5, when *one* of the D scores is zero, it is not assigned a rank and N is reduced by 1. When *two* of the D scores are tied at zero, each is given the average rank of 1.5. Each is kept in the computation, with one being assigned a plus sign and the other a minus sign. If *three* D scores are zero, one is dropped, N is reduced by 1, and the remaining two are given signed ranks of $+1.5$ and -1.5. You can generalize from these three cases to situations with four, five, or more zeros.

Wilcoxon Matched-Pairs Signed-Ranks T Test for Large Samples

When the number of pairs exceeds 50, the T statistic may be evaluated using the z test and the normal curve. The test statistic is

$$z = \frac{(T + c) - \mu_T}{\sigma_T}$$

where T = smaller sum of the signed ranks

   $c = 0.5$

   $\mu_T = \dfrac{N(N + 1)}{4}$

$$\sigma_T = \sqrt{\frac{N(N + 1)(2N + 1)}{24}}$$

$N =$ number of pairs

PROBLEMS

9. A private consultant was asked to evaluate a job-retraining program. As part of the evaluation, she determined the income of 112 individuals before and after retraining. She found a T value of 4077. Complete the analysis and draw a conclusion. Be especially careful in wording your conclusion.

10. Six industrial workers were chosen for a study of the effects of rest periods on production. Output was measured for one week before the new rest periods were instituted and again during the first week of the new schedule. Perform an appropriate nonparametric statistical test on the results shown in the table.

	Output	
Worker	Without rests	With rests
1	2240	2421
2	2069	2260
3	2132	2333
4	2095	2314
5	2162	2297
6	2203	2389

11. A political science student was interested in the differences between two sets of North Americans in attitudes toward government regulation of business. One set was made up of Canadians and the other of people from the United States. At a model United Nations session, the student managed to get 30 participants to fill out his questionnaire. High scores on this questionnaire indicated a favorable attitude toward government regulation of business. Analyze the data in the table with the appropriate nonparametric test and write a conclusion.

Canadians	United States	Canadians	United States
12	16	27	26
39	19	33	15
34	6	34	14
29	14	18	25
7	20	31	21
10	13	17	30
17	28	8	3
5	9		

12. A professor always gave his general psychology class a 50-item true/false test on the first day of class. The purpose was to measure the students' beliefs about punishment, reward, mental breakdowns, and so forth — topics that would be covered during the course. Many of the items were phrased so that they represented a common misconception ("mental breakdowns run in families and are usually caused by defective genes"). At the end of the course, the same test was given again. High scores mean lots of misconceptions. Analyze the before and after data and write a conclusion about the effects of the general psychology course on misconceptions.

Student	Before	After
1	18	4
2	14	14
3	20	10
4	6	9
5	15	10
6	17	5
7	29	16
8	5	4
9	8	8
10	10	4
11	26	15
12	17	9
13	14	10
14	12	12

13. A health specialist conducted an 8-week workshop on weight control during which all 17 of the people who completed the course lost weight. To assess the long-term effects of the workshop, she weighed the participants again 10 months later. The weight lost or gained is listed with a positive sign for those who continued to lose weight and a negative sign for those who gained some back. What can you conclude about the long-term effects of the workshop?

−5	24	0	13	9	6	−7	−3	2
−10	−16	7	12	−19	−4	8	15	

THE WILCOXON–WILCOX MULTIPLE-COMPARISONS TEST

So far in this chapter on the analysis of ranked data, you have covered both designs for the two-group case (independent and correlated samples). The next technique will analyze data from three or more independent groups. The method presented here is one that allows you to compare all possible *pairs* of groups, regardless of the number of groups in the experiment. This is the nonparametric equivalent of a one-way ANOVA followed by Tukey HSD tests.[9]

The **Wilcoxon–Wilcox multiple-comparisons test** (1964) allows you to compare all possible pairs of treatments. This is like running several Mann–Whitney tests on each pair of treatments. However, the Wilcoxon–Wilcox multiple-comparisons test keeps the α level at .05 or .01, regardless of how many pairs you have. Like the Mann–Whitney U test, this test requires independent samples.[10] (Remember that Wilcoxon devised a test identical to the Mann–Whitney U test.)

To create a Wilcoxon–Wilcox multiple-comparisons test, begin by ordering the scores from the K treatments into one overall ranking. Then, within each sample, add the ranks, which gives a ΣR for each sample. Finally, for each pair of treatments, subtract one ΣR from the other, which gives a difference. Finally, compare the absolute size of the difference to a critical value in Table K.

The rationale of the Wilcoxon–Wilcox multiple-comparisons test is that when the null hypothesis is true, the various ΣR values should be about the same. A *large* difference indicates that the two samples came from different populations. Of course, the larger K is, the greater the likelihood of large differences by chance alone, and this is taken into account in the sampling distribution that Table K is based on.

The Wilcoxon–Wilcox test can be used only when the N's for all groups are equal. A common solution to the problem of unequal N's is to reduce the too-large group(s) by throwing out one or more randomly chosen scores. A better solution is to conduct the experiment so that you have equal N's.

The data in **Table 14.6** represent the results of an experiment conducted on a solar collector by two designer-entrepreneurs. These two had designed and built a 4-foot by 8-foot solar collector they planned to market, and they wanted to know the optimal rate at which to pump water through the collector. The rule of thumb for pumping is one-half gallon per hour per square foot of collector, so they chose values of 14, 15, 16, and 17 gallons per hour for their experiment. Starting with the reservoir full of ice water, the water was pumped for 1 hour through the collector and back to the reservoir. At the end of the hour, the temperature of the water in the reservoir was measured in degrees centigrade. Then the water was replaced with ice water, the flow rate was changed, and the process repeated. The numbers in the body of Table 14.6 are the temperature measurements (to the nearest tenth of a degree).

In Table 14.6 ranks are given to each temperature, ignoring the group the temperature is in. The ranks of all those in a group are summed, producing ΣR values that range from 20 to 84 for flow rates of 14 to 17 gallons per hour. Note that you can confirm your arithmetic for the Wilcoxon–Wilcox test using the same checks you used for the Mann–Whitney U test. That is, the sum of the four group sums, 210, is equal to $N(N + 1)/2$, where N is the total number of observations. Also, the largest rank, 20, is equal to the total number of observations.

The next step is to make pairwise comparisons. With four groups, six pairwise comparisons are possible. The rate of 14 gallons per hour can be paired with 15, 16,

---

[9] The Kruskal–Wallis one-way ANOVA on ranks is a direct analogue of a one-way ANOVA. (See Howell, 1999, pp. 405–406, or McCall, 1998, pp. 411–416.) Unfortunately, tests that compare treatments (as the Tukey HSD does) are not readily available.

[10] A nonparametric test for more than two correlated samples, the Friedman rank test for correlated samples, is explained in Howell (1999, pp. 406–408).

TABLE 14.6 Temperature increases (°C) for four flow rates in a solar collector

			Flow rate (gallons/hour)				
14	Rank	15	Rank	16	Rank	17	Rank
28.1	7	28.9	3	25.1	14	24.7	15
28.8	4	27.7	8.5	25.3	13	23.5	18
29.4	1	27.7	8.5	23.7	17	22.6	19
29.0	2	28.6	5	25.9	11	21.7	20
28.3	6	26.0	10	24.2	16	25.8	12
Σ (ranks)	20		35		71		84

Check: $20 + 35 + 71 + 84 = 210$

$$\frac{N(N + 1)}{2} = \frac{20(21)}{2} = 210$$

and 17; the rate of 15 can be paired with 16 and 17; and the rate of 16 can be paired with 17. For each pair, a difference in the sum of ranks is found and the absolute value of that difference is compared with the critical value in Table K.

The upper half of **Table K** has critical values for $\alpha = .05$; the lower half has values for $\alpha = .01$. For the experiment in Table 14.6, where $K = 4$ and $N = 5$, the critical values are 48.1 ($\alpha = .05$) and 58.2 ($\alpha = .01$). To reject H_0, a difference in rank sums must be equal to or greater than the critical values.

It is convenient (and conventional) to arrange the differences in rank sums into a matrix summary table. **Table 14.7** is an example using the flow-rate data. The table displays for each pair of treatments, the difference in the sum of the ranks. Asterisks are used to indicate differences that exceed various α values. Thus, in Table 14.7, with $\alpha = .05$, rates of 14 and 16 are significantly different from each other, as are rates of 15 and 17. In addition, a rate of 14 is significantly different from a rate of 17 at the .01 level. What does all this mean for our two designer-entrepreneurs? Let's listen to their explanation to their old statistics professor.

TABLE 14.7 Summary table for differences in the sums of ranks in the flow-rate experiment

		Flow rates		
		14	15	16
Flow rates	15	15		
	16	51*	36	
	17	64**	49*	13

* $p < .05$; for $\alpha = .05$, the critical value is 48.1
** $p < .01$; for $\alpha = .01$, the critical value is 58.2

"How did the flow-rate experiment come out, fellows?" inquired the kindly old gentleman.

"OK, but we are going to have to do a follow-up experiment using different flow rates. We know that 16 and 17 gallons per hour are not as good as 14, but we don't know if 14 is optimal for our design. Fourteen was the best of the rates we tested, though. On our next experiment, we are going to test rates of 12, 13, 14, and 15."

The professor stroked his beard and nodded thoughtfully. "Typical experiment. You know more after it than you did before, . . . but not yet enough."

PROBLEMS

14. Farmer Marc, a Shakespearean devotee, delivered a plea at the County Faire, asking three different groups to lend him unshelled corn. (Perhaps you can figure out how he phrased his plea.) The numbers of bushels offered by the farmers are listed. What can you conclude about the three groups?

Friends	Romans	Countrymen
25	14	8
31	18	15
17	10	11
22	16	7
27	9	13
29	10	12

15. Many studies have investigated methods of reducing anxiety and depression. I've embedded some of their conclusions in this problem. College students who reported anxiety or depression were randomly assigned to one of four groups: no treatment (NO), relaxation (RE), solitary exercise (SE), and group exercise (GE). The numbers are the students' improvement scores after 10 weeks. Analyze the data and write a conclusion.

NO	RE	SE	GE
10	14	12	22
8	10	27	32
3	21	23	20
0	10	20	28
7	15	17	22

16. The effect of three types of leadership on group productivity and satisfaction was investigated.[11] Groups of five children were randomly constituted and assigned an authoritarian, a democratic, or a laissez-faire leader. Nine groups were formed — three with each type of leader. The groups worked for a week

[11] For a summary of a similar investigation, see Lewin (1958).

on various projects. During this time, four different measures of each child's personal satisfaction were taken. These were pooled to give one score for each child. The data are presented in the table. Test all possible comparisons with a Wilcoxon–Wilcox test.

Authoritarian	Democratic	Laissez-faire
120	108	100
102	156	69
141	92	103
90	132	76
130	161	99
153	90	126
77	105	79
97	125	114
135	146	141
121	131	82
100	107	84
147	118	120
137	110	101
128	132	62
86	100	50

17. Given the summary data in the table, test all possible comparisons. Each group had a sample of eight subjects.

Sum of ranks for six groups						
Group	1	2	3	4	5	6
ΣR	196	281	227	214	93	165

18. List the tests presented so far in this chapter and the design for which each is appropriate. Be sure you can do this from memory.

CORRELATION OF RANKED DATA

Here is a short review of what you learned about correlation in Chapter 5.

1. Correlation requires a bivariate distribution (a logical pairing of scores).
2. Correlation is a method of describing the degree of relationship between two variables — that is, the degree to which high scores on one variable are associated with low or high scores on the other variable.

3. Correlation coefficients range in value from $+1.00$ (perfect positive) to -1.00 (perfect negative). A value of .00 indicates that there is no relationship between the two variables.
4. Statements about cause and effect may not be made on the basis of a correlation coefficient alone.

In 1901, Charles Spearman (1863–1945) was "inspired by a book by Galton" and began experimenting at a little village school. He thought there was a relationship between intellect (school grades) and sensory ability (detecting musical discord). Spearman worked to find a mathematical way to express the *degree* of this relationship. He came up with a coefficient that did this, although he later found that others were ahead of him in developing such a coefficient (Spearman, 1930).

Spearman's name is attached to the coefficient that is used to show the degree of correlation between two sets of *ranked* data. He used the Greek letter ρ (rho) as the symbol for his coefficient. Later statisticians began to reserve Greek letters to indicate parameters, so the modern symbol for Spearman's statistic has become r_s, the s honoring Spearman. **Spearman r_s** is a special case of the Pearson product-moment correlation coefficient and is most often used when the number of pairs of scores is small (less than 20).

Actually, r_s is a *descriptive statistic* that could have been introduced in the first part of this book. I waited until now to introduce it because r_s is a rank-order statistic, and this is a chapter about ranks. The next section will show you how to calculate this descriptive statistic; determining the statistical significance of r_s will follow.

Calculation of r_s

The formula for r_s is

$$r_s = 1 - \frac{6\Sigma D^2}{N(N^2 - 1)}$$

where $D =$ difference in ranks of a pair of scores
$N =$ number of pairs of scores

I started this chapter with speculation about male tennis players and a made-up data set. I will end it with speculation about female tennis players, but now I have actual data.

What is the progression in women's professional tennis? Do they work their way up through the ranks, advancing their ranking year by year? Or do the younger players flash to the top and then gradually lose their ranking year by year? An r_s might lend support to one of these hypotheses.

If a rank of 1 is assigned to the oldest player, a positive r_s means that the older the player, the higher her rank (supporting the first hypothesis). A negative r_s means that the older the player, the lower her rank (supporting the second hypothesis). A zero r_s means no relationship between age and rank.

Table 14.8 shows the ten top-ranked women tennis players for 1998, their age as a rank score among the ten, and an r_s of $-.20$. Thus, the descriptive statistic lends mild support for the "youngsters flash to the top" hypothesis. But let's add inferential statistics to our analysis: Can you rule out chance as an explanation for an r_s of $-.20$?

TABLE 14.8 The top ten women tennis players in 1998; their rank in age, and the calculation of Spearman r_s

Player	Rank in tennis	Rank in age	D	D^2
Davenport (USA)	1	7	−6	36
Hingis (Switzerland)	2	10	−8	64
Novotna (Czech Republic)	3	2	1	1
Sanchez-Vicario (Spain)	4	3	1	1
Williams (USA)	5	9	−4	16
Seles (USA)	6	5	1	1
Pierce (France)	7	6	1	1
Tauziat (Fance)	8	1	7	49
Martinez (Spain)	9	4	5	25
Schnyder (Switzerland)	10	8	2	4
				$\Sigma = 198$

$$r_s = 1 - \frac{6\Sigma D^2}{N(N^2 - 1)} = 1 - \frac{6(198)}{10(99)} = 1 - 1.20 = -.20$$

error detection

For Spearman r_s, assign ranks to the scores first and then find the differences. The Wilcoxon matched-pairs signed-ranks T test is different. In it differences are found and ranks assigned to the differences.

Testing the Significance of r_s

For a sample r_s, you can test the null hypothesis that the population correlation coefficient is .00. That is, a sampling distribution of r_s values taken from a population in which the true correlation is zero will show you the probability of various r_s's. If the probability for your sample r_s is small, reject the null hypothesis and conclude that the relationship in the population that r_s came from is *not* zero. You probably recall that you conducted this same test for a Pearson product-moment r in Chapter 8.

Table L in Appendix C gives critical values for r_s for the .05 and .01 levels of significance (when the number of pairs is 16 or fewer). To use Table L, reject the hypothesis that the population correlation coefficient is zero if the obtained r_s is *equal to* or *greater than* the tabled value.

The tennis data in Table 14.8 produced $r_s = -.20$ based on ten pairs. **Table L** shows that a correlation of .648 (either positive or negative) is required for significance at the .05 level. Thus, a correlation of −.20 is not statistically significant.

Unfortunately, testing the significance of the r_s leaves you without support for either of the two initial hypotheses. At this point researchers usually (1) gather more data, (2) change their hypotheses, or (3) move on to other problems.

Notice in Table L that rather large correlations are required for significance. As with r, not much confidence can be placed in low or moderate correlation coefficients that are based on only a few pairs of scores.

For samples larger than 16, test the significance of r_s by using Table A. Note, however, that **Table A** requires df. For r_s, df is $N - 2$ (the number of pairs minus 2), which is the same formula used for r.

When $D = 0$ and Tied Ranks

In calculating r_s, you may get a D value of zero. Should a zero be dropped from further analysis as it is in a Wilcoxon matched-pairs signed-ranks T test? You can answer this question by taking a moment to decide what $D = 0$ means in an r_s correlation problem.

A zero means there is a perfect correspondence between the two ranks. If all differences were zero, the r_s would be 1.00. Thus, differences of zero should *not* be dropped when calculating an r_s.

The formula for r_s that you are working with is not designed to handle ranks that are tied. With r_s, ties are troublesome. A tedious procedure has been devised to overcome ties, but your best solution is to arrange your data collection so there are no ties. Sometimes, however, you are stuck with tied ranks, perhaps as a result of working with someone else's data. Kirk (1999) recommends assigning average ranks to ties, as you did for the three other procedures in this chapter, and then computing a Pearson r on the data.

MY FINAL WORD

In the first chapter I said that the essence of applied statistics is to let numbers stand for things of interest, manipulate the numbers according to the rules of statistics, translate the numbers back into the things, and finally describe the relationships among the things of interest. The question What things are better understood when translated into numbers? was not raised in this book, but it is important nonetheless. Here is an anecdote by E. F. Schumacher (1979), an English economist, that helps explain that importance.

> I will tell you a moment in my life when I almost missed learning something. It was during the war and I was a farm laborer and my task was before breakfast to go to yonder hill and to a field there and count the cattle. I went and I counted the cattle — there were always thirty-two — and then I went back to the bailiff, touched my cap, and said "Thirty-two, sir," and went and had my breakfast. One day when I arrived at the field an old farmer was standing at the gate, and he said, "Young man, what do you do here every morning?" I said, "Nothing much. I just count the cattle." He shook his head and said, "If you count them every day they won't flourish." I went back, I reported thirty-two, and on the way back I thought, Well, after all, I am a professional statistician, this is only a country yokel, how stupid can he get. One day I went back, I counted and counted again, there were only thirty-one. Well, I didn't want to spend all day there so I went back and reported thirty-one. The bailiff was very angry. He said, "Have your breakfast and then we'll go up there together." And we went together and we searched the place and indeed, under a bush, was a dead beast. I thought to myself, Why have I been counting them all the time? I haven't prevented this beast dying. Perhaps that's what the farmer meant. They won't flourish if you don't look and watch the quality of each individual beast. Look him in the eye. Study the sheen on his coat. Then I might have gone back and said, "Well, I don't know how many I saw but one looks mimsey."

PROBLEMS

19. For each situation described, tell whether you would use a Mann–Whitney test, a Wilcoxon matched-pairs signed-ranks test, a Wilcoxon–Wilcox multiple-comparisons test, or an r_s.

a. A limnologist (a scientist who studies freshwater streams and lakes) measured algae growth in a lake before and after the construction of a copper smelting plant to see what effect the plant had.

b. An educational psychologist compared the sociability scores of firstborn children with scores of their next-born brother or sister to see if the two groups differed in sociability.

c. A child psychologist wanted to determine the degree of relationship between eye-hand coordination scores obtained at age 6 and scores obtained from the same individuals at age 12.

d. A nutritionist randomly and evenly divided boxes of cornflakes cereal into three groups. One group was stored at 40°F, one group at 80°F, and one group alternated from day to day between the two temperatures. After 30 days, the vitamin context of each box was assayed.

e. The effect of STP gas treatment on gasoline mileage was assessed by driving six cars over a 10-mile course, adding STP, and then again driving over the same 10-mile course.

20. Fill in the cells of the table.

	Symbol of statistic (if any)	Appropriate for what design?	Reject H_0 when statistic is (greater, less) than critical value?
Mann–Whitney test			
Wilcoxon matched-pairs signed-ranks test			
Wilcoxon–Wilcox multiple-comparisons test			

21. Two members of the department of philosophy (Locke and Kant) were responsible for hiring a new professor. Each privately ranked from 1 to 10 the ten candidates who met the objective requirements (degree, specialty, and so forth). The rankings are shown in the table. Begin by writing an interpretation of a low correlation and a high correlation. Then calculate an r_s and choose the correct interpretation.

Candidates	Locke	Kant
A	7	8
B	10	10
C	3	5
D	9	9
E	1	1
F	8	7
G	5	3
H	2	4
I	6	6
J	4	2

22. Calculate an appropriate correlation coefficient for the hypothetical data in the table. Write an interpretation.

Years	Number of marriages ($\times 10,000$)	Total grain crop ($ billions)
1918	131	1.7
1919	142	2.5
1920	125	1.4
1921	129	1.7
1922	145	2.8
1923	151	3.3
1924	142	2.6
1925	160	3.7
1926	157	3.4
1927	163	3.5
1928	141	2.6
1929	138	2.6
1930	173	3.6
1931	166	3.9

23. With $N = 16$ and $\Sigma D^2 = 308$, calculate r_s and test its significance at the .05 level.

24. You are once again asked to give advice to a friend who comes to you for criticism of an experimental design. This friend has four pairs of scores obtained randomly from a population. Her intention is to calculate a correlation coefficient r_s and decide whether there is any significant correlation in the population. Give advice.

25. One last time: Review the objectives at the beginning of the chapter.

WEBSITE REFERENCES

http://psychology.wadsworth.com

Open "Choose Your Quiz" and select Spatz, 7e. Find the index of chapters, and click on Chapter 14, *Nonparametric Statistics.*

WHAT WOULD YOU RECOMMEND? chapters ten–fourteen

Here is the last set of *What would you recommend?* problems. Your task is to choose an appropriate statistical technique from among the several you have learned in the previous five chapters. No calculations are necessary. For each problem that follows, (1) recommend a statistic that will either answer the question or make the comparison, and (2) explain why you recommend that statistic.

a. One anxiety-arousing behavior is strapping on a parachute and jumping out of an airplane. Flooding and systematic desensitization are the names of two techniques that help people overcome anxiety about specific behaviors. To compare the two techniques, researchers had military trainees experience one technique or the other. Later, at the time of the first five actual jumps, each trainee's "delay time" was measured. (The assumption was that longer delay means more anxiety.) On the basis of the severely skewed distribution of delay times, each trainee was given a rank. What statistical test should be used to compare the two techniques?

b. Stanley Milgram researched the question When do people submit to demands for compliance? In one experiment, participants were told to administer a shock to another participant, who was in another room. (No one was actually shocked in this experiment.) The experimenter gave directions to half the participants over a telephone, the other half received directions in person. The dependent variable was whether or not a participant complied with the directions.

c. Forensic psychologists study factors that influence people as they judge defendents in criminal cases. Berry and McArthur (1986) supplied participants in their study with a fictitious "pretrial report" and asked them to estimate the likelihood, on a scale of .00 to 1.00, that the defendant would be convicted. The pretrial reports contained a photograph showing a person who was either baby-faced or mature-faced. In addition, half the reports indicated that the defendant was simply negligent in the alleged incident and half indicated that the defendant had been intentionally deceptive in the incident.

d. As part of a class project, two students administered a creativity test to undergraduate seniors. They also obtained each person's rank in class. In what way could the two students determine the degree of relationship between the two variables?

e. Suppose a theory predicts that the proportion of people who participate in a rather difficult exercise will drop by one-half each time the exercise is repeated. When the data were gathered, 100 people participated on the first day. On each of the next three occasions, fewer and fewer people participated. What statistical test could be used to arrive at a judgment about the theory?

f. Participants in an experiment were told to "remember what you hear." A series of sentences followed. (An example was "The plumber slips $50 into his wife's purse.") Later, when the participants were asked to recall the sentences, one group was given disposition cues (such as "generous"), one was given semantic (associative) cues (such as "pipes"), and one was given no cues at all. Recall measures ranged between 0 and 100. The recall scores for the "no cues" condition were much more variable than those for the other two conditions. The researcher's question was: Which of the three conditions produces the best recall? (See Winter, Uleman, and Cunniff, 1985.)

g. To answer a question about the effect of an empty stomach on eating, investigators in the 1930s studied dogs that ate a normal-sized meal (measured in ounces). Because of a surgical operation, however, no food reached the stomach. Thirty minutes later, the dogs were allowed to eat again. The data showed that the dogs, on the average, ate less the second time (even though their stomachs were just as empty). At both time periods, the data were positively skewed.

h. College cafeteria personnel are often dismayed by the amount of food that is left on plates. One way to reduce waste is to prepare the food using a tastier recipe. Suppose a particular vegetable is served on three different occasions, and each time a different recipe is used. As the plates are cleaned in the kitchen, the amount of the vegetable left on the plate is recorded. How might each pair of recipes be compared to determine if one recipe has resulted in significantly less waste?

i. Many programs seek to change behavior. One way to assess the effect of a program is to survey the participants about their behavior, administer the program, and give the survey again. To demonstrate the lasting value of the program, however, follow-up data some months later are required. If the data consist of each participant's three scores on the same 100-point survey (pretest, posttest, and follow-up), what statistical test is needed to determine the effect of the program?

CHOOSING TESTS AND WRITING INTERPRETATIONS

OBJECTIVES FOR CHAPTER FIFTEEN

Study the text. Work the problems as you come to them. Afterward, you should be able to:

1. List the descriptive and inferential statistical techniques that you studied in this book

2. Describe, using a table of characteristics or decision trees or both, the descriptive and inferential statistics that you studied in this book

3. Read the description of a study and identify a statistical technique that is appropriate for such data

4. Read the description of a study and recognize that you have not studied a statistical technique that is appropriate for such data

5. Given the statistical analysis of a study, write an interpretation of what the study reveals

This chapter is designed to help you consolidate the many bits and pieces of statistics that you have learned. It will help you, regardless of the number of chapters your course covered (as long as you completed through Chapter 9, Hypothesis Testing, Effect Size, and Confidence Intervals: Two-Sample Designs). First, there is a review of the major concepts of each chapter. The second section describes statistical techniques that you *did not* read about in this book. These two sections should help prepare you to apply what you have learned to the exercises in the last section of this chapter.

A REVIEW

If you have been reading for integration, the kind of reading in which you are continually asking the question, How does this stuff relate to what I learned before?, you may have clearly in your head a major thread that ties this book together from Chapter 2 onward. The name of that thread is *distributions*. It is the importance of distributions that justifies the subtitle of this book, *Tales of Distributions*. I want to summarize explicitly some of what you have learned about distributions. I hope this will help you integrate the material and facilitate your understanding of other statistical techniques that were not covered in this course (but that you might encounter in future courses or in future encounters with statistics).

In Chapter 2 you were confronted with a disorganized array of scores of one variable. You learned to organize those scores into a frequency distribution, graph the distribution, and find central tendency values. In Chapter 3 you learned about the variability of distributions — especially the standard deviation. Chapter 4 introduced the z score, effect size index, and boxplot, which are combination statistics that are useful for a Descriptive Statistics Report. In Chapter 5, where you were introduced to bivariate distributions, you learned to express the relationship between variables with a correlation coefficient and, using a linear regression equation, predict scores on one variable from scores on another. In Chapter 6 you learned about the normal distribution and how to use means and standard deviations to find probabilities associated with distributions. In Chapter 7 you were introduced to the important concept of the sampling distribution. You used two sampling distributions, the normal distribution and the t distribution, to determine probabilities about sample means and to establish confidence intervals about a sample mean. Chapter 8 presented you with the concept of hypothesis testing. You used the t distribution to test the hypothesis that a sample mean came from a population with a hypothesized mean, and you calculated the size of the effect. In Chapter 9 you studied the design of a simple two-group experiment. You applied the hypothesis-testing procedure to both independent-samples designs and correlated-samples designs, using the t distribution as a source of probabilities. You calculated effect sizes and for a second time, found confidence intervals. In Chapter 10 you learned about still another sampling distribution (F distribution) that is used when there are two or more groups in an experiment. Chapter 11 discussed an even more complex experiment, one with *two* independent variables for which probability figures could be obtained with the F distribution. The last of the three chapters on ANOVA (Chapter 12) covered the analysis of two or more correlated groups. Chapter 13 taught you to answer questions of independence and goodness of fit by using frequency counts and the chi square distribution. In Chapter 14 you analyzed data using nonparametric statistics — techniques that use distributions that do not require the assumptions that the populations of dependent-variable scores be normally distributed and have equal variances.

Look back at the preceding paragraph. You have learned a lot, even if your course did not cover *every* chapter. Allow me to offer my congratulations!

FUTURE STEPS

Several paths might be taken from here. One more or less obvious next step is to analyze an experiment with three independent variables. The three main effects and the several interactions are tested with an F distribution.

A second possibility is to study techniques for the analysis of experiments that have more than one *dependent* variable. Such techniques are called multivariate statistics. Many of these techniques are analogous to those you have studied in this book, except there are two or more dependent variables instead of just one. For example, Hotelling's T^2 is analogous to the t test; one-way multivariate analysis of variance (MANOVA) is analogous to one-way ANOVA; and higher-order MANOVA is analogous to factorial ANOVA.

Finally, there is the task of combining the results of many *separate* studies that all deal with the same topic. It is not unusual to have conflicting results among 10 or 15 or 50 studies, and researchers have been at the task of synthesizing since the beginning. Recently the task has been facilitated by a set of statistical procedures called **meta-analysis.**

CHOOSING TESTS AND WRITING INTERPRETATIONS

In Chapter 1 (page 18) I described the fourfold task of a statistics student:

1. Read a problem.
2. Decide what statistical procedure to use.
3. Perform that procedure.
4. Write an interpretation of the results

Students tell me that the parts of this task they still need work on at the end of the course are (2) deciding what statistical procedure to use and (4) writing an interpretation of the results.

Even though you may have some uncertainties, you are well on your way to being able to do all four tasks. What you may need at this point is exercises that help you "put it all together." Pick and choose from the four exercises that follow; the emphasis is heavily on the side of deciding what statistical test to use.

Exercise A: Reread all the chapters you have studied and all the tests and homework you have. If you choose to do this exercise (or part of it), spend most of your time on concepts that you recognize as especially important (such as hypothesis testing). It will take a considerable amount of time for you to do this — probably 8 to 16 hours. However, the benefits are great, according to students who have invested the time. A typical quote is "Well, during the course I would read a chapter and work the problems without too much trouble, but I didn't relate it to the earlier stuff. When I went back over the whole book, though, I got the big picture. All the pieces do fit together."

Exercise B: Construct a table with all the statistical procedures and their characteristics. List the procedures down the left side. I suggest column heads such as *Description* (brief description of procedure), *Purpose* (descriptive or inferential), *Null hypothesis* (parametric or nonparametric), *Number of independent variables* (if applicable), and so forth.

I've included my version of this table (Table 15.1), but please don't look at what I've done until you have a copy of your own. However, after you have worked for a while on your own, it wouldn't hurt to peek at the column heads I used.

Exercise C: Construct two decision trees. One should show all the descriptive statistics you have learned. To begin, you might divide these statistics into categories of univariate and bivariate distributions. Under univariate, you might list the different categories of descriptive statistics you have learned and then, under those categories,

TABLE 15.1 Summary table for elementary statistics

Procedure (text page numbers)	Purpose	Description	Null hypothesis	No. of IVs	No. of IV levels	Design
Frequency distributions (Chapter 2)	D	Picture of raw data	NA	NA	NA	NA
Boxplot (70)	D	Data picture that includes statistics	NA	NA	1, 2+	NA
Central tendency — Mean (38, 40)	D	One number to represent all	NA	NA	NA	NA
Median (39, 41)	D	One number to represent all	NA	NA	NA	NA
Mode (40, 42)	D	One number to represent all	NA	NA	NA	NA
Variability — Range (51)	D	Dispersion of scores	NA	NA	NA	NA
Standard deviation (58, 61)	D	Dispersion of scores	NA	NA	NA	NA
Variance (63)	D	Dispersion of scores	NA	NA	NA	NA
Interquartile range (52)	D	Dispersion of scores	NA	NA	NA	NA
z Score (67)	D	Relative position of one score	NA	NA	NA	NA
Regression (105)	D	Linear equation; 2 variables	NA	NA	NA	NA
Correlation — Pearson r (89)	D/I	Degree of relationship; 2 variables	P	NA	NA	Corr
Spearman r_s (319)	D/I	Degree of relationship; 2 variables	NP	NA	NA	Corr
Confidence interval (151, 200, 201)	D/I	Limits around mean or mean difference	P	1	NA	Indep
t test — One sample (166)	I	Sample $\overline{X}$ from pop. w/mean μ?	P	NA	NA	Indep
One-sample — effect size (173)	D	Assesses size of IV effect	P	1	NA	Indep
t test — Independent samples (190)	I	2 sample $\overline{X}$'s w/same μ?	P	1	2	Indep
Independent samples — effect size (199)	D	Assesses size of IV effect	P	1	2	Indep
t test — Correlated samples (194)	I	2 sample $\overline{X}$'s w/same μ?	P	1	2	Corr
Correlated samples — effect size (199)	D	Assesses size of IV effect	P	1	2	Corr
F test — Two variances (225)	I	2 sample variances w/same σ^2?	P	1	2	Indep
F test — One-way ANOVA (224)	I	≥2 sample $\overline{X}$'s w/same μ?	P	1	2+	Indep
One-way ANOVA — effect size (231)	D	Assesses size of IV effect	P	1	2+	Indep
F test — Factorial ANOVA (252)	I	2 or more IVs plus interactions	P	2+	2+	Indep
F test — Correlated measures (270)	I	≥2 sample $\overline{X}$'s w/same μ?	P	1	2+	Corr
Tukey HSD (229, 260, 271)	I	Tests all pairwise differences	P	1	2	Indep
Chi-square — goodness of fit (287)	I	Do data fit theory?	NP	NA	NA	Indep
Chi-square — independence (282)	I	Are variables independent?	NP	NA	NA	Indep
Mann–Whitney U (304)	I	2 samples from same population?	NP	1	2	Indep
Wilcoxon matched-pairs signed-ranks T (311)	I	2 samples from same population?	NP	1	2	Corr
Wilcoxon–Wilcox multiple comparisons (316)	I	Tests all pairwise differences	NP	1	2+	Indep

Key: Corr = Correlated   Indep = Independent   NP = Nonparametric
D = Descriptive   IV = Independent variable   P = Parametric
I = Inferential   NA = Not applicable

identify specific descriptive statistics. Under bivariate, choose categories under which specific descriptive statistics belong.

The second decision tree is for inferential statistics. You might begin with the categories of hypothesis testing and confidence intervals, or you might decide to divide inferential statistics into the type of statistics analyzed (means, frequency counts, and ranks). Work in issues such as the number of independent variables, whether the design is correlated samples or independent samples, and other issues that will occur to you.

There are several good ways you might organize your decision trees. My versions are Figure 15.1 and Figure 15.2, but please don't look at them now. Looking at the figures now would prevent you from going back over your text and digging out the information you learned earlier, which you now need for your decision trees. My experience has been that students who create their own decision trees learn more than those who are handed someone else's. After you have *your* trees in hand, *then* look at mine.

Exercise D: Work the two sets of problems that follow. The first set requires you to decide what statistical test or descriptive statistic will answer the question the investigator is asking. The second set requires you to interpret the results of an analyzed experiment. You will probably be glad that no number crunching is required for these problems. You won't need your calculator or computer.

However, here is a word of warning. For a few of the problems, the statistical techniques necessary for an analysis are not covered in this book. (In addition, perhaps your course did not cover all the chapters.) Thus, a part of your task is to recognize what problems you are prepared to solve and what problems would require digging into a statistics textbook.

PROBLEMS

Set A. Determine what descriptive statistic or inferential test is appropriate for each problem in this set.

1. A company has three separate divisions. Based on the capital invested, Division A made 10 cents per dollar, Division B made 20 cents per dollar, and Division C made 30 cents per dollar. How can the overall profit for the company be found?

2. Reaction time scores tend to be severely skewed. The majority of the responses are quick and there are diminishing numbers of scores in the slower categories. A student wanted to find out the effects of alcohol on reaction time, so he found the reaction time of each subject under both conditions — alcohol and no alcohol.

3. As part of a 2-week treatment for phobias, a therapist measured the general anxiety level of each client four times: before, after 1 week of treatment, at the end of treatment, and 4 months later.

4. "I want a number that will describe the typical score on this test. Most did very well, some scored in the middle, and a very few did quite poorly."

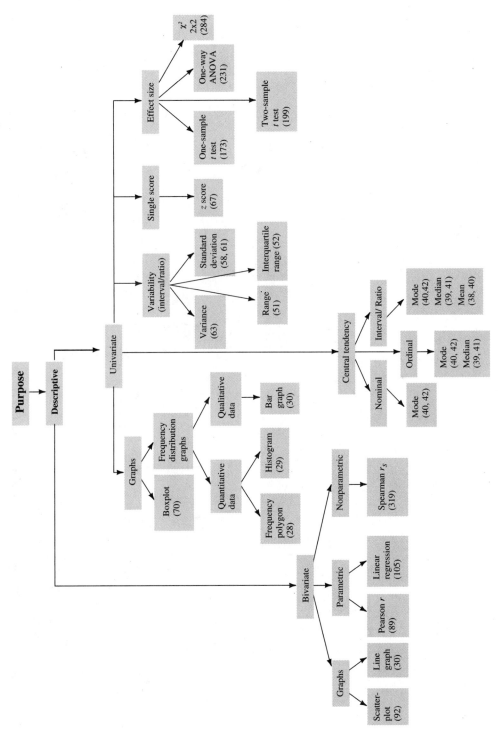

FIGURE 15.1 Decision tree for descriptive statistics (Text page numbers are in parentheses.)

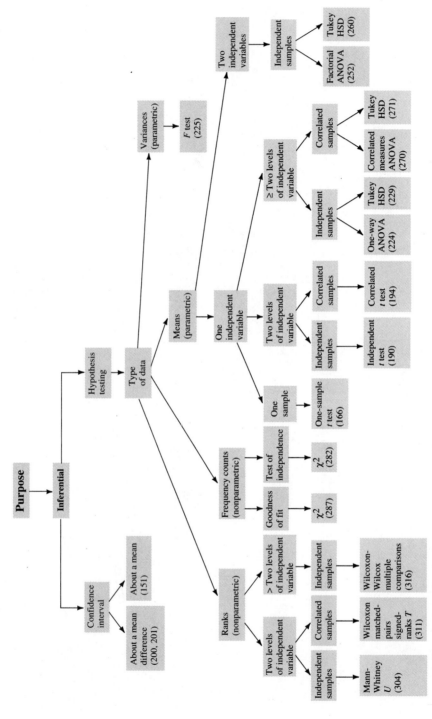

FIGURE 15.2 Decision tree for inferential statistics (Text page numbers are in parentheses.)

5. The city park benches were 16.7 inches high. The mean distance from the bottom of the shoe to the back of the knee is 18.1 inches for women. The standard deviation is 0.70 inch. What proportion of the women who use the bench will have their feet dangle (unless they sit forward on the bench)?

6. A measure of dopamine (a neurotransmitter) activity is available for 14 acute schizophrenic patients and 14 chronic schizophrenic patients. Do acute and chronic schizophrenics differ in dopamine activity?

7. A large school district needs to know the range of scores within which they can expect the mean reading achievement score of their sixth-grade students to fall. A random sample of 50 reading achievement scores is available.

8. Based on several large studies, a researcher knew that 40 percent of the public in the United States favored capital punishment, 35 percent were against it, and 25 percent expressed no strong feelings. The researcher polled 150 Canadians on this question to compare the two countries in attitudes toward capital punishment.

9. What descriptive statistic may be used to identify the most typical choice of major at your college?

10. A researcher wanted to express with a correlation coefficient the strength of the relationship between stimulus intensity and pleasantness. This researcher worked with music. She knew that both very low and very high intensities are not pleasant and that the middle range of intensity produces the highest pleasantness ratings. Would you recommend a Pearson r? An r_s?

11. A consumer testing group compared Boraxo and Tide to determine which got laundry whiter. White towels that had been subjected to a variety of filthy treatments were identified on each end and were cut in half. Each was then washed in either Boraxo or Tide. After washing, each half was tested with a photometer for the amount of light reflected.

12. An investigator wants to predict a male adult's height from his length at birth. He obtains records of both measures from a sample of male military personnel.

13. An experimenter is interested in the effect of expectations and drugs on alertness. Each participant is given an amphetamine (stimulant) or a placebo (an inert substance). In addition, half in each group are told that the drug taken was a stimulant and half that the drug was a depressant. An alertness score for each participant is obtained from a composite of measures that included a questionnaire and observation.

14. A professor told a class that she found a correlation coefficient of .20 between the alphabetical order of a person's last name and that person's cumulative point total for 50 students in a general psychology course. One student in the class wondered whether the correlation coefficient was reliable; that is, would a nonzero correlation be likely to be found in other classes? How can reliability be determined without calculating an r on another class?

15. As part of a test for some advertising copy, a company assembled 120 people who rated four toothpastes. They read several ads (including the ad being tested). Then they rated all four toothpastes again. The data to be analyzed consisted of the number of people who rated Crest highest before reading the ad and the number who rated it highest after reading the ad.

16. Most light bulb manufacturing companies claim that their 60-watt bulbs have an average life of 1000 hours. A skeptic with some skill as an electrician wired up some sockets and timers and burned 40 bulbs until they failed. The life of each bulb was recorded automatically. Is the companies' claim justified?

17. An investigator wonders if a person's education is related to his or her satisfaction in life. The investigator can measure both of these variables on a quantitative scale.

18. To find out the effect of a drug on reaction time, an investigator administered 0, 25, 50, or 75 milligrams to four groups of volunteer rats. The quickness of their response to a tipping platform was measured in seconds.

19. The experimenters (a man and a woman) staged 140 "shoplifting" incidents in a grocery store. In each case, the "shoplifter" (one experimenter) picked up a carton of cigarettes in full view of a customer and then walked out. Half the time the experimenter was well dressed and half the time sloppily dressed. Half the incidents involved a male shoplifter and half involved a female. For each incident, the experimenters (with the cooperation of the checkout person) simply noted whether the shoplifter was reported or not.

20. A psychologist wanted to illustrate for a group of seventh-graders the fact that items in the middle of a series are more difficult to learn than those at either end. The students learned an eight-item series. The psychologist found the mean number of repetitions necessary to learn each of the items.

21. A nutritionist, with the help of an anthropologist, gathered data on the incidence of cancer among 21 different cultures. The incidence data were quite skewed, with many cultures having low incidences and the rest scattered among the higher incidences. In eight of the cultures, red meat was a significant portion of the diet. For the other 13 groups, the diet consisted primarily of grains.

22. A military science teacher wondered if there was any relationship between the rank in class of graduates of West Point and their military rank in the service at age 45.

23. The boat dock operators at a lake sponsored a fishing contest. Prizes were to be awarded to the persons who caught the largest bass, the largest crappie, and the largest bream, and to the overall winner. The problem in deciding the overall winner is that the bass are by nature larger than the other two species of fish. Describe an objective way to find the overall winner that will be fair to all three kinds of contestants.

24. Yes is the answer to the question Is psychotherapy effective? Studies comparing a quantitative measure of groups who did or did not receive psychotherapy show that the therapy group improves significantly more than the nontherapy group. What statistical technique will answer the question How effective is psychotherapy?

Set B. Read each problem, look at the statistical analysis, and write an appropriate conclusion. For each conclusion, use the terms used in the problem; that is, rather than saying "The null hypothesis was rejected at the .05 level," say "Those with medium anxiety solved problems more quickly than those with high anxiety."

For some of the problems, the design of the study has flaws. There are uncontrolled extraneous variables that would make it impossible to draw precise conclusions about the experiment. It is good for you to be able to recognize such design flaws, but the task here is to interpret the statistics. Thus, don't let design flaws keep you from drawing a conclusion based on the statistic.

For all problems, use a two-tailed test with $\alpha = .05$. If the results are significant at the .01 or .001 level, report that, but treat any difference that has a probability greater than .05 as not significant.

25. Four kinds of herbicides were compared for their weed-killing characteristics. Eighty plots were randomly but evenly divided, and one herbicide was applied to each. Because the effects of two of the herbicides were quite variable, the dependent variable was reduced to ranks and a Wilcoxon–Wilcox multiple-comparisons test was run. The plot with the fewest weeds remaining was given a rank of 1. The summary table is shown here.

	A (936.5)	B (316.5)	C (1186)
B (316.5)	620		
C (1186)	249.5	869.5	
D (801)	135.5	482.5	385

26. A company that offered instruction in taking college entrance examinations correctly claimed that its methods "produced improvement that was statistically significant." A parent with some skill as a statistician obtained the information the claim was based on (a one-sample t test) and calculated an effect size. The effect size was .10.

27. In a particular recycling process, the breakeven point for each batch occurs when 54 kilograms of raw material are unusable foreign matter. An engineer developed a screening process that leaves a mean of 38 kilograms of foreign material. The upper and lower limits of a 95 percent confidence interval are 30 and 46 kilograms.

28. The interference theory of the serial position effect predicts that participants who are given extended practice will perform more poorly on the initial items of a series than will those not given extended practice. The mean numbers of errors on initial items are shown in the table. A t test with 54 df produced a value of 3.88.

	Extended practice	No extended practice
Mean number of errors	13.7	18.4

29. A developmental psychologist developed a theory that predicted the proportion of children who would, during a period of stress, cling to their mother, attack the mother, or attack a younger sibling. The stress situation was set up and the responses of 50 children recorded. The χ^2 value was 5.30.

30. An experimental psychologist at a Veterans Administration hospital obtained approval from the Institutional Review Board to conduct a study of the efficacy of Zoloft for treating depression. The experiment lasted 30 days, during which one group of depressed patients was given placebos, one was given a low dose of Zoloft, and one was given a high dose of Zoloft. At the end of the experiment, the patients were observed by two outside psychologists who rated several behaviors (eye contact, posture, verbal output, activity, and such) for degree of depression. A composite score was obtained on each patient, with low scores indicating depression. The mean scores were: placebo, 9.86; low dose, 23.00; and high dose, 9.21. An ANOVA summary table and two Tukey HSD tests are shown in the table.

Source	df	F
Between treatments	2	21.60
Error	18	
HSD (placebo vs. low) = 13.14		
HSD (placebo vs. high) = 0.65		

31. One sample of 16 third-graders had been taught to read by the "look–say" method. A second sample of 18 had been taught by the phonics method. Both were given a reading achievement test, and a Mann–Whitney U test was performed on the results. A U value of 51 was found.

32. Sociologists sometimes make a profile of the attitudes of a group by asking each individual to rate his or her tolerance of people with particular characteristics. For example, members of a group might be asked to assess, on a scale of 1–7, their tolerance of prostitutes, atheists, former mental patients, intellectuals, college graduates, and so on. The mean score for each category is then calculated and the categories are ranked from low to high. When 18 categories of people were ranked from low to high by a group of people who owned businesses and also by a group of college students, an r_s of .753 between the rankings was found.

33. In a large high school, one group of juniors took an English course that included a 9-week unit on poetry. Another group of juniors studied plays during that 9-week period. Afterward, both groups completed a questionnaire on their attitudes toward poetry. High scores mean favorable attitudes. The means and variances are presented in the table. For the t test: t (78 df) = .48.

To compare the variances, an F test was performed: $F(37, 41) = 3.62$. (See footnote 6 in Chapter 10.)

	Attitudes toward poetry	
	Studied poetry	Studied plays
Mean	23.7	21.7
Variance	51.3	16.1

34. Here is an example of the results of experiments that asked the question: Can you get more attitude change from an audience by presenting just one side of an argument and claim that it is correct, or should you present both sides and then claim that one side is correct? In this experiment, a second independent variable is also examined: level of education. The first table contains the mean change in attitude and the second is an ANOVA summary table.

	Presentation	
Level of education	One side	Both sides
Less than a high school diploma	4.3	2.7
At least 1 year of college	2.1	4.5

Source	df	F
Between presentations	1	2.01
Between education	1	1.83
Presentations × education	1	7.93
Error	44	

35. A college student was interested in the relationship between handedness and verbal ability. She gave three classes of third-grade students a test of verbal ability that she had devised and then classified each child as right-handed, left-handed, or mixed. The obtained F value was 2.63.

	Handedness		
	Left	Right	Mixed
Mean verbal ability score	21.6	25.9	31.3
N	16	36	12

36. As part of a large-scale study on alcoholism, alcoholics and nonalcoholics were classified according to when they were toilet-trained as children. The three categories were early, normal, and late. A χ^2 of 7.90 was found.

appendixes

appendix a

Arithmetic and Algebra Review

OBJECTIVES FOR APPENDIX A

After studying the text and working the problems, you should be able to:

1. **Estimate answers**
2. **Round numbers**
3. **Find answers to problems with decimals, fractions, negative numbers, proportions, percents, absolute values, $\pm$ signs, exponents, square roots, complex expressions, and simple algebraic expressions**

STATISTICAL SYMBOLS

Although, as far as I know, there has never been a clinical case of neoiconophobia,[1] some students show a mild form of this behavior. Symbols like $\overline{X}$, σ, μ, and Σ may cause a grimace, a frown, or a droopy eyelid. In severe cases, the behavior involves avoiding a statistics course entirely. I'm sure that you don't have a severe case because you have read this far. Even so, if you are a typical beginning student in statistics, symbols like σ, μ, Σ, and $\overline{X}$ are not very meaningful to you, and they may even elicit feelings of uneasiness. However, by the end of the course, you will know what these symbols mean and be able to approach them with an unruffled psyche — and perhaps even approach them joyously.

Some of the symbols will stand for concepts you are already familiar with. For example, the capital letter X with a bar over it, $\overline{X}$, stands for the mean or average. ($\overline{X}$ is pronounced "mean" or sometimes "ex-bar.") You already know about means. You just add up the scores and divide the total by the number of scores. This verbal instruction can be put into symbols: $\overline{X} = \Sigma X/N$. The greek uppercase sigma (Σ) is the instruction to add and X is the symbol for scores. $\overline{X} = \Sigma X/N$ is something you already know about, even if you have not been using these symbols.

Pay careful attention to symbols. They serve as shorthand notations for the ideas and concepts you are learning. Each time a new symbol is introduced, concentrate on it, learn it, memorize its definition and pronunciation. The more meaning a symbol

---

[1] An extreme and debilitating fear of new symbols.

has for you, the better you understand the concepts it represents and, of course, the easier the course will be.

Sometimes I will need to distinguish between two different σ's or two X's. I will use subscripts, and the results will look like σ_1 and σ_2 or X_1 and X_2. Later, subscripts other than numbers will be used to identify a symbol. You will see σ_X and $\sigma_{\bar{X}}$. The point to learn here is that subscripts are for identification purposes only; they never indicate multiplication. Thus, $\sigma_{\bar{X}}$ does not mean $(\sigma)(\bar{X})$.

Two additional comments — to encourage and to caution you. I encourage you to do more in this course than just read the text, work the problems, and pass the tests, however exciting that may be. I hope you will occasionally get beyond this elementary text and read journal articles or short portions of other statistics textbooks. I will provide recommendations in footnotes at appropriate places. The word of caution that goes with this encouragement is that reading statistics texts is like reading a Russian novel — the same characters have different names in different places. For example, the mean of a sample in some texts is symbolized M rather than $\bar{X}$. There is even more variety when it comes to symbolizing the standard deviation. If you expect such differences, it will be less difficult for you to fit the new symbol into your established scheme of understanding.

WORKING PROBLEMS

I will review the basic mathematical skills you will need to work the problems in this course. This math is not complex. The highest level of mathematical sophistication is simple algebra.

Although the mathematical reasoning is not very complex, a good bit of arithmetic is required. To be good in statistics, you need to be good at arithmetic. A wrong answer is wrong whether the error is arithmetical or logical. For the beginning statistics student, the best way to prevent arithmetic errors is to use a calculator.

Of course, calculators can do more than simple arithmetic. Answers to many of the computational problems in this book can be found using a calculator with buttons for means, standard deviations, and correlation coefficients. By using a computer program, you can find numerical answers to all the computational problems. The question to ask yourself is: How can I use these aids so that I not only get the right answer but also *understand* what is going on and can tell the story that goes with my arithmetic? Your instructor will have some ideas about how to accomplish both goals.

Here is my advice. In the beginning, use a calculator to do the arithmetic for *each* part of a problem. Write down each step so you can see the progression from formula to final answer. When you know the steps and can explain them, use the special function keys (standard deviation, correlation, and so forth) to produce final answers directly from raw scores.

Most calculators produce values that are intermediate between raw scores and a final answer. For example, suppose a problem had just three scores: 1, 2, and 3. For many problems, you will need to know the sum of the numbers (6) and the sum when each number is squared (14, which is $1 + 4 + 9$). If your calculator has a $\Sigma+$ key or a *stat* function, it will give you both the sum of the numbers and the sum of the squares with just one entry of each of the three numbers. This feature will save you time and reduce errors.

This appendix is designed to refresh your memory of arithmetic and algebra. It is divided into two parts, a pretest and a review of fundamentals. The pretest gives you problems that are similar to those you will have to work later in the course. Take the pretest and then check your answers against the answers in Appendix F. If you find that you made any mistakes, work through those sections of the review that explain the problems you missed.

As stated earlier, to be good at statistics, you need to be good at arithmetic. To be good at arithmetic, you need to know the rules and to be careful in your computations. Here are the rules; it is up to you to be careful in your computations.

PRETEST

Estimating answers (Estimate whole-number answers to the problems.)
1. $(4.02)^2$ **2.** 1.935×7.89 **3.** $31.219 \div 2.0593$

Rounding numbers (Round numbers to the nearest tenth.)
4. 6.06   **5.** 0.35   **6.** 10.348

Decimals

Add:
7. $3.12 + 6.3 + 2.004$ **8.** $12 + 8.625 + 2.316 + 4.2$

Subtract:
9. $28.76 - 8.91$ **10.** $3.2 - 1.135$

Multiply:
11. 6.2×8.06 **12.** 0.35×0.162

Divide:
13. $64.1 \div 21.25$ **14.** $0.065 \div 0.0038$

Fractions

Add:
15. $\frac{1}{8} + \frac{3}{4} + \frac{1}{2}$ **16.** $\frac{1}{3} + \frac{1}{2}$

Subtract:
17. $\frac{11}{16} - \frac{1}{2}$ **18.** $\frac{5}{9} - \frac{1}{2}$

Multiply:
19. $\frac{4}{5} \times \frac{3}{4}$ **20.** $\frac{2}{3} \times \frac{1}{4} \times \frac{3}{5}$

Divide:
21. $\frac{3}{8} \div \frac{1}{4}$ **22.** $\frac{7}{9} \div \frac{1}{4}$

Negative numbers

Add:
23. $(-5) + 16 + (-1) + (-4)$ **24.** $(-11) + (-2) + (-12) + 3$

Subtract:

25. $(-10) - (-3)$ **26.** $(-8) - (-2)$

Multiply:

27. $(-5) \times (-5)$ **28.** $(-8) \times (3)$

Divide:

29. $(-10) \div (-3)$ **30.** $(-21) \div 4$

Percents and proportions
31. 12 is what percent of 36?
32. Find 27 percent of 84.
33. What proportion of 112 is 21?
34. A proportion .40 of the tagged birds were recovered. 150 were tagged in all. How many were recovered?

Absolute value
35. $|-5|$ **36.** $|8 - 12|$

± Problems
37. $8 \pm 2 =$ **38.** $13 \pm 9 =$

Exponents
39. 4^2 **40.** 2.5^2 **41.** 0.35^2

Square roots
42. $\sqrt{9.30}$ **43.** $\sqrt{0.93}$ **44.** $\sqrt{0.093}$

Complex problems   (Round answers to two decimal places.)

45. $\dfrac{3 + 4 + 7 + 2 + 5}{5}$ **46.** $\dfrac{(5 - 3)^2 + (3 - 3)^2 + (1 - 3)^2}{3 - 1}$

47. $\dfrac{(8 - 6.5)^2 + (5 - 6.5)^2}{2 - 1}$ **48.** $\left(\dfrac{8 + 4}{8 + 4 - 2}\right)\left(\dfrac{1}{8} + \dfrac{1}{4}\right)$

49. $\left(\dfrac{3.6}{1.2}\right)^2 + \left(\dfrac{6.0}{2.4}\right)^2$ **50.** $\dfrac{(3)(4) + (6)(8) + (5)(6) + (2)(3)}{(4)(4 - 1)}$

51. $\dfrac{190 - (25^2/5)}{5 - 1}$ **52.** $\dfrac{12(50 - 20)^2}{(8)(10)(12)(11)}$

53. $\dfrac{6[(3 + 5)(4 - 1) + 2]}{5^2 - (6)(2)}$

54. For the numbers 1, 2, 3, and 4: Find the sum and the sum of the squared numbers.
55. For the numbers 2, 4, and 6: Find the sum and the sum of the squared numbers.

Simple algebra   (Solve for x.)

56. $\dfrac{x - 3}{4} = 2.5$ **57.** $\dfrac{14 - 8.5}{x} = 0.5$

58. $\dfrac{20 - 6}{2} = 4x - 3$ **59.** $\dfrac{6 - 2}{3} = \dfrac{x + 9}{5}$

REVIEW OF FUNDAMENTALS

This section will give you a quick review of the pencil-and-paper rules of arithmetic and simple algebra. I assume that you once knew all these rules but that refresher exercises will be helpful. Thus, there isn't much explanation. To obtain basic explanations, ask a teacher in your school's mathematics department to recommend one of the many "refresher" books that are available.

Definitions

Sum. The answer to an addition problem is called a sum. In Chapter 10 you will calculate a *sum of squares,* a quantity that is obtained by adding together some squared numbers.

Difference. The answer to a subtraction problem is called a difference. Much of what you will learn in statistics deals with differences and how to explain them.

Product. The answer to a multiplication problem is called a product. Chapter 5 is about the *product-moment correlation coefficient,* which requires multiplication. Multiplication problems are indicated either by a $\times$ or by parentheses. Thus, 6×4 and $(6)(4)$ call for the same operation.

Quotient. The answer to a division problem is called a quotient. The three ways to indicate a division problem are $\div$, / (slash mark), and $-$ (division line). Thus, $9 \div 4$, 9/4, and $\frac{9}{4}$ call for the same operation. It is a good idea to think of any common fraction as a division problem. The numerator is to be divided by the denominator.

Estimating Answers

It is a **very good idea** to just look at a problem and estimate the answer before you do any calculating. This is referred to as "eyeballing the data," and Edward Minium has captured its importance with Minium's First Law of Statistics: "The eyeball is the statistician's most powerful instrument." (See Minium, King, and Bear, 1993.)

Estimating answers will keep you from making gross errors such as misplacing a decimal point. For example, $\frac{31.5}{5}$ can be estimated as a little more than 6. If you estimate before you divide, you are likely to recognize that an answer of 63 or 0.63 is wrong.

The estimated answer to the problem $(21)(108)$ is 2000 since $(20)(100) = 2000$. The problem $(0.47)(0.20)$ suggests an estimated answer of 0.10 because $(\frac{1}{2})(0.20) = 0.10$. With 0.10 in mind, you are not likely to write 0.94 for the answer, which is 0.094. Estimating answers is also important if you are finding a square root. You can estimate that $\sqrt{95}$ is about 10 because $\sqrt{100} = 10$; $\sqrt{1.034}$ is about 1; $\sqrt{0.093}$ is about 0.1.

To calculate a mean, eyeball the numbers and estimate the mean. If you estimate a mean of 30 for a group of numbers that are primarily in the 20s, 30s, and 40s, a calculated mean of 60 will arouse your suspicion that you have made an error.

Rounding Numbers

There are two parts to the rule for rounding a number. If the first digit of the part that is to be dropped is less than 5, simply drop it. If the first digit of the part to be dropped is 5 or greater, increase the number to the left of it by one. These rules are built into most calculators. Here are some illustrations of these rules.

Rounding to the nearest whole number:

6.2 = 6	4.5 = 5
12.7 = 13	163.5 = 164
6.49 = 6	9.5 = 10

Rounding to the nearest hundredth:

13.614 = 13.61	12.065 = 12.07
0.049 = 0.05	4.005 = 4.01
1.097 = 1.10	0.675 = 0.68
3.6248 = 3.62	1.995 = 2.00

A reasonable question is: How many decimal places should an answer in statistics have? A good rule of thumb in statistics is to carry all operations to three decimal places and then, for the final answer, round back to two decimal places.

Sometimes this rule of thumb can get you into trouble, though. For example, if halfway through some work you have a division problem of 0.0016 ÷ 0.0074, and you dutifully round those four decimals to three (0.002 ÷ 0.007), you get an answer of 0.2857, which becomes 0.29. However, division without rounding gives you an answer of 0.2162 or 0.22. The difference between 0.22 and 0.29 may be substantial. I will often give you cues if more than two decimal places are necessary, but you always need to be alert to the problems of rounding.

Most calculators carry more decimal places in memory than they show in the display. If you have such a calculator, it will protect you from the problem of rounding too much or too soon.

PROBLEMS

1. Define (a) sum, (b) quotient, (c) product, and (d) difference.

2. Estimate answers to the expressions.

   a. $\sqrt{103.48}$ **b.** $74.16 \div 9.87$ **c.** $(11.4)^2$ **d.** $\sqrt{0.0459}$

   e. 0.41^2 **f.** 11.92×4.60 **g.** $\sqrt{0.888}$ **h.** $\sqrt{0.0098}$

3. Round the numbers to the nearest whole number.

   a. 13.9   **b.** 126.4   **c.** 9.0

   d. 0.4   **e.** 127.5   **f.** 12.51

   g. 12.49   **h.** 12.50   **i.** 9.46

4. Round the numbers to the nearest hundredth.

    a. 6.3348   **b.** 12.997   **c.** 0.050

    d. 0.965   **e.** 2.605   **f.** 0.3445

    g. 0.003   **h.** 0.015   **i.** 0.9949

Decimals

1. *Addition and subtraction of decimals.* There is only one rule about the addition and subtraction of numbers that have decimals: *Keep the decimal points in a vertical line.* The decimal point in the answer goes directly below those in the problem. This rule is illustrated in the five problems here.

Add			*Subtract*	
	0.004	6.0		
1.26	1.310	18.0	14.032	16.00
10.00	4.039	0.5	8.26	4.32
11.26	5.353	24.5	5.772	11.68

2. *Multiplication of decimals.* The basic rule for multiplying decimals is that the number of decimal places in the answer is found by adding the numbers of decimal places in the two numbers that are being multiplied. To place the decimal point in the product, count from the right.

1.3	0.21	1.47
×4.2	×0.4	×3.12
26	0.084	294
52		147
5.46		441
		4.5864

3. *Division of decimals.* Two methods have been used to teach division of decimals. The older method required the student to move the decimal in the divisor (the number you are dividing by) enough places to the right to make the divisor a whole number. The decimal in the dividend is then moved to the right the same number of places, and division is carried out in the usual way. The new decimal places are identified with carets, and the decimal place in the quotient is just above the caret in the dividend.

$$20.$$
$$0.016_\wedge \overline{)0.320_\wedge}$$
Decimal moved three places in both the divisor and the dividend

$$38.46$$
$$0.39_\wedge \overline{)15.00_\wedge}$$
Decimal moved two places in both the divisor and the dividend

$$2.072$$
$$6\overline{)12.432}$$
Divisor is already a whole number

$$.004$$
$$9.1_\wedge \overline{)0.0_\wedge 369}$$
Decimal moved one placed in both the divisor and the dividend

The newer method of teaching the division of decimals is to multiply both the divisor and the dividend by the number that will make both of them whole numbers. (Actually, this is the way the caret method works also.)

$$\frac{0.32}{0.016} \times \frac{1000}{1000} = \frac{320}{16} = 20.00$$

$$\frac{0.0369}{9.1} \times \frac{10,000}{10,000} = \frac{369}{91,000} = 0.004$$

$$\frac{15}{0.39} \times \frac{100}{100} = \frac{1500}{39} = 38.46$$

$$\frac{12.432}{6} \times \frac{1000}{1000} = \frac{12,432}{6000} = 2.072$$

Both of these methods work. Use the one you are more familiar with.

PROBLEMS

Perform the operations indicated.

5. $0.001 + 10 + 3.652 + 2.5$

6. $14.2 - 7.31$

7. 0.04×1.26

8. $143.3 + 16.92 + 2.307 + 8.1$

9. $3.06 \div 0.04$

10. $\dfrac{24}{11.75}$

11. $152.12 - 127.4$

12. $(0.5)(0.07)$

Fractions

In general, there are two ways to deal with fractions.

1. Convert the fraction to a decimal and perform the operations on the decimals.
2. Work directly with the fractions, using a set of rules for each operation. The rule for addition and subtraction is: Convert the fractions to ones with common denominators, add or subtract the numerators, and place the result over the common denominator. The rule for multiplication is: Multiply the numerators together to get the numerator of the answer, and multiply the denominators together for the denominator of the answer. The rule for division is: Invert the divisor and multiply the fractions.

For statistics problems, it is usually easier to convert the fractions to decimals and then work with the decimals. Therefore, this is the method that is illustrated. However, if you are a whiz at working directly with fractions, by all means continue with your method. To convert a fraction to a decimal, divide the lower number into the upper one. Thus, $\frac{3}{4} = 0.75$ and $\frac{13}{17} = 0.765$.

Addition of fractions

$$\frac{1}{2} + \frac{1}{4} = 0.50 + 0.25 = 0.75$$

$$\frac{13}{17} + \frac{21}{37} = 0.765 + 0.568 = 1.33$$

$$\frac{2}{3} + \frac{3}{4} = 0.667 + 0.75 = 1.42$$

Subtraction of fractions

$$\frac{1}{2} - \frac{1}{4} = 0.50 - 0.25 = 0.25$$

$$\frac{11}{12} - \frac{2}{3} = 0.917 - 0.667 = 0.25$$

$$\frac{41}{53} - \frac{17}{61} = 0.774 - 0.279 = 0.50$$

Multiplication of fractions

$$\frac{1}{2} \times \frac{3}{4} = (0.5)(0.75) = 0.38$$

$$\frac{10}{19} \times \frac{61}{90} = 0.526 \times 0.678 = 0.36$$

$$\frac{1}{11} \times \frac{2}{3} = (0.09)(0.67) = 0.06$$

Division of fractions

$$\frac{9}{21} \div \frac{13}{19} = 0.429 \div 0.684 = 0.63$$

$$14 \div \frac{1}{3} = 14 \div 0.33 = 42$$

$$\frac{7}{8} \div \frac{3}{4} = 0.875 \div 0.75 = 1.17$$

PROBLEMS

Perform the operations indicated.

13. $\dfrac{9}{10} + \dfrac{1}{2} + \dfrac{2}{5}$

14. $\dfrac{9}{20} \div \dfrac{19}{20}$

15. $\left(\dfrac{1}{3}\right)\left(\dfrac{5}{6}\right)$

16. $\dfrac{4}{5} - \dfrac{1}{6}$

17. $\dfrac{1}{3} \div \dfrac{5}{6}$

18. $\dfrac{3}{4} \times \dfrac{5}{6}$

19. $18 \div \dfrac{1}{3}$

Negative Numbers

1. *Addition of negative numbers.* (Any number without a sign is understood to be positive.)

   a. To add a series of negative numbers, add the numbers in the usual way and attach a negative sign to the total.

$$\begin{array}{r} -3 \\ -8 \\ -12 \\ \underline{-5} \\ -28 \end{array} \quad (-1) + (-6) + (-3) = -10$$

b. To add two numbers, one positive and one negative, subtract the smaller number from the larger number and attach the sign of the larger number to the result.

$$\begin{array}{r} 140 \\ -55 \\ \hline 85 \end{array} \qquad \begin{array}{r} -14 \\ 8 \\ \hline -6 \end{array} \qquad (28) + (-9) = 19$$

$$74 + (-96) = -22$$

c. To add a series of numbers, some positive and some negative, add all the positive numbers together, add all the negative numbers together (see 1a), and then combine the two sums (see 1b).

$$(-4) + (-6) + (12) + (-5) + (2) + (-9) = 14 + (-24) = -10$$
$$(-7) + (10) + (4) + (-5) = 14 + (-12) = 2$$

2. *Subtraction of negative numbers.* To subtract a negative number, change it to positive and add it.

$$\begin{array}{r} (-14) \\ -(-2) \\ \hline -12 \end{array} \qquad \begin{array}{r} 5 \\ -(-7) \\ \hline 12 \end{array} \qquad \begin{array}{r} 8 \\ -(-3) \\ \hline 11 \end{array} \qquad \begin{array}{r} (-7) \\ -(-5) \\ \hline -2 \end{array}$$

3. *Multiplication of negative numbers.* When the two numbers to be multiplied are both negative, the product is positive.

$$(-3)(-3) = 9 \qquad (-6)(-8) = 48$$

When one of the numbers is negative and the other is positive, the product is negative.

$$(-8)(3) = -24 \qquad 14 \times -2 = -28$$

4. *Division of negative numbers.* The rule in division is the same as the rule in multiplication. If the two numbers are both negative, the quotient is positive.

$$(-10) \div (-2) = 5 \qquad (-4) \div (-20) = 0.20$$

If one number is negative and the other positive, the quotient is negative.

$$(-10) \div 2 = -5 \qquad 6 \div (-18) = -0.33$$
$$14 \div (-7) = -2 \qquad (-12) \div 3 = -4$$

PROBLEMS

20. Add the numbers.
  a. $-3, 19, -14, 5, -11$ **b.** $-8, -12, -3$
  c. $-8, 11$ **d.** $3, -6, -2, 5, -7$

21. $(-8)(5)$ **22.** $(-4)(-6)$ **23.** $(4)(12)$

24. $(11)(-3)$ **25.** $(-18) - (-9)$ **26.** $14 \div (-6)$

27. $12 - (-3)$ **28.** $(-6) - (-7)$ **29.** $(-9) \div (-3)$

30. $(-10) \div 5$ **31.** $4 \div (-12)$ **32.** $(-7) - 5$

Proportions and Percents

A **proportion** is a part of a whole and can be expressed as a fraction or as a decimal. Usually proportions are expressed as decimals. If eight students in a class of 44 received A's, we may express 8 as a proportion of the whole (44). Thus, we have $\frac{8}{44}$, or 0.18. The proportion of the class that received A's is 0.18.

To convert a proportion to a percent (per one hundred), multiply by 100. Thus, $0.18 \times 100 = 18$; 18 percent of the students received A's. As you can see, proportions and percents are two ways to express the same idea.

If you know a proportion (or percent) and the size of the original whole, you can find the number that the proportion represents. If 0.28 of the students were absent due to illness and there are 50 students in all, then 0.28 of the 50 were absent: $(0.28)(50) = 14$ students who were absent. Here are some more problems. Work them with the answers covered.

1. 26 out of 31 completed the course. What proportion completed the course?

   $$\frac{26}{31} = 0.83$$

2. What percent completed the course?

   $$0.83 \times 100 = 83 \text{ percent}$$

3. What percent of 19 is 5?

   $$\frac{5}{19} = 0.26 \qquad 0.26 \times 100 = 26 \text{ percent}$$

4. If 90 percent of the population agreed and the population consisted of 210 members, how many members agreed?

   $$0.90 \times 210 = 189 \text{ members}$$

Absolute Value

The **absolute value** of a number ignores the sign of the number. Thus, the absolute value of -6 is 6. This is expressed with symbols as $|-6| = 6$. It is expressed verbally as "the absolute value of negative six is six." In a similar way, the absolute value of $4 - 7$ is 3; that is, $|4 - 7| = |-3| = 3$.

± Signs

A $\pm$ sign ("plus or minus" sign) means to both add *and* subtract. A $\pm$ problem *always* has two answers.

$$10 \pm 4 = 6, 14$$
$$8 \pm (3)(2) = 8 \pm 6 = 2, 14$$
$$\pm (4)(3) + 21 = \pm 12 + 21 = 9, 33$$
$$\pm 4 - 6 = -10, -2$$

Exponents

In the expression 5^2 ("five squared"), 2 is the exponent. The 2 means that 5 is to be multiplied by itself. Thus, $5^2 = 5 \times 5 = 25$.

In elementary statistics, the only exponent used is 2, but it will be used frequently. When a number has an exponent of 2, the number is said to be squared. The expression 4^2 ("four squared") means 4×4, and the product is 16.

$$8^2 = 8 \times 8 = 64 \qquad (0.75)^2 = (0.75)(0.75) = 0.5625$$
$$1.2^2 = (1.2)(1.2) = 1.44 \qquad 12^2 = 12 \times 12 = 144$$

Square Roots

Statistics problems often require that a square root be found. There are three possible ways to find the square root:

1. A calculator with a square root key
2. A table of square roots
3. The paper-and-pencil method

Use a calculator with a square root key. If this is not possible, find a book with a table of square roots. The third solution, the paper-and-pencil method, is so tedious and error-prone that I do not recommend it.

PROBLEMS

33. Six is what proportion of 13?   **34.** Six is what percent of 13?

35. What proportion of 25 is 18?   **36.** What percent of 115 is 85?

37. Find 72 percent of 36.   **38.** 22 is what proportion of 72?

39. If .40 of the 320 members voted against the proposal, how many voted no?

40. 22 percent of the 850 coupons were redeemed. How many were redeemed?

41. $|{-31}|$ **42.** $|21 - 25|$ **43.** $12 \pm (2)(5)$

44. $\pm(5)(6) + 10$ **45.** $\pm(2)(2) - 6$ **46.** $(2.5)^2$

47. 9^2 **48.** $(0.3)^2$

49. $\left(\dfrac{1}{4}\right)^2$

50. Find the square root of the numbers to two decimal places (or more if appropriate).

a. 625	**b.** 6.25	**c.** 0.625
d. 0.0625	**e.** 16.85	**f.** 0.003
g. 181,476	**h.** 0.25	**i.** 22.51

Complex Expressions

Two rules will take care of the complex expressions encountered in statistics.

1. Perform the operations within the parentheses. If there are brackets in the expression, perform the operations within the parentheses and then the operations within the brackets.

2. Perform the operations in the numerator separately from those in the denominator and, finally, carry out the division.

$$\frac{(8-6)^2 + (7-6)^2 + (3-6)^2}{3-1} = \frac{2^2 + 1^2 + (-3)^2}{2} = \frac{4+1+9}{2}$$

$$= \frac{14}{2} = 7.00$$

$$\left(\frac{10+12}{4+3-2}\right)\left(\frac{1}{4} + \frac{1}{3}\right) = \left(\frac{22}{5}\right)(0.25 + 0.333) = (4.40)(0.583)$$

$$= 2.57$$

$$\left(\frac{8.2}{4.1}\right)^2 + \left(\frac{4.2}{1.2}\right)^2 = (2)^2 + (3.5)^2 = 4 + 12.25 = 16.25$$

$$\frac{6[(13-10)^2 - 5]}{6(6-1)} = \frac{6(3^2 - 5)}{6(5)} = \frac{6(9-5)}{30} = \frac{6(4)}{30}$$

$$= \frac{24}{30} = 0.80$$

$$\frac{18 - \dfrac{6^2}{4}}{4-1} = \frac{18 - \dfrac{36}{4}}{3} = \frac{18-9}{3} = \frac{9}{3} = 3.00$$

Simple Algebra

To solve a simple algebra problem, isolate the unknown (x) on one side of the equal sign and combine the numbers on the other side. To do this, remember that the same number can be *added to* or *subtracted from* both sides of the equation without affecting the value of the unknown.

$$\begin{aligned} x - 5 &= 12 \\ x - 5 + 5 &= 12 + 5 \\ x &= 17 \end{aligned} \qquad \begin{aligned} x + 7 &= 9 \\ x + 7 - 7 &= 9 - 7 \\ x &= 2 \end{aligned}$$

In a similar way, you can multiply or divide both sides of the equation by the same number without affecting the value of the unknown.

$$\frac{x}{6} = 9 \qquad\qquad 11x = 30 \qquad\qquad \frac{3}{x} = 14$$

$$(6)\left(\frac{x}{6}\right) = (6)(9) \qquad \frac{11x}{11} = \frac{30}{11} \qquad (x)\left(\frac{3}{x}\right) = (14)(x)$$

$$x = 54 \qquad\qquad x = 2.73 \qquad\qquad 3 = 14x$$

$$\frac{3}{14} = \frac{14x}{14}$$

$$0.21 = x$$

I will combine some of these steps in the problems that follow. Be sure you see what operation is being performed on both sides in each step.

$$\frac{x - 2.5}{1.3} = 1.96$$

$$x - 2.5 = (1.3)(1.96)$$

$$x = 2.548 + 2.5$$

$$x = 5.048$$

$$4x - 9 = 5^2$$

$$4x = 25 + 9$$

$$x = \frac{34}{4}$$

$$x = 8.50$$

$$\frac{21.6 - 15}{x} = 0.04$$

$$6.6 = 0.04x$$

$$\frac{6.6}{0.04} = x$$

$$165 = x$$

$$\frac{14 - x}{6} = 1.9$$

$$14 - x = (1.9)(6)$$

$$14 = 11.4 + x$$

$$2.6 = x$$

PROBLEMS

Reduce these complex expressions to a single number rounded to two decimal places.

51. $\dfrac{(4 - 2)^2 + (0 - 2)^2}{6}$

52. $\dfrac{(12 - 8)^2 + (8 - 8)^2 + (5 - 8)^2 + (7 - 8)^2}{4 - 1}$

53. $\left(\dfrac{5 + 6}{3 + 2 - 2}\right)\left(\dfrac{1}{3} + \dfrac{1}{2}\right)$

54. $\left(\dfrac{13 + 18}{6 + 8 - 2}\right)\left(\dfrac{1}{6} + \dfrac{1}{8}\right)$

55. $\dfrac{8[(6 - 2)^2 - 5]}{(3)(2)(4)}$

56. $\dfrac{[(8 - 2)(5 - 1)]^2}{5(10 - 7)}$

57. $\dfrac{6}{1/2} + \dfrac{8}{1/3}$

58. $\left(\dfrac{9}{2/3}\right)^2 + \left(\dfrac{8}{3/4}\right)^2$

59. $\dfrac{10 - (6^2/9)}{8}$

60. $\dfrac{104 - (12^2/6)}{5}$

Find the value of x.

61. $\dfrac{x - 4}{2} = 2.58$

62. $\dfrac{x - 21}{6.1} = 1.04$

63. $x = \dfrac{14 - 11}{2.5}$

64. $x = \dfrac{36 - 41}{8.2}$

appendix**b**

Grouped Frequency Distributions and Central Tendency

OBJECTIVES FOR APPENDIX B

After studying the text and working the problems, you should be able to:

1. Know four conventions for constructing grouped frequency distributions
2. Arrange raw data into a grouped frequency distribution
3. Find the mean, median, and mode of a grouped frequency distribution

(In writing the material in this appendix, I have assumed that you have studied Chapter 2, Data: Organization, Graphs, and Central Tendency.) In Chapter 2 you converted a group of raw scores into a simple frequency distribution. Simple frequency distributions are quite useful; they present the data in a comprehensible form, and they facilitate drawing graphs and calculating central tendency values.

Grouped frequency distributions are equally useful. Grouped frequency distributions are used when the range of scores is too large for a simple frequency distribution. How large is too large? A rule of thumb is that grouped frequency distributions are appropriate when the range of scores is greater than 20. However, there is a fairly wide gray area around this number.

Grouped frequency distributions are used for the same purposes as simple frequency distributions. As you know from Chapter 2, the only difference is that grouped frequency distributions have class intervals.

GROUPED FREQUENCY DISTRIBUTIONS

Establishing Class Intervals

Use the conventions that follow to establish class intervals for grouped frequency distributions. These conventions make tables and graphs easier to comprehend. Many researchers use these conventions, but not all do.[1]

1. *The number of class intervals.* The number of class intervals should be 10 to 20. On the one hand, with fewer than 10 intervals, the extreme scores in the data are not as apparent because they are clustered with more frequently occurring scores. On the other hand, more than 20 class intervals often make it difficult to see the shape of the distribution.
2. *The size of i.* If i is odd, the midpoint of the class interval will be a whole number and whole numbers look better on graphs than decimal numbers. Three and five often work well as interval sizes. You may find that $i = 2$ is needed if you are to have 10 to 20 class intervals. If an i of 5 produces more than 20 class intervals, data groupers usually jump to an i of 10 or some multiple of 10. An interval size of 25 is popular.
3. *The lower limit of a class interval.* Begin each class interval with a multiple of i. For example, if the lowest score is 31 and $i = 3$ (as happened with the Personal Control scores in Table 2.4), the first class interval should be 30–32 because 30 is a multiple of 3. An exception to this convention occurs when $i = 5$. When the interval size is 5, it is usually better to use a multiple of 5 as the midpoint because multiples of 5 are easier to read on graphs.
4. *The order of the intervals.* The largest scores go at the top of the distribution. This is a convention that is not followed by some computer programs.

Converting Unorganized Scores into a Grouped Frequency Distribution

With the conventions for establishing class intervals in mind, here are the steps for converting unorganized data into a grouped frequency distribution. As an example, I will use the raw data in Table 2.1 and describe converting it into Table 2.4.

1. Find the highest and lowest scores. In Table 2.1, the highest score is 70 and the lowest score is 31.
2. Find the range of the scores by subtracting the lowest score from the highest score ($70 - 31 = 39$).
3. Determine i by a trial-and-error procedure. Remember that there are to be 10 to 20 class intervals and that the interval size should be convenient (3, 5, 10, or a multiple of 10). Dividing the range by a potential i value gives the approximate number of class intervals. Dividing the range, 39, by 5 gives 7.8, which is too few, but dividing 39 by 3 gives 13, which is between the recommended 10 and 20.

[1] Some computer programs do not follow these conventions.

4. Establish the lowest interval. Begin the interval with a multiple of i, *which may or may not be an actual raw score.* End the interval so that it contains i scores. For Table 2.4, the lowest interval is 30–32. (Note that 30 is not an actual score but is a multiple of i.) Each interval above the lowest one begins with a multiple of i. Continue building the class intervals.
5. Using the unorganized data (Table 2.1), underline each score and put a tally mark beside its class interval (Table 2.4).
6. As a check on your work, add up the frequency column. The sum should be N, the number of scores in the unorganized data.

PROBLEMS

*1. A sociology professor was deciding what statistics to present in her introduction to sociology classes. She developed a test that covered concepts such as the median, graphs, standard deviation, and correlation. She tested one class of 50 students, and on the basis of the results, planned a course syllabus for that class and the other six intro sections. Arrange the data into an appropriate rough-draft frequency distribution.

20	56	48	13	30	39	25	41	52	44
27	36	54	46	59	42	17	63	50	24
31	19	38	10	43	31	34	32	15	47
40	36	5	31	53	24	31	41	49	21
26	35	28	37	25	33	27	38	34	22

*2. The measurements that follow are weights in pounds of a sample of college men in one study. Arrange them into a grouped frequency distribution. If these data are skewed, tell the direction of the skew.

164	158	156	148	180	176	171	150	152	155	
161	168	148	175	154	155	149	149	151	160	
157	158	161	167	152	168	151	157	150	154	189

CENTRAL TENDENCY OF GROUPED FREQUENCY DISTRIBUTIONS

Mean

Finding the mean of a grouped frequency distribution involves the same arithmetic as that for a simple frequency distribution. Setting up the problem, however, requires one additional step. Look at **Table B.1,** which has four columns (compared to the three in Table 2.7). For a grouped frequency distribution, the *midpoint* of the interval represents all the scores in the interval. Thus, multiplying the midpoint by its f value includes all the scores in that interval. As you can see at the bottom of Table B.1, summing the fX column gives ΣfX, which, when divided by N, yields the mean.

TABLE B.1 A grouped frequency distribution of Personal Control scores with $i = 3$

PC scores (class interval)	Midpoint (X)	f	fX
69–71	70	1	70
66–68	67	4	268
63–65	64	1	64
60–62	61	5	305
57–59	58	9	522
54–56	55	15	825
51–53	52	18	936
48–50	49	14	686
45–47	46	15	690
42–44	43	10	430
39–41	40	4	160
36–38	37	2	74
33–35	34	0	0
30–32	31	2	62
		$\Sigma = 100$	5092

In terms of a formula,

$$\mu \text{ or } \overline{X} = \frac{\Sigma fX}{N}$$

For Table B.1,

$$\mu \text{ or } \overline{X} = \frac{\Sigma fX}{N} = \frac{5092}{100} = 50.92$$

Note that the mean of the grouped data is 50.92 but the mean of the simple frequency distribution is 51.00. The mean of grouped scores is often different, but seldom is this difference of any consequence.

Median

Finding the median of a grouped distribution is almost the same as finding the median of a simple frequency distribution. Of course, you are looking for a point that has as many frequencies above it as below it. To locate the median, use the formula

$$\text{Median location} = \frac{N + 1}{2}$$

For the data in **Table B.1,**

$$\text{Median location} = \frac{N + 1}{2} = \frac{100 + 1}{2} = 50.5$$

As before, you are looking for a point with 50 frequencies above it and 50 frequencies below it. Adding frequencies from the bottom of the distribution, you find that there are 47 scores below the interval 51–53, so the 50.5th score is in the interval 51–53. The midpoint of the interval is the median. For the grouped PC scores in Table B.1, median = 52.

Thus, to find the median of a grouped frequency distribution, locate the class interval that is the location of the middle score. The midpoint of that interval is the median.

Mode

The mode is the midpoint of the interval that has the highest frequency. In Table B.1 the highest frequency count is 18. The interval with 18 scores is 51–53. The midpoint of that interval, 52, is the mode.

PROBLEMS

3. Find the mean, median, and mode of the grouped frequency distribution you constructed from the statistics questionnaire data (problem 1).
4. Find the mean, median, and mode of the weight data in problem 2.

appendix c

TABLE A Critical values for Pearson product-moment correlation coefficients, *r**

	α levels (two-tailed test)				
df	.10	.05	.02	.01	.001
	α levels (one-tailed test)				
(df = N − 2)	.05	.025	.01	.005	.0005
1	.98769	.99692	.999507	.999877	.9999988
2	.90000	.95000	.98000	.990000	.99900
3	.8054	.8783	.93433	.95873	.99116
4	.7293	.8114	.8822	.91720	.97406
5	.6694	.7545	.8329	.8745	.95074
6	.6215	.7067	.7887	.8343	.92493
7	.5822	.6664	.7498	.7977	.8982
8	.5494	.6319	.7155	.7646	.8721
9	.5214	.6021	.6851	.7348	.8371
10	.4973	.5760	.6581	.7079	.8233
11	.4762	.5529	.6339	.6835	.8010
12	.4575	.5324	.6120	.6614	.7800
13	.4409	.5139	.5923	.6411	.7603
14	.4259	.4973	.5742	.6226	.7420
15	.4124	.4821	.5577	.6055	.7246
16	.4000	.4683	.5425	.5897	.7084
17	.3887	.4555	.5285	.5751	.6932
18	.3783	.4438	.5155	.5614	.6787
19	.3687	.4329	.5034	.5487	.6652
20	.3598	.4227	.4921	.5368	.6524
25	.3233	.3809	.4451	.4869	.5974
30	.2960	.3494	.4093	.4487	.5541
35	.2746	.3246	.3810	.4182	.5189
40	.2573	.3044	.3578	.3932	.4896
45	.2428	.2875	.3384	.3721	.4648
50	.2306	.2732	.3218	.3541	.4433
60	.2108	.2500	.2948	.3248	.4078
70	.1954	.2319	.2737	.3017	.3799
80	.1829	.2172	.2565	.2830	.3568
90	.1726	.2050	.2422	.2673	.3375
100	.1638	.1946	.2301	.2540	.3211

* To be significant the *r* obtained from the data must be equal to or **greater than** the value shown in the table.
SOURCE: From Table VII in R. A. Fisher and F. Yates, *Statistical Tables for Biological, Agricultural, and Medical Research*, Sixth Edition, published by Addison Wesley Longman Ltd. (1963). Reprinted with permission.

T A B L E B Random digits

	00–04	05–09	10–14	15–19	20–24	25–29	30–34	35–39	40–44	45–49
00	54463	22662	65905	70639	79365	67382	29085	69831	47058	08186
01	15389	85205	18850	39226	42249	90669	96325	23248	60933	26927
02	85941	40756	82414	02015	13858	78030	16269	65978	01385	15345
03	61149	69440	11286	88218	58925	03638	52862	62733	33451	77455
04	05219	81619	10651	67079	92511	59888	84502	72095	83463	75577
05	41417	98326	87719	92294	46614	50948	64886	20002	97365	30976
06	28357	94070	20652	35774	16249	75019	21145	05217	47286	76305
07	17783	00015	10806	83091	91530	36466	39981	62481	49177	75779
08	40950	84820	29881	85966	62800	70326	84740	62660	77379	90279
09	82995	64157	66164	41180	10089	41757	78258	96488	88629	37231
10	96754	17676	55659	44105	47361	34833	86679	23930	53249	27083
11	34357	88040	53364	71726	45690	66334	60332	22554	90600	71113
12	06318	37403	49927	57715	50423	67372	63116	48888	21505	80182
13	62111	52820	07243	79931	89292	84767	85693	73947	22278	11551
14	47534	09243	67879	00544	23410	12740	02540	54440	32949	13491
15	98614	75993	84460	62846	59844	14922	48730	73443	48167	34770
16	24856	03648	44898	09351	98795	18644	39765	71058	90368	44104
17	96887	12479	80621	66223	86085	78285	02432	53342	42846	94771
18	90801	21472	42815	77408	37390	76766	52615	32141	30268	18106
19	55165	77312	83666	36028	28420	70219	81369	41493	47366	41067
20	75884	12952	84318	95108	72305	64620	91318	89872	45375	85436
21	16777	37116	58550	42958	21460	43910	01175	87894	81378	10620
22	46230	43877	80207	88877	89380	32992	91380	03164	98656	59337
23	42902	66892	46134	01432	94710	23474	20423	60137	60609	13119
24	81007	00333	39693	28039	10154	95425	39220	19774	31782	49037
25	68089	01122	51111	72373	06902	74373	96199	97017	41273	21546
26	20411	67081	89950	16944	93054	87687	96693	87236	77054	33848
27	58212	13160	06468	15718	82627	76999	05999	58680	96739	63700
28	70577	42866	24969	61210	76046	67699	42054	12696	93758	03283
29	94522	74358	71659	62038	79643	79169	44741	05437	39038	13163
30	42626	86819	85651	88678	17401	03252	99547	32404	17918	62880
31	16051	33763	57194	16752	54450	19031	58580	47629	54132	60631
32	08244	27647	33851	44705	94211	46716	11738	55784	95374	72655
33	59497	04392	09419	89964	51211	04894	72882	17805	21896	83864
34	97155	13428	40293	09985	58434	91412	69124	82171	59058	82859
35	98409	66162	95763	47420	20792	61527	29441	39435	11859	41567
36	45476	84882	65109	96597	25930	66790	65706	61203	53634	22557
37	89300	69700	50741	30329	11658	23166	05400	66669	48708	02306
38	50051	95137	91631	66315	91428	12275	24816	68091	71710	33258
39	31753	85178	31310	89642	98364	92396	24617	09609	83942	22716
40	79152	53829	77250	20190	56535	18760	69942	77448	33278	48805
41	44560	38750	83635	56540	64900	42912	13953	79149	18710	68618
42	68328	83378	63369	71381	39564	95615	42451	64559	97501	65747
43	46939	38689	58625	08342	30459	85863	20781	09284	26333	91777
44	83544	86141	15707	96256	23068	13782	08467	89469	93842	55349
45	91621	00881	04900	54224	46177	55309	17852	27491	89415	23466
46	91896	67126	04151	03795	59077	11848	12630	98375	52068	60142
47	55751	62515	21108	80830	02263	29303	37204	96926	30506	09808
48	85156	87689	95493	88842	00664	55017	55539	17771	69448	87530
49	07521	56898	12236	60277	39102	62315	12239	07105	11844	01117

(continued)

SOURCE: From *Statistical Methods*, by G. W. Snedecor and W. G. Cochran, Seventh Edition. Copyright © 1980 Iowa State University Press. Reprinted by permission.

TABLE B (*continued*)

	50–54	55–59	60–64	65–69	70–74	75–79	80–84	85–89	90–94	95–99
00	59391	58030	52098	82718	87024	82848	04190	96574	90464	29065
01	99567	76364	77204	04615	27062	96621	43918	01896	83991	51141
02	10363	97518	51400	25670	98342	61891	27101	37855	06235	33316
03	86859	19558	64432	16706	99612	59798	32803	67708	15297	28612
04	11258	24591	36863	55368	31721	94335	34936	02566	80972	08188
05	95068	88628	35911	14530	33020	80428	39936	31855	34334	64865
06	54463	47237	73800	91017	36239	71824	83671	39892	60518	37092
07	16874	62677	57412	13215	31389	62233	80827	73917	82802	84420
08	92494	63157	76593	91316	03505	72389	96363	52887	01087	66091
09	15669	56689	35682	40844	53256	81872	35213	09840	34471	74441
10	99116	75486	84989	23476	52967	67104	39495	39100	17217	74073
11	15696	10703	65178	90637	63110	17622	53988	71087	84148	11670
12	97720	15369	51269	69620	03388	13699	33423	67453	43269	56720
13	11666	13841	71681	98000	35979	39719	81899	07449	47985	46967
14	71628	73130	78783	75691	41632	09847	61547	18707	65489	69944
15	40501	51089	99943	91843	41995	88931	73631	69361	05375	15417
16	22518	55576	98215	82068	10798	86211	36584	67466	69373	40054
17	75112	30485	62173	02132	14878	92879	22281	16783	86352	00077
18	80327	02671	98191	84342	90813	49268	95441	15496	20168	09271
19	60251	45548	02146	05597	48228	81366	34598	72856	66762	17002
20	57430	82270	10421	05540	43648	75888	66049	21511	47676	33444
21	73528	39559	34434	88596	54086	71693	43132	14414	79949	85193
22	25991	65959	70769	64721	86413	33475	42740	06175	82758	66248
23	78388	16638	09134	59880	63806	48472	39318	35434	24057	74739
24	12477	09965	96657	57994	59439	76330	24596	77515	09577	91871
25	83266	32883	42451	15579	38155	29793	40914	65990	16255	17777
26	76970	80876	10237	39515	79152	74798	39357	09054	73579	92359
27	37074	65198	44785	68624	98336	84481	97610	78735	46703	98265
28	83712	06514	30101	78295	54656	85417	43189	60048	72781	72606
29	20287	56862	69727	94443	64936	08366	27227	05158	50326	59566
30	74261	32592	86538	27041	65172	85532	07571	80609	39285	65340
31	64081	49863	08478	96001	18888	14810	70545	89755	59064	07210
32	05617	75818	47750	67814	29575	10526	66192	44464	27058	40467
33	26793	74951	95466	74307	13330	42664	85515	20632	05497	33625
34	65988	72850	48737	54719	52056	01596	03845	35067	03134	70322
35	27366	42271	44300	73399	21105	03280	73457	43093	05192	48657
36	56760	10909	98147	34736	33863	95256	12731	66598	50771	83665
37	72880	43338	93643	58904	59543	23943	11231	83268	65938	81581
38	77888	38100	03062	58103	47961	83841	25878	23746	55903	44115
39	28440	07819	21580	51459	47971	29882	13990	29226	23608	15873
40	63525	94441	77033	12147	51054	49955	58312	76923	96071	05813
41	47606	93410	16359	89033	89696	47231	64498	31776	05383	39902
42	52669	45030	96279	14709	52372	87832	02735	50803	72744	88208
43	16738	60159	07425	62369	07515	82721	37875	71153	21315	00132
44	59348	11695	45751	15865	74739	05572	32688	20271	65128	14551
45	12900	71775	29845	60774	94924	21810	38636	33717	67598	82521
46	75086	23527	49939	33595	13484	97588	28617	17979	70749	35234
47	99495	51434	29181	09993	38190	42553	68922	52125	91077	40197
48	26075	31671	45386	36583	93159	48599	52022	41330	60651	91321
49	13636	93596	23377	51133	95126	61496	42474	45141	46660	42338

(*continued*)

TABLE B (*continued*)

	00–04	05–09	10–14	15–19	20–24	25–29	30–34	35–39	40–44	45–49
50	64249	63664	39652	40646	97306	31741	07294	84149	46797	82487
51	26538	44249	04050	48174	65570	44072	40192	51153	11397	58212
52	05845	00512	78630	55328	18116	69296	91705	86224	29503	57071
53	74897	68373	67359	51014	33510	83048	17056	72506	82949	54600
54	20872	54570	35017	88132	25730	22626	86723	91691	13191	77212
55	31432	96156	89177	75541	81355	24480	77243	76690	42507	84362
56	66890	61505	01240	00660	05873	13568	76082	79172	57913	93448
57	48194	57790	79970	33106	86904	48119	52503	24130	72824	21627
58	11303	87118	81471	52936	08555	28420	49416	44448	04269	27029
59	54374	57325	16947	45356	78371	10563	97191	53798	12693	27928
60	64852	34421	61046	90849	13966	39810	42699	21753	76192	10508
61	16309	20384	09491	91588	97720	89846	30376	76970	23063	35894
62	42587	37065	24526	72602	57589	98131	37292	05967	26002	51945
63	40177	98590	97161	41682	84533	67588	62036	49967	01990	72308
64	82309	76128	93965	26743	24141	04838	40254	26065	07938	76236
65	79788	68243	59732	04257	27084	14743	17520	95401	55811	76099
66	40538	79000	89559	25026	42274	23489	34502	75508	06059	86682
67	64016	73598	18609	73150	62463	33102	45205	87440	96767	67042
68	49767	12691	17903	93871	99721	79109	09425	26904	07419	76013
69	76974	55108	29795	08404	82684	00497	51126	79935	57450	55671
70	23854	08480	85983	96025	50117	64610	99425	62291	86943	21541
71	68973	70551	25098	78033	98573	79848	31778	29555	61446	23037
72	36444	93600	65350	14971	25325	00427	52073	64280	18847	24768
73	03003	87800	07391	11594	21196	00781	32550	57158	58887	73041
74	17540	26188	36647	78386	04558	61463	57842	90382	77019	24210
75	38916	55809	47982	41968	69760	79422	80154	91486	19180	15100
76	64288	19843	69122	42502	48508	28820	59933	72998	99942	10515
77	86809	51564	38040	39418	49915	19000	58050	16899	79952	57849
78	99800	99566	14742	05028	30033	94889	53381	23656	75787	79223
79	92345	31890	95712	08279	91794	94068	49037	88674	35355	12267
80	90363	65162	32245	82279	79256	80834	06088	99462	56705	06118
81	64437	32242	48431	04835	39070	59702	31508	60935	22390	52246
82	91714	53662	28373	34333	55791	74758	51144	18827	10704	76803
83	20902	17646	31391	31459	33315	03444	55743	74701	58851	27427
84	12217	86007	70371	52281	14510	76094	96579	54863	78339	20839
85	45177	02863	42307	53571	22532	74921	17735	42201	80540	54721
86	28325	90814	08804	52746	47913	54577	47525	77705	95330	21866
87	29019	28776	56116	54791	64604	08815	46049	71186	34650	14994
88	84979	81353	56219	67062	26146	82567	33122	14124	46240	92973
89	50371	26347	48513	63915	11158	25563	91915	18431	92978	11591
90	53422	06825	69711	67950	64716	18003	49581	45378	99878	61130
91	67453	35651	89316	41620	32048	70225	47597	33137	31443	51445
92	07294	85353	74819	23445	68237	07202	99515	62282	53809	26685
93	79544	00302	45338	16015	66613	88968	14595	63836	77716	79596
94	64144	85442	82060	46471	24162	39500	87351	36637	42833	71875
95	90919	11883	58318	00042	52402	28210	34075	33272	00840	73268
96	06670	57353	86275	92276	77591	46924	60839	55437	03183	13191
97	36634	93976	52062	83678	41256	60948	18685	48992	19462	96062
98	75101	72891	85745	67106	26010	62107	60885	37503	55461	71213
99	05112	71222	72654	51583	05228	62056	57390	42746	39272	96659

(*continued*)

TABLE B (*continued*)

	50–54	55–59	60–64	65–69	70–74	75–79	80–84	85–89	90–94	95–99
50	32847	31282	03345	89593	69214	70381	78285	20054	91018	16742
51	16916	00041	30236	55023	14253	76582	12092	86533	92426	37655
52	66176	34047	21005	27137	03191	48970	64625	22394	39622	79085
53	46299	13335	12180	16861	38043	59292	62675	63631	37020	78195
54	22847	47839	45385	23289	47526	54098	45683	55849	51575	64689
55	41851	54160	92320	69936	34803	92479	33399	71160	64777	83378
56	28444	59497	91586	95917	68553	28639	06455	34174	11130	91994
57	47520	62378	98855	83174	13088	16561	68559	26679	06238	51254
58	34978	63271	13142	82681	05271	08822	06490	44984	49307	62717
59	37404	80416	69035	92980	49486	74378	75610	74976	70056	15478
60	32400	65482	52099	53676	74648	94148	65095	69597	52771	71551
61	89262	86332	51718	70663	11623	29834	79820	73002	84886	03591
62	86866	09127	98021	03871	27789	58444	44832	36505	40672	30180
63	90814	14833	08759	74645	05046	94056	99094	65901	32663	73040
64	19192	82756	20553	58446	55376	88914	75096	26119	83998	43816
65	77585	52593	56612	95766	10019	29531	73064	20953	53523	58136
66	23757	16364	05096	03192	62386	45389	85332	18877	55710	96459
67	45989	96257	23850	26216	23309	21526	07425	50254	19455	29315
68	92970	94243	07316	41467	64837	52406	25225	51553	31220	14032
69	74346	59596	40088	98176	17896	86900	20249	77753	·19099	48885
70	87646	41309	27636	45153	29988	94770	07255	70908	05340	99751
71	50099	71038	45146	06146	55211	99429	43169	66259	97786	59180
72	10127	46900	64984	753498	04115	33624	68774	60013	35515	62556
73	67995	81977	18984	64091	02785	27762	42529	97144	80407	64525
74	26304	80217	84934	82657	69291	35397	98714	35104	08187	48109
75	81994	41070	56642	64091	31229	02595	13513	45148	78722	30144
76	59537	34662	79631	89403	65212	09975	06118	86197	58208	16162
77	51228	10937	62396	81460	47331	91403	95007	06047	16846	64809
78	31089	37995	29577	07828	42272	54016	21950	86192	99046	84864
79	38207	97938	93459	75174	79460	55436	57206	87644	21296	43395
80	88666	31142	09474	89712	63153	62333	42212	06140	42594	43671
81	53365	56134	67582	92557	89520	33452	05134	70628	27612	33738
82	89807	74530	38004	90102	11693	90257	05500	79920	62700	43325
83	18682	81038	85662	90915	91631	22223	91588	80774	07716	12548
84	63571	32579	63942	25371	90234	94592	98475	76884	37635	33608
85	68927	56492	67799	95392	77642	54613	91853	08424	81450	76229
86	56401	63186	39389	99798	31356	89235	97036	32341	33292	73757
87	24333	95603	02359	72942	46287	95382	08452	62862	97869	71775
88	17025	84202	95199	62272	06366	16175	97577	99304	41587	03686
89	02804	08253	52133	20224	68034	50865	57868	22343	55111	03607
90	08298	03879	20995	19850	73090	13191	18963	82244	78479	99121
91	59883	01785	82403	96062	03785	03488	12970	64896	38336	30030
92	46982	06682	62864	91837	74021	89094	39952	64158	79614	78235
93	31121	47266	07661	02051	67599	24471	69843	83696	71402	76287
94	97867	56641	63416	17577	30161	87320	37752	73276	48969	41915
95	57364	86746	08415	14621	49430	22311	15836	72492	49372	44103
96	09559	26263	69511	28064	75999	44540	13337	10918	79846	54809
97	53873	55571	00608	42661	91332	63956	74087	59008	47494	99581
98	35531	19162	86406	05299	77511	24311	57257	22826	77555	05941
99	28229	88629	25695	94932	30721	16197	78742	34974	97528	45447

TABLE C Area under the normal curve between μ and z and beyond z

A	B Area between mean and	C Area beyond	A	B Area between mean and	C Area beyond	A	B Area between mean and	C Area beyond
z	z	z	z	z	z	z	z	z
0.00	.0000	.5000	0.55	.2088	.2912	1.10	.3643	.1357
0.01	.0040	.4960	0.56	.2123	.2877	1.11	.3665	.1335
0.02	.0080	.4920	0.57	.2157	.2843	1.12	.3686	.1314
0.03	.0120	.4880	0.58	.2190	.2810	1.13	.3708	.1292
0.04	.0160	.4840	0.59	.2224	.2776	1.14	.3729	.1271
0.05	.0199	.4801	0.60	.2257	.2743	1.15	.3749	.1251
0.06	.0239	.4761	0.61	.2291	.2709	1.16	.3770	.1230
0.07	.0279	.4721	0.62	.2324	.2676	1.17	.3790	.1210
0.08	.0319	.4681	0.63	.2357	.2643	1.18	.3810	.1190
0.09	.0359	.4641	0.64	.2389	.2611	1.19	.3830	.1170
0.10	.0398	.4602	0.65	.2422	.2578	1.20	.3849	.1151
0.11	.0438	.4562	0.66	.2454	.2546	1.21	.3869	.1131
0.12	.0478	.4522	0.67	.2486	.2514	1.22	.3888	.1112
0.13	.0517	.4483	0.68	.2517	.2483	1.23	.3907	.1093
0.14	.0557	.4443	0.69	.2549	.2451	1.24	.3925	.1075
0.15	.0596	.4404	0.70	.2580	.2420	1.25	.3944	.1056
0.16	.0636	.4364	0.71	.2611	.2389	1.26	.3962	.1038
0.17	.0675	.4325	0.72	.2642	.2358	1.27	.3980	.1020
0.18	.0714	.4286	0.73	.2673	.2327	1.28	.3997	.1003
0.19	.0753	.4247	0.74	.2704	.2296	1.29	.4015	.0985
0.20	.0793	.4207	0.75	.2734	.2266	1.30	.4032	.0968
0.21	.0832	.4168	0.76	.2764	.2236	1.31	.4049	.0951
0.22	.0871	.4129	0.77	.2794	.2206	1.32	.4066	.0934
0.23	.0910	.4090	0.78	.2823	.2177	1.33	.4082	.0918
0.24	.0948	.4052	0.79	.2852	.2148	1.34	.4099	.0901
0.25	.0987	.4013	0.80	.2881	.2119	1.35	.4115	.0885
0.26	.1026	.3974	0.81	.2910	.2090	1.36	.4131	.0869
0.27	.1064	.3936	0.82	.2939	.2061	1.37	.4147	.0853
0.28	.1103	.3897	0.83	.2967	.2033	1.38	.4162	.0838
0.29	.1141	.3859	0.84	.2995	.2005	1.39	.4177	.0823
0.30	.1179	.3821	0.85	.3023	.1977	1.40	.4192	.0808
0.31	.1217	.3783	0.86	.3051	.1949	1.41	.4207	.0793
0.32	.1255	.3745	0.87	.3078	.1922	1.42	.4222	.0778
0.33	.1293	.3707	0.88	.3106	.1894	1.43	.4236	.0764
0.34	.1331	.3669	0.89	.3133	.1867	1.44	.4251	.0749
0.35	.1368	.3632	0.90	.3159	.1841	1.45	.4265	.0735
0.36	.1406	.3594	0.91	.3186	.1814	1.46	.4279	.0721
0.37	.1443	.3557	0.92	.3212	.1788	1.47	.4292	.0708
0.38	.1480	.3520	0.93	.3238	.1762	1.48	.4306	.0694
0.39	.1517	.3483	0.94	.3264	.1736	1.49	.4319	.0681
0.40	.1554	.3446	0.95	.3289	.1711	1.50	.4332	.0668
0.41	.1591	.3409	0.96	.3315	.1685	1.51	.4345	.0655
0.42	.1628	.3372	0.97	.3340	.1660	1.52	.4357	.0643
0.43	.1664	.3336	0.98	.3365	.1635	1.53	.4370	.0630
0.44	.1700	.3300	0.99	.3389	.1611	1.54	.4382	.0618
0.45	.1736	.3264	1.00	.3413	.1587	1.55	.4394	.0606
0.46	.1772	.3228	1.01	.3438	.1562	1.56	.4406	.0594
0.47	.1808	.3192	1.02	.3461	.1539	1.57	.4418	.0582
0.48	.1844	.3156	1.03	.3485	.1515	1.58	.4429	.0571
0.49	.1879	.3121	1.04	.3508	.1492	1.59	.4441	.0559
0.50	.1915	.3085	1.05	.3531	.1469	1.60	.4452	.0548
0.51	.1950	.3050	1.06	.3554	.1446	1.61	.4463	.0537
0.52	.1985	.3015	1.07	.3577	.1423	1.62	.4474	.0526
0.53	.2019	.2981	1.08	.3599	.1401	1.63	.4484	.0516
0.54	.2054	.2946	1.09	.3621	.1379	1.64	.4495	.0505

(continued)

TABLE C (*continued*)

A	B Area between mean and	C Area beyond	A	B Area between mean and	C Area beyond	A	B Area between mean and	C Area beyond
z	z	z	z	z	z	z	z	z
1.65	.4505	.0495	2.22	.4868	.0132	2.79	.4974	.0026
1.66	.4515	.0485	2.23	.4871	.0129	2.80	.4974	.0026
1.67	.4525	.0475	2.24	.4875	.0125	2.81	.4975	.0025
1.68	.4535	.0465	2.25	.4878	.0122	2.82	.4976	.0024
1.69	.4545	.0455	2.26	.4881	.0119	2.83	.4977	.0023
1.70	.4554	.0446	2.27	.4884	.0116	2.84	.4977	.0023
1.71	.4564	.0436	2.28	.4887	.0113	2.85	.4978	.0022
1.72	.4573	.0427	2.29	.4890	.0110	2.86	.4979	.0021
1.73	.4582	.0418	2.30	.4893	.0107	2.87	.4979	.0021
1.74	.4591	.0409	2.31	.4896	.0104	2.88	.4980	.0020
1.75	.4599	.0401	2.32	.4898	.0102	2.89	.4981	.0019
1.76	.4608	.0392	2.33	.4901	.0099	2.90	.4981	.0019
1.77	.4616	.0384	2.34	.4904	.0096	2.91	.4982	.0018
1.78	.4625	.0375	2.35	.4906	.0094	2.92	.4982	.0018
1.79	.4633	.0367	2.36	.4909	.0091	2.93	.4983	.0017
1.80	.4641	.0359	2.37	.4911	.0089	2.94	.4984	.0016
1.81	.4649	.0351	2.38	.4913	.0087	2.95	.4984	.0016
1.82	.4656	.0344	2.39	.4916	.0084	2.96	.4985	.0015
1.83	.4664	.0336	2.40	.4918	.0082	2.97	.4985	.0015
1.84	.4671	.0329	2.41	.4920	.0080	2.98	.4986	.0014
1.85	.4678	.0322	2.42	.4922	.0078	2.99	.4986	.0014
1.86	.4686	.0314	2.43	.4925	.0075	3.00	.4987	.0013
1.87	.4693	.0307	2.44	.4927	.0073	3.01	.4987	.0013
1.88	.4699	.0301	2.45	.4929	.0071	3.02	.4987	.0013
1.89	.4706	.0294	2.46	.4931	.0069	3.03	.4988	.0012
1.90	.4713	.0287	2.47	.4932	.0068	3.04	.4988	.0012
1.91	.4719	.0281	2.48	.4934	.0066	3.05	.4989	.0011
1.92	.4726	.0274	2.49	.4936	.0064	3.06	.4989	.0011
1.93	.4732	.0268	2.50	.4938	.0062	3.07	.4989	.0011
1.94	.4738	.0262	2.51	.4940	.0060	3.08	.4990	.0010
1.95	.4744	.0256	2.52	.4941	.0059	3.09	.4990	.0010
1.96	.4750	.0250	2.53	.4943	.0057	3.10	.4990	.0010
1.97	.4756	.0244	2.54	.4945	.0055	3.11	.4991	.0009
1.98	.4761	.0239	2.55	.4946	.0054	3.12	.4991	.0009
1.99	.4767	.0233	2.56	.4948	.0052	3.13	.4991	.0009
2.00	.4772	.0228	2.57	.4949	.0051	3.14	.4992	.0008
2.01	.4778	.0222	2.58	.4951	.0049	3.15	.4992	.0008
2.02	.4783	.0217	2.59	.4952	.0048	3.16	.4992	.0008
2.03	.4788	.0212	2.60	.4953	.0047	3.17	.4992	.0008
2.04	.4793	.0207	2.61	.4955	.0045	3.18	.4993	.0007
2.05	.4798	.0202	2.62	.4956	.0044	3.19	.4993	.0007
2.06	.4803	.0197	2.63	.4957	.0043	3.20	.4993	.0007
2.07	.4808	.0192	2.64	.4959	.0041	3.21	.4993	.0007
2.08	.4812	.0188	2.65	.4960	.0040	3.22	.4994	.0006
2.09	.4817	.0183	2.66	.4961	.0039	3.23	.4994	.0006
2.10	.4821	.0179	2.67	.4962	.0038	3.24	.4994	.0006
2.11	.4826	.0174	2.68	.4963	.0037	3.25	.4994	.0006
2.12	.4830	.0170	2.69	.4964	.0036	3.30	.4995	.0005
2.13	.4834	.0166	2.70	.4965	.0035	3.35	.4996	.0004
2.14	.4838	.0162	2.71	.4966	.0034	3.40	.4997	.0003
2.15	.4842	.0158	2.72	.4967	.0033	3.45	.4997	.0003
2.16	.4846	.0154	2.73	.4968	.0032	3.50	.4998	.0002
2.17	.4850	.0150	2.74	.4969	.0031	3.60	.4998	.0002
2.18	.4854	.0146	2.75	.4970	.0030	3.70	.4999	.0001
2.19	.4857	.0143	2.76	.4971	.0029	3.80	.4999	.0001
2.20	.4861	.0139	2.77	.4972	.0028	3.90	.49995	.00005
2.21	.4864	.0136	2.78	.4973	.0027	4.00	.49997	.00003

T A B L E D The *t* distribution*

| | Confidence interval percents (two-tailed) | | | | | |
	80%	90%	95%	98%	99%	99.9%
	α level for two-tailed test					
	.20	.10	.05	.02	.01	.001
	α level for one-tailed test					
df	.10	.05	.025	.01	.005	.0005
1	3.078	6.314	12.706	31.821	63.657	636.619
2	1.886	2.920	4.303	6.965	9.925	31.598
3	1.638	2.353	3.182	4.541	5.841	12.924
4	1.533	2.132	2.776	3.747	4.604	8.610
5	1.476	2.015	2.571	3.365	4.032	6.869
6	1.440	1.943	2.447	3.143	3.707	5.959
7	1.415	1.895	2.365	2.998	3.499	5.408
8	1.397	1.860	2.306	2.896	3.355	5.041
9	1.383	1.833	2.262	2.821	3.250	4.781
10	1.372	1.812	2.228	2.764	3.169	4.587
11	1.363	1.796	2.201	2.718	3.106	4.437
12	1.356	1.782	2.179	2.681	3.055	4.318
13	1.350	1.771	2.160	2.650	3.012	4.221
14	1.345	1.761	2.145	2.624	2.977	4.140
15	1.341	1.753	2.131	2.602	2.947	4.073
16	1.337	1.746	2.120	2.583	2.921	4.015
17	1.333	1.740	2.110	2.567	2.898	3.965
18	1.330	1.734	2.101	2.552	2.878	3.922
19	1.328	1.729	2.093	2.539	2.861	3.883
20	1.325	1.725	2.086	2.528	2.845	3.850
21	1.323	1.721	2.080	2.518	2.831	3.819
22	1.321	1.717	2.074	2.508	2.819	3.792
23	1.319	1.714	2.069	2.500	2.807	3.767
24	1.318	1.711	2.064	2.492	2.797	3.745
25	1.316	1.708	2.060	2.485	2.787	3.725
26	1.315	1.706	2.056	2.479	2.779	3.707
27	1.314	1.703	2.052	2.474	2.771	3.690
28	1.313	1.701	2.048	2.467	2.763	3.674
29	1.311	1.699	2.045	2.462	2.756	3.659
30	1.310	1.697	2.042	2.457	2.750	3.646
40	1.303	1.684	2.021	2.423	2.704	3.551
60	1.296	1.671	2.000	2.390	2.660	3.460
120	1.289	1.658	1.980	2.358	2.617	3.373
∞	1.282	1.645	1.960	2.326	2.576	3.291

* To be significant the *t* obtained from the data must be equal to or **greater than** the value shown in the table.

SOURCE: From Table III in R. A. Fisher and F. Yates, *Statistical Tables for Biological, Agricultural, and Medical Research,* Sixth Edition, published by Addison Wesley Longman Ltd. (1974). Reprinted with permission.

TABLE E Chi square distribution*

df	α levels				
	.10	.05	.02	.01	.001
1	2.71	3.84	5.41	6.64	10.81
2	4.60	5.99	7.82	9.21	13.82
3	6.25	7.82	9.84	11.34	16.27
4	7.78	9.49	11.67	13.28	18.46
5	9.24	11.07	13.39	15.09	20.52
6	10.64	12.59	15.03	16.81	22.46
7	12.02	14.07	16.62	18.48	24.32
8	13.36	15.51	18.17	20.09	26.12
9	14.68	16.92	19.68	21.67	27.88
10	15.99	18.31	21.16	23.21	29.59
11	17.28	19.68	22.62	24.72	31.26
12	18.55	21.03	24.05	26.22	32.91
13	19.81	22.36	25.47	27.69	34.53
14	21.06	23.68	26.87	29.14	36.12
15	22.31	25.00	28.26	30.58	37.70
16	23.54	26.30	29.63	32.00	39.25
17	24.77	27.59	31.00	33.41	40.79
18	25.99	28.87	32.35	34.80	42.31
19	27.20	30.14	33.69	36.19	43.82
20	28.41	31.41	35.02	37.57	45.32
21	29.62	32.67	36.34	38.93	46.80
22	30.81	33.92	37.66	40.29	48.27
23	32.01	35.17	38.97	41.64	49.73
24	33.20	36.42	40.27	42.98	51.18
25	34.38	37.65	41.57	44.31	52.62
26	35.56	38.88	42.86	45.64	54.05
27	36.74	40.11	44.14	46.96	55.48
28	37.92	41.34	45.42	48.28	56.89
29	39.09	42.56	46.69	49.59	58.30
30	40.26	43.77	47.96	50.89	59.70

* To be significant the χ^2 obtained from the data must be equal to or **greater than** the value shown in the table.

SOURCE: From Table IV in R. A. Fisher and F. Yates, *Statistical Tables for Biological, Agricultural, and Medical Research,* Sixth Edition, published by Addison Wesley Longman Ltd. (1974). Reprinted with permission.

TABLE F The *F* distribution*

α levels of .05 (lightface) and .01 (**boldface**) for the distribution of *F*

Degrees of freedom (for the numerator of *F* ratio)

Each cell lists the .05 value (lightface) and the .01 value (**boldface**). The first column is the degrees of freedom for the denominator of the *F* ratio.

den. df	1	2	3	4	5	6	7	8	9	10	11	12	14	16	20	24	30	40	50	75	100	200	500	∞
1	161 / **4,052**	200 / **4,999**	216 / **5,403**	225 / **5,625**	230 / **5,764**	234 / **5,859**	237 / **5,928**	239 / **5,981**	241 / **6,022**	242 / **6,056**	243 / **6,082**	244 / **6,106**	245 / **6,142**	246 / **6,169**	248 / **6,208**	249 / **6,234**	250 / **6,258**	251 / **6,286**	252 / **6,302**	253 / **6,323**	253 / **6,334**	254 / **6,352**	254 / **6,361**	254 / **6,366**
2	18.51 / **98.49**	19.00 / **99.00**	19.16 / **99.17**	19.25 / **99.25**	19.30 / **99.30**	19.33 / **99.33**	19.36 / **99.34**	19.37 / **99.36**	19.38 / **99.38**	19.39 / **99.40**	19.40 / **99.41**	19.41 / **99.42**	19.42 / **99.43**	19.43 / **99.44**	19.44 / **99.45**	19.45 / **99.46**	19.46 / **99.47**	19.47 / **99.48**	19.47 / **99.49**	19.48 / **99.49**	19.49 / **99.49**	19.49 / **99.49**	19.50 / **99.50**	19.50 / **99.50**
3	10.13 / **34.12**	9.55 / **30.82**	9.28 / **29.46**	9.12 / **28.71**	9.01 / **28.24**	8.94 / **27.91**	8.88 / **27.67**	8.84 / **27.49**	8.81 / **27.34**	8.78 / **27.23**	8.76 / **27.13**	8.74 / **27.05**	8.71 / **29.92**	8.69 / **26.83**	8.66 / **26.69**	8.64 / **26.60**	8.62 / **26.50**	8.60 / **26.41**	8.58 / **26.35**	8.57 / **26.27**	8.56 / **26.23**	8.54 / **26.18**	8.54 / **26.14**	8.53 / **26.12**
4	7.71 / **21.20**	6.94 / **18.00**	6.59 / **16.69**	6.39 / **15.98**	6.26 / **15.52**	6.16 / **15.21**	6.09 / **14.98**	6.04 / **14.80**	6.00 / **14.66**	5.96 / **14.54**	5.93 / **14.45**	5.91 / **14.37**	5.87 / **14.24**	5.84 / **14.15**	5.80 / **14.02**	5.77 / **13.93**	5.74 / **13.83**	5.71 / **13.74**	5.70 / **13.69**	5.68 / **13.61**	5.66 / **13.57**	5.66 / **13.52**	5.64 / **13.48**	5.63 / **13.46**
5	6.61 / **16.26**	5.79 / **13.27**	5.41 / **12.06**	5.19 / **11.39**	5.05 / **10.97**	4.95 / **10.67**	4.88 / **10.45**	4.82 / **10.27**	4.78 / **10.15**	4.74 / **10.05**	4.70 / **9.96**	4.68 / **9.89**	4.64 / **9.77**	4.60 / **9.68**	4.56 / **9.55**	4.53 / **9.47**	4.50 / **9.38**	4.46 / **9.29**	4.44 / **9.24**	4.42 / **9.17**	4.40 / **9.13**	4.38 / **9.07**	4.37 / **9.04**	4.36 / **9.02**
6	5.99 / **13.74**	5.14 / **10.92**	4.76 / **9.78**	4.53 / **9.15**	4.39 / **8.75**	4.28 / **8.47**	4.21 / **8.26**	4.15 / **8.10**	4.10 / **7.98**	4.06 / **7.87**	4.03 / **7.79**	4.00 / **7.72**	3.96 / **7.60**	3.92 / **7.52**	3.87 / **7.39**	3.84 / **7.31**	3.81 / **7.23**	3.77 / **7.14**	3.75 / **7.09**	3.72 / **7.02**	3.71 / **6.99**	3.69 / **6.94**	3.68 / **6.90**	3.67 / **6.88**
7	5.59 / **12.25**	4.74 / **9.55**	4.35 / **8.45**	4.12 / **7.85**	3.97 / **7.46**	3.87 / **7.19**	3.79 / **7.00**	3.73 / **6.84**	3.68 / **6.71**	3.63 / **6.62**	3.60 / **6.54**	3.57 / **6.47**	3.52 / **6.35**	3.49 / **6.27**	3.44 / **6.15**	3.41 / **6.07**	3.38 / **5.98**	3.34 / **5.90**	3.32 / **5.85**	3.29 / **5.78**	3.28 / **5.75**	3.25 / **5.70**	3.24 / **5.67**	3.23 / **5.65**
8	5.32 / **11.26**	4.46 / **8.65**	4.07 / **7.59**	3.84 / **7.01**	3.69 / **6.63**	3.58 / **6.37**	3.50 / **6.19**	3.44 / **6.03**	3.39 / **5.91**	3.34 / **5.82**	3.31 / **5.74**	3.28 / **5.67**	3.23 / **5.56**	3.20 / **5.48**	3.15 / **5.36**	3.12 / **5.28**	3.08 / **5.20**	3.05 / **5.11**	3.03 / **5.06**	3.00 / **5.00**	2.98 / **4.96**	2.96 / **4.91**	2.94 / **4.88**	2.93 / **4.86**
9	5.12 / **10.56**	4.26 / **8.02**	3.86 / **6.99**	3.63 / **6.42**	3.48 / **6.06**	3.37 / **5.80**	3.29 / **5.62**	3.23 / **5.47**	3.18 / **5.35**	3.13 / **5.26**	3.10 / **5.18**	3.07 / **5.11**	3.02 / **5.00**	2.98 / **4.92**	2.93 / **4.80**	2.90 / **4.73**	2.86 / **4.64**	2.82 / **4.56**	2.80 / **4.51**	2.77 / **4.45**	2.76 / **4.41**	2.73 / **4.36**	2.72 / **4.33**	2.71 / **4.31**
10	4.96 / **10.04**	4.10 / **7.56**	3.71 / **6.55**	3.48 / **5.99**	3.33 / **5.64**	3.22 / **5.39**	3.14 / **5.21**	3.07 / **5.06**	3.02 / **4.95**	2.97 / **4.85**	2.94 / **4.78**	2.91 / **4.71**	2.86 / **4.60**	2.82 / **4.52**	2.77 / **4.41**	2.74 / **4.33**	2.70 / **4.25**	2.67 / **4.17**	2.64 / **4.12**	2.61 / **4.05**	2.59 / **4.01**	2.56 / **3.96**	2.55 / **3.93**	2.54 / **3.91**
11	4.84 / **9.65**	3.98 / **7.20**	3.59 / **6.22**	3.36 / **5.67**	3.20 / **5.32**	3.09 / **5.07**	3.01 / **4.88**	2.95 / **4.74**	2.90 / **4.63**	2.86 / **4.54**	2.82 / **4.46**	2.79 / **4.40**	2.74 / **4.29**	2.70 / **4.21**	2.65 / **4.10**	2.61 / **4.02**	2.57 / **3.94**	2.53 / **3.86**	2.50 / **3.80**	2.47 / **3.74**	2.45 / **3.70**	2.42 / **3.66**	2.41 / **3.62**	2.40 / **3.60**
12	4.75 / **9.33**	3.88 / **6.93**	3.49 / **5.95**	3.26 / **5.41**	3.11 / **5.06**	3.00 / **4.82**	2.92 / **4.65**	2.85 / **4.50**	2.80 / **4.39**	2.76 / **4.30**	2.72 / **4.22**	2.69 / **4.16**	2.64 / **4.05**	2.60 / **3.98**	2.54 / **3.86**	2.50 / **3.78**	2.46 / **3.70**	2.42 / **3.61**	2.40 / **3.56**	2.36 / **3.49**	2.35 / **3.46**	2.32 / **3.41**	2.31 / **3.38**	2.30 / **3.36**
13	4.67 / **9.07**	3.80 / **6.70**	3.41 / **5.74**	3.18 / **5.20**	3.02 / **4.86**	2.92 / **4.62**	2.84 / **4.44**	2.77 / **4.30**	2.72 / **4.19**	2.67 / **4.10**	2.63 / **4.02**	2.60 / **3.96**	2.55 / **3.85**	2.51 / **3.78**	2.46 / **3.67**	2.42 / **3.59**	2.38 / **3.51**	2.34 / **3.42**	2.32 / **3.37**	2.28 / **3.30**	2.26 / **3.27**	2.24 / **3.21**	2.22 / **3.18**	2.21 / **3.16**

(continued)

Degrees of freedom (for the denominator of *F* ratio)

*To be significant the *F* obtained from the data must be equal to or **greater than** the value shown in the table.
SOURCE: From *Statistical Methods*, by G. W. Snedecor and W. W. Cochran, Seventh Edition. Copyright © 1980 Iowa State University Press. Reprinted with permission.

TABLE F (continued)

Each cell shows the 0.05 value (top) and the 0.01 value (bold, bottom).

Degrees of freedom (for the denominator of F ratio)	1	2	3	4	5	6	7	8	9	10	11	12	14	16	20	24	30	40	50	75	100	200	500	∞
14	4.60 / **8.86**	3.74 / **6.51**	3.34 / **5.56**	3.11 / **5.03**	2.96 / **4.69**	2.85 / **4.46**	2.77 / **4.28**	2.70 / **4.14**	2.65 / **4.03**	2.60 / **3.94**	2.56 / **3.86**	2.53 / **3.80**	2.48 / **3.70**	2.44 / **3.62**	2.39 / **3.51**	2.35 / **3.43**	2.31 / **3.34**	2.27 / **3.26**	2.24 / **3.21**	2.21 / **3.14**	2.19 / **3.11**	2.16 / **3.06**	2.14 / **3.02**	2.13 / **3.00**
15	4.54 / **8.68**	3.68 / **6.36**	3.29 / **5.52**	3.06 / **4.89**	2.90 / **4.56**	2.79 / **4.32**	2.70 / **4.14**	2.64 / **4.00**	2.59 / **3.89**	2.55 / **3.80**	2.51 / **3.73**	2.48 / **3.67**	2.43 / **3.56**	2.39 / **3.48**	2.33 / **3.36**	2.29 / **3.29**	2.25 / **3.20**	2.21 / **3.12**	2.18 / **3.07**	2.15 / **3.00**	2.12 / **2.97**	2.10 / **2.92**	2.08 / **2.89**	2.07 / **2.87**
16	4.49 / **8.53**	3.63 / **6.23**	3.24 / **5.29**	3.01 / **4.77**	2.85 / **4.44**	2.74 / **4.20**	2.66 / **4.03**	2.59 / **3.89**	2.54 / **3.78**	2.49 / **3.69**	2.45 / **3.61**	2.42 / **3.55**	2.37 / **3.45**	2.33 / **3.37**	2.28 / **3.25**	2.24 / **3.18**	2.20 / **3.10**	2.16 / **3.01**	2.13 / **2.96**	2.09 / **2.89**	2.07 / **2.86**	2.04 / **2.80**	2.02 / **2.77**	2.01 / **2.75**
17	4.45 / **8.40**	3.59 / **6.11**	3.20 / **5.18**	2.96 / **4.67**	2.81 / **4.34**	2.70 / **4.10**	2.62 / **3.93**	2.55 / **3.79**	2.50 / **3.68**	2.45 / **3.59**	2.41 / **3.52**	2.38 / **3.45**	2.33 / **3.35**	2.29 / **3.27**	2.23 / **3.16**	2.19 / **3.08**	2.15 / **3.00**	2.11 / **2.92**	2.08 / **2.86**	2.04 / **2.79**	2.02 / **2.76**	1.99 / **2.70**	1.97 / **2.67**	1.96 / **2.65**
18	4.41 / **8.28**	3.55 / **6.01**	3.16 / **5.09**	2.93 / **4.58**	2.77 / **4.25**	2.66 / **4.01**	2.58 / **3.85**	2.51 / **3.71**	2.46 / **3.60**	2.41 / **3.51**	2.37 / **3.44**	2.34 / **3.37**	2.29 / **3.27**	2.25 / **3.19**	2.19 / **3.07**	2.15 / **3.00**	2.11 / **2.91**	2.07 / **2.83**	2.04 / **2.78**	2.00 / **2.71**	1.98 / **2.68**	1.95 / **2.62**	1.93 / **2.59**	1.92 / **2.57**
19	4.38 / **8.18**	3.52 / **5.93**	3.13 / **5.01**	2.90 / **4.50**	2.74 / **4.17**	2.63 / **3.94**	2.55 / **3.77**	2.48 / **3.63**	2.43 / **3.52**	2.38 / **3.43**	2.34 / **3.36**	2.31 / **3.30**	2.26 / **3.19**	2.21 / **3.12**	2.15 / **3.00**	2.11 / **2.92**	2.07 / **2.84**	2.02 / **2.76**	2.00 / **2.70**	1.96 / **2.63**	1.94 / **2.60**	1.91 / **2.54**	1.90 / **2.51**	1.88 / **2.49**
20	4.35 / **8.10**	3.49 / **5.85**	3.10 / **4.94**	2.87 / **4.43**	2.71 / **4.10**	2.60 / **3.87**	2.52 / **3.71**	2.45 / **3.56**	2.40 / **3.45**	2.35 / **3.37**	2.31 / **3.30**	2.28 / **3.23**	2.23 / **3.13**	2.18 / **3.05**	2.12 / **2.94**	2.08 / **2.86**	2.04 / **2.77**	1.99 / **2.69**	1.96 / **2.63**	1.92 / **2.56**	1.90 / **2.53**	1.87 / **2.47**	1.85 / **2.44**	1.84 / **2.42**
21	4.32 / **8.02**	3.47 / **5.78**	3.07 / **4.87**	2.84 / **4.37**	2.68 / **4.04**	2.57 / **3.81**	2.49 / **3.65**	2.42 / **3.51**	2.37 / **3.40**	2.32 / **3.31**	2.28 / **3.24**	2.25 / **3.17**	2.20 / **3.07**	2.15 / **2.99**	2.09 / **2.88**	2.05 / **2.80**	2.00 / **2.72**	1.96 / **2.63**	1.93 / **2.58**	1.89 / **2.51**	1.87 / **2.47**	1.84 / **2.42**	1.82 / **2.38**	1.81 / **2.36**
22	4.30 / **7.94**	3.44 / **5.72**	3.05 / **4.82**	2.82 / **4.31**	2.66 / **3.99**	2.55 / **3.76**	2.47 / **3.59**	2.40 / **3.45**	2.35 / **3.35**	2.30 / **3.26**	2.26 / **3.18**	2.23 / **3.12**	2.18 / **3.02**	2.13 / **2.94**	2.07 / **2.83**	2.03 / **2.75**	1.98 / **2.67**	1.93 / **2.58**	1.91 / **2.53**	1.87 / **2.46**	1.84 / **2.42**	1.81 / **2.37**	1.80 / **2.33**	1.78 / **2.31**
23	4.28 / **7.88**	3.42 / **5.66**	3.03 / **4.76**	2.80 / **4.26**	2.64 / **3.94**	2.53 / **3.71**	2.45 / **3.54**	2.38 / **3.41**	2.32 / **3.30**	2.28 / **3.21**	2.24 / **3.14**	2.20 / **3.07**	2.14 / **2.97**	2.10 / **2.89**	2.04 / **2.78**	2.00 / **2.70**	1.96 / **2.62**	1.91 / **2.53**	1.88 / **2.48**	1.84 / **2.41**	1.82 / **2.37**	1.79 / **2.32**	1.77 / **2.28**	1.76 / **2.26**
24	4.26 / **7.82**	3.40 / **5.61**	3.01 / **4.72**	2.78 / **4.22**	2.62 / **3.90**	2.51 / **3.67**	2.43 / **3.50**	2.36 / **3.36**	2.30 / **3.25**	2.26 / **3.17**	2.22 / **3.09**	2.18 / **3.03**	2.13 / **2.93**	2.09 / **2.85**	2.02 / **2.74**	1.98 / **2.66**	1.94 / **2.58**	1.89 / **2.49**	1.86 / **2.44**	1.82 / **2.36**	1.80 / **2.33**	1.76 / **2.27**	1.74 / **2.23**	1.73 / **2.21**
25	4.24 / **7.77**	3.38 / **5.57**	2.99 / **4.68**	2.76 / **4.18**	2.60 / **3.86**	2.49 / **3.63**	2.41 / **3.46**	2.34 / **3.32**	2.28 / **3.21**	2.24 / **3.13**	2.20 / **3.05**	2.16 / **2.99**	2.11 / **2.89**	2.06 / **2.81**	2.00 / **2.70**	1.96 / **2.62**	1.92 / **2.54**	1.87 / **2.45**	1.84 / **2.40**	1.80 / **2.32**	1.77 / **2.29**	1.74 / **2.23**	1.72 / **2.19**	1.71 / **2.17**
26	4.22 / **7.72**	3.37 / **5.53**	2.98 / **4.64**	2.74 / **4.14**	2.59 / **3.82**	2.47 / **3.59**	2.39 / **3.42**	2.32 / **3.29**	2.27 / **3.17**	2.22 / **3.09**	2.18 / **3.02**	2.15 / **2.96**	2.10 / **2.86**	2.05 / **2.77**	1.99 / **2.66**	1.95 / **2.58**	1.90 / **2.50**	1.85 / **2.41**	1.82 / **2.36**	1.78 / **2.28**	1.76 / **2.25**	1.72 / **2.19**	1.70 / **2.15**	1.69 / **2.13**

Degrees of freedom (for the numerator of F ratio)

(continued)

T A B L E F *(continued)*

Degrees of freedom (for the numerator of F ratio)

Each cell shows the value for the .05 level (light) over the .01 level (bold).

df (denom.)	1	2	3	4	5	6	7	8	9	10	11	12	14	16	20	24	30	40	50	75	100	200	500	∞	df
27	4.21 / **7.68**	3.35 / **5.49**	2.96 / **4.60**	2.73 / **4.11**	2.57 / **3.79**	2.46 / **3.56**	2.37 / **3.39**	2.30 / **3.26**	2.25 / **3.14**	2.20 / **3.06**	2.16 / **2.98**	2.13 / **2.93**	2.08 / **2.83**	2.03 / **2.74**	1.97 / **2.63**	1.93 / **2.55**	1.88 / **2.47**	1.84 / **2.38**	1.80 / **2.33**	1.76 / **2.25**	1.74 / **2.21**	1.71 / **2.16**	1.68 / **2.12**	1.67 / **2.10**	27
28	4.20 / **7.64**	3.34 / **5.45**	2.95 / **4.57**	2.71 / **4.07**	2.56 / **3.76**	2.44 / **3.53**	2.36 / **3.36**	2.29 / **3.23**	2.24 / **3.11**	2.19 / **3.03**	2.15 / **2.95**	2.12 / **2.90**	2.06 / **2.80**	2.02 / **2.71**	1.96 / **2.60**	1.91 / **2.52**	1.87 / **2.44**	1.81 / **2.35**	1.78 / **2.30**	1.75 / **2.22**	1.72 / **2.18**	1.69 / **2.13**	1.67 / **2.09**	1.65 / **2.06**	28
29	4.18 / **7.60**	3.33 / **5.42**	2.93 / **4.54**	2.70 / **4.04**	2.54 / **3.73**	2.43 / **3.50**	2.35 / **3.33**	2.28 / **3.20**	2.22 / **3.08**	2.18 / **3.00**	2.14 / **2.92**	2.10 / **2.87**	2.05 / **2.77**	2.00 / **2.68**	1.94 / **2.57**	1.90 / **2.49**	1.85 / **2.41**	1.80 / **2.32**	1.77 / **2.27**	1.73 / **2.19**	1.71 / **2.15**	1.68 / **2.10**	1.65 / **2.06**	1.64 / **2.03**	29
30	4.17 / **7.56**	3.32 / **5.39**	2.92 / **4.51**	2.69 / **4.02**	2.53 / **3.70**	2.42 / **3.47**	2.34 / **3.30**	2.27 / **3.17**	2.21 / **3.06**	2.16 / **2.98**	2.12 / **2.90**	2.09 / **2.84**	2.04 / **2.74**	1.99 / **2.66**	1.93 / **2.55**	1.89 / **2.47**	1.84 / **2.38**	1.79 / **2.29**	1.76 / **2.24**	1.72 / **2.16**	1.69 / **2.13**	1.66 / **2.07**	1.64 / **2.03**	1.62 / **2.01**	30
32	4.15 / **7.50**	3.30 / **5.34**	2.90 / **4.46**	2.67 / **3.97**	2.51 / **3.66**	2.40 / **3.42**	2.32 / **3.25**	2.25 / **3.12**	2.19 / **3.01**	2.14 / **2.94**	2.10 / **2.86**	2.07 / **2.80**	2.02 / **2.70**	1.97 / **2.62**	1.91 / **2.51**	1.86 / **2.42**	1.82 / **2.34**	1.76 / **2.25**	1.74 / **2.20**	1.69 / **2.12**	1.67 / **2.08**	1.64 / **2.02**	1.61 / **1.98**	1.59 / **1.96**	32
34	4.13 / **7.44**	3.28 / **5.29**	2.88 / **4.42**	2.65 / **3.93**	2.49 / **3.61**	2.38 / **3.38**	2.30 / **3.21**	2.23 / **3.08**	2.17 / **2.97**	2.12 / **2.89**	2.08 / **2.82**	2.05 / **2.76**	2.00 / **2.66**	1.95 / **2.58**	1.89 / **2.47**	1.84 / **2.38**	1.80 / **2.30**	1.74 / **2.21**	1.71 / **2.15**	1.67 / **2.08**	1.64 / **2.04**	1.61 / **1.98**	1.59 / **1.94**	1.57 / **1.91**	34
36	4.11 / **7.39**	3.26 / **5.24**	2.86 / **4.38**	2.63 / **3.89**	2.48 / **3.58**	2.36 / **3.35**	2.28 / **3.18**	2.21 / **3.04**	2.15 / **2.94**	2.10 / **2.86**	2.06 / **2.78**	2.03 / **2.72**	1.98 / **2.62**	1.93 / **2.54**	1.87 / **2.43**	1.82 / **2.35**	1.78 / **2.26**	1.72 / **2.17**	1.69 / **2.12**	1.65 / **2.04**	1.62 / **2.00**	1.59 / **1.94**	1.56 / **1.90**	1.55 / **1.87**	36
38	4.10 / **7.35**	3.25 / **5.21**	2.85 / **4.34**	2.62 / **3.86**	2.46 / **3.54**	2.35 / **3.32**	2.26 / **3.15**	2.19 / **3.02**	2.14 / **2.91**	2.09 / **2.82**	2.05 / **2.75**	2.02 / **2.69**	1.96 / **2.59**	1.92 / **2.51**	1.85 / **2.40**	1.80 / **2.32**	1.76 / **2.22**	1.71 / **2.14**	1.67 / **2.08**	1.63 / **2.00**	1.60 / **1.97**	1.57 / **1.90**	1.54 / **1.86**	1.53 / **1.84**	38
40	4.08 / **7.31**	3.23 / **5.18**	2.84 / **4.31**	2.61 / **3.83**	2.45 / **3.51**	2.34 / **3.29**	2.25 / **3.12**	2.18 / **2.99**	2.12 / **2.88**	2.07 / **2.80**	2.04 / **2.73**	2.00 / **2.66**	1.95 / **2.56**	1.90 / **2.49**	1.84 / **2.37**	1.79 / **2.29**	1.74 / **2.20**	1.69 / **2.11**	1.66 / **2.05**	1.61 / **1.97**	1.59 / **1.94**	1.55 / **1.88**	1.53 / **1.84**	1.51 / **1.81**	40
42	4.07 / **7.25**	3.22 / **5.15**	2.83 / **4.29**	2.59 / **3.80**	2.44 / **3.49**	2.32 / **3.26**	2.24 / **3.10**	2.17 / **2.96**	2.11 / **2.86**	2.06 / **2.77**	2.02 / **2.70**	1.99 / **2.64**	1.94 / **2.54**	1.89 / **2.46**	1.82 / **2.35**	1.78 / **2.26**	1.73 / **2.17**	1.68 / **2.08**	1.64 / **2.02**	1.60 / **1.94**	1.57 / **1.91**	1.54 / **1.85**	1.51 / **1.80**	1.49 / **1.78**	42
44	4.06 / **7.24**	3.21 / **5.12**	2.82 / **4.26**	2.58 / **3.78**	2.43 / **3.46**	2.31 / **3.24**	2.23 / **3.07**	2.16 / **2.94**	2.10 / **2.84**	2.05 / **2.75**	2.01 / **2.68**	1.98 / **2.62**	1.92 / **2.52**	1.88 / **2.44**	1.81 / **2.32**	1.76 / **2.24**	1.72 / **2.15**	1.66 / **2.06**	1.63 / **2.00**	1.58 / **1.92**	1.56 / **1.88**	1.52 / **1.82**	1.50 / **1.78**	1.48 / **1.75**	44
46	4.05 / **7.21**	3.20 / **5.10**	2.81 / **4.24**	2.57 / **3.76**	2.42 / **3.44**	2.30 / **3.22**	2.22 / **3.05**	2.14 / **2.92**	2.09 / **2.82**	2.04 / **2.73**	2.00 / **2.66**	1.97 / **2.60**	1.91 / **2.50**	1.87 / **2.42**	1.80 / **2.30**	1.75 / **2.22**	1.71 / **2.13**	1.65 / **2.04**	1.62 / **1.98**	1.57 / **1.90**	1.54 / **1.86**	1.51 / **1.80**	1.48 / **1.76**	1.46 / **1.72**	46
48	4.04 / **7.19**	3.19 / **5.08**	2.80 / **4.22**	2.56 / **3.74**	2.41 / **3.42**	2.30 / **3.20**	2.21 / **3.04**	2.14 / **2.90**	2.08 / **2.80**	2.03 / **2.71**	1.99 / **2.64**	1.96 / **2.58**	1.90 / **2.48**	1.86 / **2.40**	1.79 / **2.28**	1.74 / **2.20**	1.70 / **2.11**	1.64 / **2.02**	1.61 / **1.96**	1.56 / **1.88**	1.53 / **1.84**	1.50 / **1.78**	1.47 / **1.73**	1.45 / **1.70**	48

Degrees of freedom (for the denominator of F ratio)

(continued)

TABLE F (continued)

Degrees of freedom (for the numerator of F ratio)

df (denom)	1	2	3	4	5	6	7	8	9	10	11	12	14	16	20	24	30	40	50	75	100	200	500	∞
50	4.03 **7.17**	3.18 **5.06**	2.79 **4.20**	2.56 **3.72**	2.40 **3.41**	2.29 **3.18**	2.20 **3.02**	2.13 **2.88**	2.07 **2.78**	2.02 **2.70**	1.98 **2.62**	1.95 **2.56**	1.90 **2.46**	1.85 **2.39**	1.78 **2.26**	1.74 **2.18**	1.69 **2.10**	1.63 **2.00**	1.60 **1.94**	1.55 **1.86**	1.52 **1.82**	1.48 **1.76**	1.46 **1.71**	1.44 **1.68**
55	4.02 **7.12**	3.17 **5.01**	2.78 **4.16**	2.54 **3.68**	2.38 **3.37**	2.27 **3.15**	2.18 **2.98**	2.11 **2.85**	2.05 **2.75**	2.00 **2.66**	1.97 **2.59**	1.93 **2.53**	1.88 **2.43**	1.83 **2.35**	1.76 **2.23**	1.72 **2.15**	1.67 **2.06**	1.61 **1.96**	1.58 **1.90**	1.52 **1.82**	1.50 **1.78**	1.46 **1.71**	1.43 **1.66**	1.41 **1.64**
60	4.00 **7.08**	3.15 **4.98**	2.76 **4.13**	2.52 **3.65**	2.37 **3.34**	2.25 **3.12**	2.17 **2.95**	2.10 **2.82**	2.04 **2.72**	1.99 **2.63**	1.95 **2.56**	1.92 **2.50**	1.86 **2.40**	1.81 **2.32**	1.75 **2.20**	1.70 **2.12**	1.65 **2.03**	1.59 **1.93**	1.56 **1.87**	1.50 **1.79**	1.48 **1.74**	1.44 **1.68**	1.41 **1.63**	1.39 **1.60**
65	3.99 **7.04**	3.14 **4.95**	2.75 **4.10**	2.51 **3.62**	2.36 **3.31**	2.24 **3.09**	2.15 **2.93**	2.08 **2.79**	2.02 **2.70**	1.98 **2.61**	1.94 **2.54**	1.90 **2.47**	1.85 **2.37**	1.80 **2.30**	1.73 **2.18**	1.68 **2.09**	1.63 **2.00**	1.57 **1.90**	1.54 **1.84**	1.49 **1.76**	1.46 **1.71**	1.42 **1.64**	1.39 **1.60**	1.37 **1.56**
70	3.98 **7.01**	3.13 **4.92**	2.74 **4.08**	2.50 **3.60**	2.35 **3.29**	2.23 **3.07**	2.14 **2.91**	2.07 **2.77**	2.01 **2.67**	1.97 **2.59**	1.93 **2.51**	1.89 **2.45**	1.84 **2.35**	1.79 **2.28**	1.72 **2.15**	1.67 **2.07**	1.62 **1.98**	1.56 **1.88**	1.53 **1.82**	1.47 **1.74**	1.45 **1.69**	1.40 **1.62**	1.37 **1.56**	1.35 **1.53**
80	3.96 **6.96**	3.11 **4.88**	2.72 **4.04**	2.48 **3.56**	2.33 **3.25**	2.21 **3.04**	2.12 **2.87**	2.05 **2.74**	1.99 **2.64**	1.95 **2.55**	1.91 **2.48**	1.88 **2.41**	1.82 **2.32**	1.77 **2.24**	1.70 **2.11**	1.65 **2.03**	1.60 **1.94**	1.54 **1.84**	1.51 **1.78**	1.45 **1.70**	1.42 **1.65**	1.38 **1.57**	1.35 **1.52**	1.32 **1.49**
100	3.94 **6.90**	3.09 **4.82**	2.70 **3.98**	2.46 **3.51**	2.30 **3.20**	2.19 **2.99**	2.10 **2.82**	2.03 **2.69**	1.97 **2.59**	1.92 **2.51**	1.88 **2.43**	1.85 **2.36**	1.79 **2.26**	1.75 **2.19**	1.68 **2.06**	1.63 **1.98**	1.57 **1.89**	1.51 **1.79**	1.48 **1.73**	1.42 **1.64**	1.39 **1.59**	1.34 **1.51**	1.30 **1.46**	1.28 **1.43**
125	3.92 **6.84**	3.07 **4.78**	2.68 **3.94**	2.44 **3.47**	2.29 **3.17**	2.17 **2.95**	2.08 **2.79**	2.01 **2.65**	1.95 **2.56**	1.90 **2.47**	1.86 **2.40**	1.83 **2.33**	1.77 **2.23**	1.72 **2.15**	1.65 **2.03**	1.60 **1.94**	1.55 **1.85**	1.49 **1.75**	1.45 **1.68**	1.39 **1.59**	1.36 **1.54**	1.31 **1.46**	1.27 **1.40**	1.25 **1.37**
150	3.91 **6.81**	3.06 **4.75**	2.67 **3.91**	2.43 **3.44**	2.27 **3.14**	2.16 **2.92**	2.07 **2.76**	2.00 **2.62**	1.94 **2.53**	1.89 **2.44**	1.85 **2.37**	1.82 **2.30**	1.76 **2.20**	1.71 **2.12**	1.64 **2.00**	1.59 **1.91**	1.54 **1.83**	1.47 **1.72**	1.44 **1.66**	1.37 **1.56**	1.34 **1.51**	1.29 **1.43**	1.25 **1.37**	1.22 **1.33**
200	3.89 **6.76**	3.04 **4.71**	2.65 **3.88**	2.41 **3.41**	2.26 **3.11**	2.14 **2.90**	2.05 **2.73**	1.98 **2.60**	1.92 **2.50**	1.87 **2.41**	1.83 **2.34**	1.80 **2.28**	1.74 **2.17**	1.69 **2.09**	1.62 **1.97**	1.57 **1.88**	1.52 **1.79**	1.45 **1.69**	1.42 **1.62**	1.35 **1.53**	1.32 **1.48**	1.26 **1.39**	1.22 **1.33**	1.19 **1.28**
400	3.86 **6.70**	3.02 **4.66**	2.62 **3.83**	2.39 **3.36**	2.23 **3.06**	2.12 **2.85**	2.03 **2.69**	1.96 **2.55**	1.90 **2.46**	1.85 **2.37**	1.81 **2.29**	1.78 **2.23**	1.72 **2.12**	1.67 **2.04**	1.60 **1.92**	1.54 **1.84**	1.49 **1.74**	1.42 **1.64**	1.38 **1.57**	1.32 **1.47**	1.28 **1.42**	1.22 **1.32**	1.16 **1.24**	1.13 **1.19**
1000	3.85 **6.66**	3.00 **4.62**	2.61 **3.80**	2.38 **3.34**	2.22 **3.04**	2.10 **2.82**	2.02 **2.66**	1.95 **2.53**	1.89 **2.43**	1.84 **2.34**	1.80 **2.26**	1.76 **2.20**	1.70 **2.09**	1.65 **2.01**	1.58 **1.89**	1.53 **1.81**	1.47 **1.71**	1.41 **1.61**	1.36 **1.54**	1.30 **1.44**	1.26 **1.38**	1.19 **1.28**	1.13 **1.19**	1.108 **1.11**
∞	3.84 **6.64**	2.99 **4.60**	2.60 **3.78**	2.37 **3.32**	2.21 **3.02**	2.09 **2.80**	2.01 **2.64**	1.94 **2.51**	1.88 **2.41**	1.83 **2.32**	1.79 **2.24**	1.75 **2.18**	1.69 **2.07**	1.64 **1.99**	1.57 **1.87**	1.52 **1.79**	1.46 **1.69**	1.40 **1.59**	1.35 **1.52**	1.28 **1.41**	1.24 **1.36**	1.17 **1.25**	1.11 **1.15**	1.00 **1.00**

Degrees of freedom (for the denominator of F ratio)

TABLE G Critical values of the studentized range statistic (for Tukey HSD tests)

$\alpha = .05$

df_{error}	\multicolumn{14}{c}{Number of levels of the independent variable}
	2	3	4	5	6	7	8	9	10	11	12	13	14	15
1	17.97	26.98	32.82	37.07	40.41	43.12	45.40	47.36	49.07	50.59	51.96	53.20	54.33	55.36
2	6.08	8.33	9.80	10.88	11.74	12.44	13.03	13.54	13.99	14.39	14.75	15.08	15.38	15.65
3	4.50	5.91	6.82	7.50	8.04	8.48	8.85	9.18	9.46	9.72	9.95	10.15	10.35	10.53
4	3.93	5.04	5.76	6.29	6.71	7.05	7.35	7.60	7.83	8.03	8.21	8.37	8.52	8.66
5	3.64	4.60	5.22	5.67	6.03	6.33	6.58	6.80	7.00	7.17	7.32	7.47	7.60	7.72
6	3.46	4.34	4.90	5.31	5.63	5.90	6.12	6.32	6.49	6.65	6.79	6.92	7.03	7.14
7	3.34	4.16	4.68	5.06	5.36	5.61	5.82	6.00	6.16	6.30	6.43	6.55	6.66	6.76
8	3.26	4.04	4.53	4.89	5.17	5.40	5.60	5.77	5.92	6.05	6.18	6.29	6.39	6.48
9	3.20	3.95	4.42	4.76	5.02	5.24	5.43	5.60	5.74	5.87	5.98	6.09	6.19	6.28
10	3.15	3.88	4.33	4.65	4.91	5.12	5.30	5.46	5.60	5.72	5.83	5.94	6.03	6.11
11	3.11	3.82	4.26	4.57	4.82	5.03	5.20	5.35	5.49	5.60	5.71	5.81	5.90	5.98
12	3.08	3.77	4.20	4.51	4.75	4.95	5.12	5.26	5.40	5.51	5.62	5.71	5.79	5.88
13	3.06	3.74	4.15	4.45	4.69	4.88	5.05	5.19	5.32	5.43	5.53	5.63	5.71	5.79
14	3.03	3.70	4.11	4.41	4.64	4.83	4.99	5.13	5.25	5.36	5.46	5.55	5.64	5.71
15	3.01	3.67	4.08	4.37	4.60	4.78	4.94	5.08	5.20	5.31	5.40	5.49	5.57	5.65
16	3.00	3.65	4.05	4.33	4.56	4.74	4.90	5.03	5.15	5.26	5.35	5.44	5.52	5.59
17	2.98	3.63	4.02	4.30	4.52	4.70	4.86	4.99	5.11	5.21	5.31	5.39	5.47	5.54
18	2.97	3.61	4.00	4.28	4.50	4.67	4.82	4.96	5.07	5.17	5.27	5.35	5.43	5.50
19	2.96	3.59	3.98	4.25	4.47	4.64	4.79	4.92	5.04	5.14	5.23	5.32	5.39	5.46
20	2.95	3.58	3.96	4.23	4.44	4.62	4.77	4.90	5.01	5.11	5.20	5.28	5.36	5.43
24	2.92	3.53	3.90	4.17	4.37	4.54	4.68	4.81	4.92	5.01	5.10	5.18	5.25	5.32
30	2.89	3.49	3.84	4.10	4.30	4.46	4.60	4.72	4.82	4.92	5.00	5.08	5.15	5.21
40	2.86	3.44	3.79	4.04	4.23	4.39	4.52	4.64	4.74	4.82	4.90	4.98	5.04	5.11
60	2.83	3.40	3.74	3.98	4.16	4.31	4.44	4.55	4.65	4.73	4.81	4.88	4.94	5.00
120	2.80	3.36	3.69	3.92	4.10	4.24	4.36	4.47	4.56	4.64	4.71	4.78	4.84	4.90
∞	2.77	3.31	3.63	3.86	4.03	4.17	4.29	4.39	4.47	4.55	4.62	4.68	4.74	4.80

(continued)

SOURCE: From "Tables of Range and Studentized Range," by M. L. Harter, 1960, *Annals of Mathematical Statistics, 31*, 1122–1147. Copyright © 1960 The Institute of Mathematical Statistics. Reprinted with permission.

T A B L E   G · (continued)

$\alpha = .01$

df_{error}						Number of levels of the independent variable								
	2	3	4	5	6	7	8	9	10	11	12	13	14	15
1	90.03	135.00	164.30	185.60	202.20	215.80	227.20	237.00	245.60	253.20	260.00	266.20	271.80	277.00
2	14.04	19.02	22.29	24.72	26.63	28.20	29.53	30.68	31.69	32.59	33.40	34.13	34.81	35.43
3	8.26	10.62	12.17	13.33	14.24	15.00	15.64	12.60	16.69	17.13	17.53	17.89	18.22	18.52
4	6.51	8.12	9.17	9.96	10.58	11.10	11.55	11.93	12.27	12.57	12.84	13.09	13.32	13.53
5	5.70	6.98	7.80	8.42	8.91	9.32	9.67	9.97	10.24	10.48	10.70	10.89	11.08	11.24
6	5.24	6.33	7.03	7.56	7.97	8.32	8.62	8.87	9.10	9.30	9.48	9.65	9.81	9.95
7	4.95	5.92	6.54	7.00	7.37	7.68	7.94	8.17	8.37	8.55	8.71	8.86	9.00	9.12
8	4.75	5.64	6.20	6.62	6.96	7.24	7.47	7.68	7.86	8.03	8.18	8.31	8.44	8.55
9	4.60	5.43	5.96	6.35	6.66	6.92	7.13	7.32	7.50	7.65	7.78	7.91	8.02	8.13
10	4.48	5.27	5.77	6.14	6.43	6.67	6.88	7.06	7.21	7.36	7.48	7.60	7.71	7.81
11	4.39	5.15	5.62	5.97	6.25	6.48	6.67	6.84	6.99	7.13	7.25	7.36	7.46	7.56
12	4.32	5.05	5.50	5.84	6.10	6.32	6.51	6.67	6.81	6.94	7.06	7.17	7.26	7.36
13	4.26	4.96	5.40	5.73	5.98	6.19	6.37	6.53	6.67	6.79	6.90	7.01	7.10	7.19
14	4.21	4.90	5.32	5.63	5.88	6.08	6.26	6.41	6.54	6.66	6.77	6.87	6.96	7.05
15	4.17	4.84	5.25	5.56	5.80	5.99	6.16	6.31	6.44	6.56	6.66	6.76	6.84	6.93
16	4.13	4.79	5.19	5.49	5.72	5.92	6.08	6.22	6.35	6.46	6.56	6.66	6.74	6.82
17	4.10	4.74	5.14	5.43	5.66	5.85	6.01	6.15	6.27	6.38	6.48	6.57	6.66	6.73
18	4.07	4.70	5.09	5.38	5.60	5.79	5.94	6.08	6.20	6.31	6.41	6.50	6.58	6.66
19	4.05	4.67	5.05	5.33	5.55	5.74	5.89	6.02	6.14	6.25	6.34	6.43	6.51	6.58
20	4.02	4.64	5.02	5.29	5.51	5.69	5.84	5.97	6.09	6.19	6.28	6.37	6.45	6.52
24	3.96	4.55	4.91	5.17	5.37	5.54	5.69	5.81	5.92	6.02	6.11	6.19	6.26	6.33
30	3.89	4.46	4.80	5.05	5.24	5.40	5.54	5.65	5.76	5.85	5.93	6.01	6.08	6.14
40	3.82	4.37	4.70	4.93	5.11	5.26	5.39	5.50	5.60	5.69	5.76	5.84	5.90	5.96
60	3.76	4.28	4.60	4.82	4.99	5.13	5.25	5.36	5.45	5.53	5.60	5.67	5.73	5.78
120	3.70	4.20	3.50	4.71	4.87	5.01	5.12	5.21	5.30	5.38	5.44	5.51	5.56	5.61
∞	3.64	4.12	4.40	4.60	4.76	4.88	4.99	5.08	5.16	5.23	5.29	5.35	5.40	5.45

TABLE H Critical values for the Mann–Whitney U test*

One-tailed test
$\alpha = .01$ (lightface)
$\alpha = .005$ (**boldface**)

Two-tailed test
$\alpha = .02$ (lightface)
$\alpha = .01$ (**boldface**)

N_2＼N_1	1	2	3	4	5	6	7	8	9	10	11	12	13	14	15	16	17	18	19	20
1	—	—	—	—	—	—	—	—	—	—	—	—	—	—	—	—	—	—	—	—
2	—	—	—	—	—	—	—	—	—	—	—	—	0	0	0	0	0	0	1	1
	—	—	—	—	—	—	—	—	—	—	—	—	—	—	—	—	—	—	**0**	**0**
3	—	—	—	—	—	—	0	0	1	1	1	2	2	2	3	3	4	4	4	5
	—	—	—	—	—	—	—	—	**0**	**0**	**0**	**1**	**1**	**1**	**2**	**2**	**2**	**2**	**3**	**3**
4	—	—	—	—	0	1	1	2	3	3	4	5	5	6	7	7	8	9	9	10
	—	—	—	—	—	**0**	**0**	**1**	**1**	**2**	**2**	**3**	**3**	**4**	**5**	**5**	**6**	**6**	**7**	**8**
5	—	—	—	0	1	2	3	4	5	6	7	8	9	10	11	12	13	14	15	16
	—	—	—	—	**0**	**1**	**1**	**2**	**3**	**4**	**5**	**6**	**7**	**7**	**8**	**9**	**10**	**11**	**12**	**13**
6	—	—	—	1	2	3	4	6	7	8	9	11	12	13	15	16	18	19	20	22
	—	—	—	**0**	**1**	**2**	**3**	**4**	**5**	**6**	**7**	**9**	**10**	**11**	**12**	**13**	**15**	**16**	**17**	**18**
7	—	—	0	1	3	4	6	7	9	11	12	14	16	17	19	21	23	24	26	28
	—	—	—	**0**	**1**	**3**	**4**	**6**	**7**	**9**	**10**	**12**	**13**	**15**	**16**	**18**	**19**	**21**	**22**	**24**
8	—	—	0	2	4	6	7	9	11	13	15	17	20	22	24	26	28	30	32	34
	—	—	—	**1**	**2**	**4**	**6**	**7**	**9**	**11**	**13**	**15**	**17**	**18**	**20**	**22**	**24**	**26**	**29**	**30**
9	—	—	1	3	5	7	9	11	14	16	18	21	23	26	28	31	33	36	38	40
	—	—	**0**	**1**	**3**	**5**	**7**	**9**	**11**	**13**	**16**	**18**	**20**	**22**	**24**	**27**	**29**	**31**	**33**	**36**
10	—	—	1	3	6	8	11	13	16	19	22	24	27	30	33	36	38	41	44	47
	—	—	**0**	**2**	**4**	**6**	**9**	**11**	**13**	**16**	**18**	**21**	**24**	**26**	**29**	**31**	**34**	**37**	**39**	**42**
11	—	—	1	4	7	9	12	15	18	22	25	28	31	34	37	41	44	47	50	53
	—	—	**0**	**2**	**5**	**7**	**10**	**13**	**16**	**18**	**21**	**24**	**27**	**30**	**33**	**36**	**39**	**42**	**45**	**48**
12	—	—	2	5	8	11	14	17	21	24	28	31	35	38	42	46	49	53	56	60
	—	—	**1**	**3**	**6**	**9**	**12**	**15**	**18**	**21**	**24**	**27**	**31**	**34**	**37**	**41**	**44**	**47**	**51**	**54**
13	—	0	2	5	9	12	16	20	23	27	31	35	39	43	47	51	55	59	63	67
	—	—	**1**	**3**	**7**	**10**	**13**	**17**	**20**	**24**	**27**	**31**	**34**	**38**	**42**	**45**	**49**	**53**	**56**	**60**
14	—	0	2	6	10	13	17	22	26	30	34	38	43	47	51	56	60	65	69	73
	—	—	**1**	**4**	**7**	**11**	**15**	**18**	**22**	**26**	**30**	**34**	**38**	**42**	**46**	**50**	**54**	**58**	**63**	**67**
15	—	0	3	7	11	15	19	24	28	33	37	42	47	51	56	61	66	70	75	80
	—	—	**2**	**5**	**8**	**12**	**16**	**20**	**24**	**29**	**33**	**37**	**42**	**46**	**51**	**55**	**60**	**64**	**69**	**73**
16	—	0	3	7	12	16	21	26	31	36	41	46	51	56	61	66	71	76	82	87
	—	—	**2**	**5**	**9**	**13**	**18**	**22**	**27**	**31**	**36**	**41**	**45**	**50**	**55**	**60**	**65**	**70**	**74**	**79**
17	—	0	4	8	13	18	23	28	33	38	44	49	55	60	66	71	77	82	88	93
	—	—	**2**	**6**	**10**	**15**	**19**	**24**	**29**	**34**	**39**	**44**	**49**	**54**	**60**	**65**	**70**	**75**	**81**	**86**
18	—	0	4	9	14	19	24	30	36	41	47	53	59	65	70	76	82	88	94	100
	—	—	**2**	**6**	**11**	**16**	**21**	**26**	**31**	**37**	**42**	**47**	**53**	**58**	**64**	**70**	**75**	**81**	**87**	**92**
19	—	1	4	9	15	20	26	32	38	44	50	56	63	69	75	82	88	94	101	107
	—	**0**	**3**	**7**	**12**	**17**	**22**	**28**	**33**	**39**	**45**	**51**	**56**	**63**	**69**	**74**	**81**	**87**	**93**	**99**
20	—	1	5	10	16	22	28	34	40	47	53	60	67	73	80	87	93	100	107	114
	—	**0**	**3**	**8**	**13**	**18**	**24**	**30**	**36**	**42**	**48**	**54**	**60**	**67**	**73**	**79**	**86**	**92**	**99**	**105**

(*continued*)

* To be significant the U obtained from data must be equal to or **less than** the value shown in the table. Dashes in the body of the table indicate that no decision is possible at the stated level of significance.

SOURCE: From *Elementary Statistics,* Second Edition, by R. E. Kirk, Brooks/Cole Publishing, 1984.

TABLE H (continued)

One-tailed test	Two-tailed test
α = .05 (lightface)	α = .10 (lightface)
α = .025 (**boldface**)	α = .05 (**boldface**)

N_2 \ N_1	1	2	3	4	5	6	7	8	9	10	11	12	13	14	15	16	17	18	19	20
1	—	—	—	—	—	—	—	—	—	—	—	—	—	—	—	—	—	—	0	0
2	—	—	—	—	0	0	0	1	1	1	1	2	2	2	3	3	3	4	4	4
	—	—	—	—	—	—	—	**0**	**0**	**0**	**0**	**1**	**1**	**1**	**1**	**1**	**2**	**2**	**2**	**2**
3	—	—	0	0	1	2	2	3	3	4	5	5	6	7	7	8	9	9	10	11
	—	—	—	—	**0**	**1**	**1**	**2**	**2**	**3**	**3**	**4**	**4**	**5**	**5**	**6**	**6**	**7**	**7**	**8**
4	—	—	0	1	2	3	4	5	6	7	8	9	10	11	12	14	15	16	17	18
	—	—	—	**0**	**1**	**2**	**3**	**4**	**4**	**5**	**6**	**7**	**8**	**9**	**10**	**11**	**11**	**12**	**13**	**13**
5	—	0	1	2	4	5	6	8	9	11	12	13	15	16	18	19	20	22	23	25
	—	—	**0**	**1**	**2**	**3**	**5**	**6**	**7**	**8**	**9**	**11**	**12**	**13**	**14**	**15**	**17**	**18**	**19**	**20**
6	—	0	2	3	5	7	8	10	12	14	16	17	19	21	23	25	26	28	30	32
	—	—	**1**	**2**	**3**	**5**	**6**	**8**	**10**	**11**	**13**	**14**	**16**	**17**	**19**	**21**	**22**	**24**	**25**	**27**
7	—	0	2	4	6	8	11	13	15	17	19	21	24	26	28	30	33	35	37	39
	—	—	**1**	**3**	**5**	**6**	**8**	**10**	**12**	**14**	**16**	**18**	**20**	**22**	**24**	**26**	**28**	**30**	**32**	**34**
8	—	1	3	5	8	10	13	15	18	20	23	26	28	31	33	36	39	41	44	47
	—	**0**	**2**	**4**	**6**	**8**	**10**	**13**	**15**	**17**	**19**	**22**	**24**	**26**	**29**	**31**	**34**	**36**	**38**	**41**
9	—	1	3	6	9	12	15	18	21	24	27	30	33	36	39	42	45	48	51	54
	—	**0**	**2**	**4**	**7**	**10**	**12**	**15**	**17**	**20**	**23**	**26**	**28**	**31**	**34**	**37**	**39**	**42**	**45**	**48**
10	—	1	4	7	11	14	17	20	24	27	31	34	37	41	44	48	51	55	58	62
	—	**0**	**3**	**5**	**8**	**11**	**14**	**17**	**20**	**23**	**26**	**29**	**33**	**36**	**39**	**42**	**45**	**48**	**52**	**55**
11	—	1	5	8	12	16	19	23	27	31	34	38	42	46	50	54	57	61	65	69
	—	**0**	**3**	**6**	**9**	**13**	**16**	**19**	**23**	**26**	**30**	**33**	**37**	**40**	**44**	**47**	**51**	**55**	**58**	**62**
12	—	2	5	9	13	17	21	26	30	34	38	42	47	51	55	60	64	68	72	77
	—	**1**	**4**	**7**	**11**	**14**	**18**	**22**	**26**	**29**	**33**	**37**	**41**	**45**	**49**	**53**	**57**	**61**	**65**	**69**
13	—	2	6	10	15	19	24	28	33	37	42	47	51	56	61	65	70	75	80	84
	—	**1**	**4**	**8**	**12**	**16**	**20**	**24**	**28**	**33**	**37**	**41**	**45**	**50**	**54**	**59**	**63**	**67**	**72**	**76**
14	—	2	7	11	16	21	26	31	36	41	46	51	56	61	66	71	77	82	87	92
	—	**1**	**5**	**9**	**13**	**17**	**22**	**26**	**31**	**36**	**40**	**45**	**50**	**55**	**59**	**64**	**67**	**74**	**78**	**83**
15	—	3	7	12	18	23	28	33	39	44	50	55	61	66	72	77	83	88	94	100
	—	**1**	**5**	**10**	**14**	**19**	**24**	**29**	**34**	**39**	**44**	**49**	**54**	**59**	**64**	**70**	**75**	**80**	**85**	**90**
16	—	3	8	14	19	25	30	36	42	48	54	60	65	71	77	83	89	95	101	107
	—	**1**	**6**	**11**	**15**	**21**	**26**	**31**	**37**	**42**	**47**	**53**	**59**	**64**	**70**	**75**	**81**	**86**	**92**	**98**
17	—	3	9	15	20	26	33	39	45	51	57	64	70	77	83	89	96	102	109	115
	—	**2**	**6**	**11**	**17**	**22**	**28**	**34**	**39**	**45**	**51**	**57**	**63**	**67**	**75**	**81**	**87**	**93**	**99**	**105**
18	—	4	9	16	22	28	35	41	48	55	61	68	75	82	88	95	102	109	116	123
	—	**2**	**7**	**12**	**18**	**24**	**30**	**36**	**42**	**48**	**55**	**61**	**67**	**74**	**80**	**86**	**93**	**99**	**106**	**112**
19	0	4	10	17	23	30	37	44	51	58	65	72	80	87	94	101	109	116	123	130
	—	**2**	**7**	**13**	**19**	**25**	**32**	**38**	**45**	**52**	**58**	**65**	**72**	**78**	**85**	**92**	**99**	**106**	**113**	**119**
20	0	4	11	18	25	32	39	47	54	62	69	77	84	92	100	107	115	123	130	138
	—	**2**	**8**	**13**	**20**	**27**	**34**	**41**	**48**	**55**	**62**	**69**	**76**	**83**	**90**	**98**	**105**	**112**	**119**	**127**

TABLE J Critical values for the Wilcoxon matched-pairs signed-ranks T test*

No. of pairs N	α levels for a one-tailed test .05	.025	.01	.005	N	α levels for a one-tailed test .05	.025	.01	.005
	α levels for a two-tailed test .10	.05	.02	.01		α levels for a two-tailed test .10	.05	.02	.01
5	0	—	—	—	28	130	116	101	91
6	2	0	—	—	29	140	126	110	100
7	3	2	0	—	30	151	137	120	109
8	5	3	1	0	31	163	147	130	118
9	8	5	3	1	32	175	159	140	128
10	10	8	5	3	33	187	170	151	138
11	13	10	7	5	34	200	182	162	148
12	17	13	9	7	35	213	195	173	159
13	21	17	12	9	36	227	208	185	171
14	25	21	15	12	37	241	221	198	182
15	30	25	19	15	38	256	235	211	194
16	35	29	23	19	39	271	249	224	207
17	41	34	27	23	40	286	264	238	220
18	47	40	32	27	41	302	279	252	233
19	53	46	37	32	42	319	294	266	247
20	60	52	43	37	43	336	310	281	261
21	67	58	49	42	44	353	327	296	276
22	75	65	55	48	45	371	343	312	291
23	83	73	62	54	46	389	361	328	307
24	91	81	69	61	47	407	378	345	322
25	100	89	76	68	48	426	396	362	339
26	110	98	84	75	49	446	415	379	355
27	119	107	92	83	50	466	434	397	373

* To be significant the T obtained from the data must be equal to or **less than** the value shown in the table.
SOURCE: From *Elementary Statistics,* Second Edition, by R. E. Kirk, Brooks/Cole Publishing, 1984.

TABLE K Critical differences for the Wilcoxon–Wilcox multiple-comparisons test*
(Column N represents the number in one group.)

				$\alpha = .05$ (two-tailed)				
N	K = 3	K = 4	K = 5	K = 6	K = 7	K = 8	K = 9	K = 10
1	3.3	4.7	6.1	7.5	9.0	10.5	12.0	13.5
2	8.8	12.6	16.5	20.5	24.7	28.9	33.1	37.4
3	15.7	22.7	29.9	37.3	44.8	52.5	60.3	68.2
4	23.9	34.6	45.6	57.0	68.6	80.4	92.4	104.6
5	33.1	48.1	63.5	79.3	95.5	112.0	128.8	145.8
6	43.3	62.9	83.2	104.0	125.3	147.0	169.1	191.4
7	54.4	79.1	104.6	130.8	157.6	184.9	212.8	240.9
8	66.3	96.4	127.6	159.6	192.4	225.7	259.7	294.1
9	78.9	114.8	152.0	190.2	229.3	269.1	309.6	350.6
10	92.3	134.3	177.8	222.6	268.4	315.0	362.4	410.5
11	106.3	154.8	205.0	256.6	309.4	363.2	417.9	473.3
12	120.9	176.2	233.4	292.2	352.4	413.6	476.0	539.1
13	136.2	198.5	263.0	329.3	397.1	466.2	536.5	607.7
14	152.1	221.7	293.8	367.8	443.6	520.8	599.4	679.0
15	168.6	245.7	325.7	407.8	491.9	577.4	664.6	752.8
16	185.6	270.6	358.6	449.1	541.7	635.9	732.0	829.2
17	203.1	269.2	392.6	491.7	593.1	696.3	801.5	907.9
18	221.2	322.6	427.6	535.5	646.1	758.5	873.1	989.0
19	239.8	349.7	463.6	580.6	700.5	822.4	946.7	1072.4
20	258.8	377.6	500.5	626.9	756.4	888.1	1022.3	1158.1
21	278.4	406.1	538.4	674.4	813.7	955.4	1099.8	1245.9
22	298.4	435.3	577.2	723.0	872.3	1024.3	1179.1	1335.7
23	318.9	465.2	616.9	772.7	932.4	1094.8	1260.3	1427.7
24	339.8	495.8	657.4	823.5	993.7	1166.8	1343.2	1521.7
25	361.1	527.0	698.8	875.4	1056.3	1240.4	1427.9	1617.6

				$\alpha = .01$ (two-tailed)				
N	K = 3	K = 4	K = 5	K = 6	K = 7	K = 8	K = 9	K = 10
1	4.1	5.7	7.3	8.9	10.5	12.2	13.9	15.6
2	10.9	15.3	19.7	24.3	28.9	33.6	38.3	43.1
3	19.5	27.5	35.7	44.0	52.5	61.1	69.8	78.6
4	29.7	41.9	54.5	67.3	80.3	93.6	107.0	120.6
5	41.2	58.2	75.8	93.6	111.9	130.4	149.1	168.1
6	53.9	76.3	99.3	122.8	146.7	171.0	195.7	220.6
7	67.6	95.8	124.8	154.4	184.6	215.2	246.3	277.7
8	82.4	116.8	152.2	188.4	225.2	262.6	300.6	339.0
9	98.1	139.2	181.4	224.5	268.5	313.1	358.4	404.2
10	114.7	162.8	212.2	262.7	314.2	366.5	419.5	473.1
11	132.1	187.6	244.6	302.9	362.2	422.6	483.7	545.6
12	150.4	213.5	278.5	344.9	412.5	481.2	551.0	621.4
13	169.4	240.6	313.8	388.7	464.9	542.4	621.0	700.5
14	189.1	268.7	350.5	434.2	519.4	606.0	693.8	782.6
15	209.6	297.6	388.5	481.3	575.8	671.9	769.3	867.7
16	230.7	327.9	427.9	530.1	634.2	740.0	847.3	955.7
17	252.5	359.0	468.4	580.3	694.4	810.2	927.8	1046.5
18	275.0	391.0	510.2	632.1	756.4	882.6	1010.6	1140.0
19	298.1	423.8	553.1	685.4	820.1	957.0	1095.8	1236.2
20	321.8	457.6	597.2	740.0	885.5	1033.3	1183.3	1334.9
21	346.1	492.2	642.4	796.0	952.6	1111.6	1273.0	1436.0
22	371.0	527.6	688.7	853.4	1021.3	1191.8	1364.8	1539.7
23	396.4	563.8	736.0	912.1	1091.5	1273.8	1458.8	1645.7
24	422.4	600.9	784.4	972.1	1163.4	1357.6	1554.8	1754.0
25	449.0	638.7	833.8	1033.3	1236.7	1443.2	1652.8	1864.6

* To be significant the difference obtained from the data must be equal to or **greater than** the tabled value.
SOURCE: From *Some Rapid Approximate Statistical Procedures*, by F. Wilcoxon and R. Wilcox. Copyright ©
1964 Lederle Laboratories, a division of American Cyanamid Co. Reprinted with permission.

TABLE L Critical values for Spearman r_s*

No. of pairs N	One-tailed test		Two-tailed test	
	.01	.05	.01	.05
4	—	1.000	—	—
5	1.000	.900	—	1.000
6	.943	.829	1.000	.886
7	.893	.714	.929	.786
8	.833	.643	.881	.738
9	.783	.600	.833	.700
10	.745	.564	.794	.648
11	.709	.536	.755	.618
12	.677	.503	.727	.587
13	.648	.484	.703	.560
14	.626	.464	.679	.538
15	.604	.446	.654	.521
16	.582	.429	.635	.503

For samples larger than 16, use Table A.

* To be significant, the r_s obtained from the data must be equal to or **greater than** the value shown in the table.

SOURCE: From "Testing the Significance of Kendall's T and Spearman's r_s" by M. Nijsse, 1988, *Psychological Bulletin, 103,* 235–237. Copyright © 1988 American Psychological Association. Reprinted by permission.

appendix **d**

Glossary of Words

***a priori* tests**   A category of multiple comparison tests; planned before examining the data

abscissa   The horizontal, or X, axis of a graph.

absolute value   The number without consideration of its algebraic sign.

alternative hypothesis   A hypothesis about population parameters that is accepted if the null hypothesis is rejected; symbolized H_1.

analysis of variance (ANOVA)   An inferential statistics technique for comparing means, comparing variances, and assessing interactions.

asymptotic   A line that continually approaches but never reaches a specified limit.

bar graph   A graph of the frequency distribution for nominal or qualitative data.

biased sample   A sample selected in such a way that not all samples from the population have an equal chance of being chosen.

bimodal distribution   A distribution with two modes.

binomial distribution   A distribution of the frequency of events that have two possible outcomes.

bivariate distribution   A joint distribution of two variables, the individual scores of which are paired in some logical way.

boxplot   A graphic that shows the mean, median, range, interquartile range, and skew.

cell   The portion of an ANOVA data array table containing the scores of participants who are treated alike.

Central Limit Theorem   The sampling distribution of the mean approaches a normal curve as N gets larger. This normal curve has a mean equal to μ and a standard deviation equal to $\sigma/\sqrt{N}$.

central tendency   The mean, median, or mode; a statistic that best represents the scores in a distribution.

chi square distribution   A theoretical sampling distribution of chi square values. Chi square distributions vary with degrees of freedom.

class interval   A range of scores in a grouped frequency distribution.

coefficient of determination   A squared correlation coefficient; an estimate of common variance of the two variables.

common variance   Variance held in common by two variables. Common variance is assumed to be related to the same factors.

confidence interval An interval of scores that is expected, with specified confidence, to contain a parameter.

control group A baseline group to which other groups are compared.

correlated-measures ANOVA A technique that tests the hypothesis that two or more population means in a correlated-samples design are equal.

correlated-samples design An experimental design in which measures from different groups can be logically paired.

correlation coefficient A descriptive statistic calculated on bivariate data; expresses the degree of relationship between two variables.

critical region Synonym for rejection region.

critical value The value from a sampling distribution against which a computed statistic is compared to determine whether the null hypothesis may be rejected.

degrees of freedom A concept in mathematical statistics that determines which distribution from a family of distributions is appropriate for a particular set of sample data.

dependent variable The observed variable that is expected to change as a result of changes in the independent variable in an experiment.

descriptive statistic A number that expresses some particular characteristic of a set of data. Graphs and tables are sometimes included in this category. (Congratulations to you if you have looked up this entry after reading footnote 1 in Chapter 1. Very few students make the effort to check out their authors' claims, as you just did. You have one of the makings of a scholar.)

deviation score A raw score minus the mean of the distribution it is in.

dichotomous variable A variable that takes two, and only two, values.

distribution-free statistics Inferential statistical methods that do not assume a particular population distribution.

effect size (index) The amount of difference that a variable makes.

empirical distribution A set of scores that come from actual observations.

epistemology The study or theory of the nature of knowledge.

error estimate (or error term) Variance due to factors not controlled in the experiment; within-group or within-cell variance.

expected frequency Value in a chi square analysis that is derived from the null hypothesis; theoretical frequency.

expected value The mean value of a random variable over an infinite number of samplings. The expected value of a statistic is the mean of the sampling distribution of the statistic.

experimental group A group that receives a treatment in an experiment and whose dependent-variable scores are compared with those of a control group.

extraneous variable A variable other than the independent variable that may affect the dependent variable.

***F* distribution** A theoretical sampling distribution of F values. There is a different F distribution for each combination of degrees of freedom.

***F* test** A method of determining the significance of the difference among two or more means or between two variances.

factor Independent variable.

factorial ANOVA An experimental design with two or more independent variables; allows F tests for main effects and interactions.

frequency polygon A graph of connected points with quantitative scores on the X axis and frequencies on the Y axis. A point on the graph represents a score and its frequency.

goodness of fit Degree to which observed data coincide with theoretical expectations.

grand mean The mean of all the scores in an experiment.

grouped frequency distribution An arrangement of scores from highest to lowest. Scores are grouped into equal-sized ranges called class intervals. Includes frequencies for each interval; may include midpoints of class intervals.

histogram A graph of contiguous bars; quantitative scores are on the X axis and frequencies on the Y axis.

hypothesis testing The process of hypothesizing a parameter, comparing (or testing) the parameter with a statistic from a sample, and deciding whether or not the parameter is reasonable.

independent The occurrence or variation of one event does not affect the occurrence or variation of a second event.

independent-samples design An experimental design using samples whose dependent-variable scores cannot logically be paired.

independent variable A variable controlled by the researcher; changes in this variable may produce changes in the dependent variable.

inferential statistics A method of reaching conclusions about unmeasurable populations that uses sample evidence and probability.

inflection point A point on a curve that separates an arc that is concave upward from the part that is concave downward, and vice versa.

interaction The situation in which the effect of one variable on the dependent variable depends on which level of a second variable is operative.

interpolation A method for determining a value that is between two other values.

interquartile range A range of scores that contains the middle 50 percent of a distribution.

interval scale A measurement scale in which equal differences between numbers represent equal differences in the thing measured. The zero point is arbitrarily defined.

J-curve A severely skewed distribution with the mode at one extreme.

least squares Method of fitting a regression line such that the sums of the squared deviations from the straight regression line is a minimum.

level One value of the independent variable.

Likert scale A method of measuring attitudes; subjects indicate degree of approval or disapproval on a 5- or 7-point scale.

line graph A graph that shows the relationship between two variables.

lower limit The bottom of the range of possible values that a measurement on a quantitative variable can take.

main effect The deviations of two or more treatment means from the grand mean.

Mann–Whitney U test A nonparametric test of the hypothesis that the populations that two independent samples came from are the same.

matched pairs A correlated-samples design in which individuals are paired by the researcher before the experiment.

mean   The arithmetic average; the sum of the scores divided by the number of scores.

mean square   An ANOVA term for the variance; a sum of squares divided by its degrees of freedom.

median   The point that divides a distribution of scores into equal halves, so that half the scores are above the median and half are below it.

meta-analysis   A technique for reaching conclusions when the data consist of many separate studies done by different investigators.

mode   The score that occurs most frequently in a distribution.

multiple-comparisons tests   Tests of differences between treatment means or combination of means.

multiple correlation   A correlation method that combines intercorrelations among more than two variables into a single statistic.

natural pairs   A correlated-samples design in which pairing occurs naturally, prior to the experiment.

nominal scale   A scale of measurement in which numbers are used simply as names and not as quantities.

nonparametric statistics   Statistical methods that do not require assumptions about the parameters of the sampled populations.

normal distribution (or normal curve)   A bell-shaped, theoretical distribution that predicts the frequency of occurrence of chance events. An empirical distribution of similar shape.

NS   Not statistically significant.

null hypothesis   A hypothesis about a population or the relationship between populations; symbolized H_0.

observed frequency   The actual count of events in a category in a chi square analysis.

one-sample t test   A statistical test of the hypothesis that a sample with mean $\overline{X}$ came from a population with mean μ.

one-tailed test of significance   A statistical test in which the rejection region is in only one tail of the distribution.

one-way ANOVA   A technique that tests the hypothesis that two or more population means in an independent-samples design are equal.

operational definition   A definition that tells how to measure a variable.

ordinal scale   Scale of measurement in which numbers are ranks but equal differences between numbers do not represent equal differences between the things measured.

ordinate   The vertical, or Y, axis of a graph.

outlier   A score that is very high or very low in relation to the other scores in the distribution.

parameter   A numerical or nominal characteristic of a population.

partial correlation   Technique that allows the separation or partialing out of the effects of one variable from the correlation of two other variables.

phi (ϕ)   An effect size index for a 2×2 chi square test of independence.

population   All measurements of a specified group.

post hoc **tests**   A category of multiple-comparisons tests; selected after examining the data.

power Power $= 1 - \beta$. Power is the probability of rejecting a false null hypothesis.

proportion A part of a whole.

qualitative variable A variable that exists in different kinds rather than different amounts.

quantification The idea that translating a phenomenon into numbers will promote a better understanding of the phenomenon.

quantitative variable A variable that exists in different amounts.

random sample A subset of a population chosen in such a way that all samples of the specified size have an equal probability of being selected.

range The upper limit of the highest score minus the lower limit of the lowest score.

ratio scale A scale that has all the characteristics of an interval scale; in addition, zero means that none of the thing measured is present.

raw score A score as it is obtained by observation or in an experiment.

rectangular distribution A distribution in which all scores have the same frequency; also called a uniform distribution.

regression coefficients The values a (point where the regression line intersects the Y axis) and b (slope of the regression line).

regression equation An equation used to predict particular values of Y for specific values of X.

regression line The "line of best fit" that runs through a scatterplot.

rejection region The area of a sampling distribution that includes values of the test statistic that lead to a decision to reject the null hypothesis.

reliability The dependability or test–retest consistency of a measure.

repeated measures An experimental design in which more than one dependent-variable measure is taken on each subject.

residual variance ANOVA term for variability due to unknown or uncontrolled variables.

sample A subset of a population.

sampling distribution A theoretical distribution of a statistic based on all possible random samples drawn from the same population; used to determine probabilities.

scatterplot A graph of the scores of a bivariate frequency distribution.

significance level A probability chosen as the criterion for rejecting the null hypothesis; α.

simple frequency distribution Scores arranged from highest to lowest, with the frequency of each score in a column beside the score.

skewed distribution An asymmetrical distribution; may be positive or negative.

Spearman r_s A correlation coefficient for two sets of ranked data.

standard deviation A measure of the amount of variability around the mean of the distribution.

standard error The standard deviation of a sampling distribution.

standard error of a difference The standard deviation of a sampling distribution of differences between means.

standard error of estimate The standard deviation of the differences between predicted outcomes and actual outcomes.

standard score   A score expressed in standard deviation units.

statistic   A numerical or nominal characteristic of a sample.

statistically significant   A difference so large that chance is not a plausible explanation.

sum of squares   The sum of the squared deviations from the mean; the numerator of the formula for the standard deviation.

t distribution   Theoretical distribution used to determine probabilities when σ is not known or samples are small.

t test   Test of a null hypothesis that uses the t distribution.

T value   Statistic used in the Wilcoxon matched-pairs signed-ranks T test.

theoretical distribution   Hypothesized scores based on mathematical formulas and logic.

treatment   One level of the independent variable.

truncated range   The range of the sample is much smaller than the range of its population.

Tukey Honestly Significant Difference (HSD) test   Significance test for all possible pairs of treatments in a multitreatment experiment.

two-tailed test of significance   Any statistical test in which the rejection region is in both two tails of the sampling distribution.

Type I error   Rejection of the null hypothesis when it is true.

Type II error   Retention of the null hypothesis when it is false.

U value   Statistic used in the Mann–Whitney U test.

univariate distribution   A frequency distribution of one variable.

upper limit   The top of the range of possible values that a measurement on a quantitative variable can take.

variability   Characterized by having more than one value.

variable   Something that exists in more than one amount or in more than one form.

variance   The square of the standard deviation; also mean square in ANOVA.

very good idea   Examine a statistics problem until you understand it well enough to estimate the answer.

Wilcoxon matched-pairs signed-ranks T test   A nonparametric test of the hypothesis that the populations that two correlated samples came from are the same.

Wilcoxon rank-sum test   A nonparametric test of the difference between two independent samples.

Wilcoxon–Wilcox multiple-comparisons test   A nonparametric test of independent samples in which all possible pairs of treatments are tested.

Yates' correction   A correction for a 2×2 chi square with one small expected frequency.

z score   A score expressed in standard deviation units; used to compare the relative standing of scores in two different distributions.

z test   An inferential statistics test that uses the normal curve as a sampling distribution.

appendix e

Glossary of Formulas

Analysis of Variance

degrees of freedom in one-way ANOVA

$$df_{treat} = K - 1$$
$$df_{error} = N_{tot} - K$$
$$df_{tot} = N_{tot} - 1$$

degrees of freedom in factorial ANOVA

$$df_A = A - 1$$
$$df_B = B - 1$$
$$df_{AB} = (A - 1)(B - 1)$$
$$df_{error} = N_{tot} - (A)(B)$$
$$df_{tot} = N_{tot} - 1$$

degrees of freedom in one-factor correlated-measures ANOVA

$$df_{subjects} = N_s - 1$$
$$df_{treat} = N_t - 1$$
$$df_{error} = (N_s - 1)(N_t - 1)$$
$$df_{tot} = N_{tot} - 1$$

F value in one-way ANOVA and correlated-measures ANOVA

$$F = \frac{MS_{treat}}{MS_{error}}$$

F values in factorial ANOVA

$$F_A = \frac{MS_A}{MS_{error}}$$

$$F_B = \frac{MS_B}{MS_{error}}$$

$$F_{AB} = \frac{MS_{AB}}{MS_{error}}$$

mean square

$$MS = \frac{SS}{df}$$

total sum of squares

$$SS_{tot} = \Sigma X_{tot}^2 - \frac{(\Sigma X_{tot})^2}{N_{tot}}$$

between-treatments sum of squares for one-way and factorial ANOVA

$$SS_{treat} = \Sigma\left[\frac{(\Sigma X_t)^2}{N_t}\right] - \frac{(\Sigma X_{tot})^2}{N_{tot}}$$

between-treatments sum of squares for correlated-measures ANOVA

$$SS_{treat} = \Sigma\left[\frac{(\Sigma X_t)^2}{N_s}\right] - \frac{(\Sigma X_{tot})^2}{N_{tot}}$$

between-cells sum of squares

$$SS_{cells} = \Sigma\left[\frac{(\Sigma X_c)^2}{N_c}\right] - \frac{(\Sigma X_{tot})^2}{N_{tot}}$$

between-subjects sum of squares

$$SS_{subjects} = \Sigma\left[\frac{(\Sigma X_s)^2}{N_t}\right] - \frac{(\Sigma X_{tot})^2}{N_{tot}}$$

error sum of squares for one-way ANOVA

$$SS_{error} = \Sigma\left[\Sigma X_t^2 - \frac{(\Sigma X_t)^2}{N_t}\right]$$

error sum of squares for factorial ANOVA

$$SS_{error} = \Sigma\left[\Sigma X_c^2 - \frac{(\Sigma X_c)^2}{N_c}\right]$$

error sum of squares for correlated-measures ANOVA

$$SS_{error} = SS_{tot} - SS_{subjects} - SS_{treat}$$

sum of squares for the interaction effect in factorial ANOVA

$$SS_{AB} = N_c\Sigma[(\bar{X}_{AB} - \bar{X}_A - \bar{X}_B + \bar{X}_{tot})^2]$$

Check:

$$SS_{AB} = SS_{cells} - SS_A - SS_B$$

sum of squares for a main effect in factorial ANOVA

$$SS = \frac{(\Sigma X_1)^2}{N_1} + \frac{(\Sigma X_2)^2}{N_2} + \cdots$$
$$+ \frac{(\Sigma X_K)^2}{N_K} - \frac{(\Sigma X_{tot})^2}{N_{tot}}$$

where 1 and 2 denote levels of a factor and K denotes the last level of a factor

Chi Square

basic formula

$$\chi^2 = \Sigma\left[\frac{(O - E)^2}{E}\right]$$

degrees of freedom for a chi square table

$$df = (R - 1)(C - 1)$$

where R = number of rows
C = number of columns

shortcut formula for a 2 × 2 table

$$\chi^2 = \frac{N(AD - BC)^2}{(A + B)(C + D)(A + C)(B + D)}$$

where $A, B, C,$ and D designate the four cells of the table, moving left to right across the top row and then across the bottom row

Confidence Intervals

about a sample mean

$$\text{LL} = \bar{X} - ts_{\bar{X}}$$
$$\text{UL} = \bar{X} + ts_{\bar{X}}$$

about a mean difference
(independent samples)

$$\text{LL} = (\bar{X}_1 - \bar{X}_2) - t(s_{\bar{X}_1 - \bar{X}_2})$$
$$\text{UL} = (\bar{X}_1 - \bar{X}_2) + t(s_{\bar{X}_1 - \bar{X}_2})$$

about a mean difference
(correlated samples)

$$\text{LL} = (\bar{X} - \bar{Y}) - t(s_{\bar{D}})$$
$$\text{UL} = (\bar{X} - \bar{Y}) + t(s_{\bar{D}})$$

Correlation

Pearson product-moment
definition formula

$$r = \frac{\Sigma(z_X z_Y)}{N}$$

computation formulas

$$r = \frac{\dfrac{\Sigma XY}{N} - (\bar{X})(\bar{Y})}{(S_X)(S_Y)}$$

$$r = \frac{N\Sigma XY - (\Sigma X)(\Sigma Y)}{\sqrt{[N\Sigma X^2 - (\Sigma X)^2][N\Sigma Y^2 - (\Sigma Y)^2]}}$$

testing significance from
.00 (or use Table A)

$$t = (r)\sqrt{\frac{N - 2}{1 - r^2}}$$
$$df = N - 2$$

Spearman r_s computation
formula

$$r_s = 1 - \frac{6\Sigma D^2}{N(N^2 - 1)}$$

Degrees of Freedom

See specific statistical tests.

Deviation Score

$$X - \bar{X} \quad \text{or} \quad X - \mu$$

Effect Size Index

one-sample design

$$d = \frac{|\bar{X} - \mu|}{\hat{s}}$$

two independent samples,
$N_1 = N_2$

$$d = \frac{|\bar{X}_1 - \bar{X}_2|}{\sqrt{N_1}(s_{\bar{X}_1 - \bar{X}_2})}$$

two independent samples,
$N_1 \neq N_2$

$$d = \frac{|\bar{X}_1 - \bar{X}_2|}{\sqrt{\dfrac{\hat{s}_1^2(df_1) + \hat{s}_2^2(df_2)}{df_1 + df_2}}}$$

two correlated samples

$$d = \frac{|\overline{X} - \overline{Y}|}{\sqrt{N}(s_{\overline{D}})}$$

one-way ANOVA

$$f = \frac{\hat{s}_{\text{treat}}}{\hat{s}_{\text{error}}} = \frac{\sqrt{\dfrac{K-1}{N_{\text{tot}}}(MS_{\text{treat}} - MS_{\text{error}})}}{\sqrt{MS_{\text{error}}}}$$

chi square (2×2)

$$\phi = \sqrt{\frac{\chi^2}{N}}$$

Mann–Whitney U Test

value for U

$$U_1 = (N_1)(N_2) + \frac{N_1(N_1 + 1)}{2} - \Sigma R_1$$

where $\Sigma R_1 =$ sum of the ranks of the N_1 group

testing significance for
  larger samples, $N \geq 21$

$$z = \frac{(U_1 + c) - \mu_U}{\sigma_U}$$

where $c = 0.5$

$$\mu_U = \frac{N_1 N_2}{2}$$

$$\sigma_U = \sqrt{\frac{(N_1)(N_2)(N_1 + N_2 + 1)}{12}}$$

Mean

from raw data

$$\mu \text{ or } \overline{X} = \frac{\Sigma X}{N}$$

from a frequency distribution

$$\mu \text{ or } \overline{X} = \frac{\Sigma f X}{N}$$

of a set of means

$$\overline{\overline{X}} = \frac{\Sigma(N_1 \overline{X}_1 + N_2 \overline{X}_2 + \cdots + N_K \overline{X}_K)}{\Sigma N}$$

where N_1, N_2, and so on are the numbers of
scores associated with their respective means
and K is the last of the means being averaged

Range

of a distribution

$$\text{Range} = X_{\text{H}} - X_{\text{L}}$$

where $X_{\text{H}} =$ highest score
$X_{\text{L}} =$ lowest score

interquartile range

$$\text{IQR} = 75\text{th percentile} - 25\text{th percentile}$$

Regression

for predicting Y from X

$$\hat{Y} = r\frac{S_Y}{S_X}(X - \bar{X}) + \bar{Y}$$

for a straight line

$$\hat{Y} = a + bX$$

where a = value at the Y intercept

b = slope of the regression line

the Y intercept of a regression line

$$a = \bar{Y} - b\bar{X}$$

the slope of a regression line

$$b = r\frac{S_Y}{S_X}$$

$$b = \frac{N\Sigma XY - (\Sigma X)(\Sigma Y)}{N\Sigma X^2 - (\Sigma X)^2}$$

Standard Deviation of a Population or Sample (for description)

by the raw-score method from ungrouped data

$$\sigma \ \text{ or } \ S = \sqrt{\frac{\Sigma X^2 - \frac{(\Sigma X)^2}{N}}{N}}$$

by the raw-score method from grouped data

$$\sigma \ \text{ or } \ S = \sqrt{\frac{\Sigma f X^2 - \frac{(\Sigma f X)^2}{N}}{N}}$$

by the deviation-score method from ungrouped data

$$S = \sqrt{\frac{\Sigma(X - \bar{X})^2}{N}}$$

Standard Deviation of a Sample (an estimator of σ)

by the raw-score method from ungrouped data

$$\hat{s} = \sqrt{\frac{\Sigma X^2 - \frac{(\Sigma X)^2}{N}}{N - 1}}$$

$$\hat{s} = \sqrt{\frac{N\Sigma X^2 - (\Sigma X)^2}{N(N - 1)}}$$

by the raw-score method from grouped data

$$\hat{s} = \sqrt{\frac{\Sigma f X^2 - \frac{(\Sigma f X)^2}{N}}{N - 1}}$$

$$\hat{s} = \sqrt{\frac{N\Sigma f X^2 - (\Sigma f X)^2}{N(N - 1)}}$$

by the deviation-score method
from ungrouped data

$$\hat{s} = \sqrt{\frac{\Sigma(X - \bar{X})^2}{N - 1}}$$

for correlated samples

$$\hat{s}_D = \sqrt{\frac{\Sigma D^2 - \dfrac{(\Sigma D)^2}{N}}{N - 1}}$$

Standard Error

of the mean
   where the population
   standard deviation is known

$$\sigma_{\bar{X}} = \frac{\sigma}{\sqrt{N}}$$

estimated from a single sample

$$s_{\bar{X}} = \frac{\hat{s}}{\sqrt{N}}$$

of a difference between means
   for equal-N samples

$$s_{\bar{X}_1 - \bar{X}_2} = \sqrt{s_{\bar{X}_1}^2 + s_{\bar{X}_2}^2}$$

$$= \sqrt{\left(\frac{\hat{s}_1}{\sqrt{N_1}}\right)^2 + \left(\frac{\hat{s}_2}{\sqrt{N_2}}\right)^2}$$

$$= \sqrt{\frac{\Sigma X_1^2 - \dfrac{(\Sigma X_1)^2}{N_1} + \Sigma X_2^2 - \dfrac{(\Sigma X_2)^2}{N_2}}{N_1(N_2 - 1)}}$$

for samples with
   unequal N's

$$s_{\bar{X}_1 - \bar{X}_2} = \sqrt{\left(\frac{\Sigma X_1^2 - \dfrac{(\Sigma X_1)^2}{N_1} + \Sigma X_2^2 - \dfrac{(\Sigma X_2)^2}{N_2}}{N_1 + N_2 - 2}\right)\left(\frac{1}{N_1} + \frac{1}{N_2}\right)}$$

for correlated samples by the
   direct-difference method

$$s_{\bar{D}} = \frac{\hat{s}_D}{\sqrt{N}}$$

$$\text{where } \hat{s}_D = \sqrt{\frac{\Sigma D^2 - \dfrac{(\Sigma D)^2}{N}}{N - 1}}$$

for correlated samples
   when r is known

$$s_{\bar{D}} = \sqrt{s_{\bar{X}}^2 + s_{\bar{Y}}^2 - 2r_{XY}(s_{\bar{X}})(s_{\bar{Y}})}$$

t Test

as a test for whether a sample
   mean came from a
   population with a mean μ

$$t = \frac{\bar{X} - \mu}{s_{\bar{X}}}$$

$$df = N - 1$$

for correlated samples
where r is known

$$t = \frac{\overline{X} - \overline{Y}}{\sqrt{s_{\overline{X}}^2 + s_{\overline{Y}}^2 - 2r_{XY}(s_{\overline{X}})(s_{\overline{Y}})}}$$

for correlated samples
where the direct-difference
method is used

$$t = \frac{\overline{X} - \overline{Y}}{s_{\overline{D}}}$$

df = number of pairs minus one

for independent samples

$$t = \frac{\overline{X}_1 - \overline{X}_2}{s_{\overline{X}_1 - \overline{X}_2}}$$

$$df = N_1 + N_2 - 2$$

for testing whether a correlation
coefficient is significantly
different from .00

$$t = (r)\sqrt{\frac{N - 2}{1 - r^2}}$$

df = number of pairs minus two

Tukey HSD

$$\text{HSD} = \frac{\overline{X}_1 - \overline{X}_2}{s_{\overline{X}}}$$

for one-way and factorial
  ANOVA

where $s_{\overline{X}}$ (for $N_1 = N_2 = N_t$) $= \sqrt{\dfrac{MS_{\text{error}}}{N_t}}$

$$s_{\overline{X}} \text{ (for } N_1 \neq N_2) = \sqrt{\frac{MS_{\text{error}}}{2}\left(\frac{1}{N_1} + \frac{1}{N_2}\right)}$$

for correlated-measures
  ANOVA

$$s_{\overline{X}} = \sqrt{\frac{MS_{\text{error}}}{N_t}}$$

Variance

Use the formulas for the standard deviation.
For $\hat{s}^2$ and σ^2, square $\hat{s}$ and σ.

Wilcoxon Matched-Pairs Signed-Ranks T Test

value for T

T = smaller sum of the signed ranks

testing significance for
  larger samples, $N \geq 50$

$$z = \frac{(T + c) - \mu_T}{\sigma_T}$$

where $c = 0.5$

$$\mu_T = \frac{N(N + 1)}{4}$$

$$\sigma_T = \sqrt{\frac{N(N + 1)(2N + 1)}{24}}$$

N = number of pairs

z Score Formulas

descriptive
$$z = \frac{X - \overline{X}}{S}; \quad \text{also } z = \frac{X - \mu}{\sigma}$$

probability of a sample mean
$$z = \frac{\overline{X} - \mu}{\sigma_{\overline{X}}}$$

(See also Mann–Whitney U test and Wilcoxon Matched-Pairs Signed-Ranks T test.)

appendix f

Answers to Problems

chapter one

1. **a.** 7.5–8.5   **b.** Qualitative   **c.** 9.995–10.005
   d. Qualitative   **e.** 4.45–4.55   **f.** 2.945–2.955
2. Many paragraphs can qualify as good answers to this question. Your paragraph should include variations of these definitions:
   Population: a defined set of scores that is of interest to an investigator
   Sample: some subset of a population
   Statistic: a numerical or nominal characteristic of a sample
   Parameter: a numerical or nominal characteristic of a population
3. **a.** Inferential; sample data are used to predict a future event.
   b. Descriptive; the population was available for counting.
   c. Descriptive; all the votes were counted and no inference was made.
   d. Inferential; a conclusion about the generative approach is based on sample data.
4. Nominal, ordinal, interval, and ratio
5. *Nominal:* Different numbers are assigned to different classes of things.
   Ordinal: Nominal properties, plus the numbers carry information about greater than and less than.
   Interval: Ordinal properties, plus the distance between units is equal.
   Ratio: Interval properties, plus a zero point that means "none of the thing being measured."
6. **a.** Ordinal   **b.** Ratio   **c.** Nominal   **d.** Nominal
   e. Ordinal   **f.** Ratio   **g.** Ordinal
7. **a.** *Dependent variable:* rating of qualification; *independent variable:* gender; *number of levels:* two; *names of levels:* female and male
   b. *Dependent variable:* narcissism; *independent variable:* birth order; *number of levels:* three; *names of levels:* first, second, and later born
   c. *Dependent variable:* jail sentence; *independent variable:* occupation; *number of levels:* three; *names of levels:* vice president, janitor, and unspecified

8. a. *Dependent variable:* number of suggestions complied with
Independent variable, number of levels, and their names: hypnosis; two; yes and no

　　i. *Nominal variables:* hypnosis, suggestions; *statistic:* mean number of suggestions complied with

　　ii. Barber's study shows that hypnotized participants do not acquire powers greater than normal (see Barber, 1976).

b. *Dependent variable:* answer to the question Did you see a barn?
Independent variable, number of levels, and their names: mention of the barn in the question about how fast the car was going; two; mentioned and not mentioned

　　i. Several answers can be correct here. The most general population answer is "people" or "people's memory." Less general is "the memory of people given misinformation about an event." One parameter is the percent of all people who say they see a barn even though no barn was in the film.

　　ii. If people are given misinformation after they have witnessed an event, they sometimes incorporate the misinformation into their memory. [See Loftus's book, *Eyewitness Testimony* (1979), which relates this phenomenon to courts of law.]

c. *Dependent variable:* weight (grams) of crackers consumed
Independent variable, number of levels, and their names: the time shown on the clock on the wall; three; slow, correct, and fast

　　i. Weight (grams)

　　ii. Eating by obese men is affected by what time they think it is; that is, consumption by obese men does not depend entirely on internal physiological cues of hunger (see Schachter and Gross, 1968).

9. Besides differing in the textbook used, the two classes also differed in (1) the professor, one of whom might be a better teacher; (2) the time the class met; and (3) the length of the class period. A 10:00 A.M. class probably has younger students who have fewer outside commitments. Students probably concentrate less during a 3-hour class than they do in three 1-hour classes. Any of these extraneous variables might be responsible for the difference in scores. Also, for the comparison to be valid, the investigator would need assurance that the comprehensive test was appropriate for both textbooks.

chapter **two**

1.

	Women				Men	
Height (in.)	Tally marks	f		Height (in.)	Tally marks	f
72	/	1		77	/	1
71		0		76	/	1
70	/	1		75	/	1
69	//	2		74	/	1
68		0		73	////	4
67	////	5		72	//// /	6
66	//// //	7		71	//// //	7
65	//// ////	10		70	//// /	6
64	//// ////	9		69	//// ///	8
63	////	5		68	//// /	6
62	////	5		67	//	2
61	///	3		66	//	2
60	/	1		65	///	3
59	/	1		64	/	1
				63		0
				62	/	1

2. The order of candidates is arbitrary; any order is correct.

Candidate	f
Attila	13
Bolivar	5
Gandhi	19
Lenin	8
Mao	11

3.

Temperature intervals	Tally marks	f
99.3–99.5	///	3
99.0–99.2	//	2
98.7–98.9	//// /	6
98.4–98.6	////	5
98.1–98.3	//// /	6
97.8–98.0	//// ////	10
97.5–97.7	//	2
97.2–97.4	///	3
96.9–97.1	//	2
96.6–96.8		0
96.3–96.5	/	1

4.

Number of sentences heard before	Tally marks	*f*
16	//	2
15		0
14	//	2
13	///	3
12	///	3
11	~~////~~ ////	9
10	~~////~~ /	6
9	~~////~~ //	7
8	////	4
7	//	2
6	/	1

Interpretation: Here are some points that would be appropriate in your interpretation statement: (1) Surprisingly, no one recognized that all the statements were new. (2) All the students thought they had heard more than a fourth of the statements before. (3) Most students thought they had heard about half the statements before, but, as the number of statements increased or decreased from half, fewer and fewer thought that.

5. a. The midpoint of the interval 54–56

 b. A frequency count; for example, there were 15 scores of 55

 c. 14

6.

7. A bar graph is appropriate.

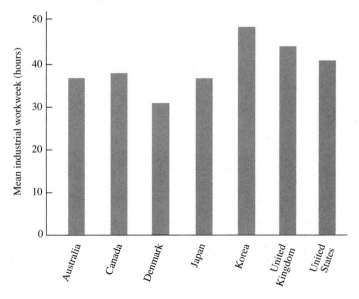

8. They should be graphed as frequency polygons. Note that the midpoint of the class interval is used as the *X*-axis score.

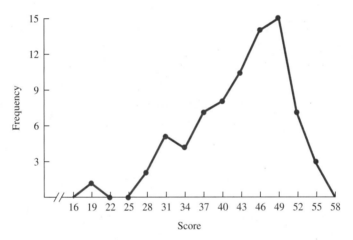

9. Part a is positively skewed; part b is negatively skewed.
10. A bar graph is the proper graph for the qualitative data in problem 2.

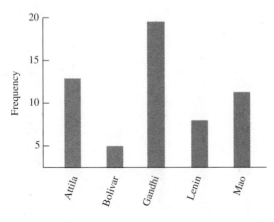

11. A frequency polygon has a frequency count on the Y axis; the variable that is being counted is on the X axis. A line graph has some variable other than frequency on the Y axis.

12.

Elevation (feet)

Average temperature in January is affected by elevation as well as by latitude. The higher the elevation, the lower the temperature, although the relationship is not a straight line.

13. Check your sketches against Figures 2.6 and 2.9.

14. Right

15. a. Positively skewed **b.** Symmetrical **c.** Positively skewed
d. Positively skewed **e.** Negatively skewed **f.** Symmetrical

16. To find the median, arrange the scores in descending order.
a. 5 **b.** 12.5, halfway between 12 and 13
c. 9.5, halfway between 8 and 11

17. It may help to arrange the scores in descending order. Only distribution c has two modes. They are 14 and 18.

18. $Mean = \dfrac{\Sigma X}{N} = \dfrac{409}{39} = 10.49$ sentences

$$\text{Median location} = \frac{N + 1}{2} = \frac{39 + 1}{2} = 20$$

Counting from the bottom of the distribution, you find 20 scores of 10 or less. $Median = 10$ sentences. Counting from the top of the distribution, you find 19 scores of 11 or more. The location of the 20th score is at 10. $Median = 10$ sentences. There are 9 scores of 11. $Mode = 11$ sentences.

People are poor at recognizing whether or not they have heard a particular sentence if what they hear has concepts similar to material they heard earlier. On the average, more than half the sentences never heard before were identified as having been heard.

19. a. For these nominal data, only the mode is appropriate. The mode is Gandhi, who accounted for 34 percent of the signs.
b. Because only one precinct was covered and the student's interest was city-wide, this is a sample mode, a statistic.
c. Because there are more Gandhi yard signs than signs for any other candidate, I expect Gandhi to get the most votes.

20.

Women

Height (in.)	Tally marks	f	fX
72	/	1	72
71		0	
70	/	1	70
69	//	2	138
67	̶H̶H̶	5	335
66	̶H̶H̶ //	7	462
65	̶H̶H̶ ̶H̶H̶	10	650
64	̶H̶H̶ ////	9	576
63	̶H̶H̶	5	315
62	̶H̶H̶	5	310
61	///	3	183
60	/	1	60
59	/	1	59
		50	3230

$$Mean = \overline{X}_{women} = \frac{\Sigma X}{N} = \frac{3230}{50} = 64.6 \text{ inches}$$

$$\text{Median location} = \frac{N + 1}{2} = \frac{50 + 1}{2} = 25.5$$

Counting from the bottom of the distribution, you find 24 scores below 65 inches, so the median is located among the 10 scores of 65. *Median* = 65 inches. The most frequently occurring score (10 times) is 65 inches. *Mode* = 65 inches, which is found for 20 percent of the women.

Men

Height (in.)	Tally marks	f	fX
77	/	1	77
76	/	1	76
75	/	1	75
74	/	1	74
73	////	4	292
72	̶H̶H̶ /	6	432
71	̶H̶H̶ //	7	497
70	̶H̶H̶ /	6	420
69	̶H̶H̶ ///	8	552
68	̶H̶H̶ /	6	408
67	//	2	134
66	//	2	132
65	///	3	195
64	/	1	64
63		0	
62	/	1	62
		50	3490

$$Mean = \overline{X}_{men} = \frac{\Sigma X}{N} = \frac{3490}{50} = 69.80 \text{ inches}$$

$$\text{Median location} = \frac{N + 1}{2} = \frac{50 + 1}{2} = 25.5$$

Counting from the top of the distribution, you find 21 scores above 70 inches. The median is among the 6 scores of 70. *Median* = 70 inches. A height of 69 inches occurs most frequently. *Mode* = 69 inches, which is the height of 16 percent of the men.

21. It is easy to infer that the interest is in the oral body temperature of human beings. Thus, these are sample data.

22. a. Median location = $\frac{N + 1}{2} = \frac{19 + 1}{2} = 10$

Counting from the *bottom,* you reach 7 when you include the score of 12. The 10th score is located among the 5 scores of 13. *Median* = 13.

b. Median location = $\frac{N + 1}{2} = \frac{20 + 1}{2} = 10.5$

Arrange the scores in descending order. Counting from the top, you reach 6 when you include the score of 2. The location of the 10.5th score is among the 5 scores of 1. *Median* = 1.

c. Median location = $\frac{N + 1}{2} = \frac{6 + 1}{2} = 3.5$

The median is the average of 26 and 21. *Median* = 23.5.

23. $\Sigma(X - \overline{X}) = 0$ and $\Sigma(X - \overline{X})^2$ is a minimum.

24. a. The mode is appropriate because the observations are of a nominal variable.
   b. The median or mode is appropriate because helpfulness is measured with an ordinal variable.
   c. The median is appropriate for data with an open-ended category.
   d. The median is the appropriate measure. It is conventional to use the median for income data because the distribution is often severely skewed. (About half of the frequencies are in the $0–$18,000 range.)
   e. The mode is appropriate because these are nominal data.
   f. The mean is appropriate because the data are not severely skewed.

25. $74 \times 12 = 888$

$69 \times 31 = 2139$

$75 \times \underline{17} = \underline{1275}$

$\Sigma 60 = 4302 \qquad$ *Overall mean* $(\overline{\overline{X}}) = \frac{4302}{60} = 71.7$

26. For part a, mean = 21.80 and median = 20. The mean is larger, so the distribution is positively skewed.

For part b, mean = 43.19 and median = 46. The mean is smaller, so the distribution is negatively skewed.

27. This is not correct. To find his lifetime batting average he would need to add his hits for the three years and divide by the total number of at-bats. From the description of the problem, it appears that his average is less than .317.

28.

Body temperature	f	fX
99.5	2	199.0
99.4	1	99.4
99.2	1	99.2
99.1	1	99.1
98.9	2	197.8
98.8	2	197.6
98.7	2	197.4
98.6	2	197.2
98.5	1	98.5
98.4	2	196.8
98.3	3	294.9
98.2	2	196.4
98.1	1	98.1
98.0	3	294.0
97.9	5	489.5
97.8	2	195.6
97.7	1	97.7
97.5	1	97.5
97.4	2	194.8
97.2	1	97.2
97.0	1	97.0
96.9	1	96.9
96.4	1	96.4
	$\Sigma = 40$	3928.0

$$Mean = \overline{X} = \frac{\Sigma X}{N} = \frac{3928.0}{40} = 98.2°F$$

$$Median\ location = \frac{N + 1}{2} = \frac{40 + 1}{2} = 20.5$$

If you count from the bottom of the distribution, there are 19 scores below 98.2. The 20.5th score is in the two scores of 98.2. *Median* = 98.2°F. The most frequently occurring score is 97.9, which is found 5 times. *Mode* = 97.9°F, which occurs in 12.5 percent of the cases.

Oral body temperature (°F)

The central tendency statistics show that the mean and median are 98.2°F and the mode is 97.9°F. All three of these measures are below 98.6°F, the figure usually considered "normal." The frequency polygon of the grouped frequency distribution shows that oral body temperatures are distributed in a fairly symmetrical, bell-shaped fashion.

29. What is your response? Do you know this stuff? Can you work the problems?

chapter three

1. **a.** Range = 17 − 1 = 16   **b.** Range = 0.45 − 0.30 = 0.15

2. With $N = 100$, the 25th percentile score is the 25th score from the bottom. There are 21 scores less than 46 and 5 scores of 46. The 25th percentile score is 46. The 75th percentile score is 25 frequencies from the top. There are 23 scores greater than 55 and 9 scores of 55. The 75th percentile score is 55. The interquartile range is 55 − 46 = 9. Thus, 50 percent of the PC scores are in the 10-point range of 46–55.

3. *Women:* Because $0.25 \times 50 = 12.5$, the 25th percentile score is 12.5 frequencies up from the bottom and the 75th percentile score is 12.5 frequencies down from the top. The 25th percentile score is 63 inches (there are 5 frequencies of 63 and 10 less than 63). The 75th percentile score is 66 inches (there are 7 frequencies of 66 and 9 greater than 66). Thus, the interquartile range of heights of women is 3 inches; 50 percent are 63 to 66 inches tall.

   Men: With $N = 50$ for men also, the same 12.5 frequencies from the bottom and top identify the 25th and 75th percentile scores. The 25th percentile score is among the 6 scores of 68 inches (there are 9 less than 68). The 75th percentile score is among the 6 scores of 72 inches (there are 8 greater than 72). Thus, the interquartile range is 4 inches for men; 50 percent fall in the range from 68 to 72 inches.

4. σ is used to describe the variability of a population.
   $\hat{s}$ is used to estimate μ from a sample of the population.
   S is used to describe the variability of a sample when you have no desire to estimate σ.

5. a.

X	$X - \overline{X}$	$(X - \overline{X})^2$
7	2	4
6	1	1
5	0	0
2	−3	9
$\Sigma = 20$	0	14

$\overline{X} = 5$

$$S = \sqrt{\frac{\Sigma(X - \overline{X})^2}{N}} = \sqrt{\frac{14}{4}} = \sqrt{3.5} = 1.87$$

b.

X	$X - \overline{X}$	$(X - \overline{X})^2$
14	3.8	14.44
11	0.8	0.64
10	−0.2	0.04
8	−2.2	4.84
8	−2.2	4.84
$\Sigma = 151$	0	24.80

$\overline{X} = 10.2$

$$S = \sqrt{\frac{\Sigma(X - \overline{X})^2}{N}} = \sqrt{\frac{24.80}{5}} = \sqrt{4.96} = 2.23$$

c.

X	$X - \overline{X}$	$(X - \overline{X})^2$
107	2	4
106	1	1
105	0	0
102	−3	9
$\Sigma = 420$	0	14

$\overline{X} = 105$

$$S = \sqrt{\frac{\Sigma(X - \overline{X})^2}{N}} = \sqrt{\frac{14}{4}} = \sqrt{3.5} = 1.87$$

6. Although the numbers in problem 5c are much greater than those in problem 5a, the standard deviations are the same. Thus, the size of the numbers does not give you any information about the variability of the numbers.

7. Yes; the larger the numbers, the larger the mean.

8.

City	Mean	Standard deviation
San Francisco	56.75°F	3.96°F
Albuquerque	56.75°F	16.24°F

Although the mean temperature is the same for the two cities, Albuquerque has a wider variety of temperatures.

9. In eyeballing data for variability, use the range as a quick index.
   a. The second distribution is more variable.
   b. Equal variability
   c. The first distribution is more variable.
   d. Equal variability
   e. The second distribution is more variable; however, most of the variability is due to one extreme score, 15.
10. The second distribution (b) is more variable than the first (a).

a.

X	X^2	$X - \bar{X}$	$(X - \bar{X})^2$
5	25	2.5	6.25
4	16	1.5	2.25
3	9	0.5	0.25
2	4	−0.5	0.25
1	1	−1.5	2.25
0	0	−2.5	6.25
$\Sigma = 15$	55	0	17.50

$$S = \sqrt{\frac{\Sigma X^2 - \dfrac{(\Sigma X)^2}{N}}{N}} = \sqrt{\frac{55 - \dfrac{(15)^2}{6}}{6}} = \sqrt{2.917} = 1.71$$

$$S = \sqrt{\frac{\Sigma(X - \bar{X})^2}{N}} = \sqrt{\frac{17.50}{6}} = \sqrt{2.917} = 1.71$$

b.

X	X^2	$X - \bar{X}$	$(X - \bar{X})^2$
5	25	2.5	6.25
5	25	2.5	6.25
5	25	2.5	6.25
0	0	−2.5	6.25
0	0	−2.5	6.25
0	0	−2.5	6.25
$\Sigma = 15$	75	0	37.50

$$S = \sqrt{\frac{\Sigma X^2 - \dfrac{(\Sigma X)^2}{N}}{N}} = \sqrt{\frac{75 - \dfrac{(15)^2}{6}}{6}} = \sqrt{6.25} = 2.50$$

$$S = \sqrt{\frac{\Sigma(X - \bar{X})^2}{N}} = \sqrt{\frac{37.50}{6}} = \sqrt{6.25} = 2.50$$

How did your guess compare to your computation?

11. For problem 10a, $\dfrac{5}{1.71} = 2.92$. For problem 10b, $\dfrac{5}{2.50} = 2.00$. Yes, these results are between 2 and 5.

12. a. $\sigma = \sqrt{\dfrac{\Sigma X^2 - \dfrac{(\Sigma X)^2}{N}}{N}} = \sqrt{\dfrac{262 - \dfrac{(34)^2}{5}}{5}} = 2.48$

b. $\sigma = \sqrt{\dfrac{294 - \dfrac{(38)^2}{5}}{5}} = 1.02$

13. For $\hat{s} = \$0.02$: Because you spend almost exactly \$2.17 each day in the Student Center, each day you don't go to the center will save you \$2.17.

    For $\hat{s} = \$2.50$: Because you spend widely varying amounts in the Student Center, restrict your buying of the large-ticket items, which will save you a bundle each time you spend. (The actual value of $\hat{s}$ for the data in Table 2.5 is \$2.69.)

14. The variance is the square of the standard deviation.

15. Because the researcher wanted to estimate the standard deviation of the freshman class scores from a sample, $\hat{s}$ is the correct standard deviation.

$$\hat{s} = \sqrt{\dfrac{\Sigma X^2 - \dfrac{(\Sigma X)^2}{N}}{N - 1}} = \sqrt{\dfrac{5064 - \dfrac{(304)^2}{21}}{20}} = \sqrt{33.16} = 5.76$$

$\hat{s}^2 = 33.162$

16.

Height (in.)	f	fX	fX²
		Females	
72	1	72	5,184
70	1	70	4,900
69	2	138	9,522
67	5	335	22,445
66	7	462	30,492
65	10	650	42,250
64	9	576	36,864
63	5	315	19,845
62	5	310	19,220
61	3	183	11,163
60	1	60	3,600
59	1	59	3,481
		$\Sigma = 3230$	208,966

$$\hat{s} = \sqrt{\dfrac{208{,}966 - \dfrac{3230^2}{50}}{49}} = 2.51 \text{ inches}$$

	Males		
Height (in.)	f	fX	fX^2
77	1	77	5,929
76	1	76	5,776
75	1	75	5,625
74	1	74	5,476
73	4	292	21,316
72	6	432	31,104
71	7	497	35,287
70	6	420	29,400
69	8	552	38,088
68	6	408	27,744
67	2	134	8,978
66	2	132	8,712
65	3	195	12,675
64	1	64	4,096
62	1	62	3,844
		$\Sigma = 3490$	244,050

$$\hat{s} = \sqrt{\frac{244,050 - \dfrac{3490^2}{50}}{49}} = 3.02 \text{ inches}$$

Interpretation: The heights of men are more variable than the heights of women.

17. Before: $\hat{s} = \sqrt{1.429} = 1.20$, $\overline{X} = 5.0$
    After: $\hat{s} = \sqrt{14.286} = 3.78$, $\overline{X} = 5.0$

It appears that students were homogeneous and neutral before studying poetry. After studying poetry for 9 weeks, some students were turned on and some were turned off; they were no longer neutral.

18. $\hat{s}$ is the more appropriate standard deviation; generalization to the manufacturing process is expected.

$$\text{Process A:} \quad \hat{s} = \sqrt{\frac{18 - \dfrac{0^2}{6}}{5}} = 1.90 \text{ millimeters}$$

$$\text{Process B:} \quad \hat{s} = \sqrt{\frac{12 - \dfrac{0^2}{6}}{5}} = 1.55 \text{ millimeters}$$

Process B produces more consistent doodads. Note that both processes have an average error of zero.

19.

$$\hat{s} = \sqrt{\frac{\Sigma fX^2 - \dfrac{(\Sigma fX)^2}{N}}{N-1}} = \sqrt{\frac{385,749.24 - \dfrac{(3928)^2}{40}}{39}} = \sqrt{0.504} = 0.71°\text{F}$$

20. Range $= 16 - 6 = 10$ sentences. The 25th percentile is 9 sentences; the 75th percentile is 12. The interquartile range is 3 sentences ($12 - 9 = 3$).

Bransford and Franks are interested in the way human beings process information and remember it. Thus, these data are a sample and the purpose is to generalize to all humans. $\hat{s}$ is the appropriate standard deviation, and $\hat{s}^2$ is the appropriate variance.

Number of sentences			
heard before	f	fX	fX^2
16	2	32	512
15	0	0	0
14	2	28	392
13	3	39	507
12	3	36	432
11	9	99	1089
10	6	60	600
9	7	63	567
8	4	32	256
7	2	14	98
6	1	6	36
	$\Sigma = 39$	409	4489

$$\hat{s} = \sqrt{\frac{\Sigma f X^2 - \dfrac{(\Sigma f X)^2}{N}}{N - 1}} = \sqrt{\frac{4489 - \dfrac{(409)^2}{39}}{38}} = \sqrt{5.256} = 2.29 \text{ sentences}$$

$\hat{s}^2 = 5.26$

chapter**four**

1. Zero

2. $\Sigma z = 0$. Because $\Sigma(X - \overline{X}) = 0$, it follows that $\dfrac{\Sigma(X - \overline{X})}{S} = 0$.

3. Negative z scores are preferable when lower scores are preferable to higher scores. Examples include measures such as errors, pollution levels, and race times.

4. Hattie: $z = \dfrac{37 - 39.5}{1.803} = -1.39$;   Missy: $z = \dfrac{24 - 26.25}{1.479} = -1.52$

Of course, in timed events, the more negative the z score, the better the score. Missy's -1.52 is superior to Hattie's -1.39.

5. Tobe's apple: $z = \dfrac{9 - 5}{1} = 4.00$;   Zeke's orange: $z = \dfrac{10 - 6}{1.2} = 3.33$

Tobe's z score is larger, so the answer to Hamlet must be a resounding "To be." Notice that each fruit varies from its group mean by the same amount. It is the smaller variability of the apple weights that makes Tobe's fruit a winner.

6. First test: $z = \dfrac{79 - 67}{4} = 3.00$;   second test: $z = \dfrac{125 - 105}{15} = 1.33$;

third test: $z = \dfrac{51 - 45}{3} = 2.00$

Ableson's performance was poorest on the second test.

7. Distribution X, with a range of 7 and an interquartile range of 2, is positively skewed with the mean at about 3 and the median at about $1\frac{1}{2}$. Distribution Y has a range of 6 and an interquartile range of 2. It is negatively skewed with the mean near $6\frac{1}{2}$ and the median near $7\frac{1}{2}$. Distribution Z, with a range of 3 and an interquartile range of 1, is a symmetrical distribution with the mean and median at about $6\frac{1}{2}$.

8. Distribution X and Y are most different. More than 50 percent of the scores in Y (median $= 7\frac{1}{2}$) are greater than the highest score in X (7). This isn't true for Distribution Z.

9.

10. For normal oral body temperatures: highest $= 99.5°F$; lowest $= 96.4°F$; 25th percentile $= 97.8°F$; 75th percentile $= 98.7°F$; median $= 98.2°F$

11. $d = \dfrac{|\mu_1 - \mu_2|}{\sigma} = \dfrac{|\bar{X}_1 - \bar{X}_2|}{\sigma}$

a. $d = .50$; a d of .50 is a medium effect size index.
b. $d = .90$ A d of .90 is a quite large effect size, greater even than the .80 that is considered large.
c. $d = .30$; a d of .30 is intermediate in size between values that are considered small and medium; I'd call it "somewhat larger than a small effect size." (Note that the means are the same in problems b and c. The data's variability is always important.)

12. $d = \dfrac{|\mu_1 - \mu_2|}{\sigma} = \dfrac{|\bar{X}_1 - \bar{X}_2|}{\sigma} = \dfrac{|469 - 456|}{110} = 0.12$

Although first-born children score higher on cognitive ability than second-born children, the size of the effect is quite small, being about half the size needed to be designated as "small."

13. For cabbages: $\overline{X} = \dfrac{\Sigma X}{N} = \dfrac{300}{3} = 100$ pounds

$$S = \sqrt{\dfrac{\Sigma X^2 - \dfrac{(\Sigma X)^2}{N}}{N}} = \sqrt{\dfrac{30{,}200 - \dfrac{(300)^2}{3}}{3}} = 8.165 \text{ pounds}$$

For pumpkins: $\overline{X} = \dfrac{\Sigma X}{N} = \dfrac{2700}{3} = 900$ pounds

$$S = \sqrt{\dfrac{\Sigma X^2 - \dfrac{(\Sigma X)^2}{N}}{N}} = \sqrt{\dfrac{2{,}450{,}000 - \dfrac{(2700)^2}{3}}{3}} = 81.65 \text{ pounds}$$

Lavery's cabbage: $z = \dfrac{X - \overline{X}}{S} = \dfrac{124 - 100}{8.165} = 2.94$

Burke's pumpkin: $z = \dfrac{X - \overline{X}}{S} = \dfrac{1092 - 900}{81.65} = 2.35$

Lavery's cabbage is the overall winner.

14. Compile the raw data into simple frequency distributions:

Psychotherapy	f	Control	f
34	1	22	1
28	1	21	1
25	1	18	1
23	1	13	1
18	1	10	1
15	1	9	1
13	2	5	1
11	1	4	2
10	2	3	2
9	1	2	1
7	2	1	1
5	1	0	1
0	1	−2	1
−2	1	−3	1
−5	1	−7	1
−10	1	−9	1
−15	1	−12	1
		−22	1
	$\Sigma = 20$		$\Sigma = 20$

$$\bar{X}_{\text{therapy}} = \frac{\Sigma X}{N} = \frac{196}{20} = 9.8$$

$$\bar{X}_{\text{control}} = \frac{\Sigma X}{N} = \frac{60}{20} = 3.0$$

$$\text{Median location} = \frac{N+1}{2} = \frac{20+1}{2} = 10.5$$

Median$_{\text{therapy}}$: The 10.5th score is between the two scores of 10. Median$_{\text{therapy}}$ = 10.

Median$_{\text{control}}$: The 10.5th score is between the two scores of 3. Median$_{\text{control}}$ = 3.

Therapy: 25th percentile: $0.25 \times N = 5$. There are 5 scores below 5, so 5 is the 25th percentile.

75th percentile: There are 5 scores above 15; 15 is the 75th percentile.

Control: 25th percentile: $0.25 \times N = 5$. There are 5 scores below -2, so -2 is the 25th percentile.

75th percentile: There are 5 scores above 9; 9 is the 75th percentile.

$$d = \frac{|\mu_1 - \mu_2|}{\sigma} = \frac{|\bar{X}_1 - \bar{X}_2|}{\sigma} = \frac{|9.8 - 3.0|}{10} = |0.68| = 0.68$$

Change in psychological health

Psychotherapy helps. On the average, the psychological health of those who received psychotherapy was 9.8 points higher after treatment. For those not receiving treatment, scores improved only 3.0 points. The difference in the two means produced an effect size index of 0.68. Such an effect size is intermediate between "large" and "medium." Both distributions of scores were approximately symmetrical.

chapter**five**

1. A bivariate distribution has two variables. The scores of the variables are paired in some logical way.

2. Variation in one variable is accompanied by predictable variation in the other; the statement says nothing about direction. The variables may vary in the same direction (positive correlation) or in the opposite direction (negative correlation).

3. In positive correlation, as X increases, Y increases. (It is equivalent to say that as X decreases, Y decreases.) In negative correlation, as X increases, Y decreases. (An equivalent statement is: As X decreases, Y increases in negative correlation.)

4. a. Yes; positive; taller people usually weigh more than shorter people.

  b. No; these scores cannot be correlated because there is no basis for pairing a particular oak tree with a particular pine tree.

  c. Yes, negative; as temperatures go up, less heat is needed and fuel bills go down.

  d. Yes, positive; people with higher IQs score higher on reading comprehension tests.

  e. Yes, positive; those who score well on quiz 1 are likely to score well on quiz 2. The fact that some students are in section 1 and some are in section 2 can be ignored.

  f. No; there is no basis for pairing the score of a student in section 1 with that of a student in section 2.

5.

	Fathers	Daughters
Mean	67.50	62.50
S	3.15	2.22
Σ (score)	405	375
Σ (score)^2	27,397	23,467

$$\Sigma XY = 25{,}334$$
$$r = .513$$

r by the blanched formula:

$$r = \frac{\dfrac{25{,}334}{6} - (67.5)(62.5)}{(3.15)(2.22)} = \frac{4222.33 - 4218.75}{6.99} = .512$$

By the raw-score formula:

$$r = \frac{(6)(25{,}334) - (405)(375)}{\sqrt{[(6)(27{,}397) - (405)^2][(6)(23{,}467) - (375)^2]}} = .513$$

6. $\overline{X} = 46.18 \qquad \overline{Y} = 30.00$

$$S_X = \sqrt{\dfrac{87{,}373 - \dfrac{(1755)^2}{38}}{38}} = \sqrt{166.308} = 12.90$$

$$S_Y = \sqrt{\dfrac{37{,}592 - \dfrac{(1140)^2}{38}}{38}} = \sqrt{89.263} = 9.45$$

$$r = \dfrac{\dfrac{55{,}300}{38} - (46.18)(30)}{(12.90)(9.45)} = \dfrac{1455.26 - 1385.40}{121.91} = \dfrac{69.86}{121.91} = .57$$

7. $r = \dfrac{(50)(175{,}711) - (202)(41{,}048)}{\sqrt{[(50)(1740) - (202)^2][(50)(35{,}451{,}830) - (41{,}048)^2]}} = .25$

$$r = \dfrac{\dfrac{175{,}711}{50} - (4.04)(820.96)}{(4.299)(187.247)} = .25$$

8. a. $-.60$ **b.** $-.95$ **c.** $.50$ **d.** $.25$

Would you be interested in knowing how statistics teachers did on this question? I asked 31 statistics teachers to estimate r for these four scatterplots. These are the means and standard deviations of their estimates:

a. $\overline{X} = -.50;\ \hat{s} = .19$ **b.** $\overline{X} = -.87;\ \hat{s} = .06$
c. $\overline{X} = .42;\ \hat{s} = .17$ **d.** $\overline{X} = .13;\ \hat{s} = .11$

These data suggest to me that the question is a difficult one.

9. Coefficient of determination $= .32$. The two measures of self-esteem have about 32 percent of their variance in common. Although the two measures are to some extent measuring the same traits, they are, in large part, measuring different traits.

10. $r^2 = (.25)^2 = .0625$. This means that 6.25 percent of the variance in expenditure per pupil is related to population variability; 93.75 percent of the variance is related to factors other than population size. Thus, to "explain" the differences among the states, factors other than population differences are needed.

11. Your shortest answer here is *no*. Although you have two distributions, you do not have a bivariate distribution. There is no logical pairing of a height score for a woman with a height score for a man.

12. $(.10)^2 = .01$, or 1 percent; $(.40)^2 = .16$, or 16 percent. Note that a fourfold increase in the correlation coefficient produces a 16-fold increase in the common variance.

13.

	Cigarette consumption	Death rate
Mean	604.27	205.00
S	367.865	113.719
Σ (score)	6,647	2,255
Σ (score)^2	5,505,173	604,527

$$\Sigma\, XY = 1,705,077$$
$$N = 11$$
$$r = .74$$

With $r = .74$, these data show that there is a fairly strong relationship between per capita cigarette consumption and male death rate 20 years later.

14. a. Vocational interests tend to remain similar from age 20 to age 40. The effect size index (.70) is very large.

　　b. Identical twins raised together have very similar IQs. The effect size index (.86) is just huge.

　　c. There is a slight tendency for IQ to be lower as family size increases. An effect size index of .30 is a medium effect.

　　d. There is a slight tendency for taller men to have higher IQs than shorter men. The effect size index (.22) is midway between small and medium.

　　e. The lower a person's income level is, the greater the probability that he or she will be diagnosed as psychotic. An effect size index of .83 is huge.

15. a. There is some tendency for children with more older siblings to accept less credit or blame for their own successes and failures than children with fewer older siblings. The effect size index is somewhat larger than medium.

　　b. Nothing; correlation coefficients do not permit you to make cause-and-effect statements.

　　c. The coefficient of determination is .1369 ($-.37^2$), so you can say that about 14 percent of the variance in acceptance of responsibility is predictable from knowledge of the number of older siblings; 86 percent is not. Thus, knowing the number of siblings is only somewhat helpful in predicting degree of acceptance of responsibility.

16. a. Because $(.97)^2 = .94$, these two tests have 94 percent of their variance in common. They must be measuring essentially the same trait. There is a very strong tendency for persons who score high on one test to score high on the other.

　　b. No; this is a cause-and-effect statement and it is not justified on the basis of correlational data.

17. A Pearson *r* is not a useful statistic for data such as these because the relationship is a curved one. You might note that the scatterplot is similar to the curved relationship you saw in Figure 2.5.

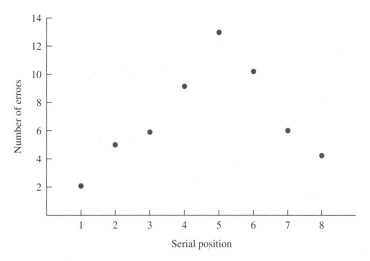

18. a. $b = r\dfrac{S_Y}{S_X} = (.513)\dfrac{2.217}{3.149} = (.513)(.704) = .3612$

$a = \overline{Y} - b\overline{X} = 62.5 - (.3612)(67.5) = 38.12$

b. To draw the regression line, I used two points: $(\overline{X}, \overline{Y})$ and $(62, 60.5)$. I got the point $(62, 60.5)$ by using the formula $\hat{Y} = a + bX$. Other X values would work just as well.

19. a. $b = r\dfrac{S_Y}{S_X} = (.57)\dfrac{9.45}{12.90} = (.57)(.73) = .42$

$a = \overline{Y} - b\overline{X} = 30 - (.42)(46.18) = 30 - 19.40 = 10.60$

b. $\hat{Y} = a + bX = 10.60 + (.42)(42) = 10.60 + 17.64 = 28.24$

20.

	Advertising X	Sales Y
Mean	4.286	108.571
S	1.030	20.304
Σ (score)	30	760
Σ (score)²	136	85,400
	ΣXY = 3360	
	r = .703	

a. $\hat{Y} = a + bX = 49.176 + 13.858X$

b. One point that I used for the regression line on the scatterplot was for the two means, $\overline{X} = 4.29$ and $\overline{Y} = 108.57$. The other point was $X = 6.00$ and $Y = 132.33$. Other X values would work just as well.

c.

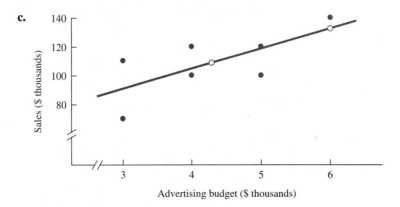

Advertising budget ($ thousands)

d. $\hat{Y} = a + bX$. For $X = 10$, $\hat{Y} = 187.756$ or, in terms of sales, $187,756.

e. You were warned that the interpretation of r can be a tricky business. This problem leads you to one of those common errors I mentioned. With $r = .70$, you have a fair degree of confidence that the prediction will hold *in general*. However, if your thinking (and your answer) was that by increasing advertising, sales would increase, you made the error of inferring a causal relationship from a correlation coefficient. As a matter of fact, some analysts say that sales *cause* advertising; they say that advertising budgets are determined by *last year's* sales. In any case, don't infer more than the data permit. These data allow you to estimate with some confidence what sales were by knowing the amount spent for advertising. In this problem, the percent of sales spent on advertising (just less than 4 percent) is about the same as the percent spent nationally in the United States.

   Also, $10,000 for advertising is much more than any amount used to find the regression line. It is not safe to predict outcomes for points far from the data.

21. $\hat{Y} = (.80)\left(\dfrac{15}{16}\right)(65 - 100) + 100 = 73.75 = 74$

22. Designate the year X and the number of graduates Y.

$\Sigma = 555$ $\Sigma Y = 68.10$ $N = 6$
$\Sigma X^2 = 51,355$ $\Sigma Y^2 = 774.47$ $\Sigma XY = 6303.80$
$S_X = 1.708$ $S_Y = 0.506$
$r = .878$ $a = -12.700$ $b = 0.260$

For the number of graduates in the year 2000,

$$\hat{Y} = a + bX = -12.70 + .26(100) = 13.30$$

Adjusting the decimals, you may predict that the number of bachelor's degree graduates in 2000 is 1,330,000. Depending on when you are working this problem, the accuracy of this prediction can be checked in the *Statistical Abstract of the United States.*

24. On the big important questions like this one, you are on your own!

WHAT WOULD YOU RECOMMEND? chapters one–five

a. The mode is appropriate because this is a distribution of nominal data.
b. z scores can be used.
c. A regression analysis produces a prediction.
d. The median is appropriate because the distribution is severely skewed (the class interval with zero has the highest frequency). The range, interquartile range, and standard deviation would all be informative. The appropriate standard deviation is σ because the 50 states represent the entire population of states.
e. The applicant with the more consistent judgments is the better candidate. The range, interquartile range, and standard deviation would all be helpful.
f. A line graph is appropriate for the relationship that is described. A Pearson product-moment correlation coefficient is not appropriate because the relationship is curvilinear (much forgetting at first, and much less after 4 days).
g. The median is the appropriate central tendency statistic when one or more of the categories is open-ended.
h. A correlation coefficient will describe this relationship. What direction and size would you guess for this coefficient? The answer (based on several studies) is $-.10$.

chapter**six**

1. There are seven cards between the 3 and jack, each with a probability of $\frac{4}{52} = .077$. So $(7)(.077) = .539$.
2. The probability of drawing a 7 is $\frac{4}{52}$, and there are 52 opportunities to get a 7. Thus, $(52)(\frac{4}{52}) = 4$.
3. There are four cards that are higher than a jack or lower than a 3. Each has a probability of $\frac{4}{52}$. Thus, $(4)(\frac{4}{52}) = \frac{16}{52} = .308$.
4. The probability of a 5 or 6 is $\frac{4}{52} + \frac{4}{52} = \frac{8}{52} = .154$. In 78 draws, $(78)(.154) = 12$ cards that are 5s or 6s.
5. a. .7500; adding the probability of one head (.3750) to that of two heads (.3750), you get .7500.
    b. $.1250 + .1250 = .2500$

6. 16(.1250) = 2 times
7. a. .0668
   b. The test scores are normally distributed.
8. a. .0832
   b. .2912
   c. .4778
9. Empirical; the scores are based on observations.
10. $z = (X - \mu)/\sigma$. For IQ = 55, z score = −3.00; for IQ = 110, z score = .67; for IQ = 103, z score = .20; for IQ = 100, z score = .00.
11. A quick check of your answer can be made by comparing it with the proportion of students who have IQs of 120 or higher, .0918. The proportion with IQs of 70 or lower will be less than the proportion with IQs of 120 or higher. Is your calculated proportion less? The figure shows a normal distribution in which the proportion of the population with IQs of 70 or lower is shaded. For IQ = 70, z = (70 − 100)/15 = −30/15 = −2.00. The proportion beyond z = 2.00 is .0228 (column C in Table C). Thus, .0228 or 2.28 percent of the population is expected to be in special education classes.

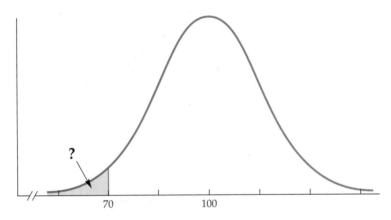

12. (.0228)(4000) = 91.2, or 91 students.
13. $z = (110 - 100)/15 = 10/15 = .67$. The proportion beyond z = .67 is .2514, which is the expected proportion of people with IQs of 110 or higher.
14. a. .2514 × 250 = 62.85, or 63 students
    b. 250 − 62.85 = 187.15, or 187 students
    c. $\frac{1}{2}$ × 250 = 125. I hope you were able to get this one immediately by thinking about the symmetrical nature of the normal distribution.
15. The z score that includes a proportion of .02 is 2.06. If z = 2.06, then X = 100 + (2.06)(15) = 100 + 30.9 = 130.9. In fact, Mensa requires an IQ of 130 on tests that have a standard deviation of 15.
16. a. The z score you need is 1.65. (1.64 will include more than the tallest 5 percent.) 1.65 = (X − 64.6)/2.5; X = 64.6 + 4.1 = 68.7 inches.
    b. $z = (58 - 64.6)/2.5 = -6.6/2.5 = -2.64$. The proportion excluded is .0041.

17. a. $z = (60 - 69.8)/3.0 = -9.8/3.0 = -3.27$. The proportion excluded is .0006.
   b. $z = (62 - 69.8)/3.0 = -7.8/3.0 = -2.60$. The proportion taller than Napoleon is .9953 (.5000 + .4953).

18. a. $z = \dfrac{3.20 - 3.11}{.05} = 1.80$, proportion $= .0359$

   b. The z score corresponding to a proportion of .1000 is 1.28. Thus,

$$X = 3.11 + (1.28)(.05)$$
$$= 3.11 + .064$$
$$= 3.17$$
$$X = 3.11 - .064$$
$$= 3.05$$

   Thus, the middle 80 percent of the pennies weigh between 3.05 and 3.17 grams.

19.

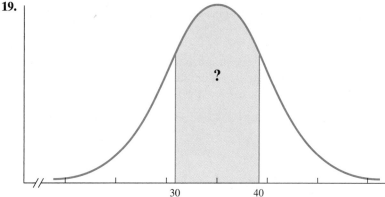

   a. $z = \dfrac{30 - 35}{6} = -.83$, proportion $= .2967$; $z = \dfrac{40 - 35}{6} = .83$, proportion $=$ .2967; $(2)(.2967) = .5934 =$ the proportion of students with scores between 30 and 40.
   b. The probability is also .5934.

20. No, because the distance between 30 and 40 straddles the mean, where scores that occur frequently are found. The distance between 20 and 30 is all in one tail of the curve, where scores occur less frequently.

21.

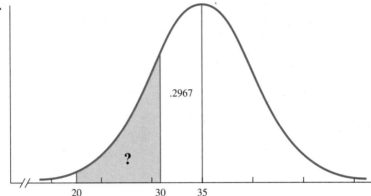

$$z = \frac{20 - 35}{6} = -2.50, \text{ proportion} = .4938$$

$$z = \frac{30 - 35}{6} = -.83, \text{ proportion} = .2967$$

.4938 − .2967 = .1971. The proportion of students with scores between 35 and 30 is subtracted from the proportion with scores between 35 and 20. There are also other correct ways to set up this problem.

22. 800 people × .1971 = 157.7 = 158 people

23.

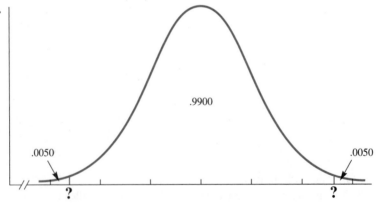

The z score that is conventionally used for a proportion of .0050 is 2.58 (column C). Thus,

$$X = \mu + (z)(\sigma)$$
$$= 100 + (2.58)(15)$$
$$= 138.7 \text{ or } 139$$

and

$$X = 100 + (-2.58)(15)$$
$$= 61.3 \text{ or } 61$$

IQ scores of 139 or higher or 61 or lower are achieved by 1 percent of the population.

24. Because .01 of the curve lies outside the limits of the scores 61 and 139, the probability is .01. The probability is .005 that the randomly selected person has an IQ of 139 or higher and .005 that it is 61 or lower.

25. $z = \dfrac{.85 - .99}{.17} = \dfrac{-.14}{.17} = -.82$

The proportion associated with $z = -.82$ is .2061 (column C), and $(.2061)(185,822) = 38,298$ truck drivers. Because the normal curve corresponds only approximately to the empirical curve, the usual procedure is to report this result as "About 38,000 truck drivers would get raises."

26. This problem has two parts: determining the number of trees over 8 inches DBH on 1 acre and multiplying by 100:

$$z = \frac{8 - 13.68}{4.83} = -1.18, \text{ proportion} = .5000 + .3810 = .8810$$

.8810(199) = 175.319 trees on 1 acre
175.319(100) = 17,532 trees on 100 acres (about 17,500 trees)

27. $z = \dfrac{300 - 268}{14} = \dfrac{32}{14} = 2.29$

The proportion of gestation periods expected to last 10 months or longer is .0110. In modern obstetrical care, pregnancies are normally brought to term (by drugs or surgery) earlier than 300 days.

28. Converting this problem to inches, you find the z score is

$$z = \frac{80 - 69.8}{3.0} = 3.40$$

The proportion of the curve above 3.40 is .0003, so only 3 in 10,000 have to duck to enter a residential room.

29. $\mu = \dfrac{3600}{100} = 36.0$ inches

$$\sigma = \sqrt{\frac{130,000 - \dfrac{(3600)^2}{100}}{100}} = 2.00 \text{ inches}$$

a. $z = \dfrac{32 - 36}{2} = -2.00$, proportion $= .5000 + .4772 = .9772$

b. $z = \dfrac{39 - 36}{2} = 1.50$, proportion $= .5000 + .4332 = .9332$

.9332(300) = 279.96 or 280 hobbits

c. $z = \dfrac{46 - 36}{2} = 5.00$, proportion $< .00003$ or, practically speaking, zero chapter **seven**

1. $\bar{X} = \dfrac{\Sigma X}{N} = \dfrac{68}{8} = 8.50$

2. For this problem, it is not the numbers in your sample that are "right" or "wrong" but the method you used to get them. Using the table of random numbers, create an identifying number (01–20) for every population member. (Table 7.2 will work fine.) Haphazardly find a starting place in Table B and record in sequence 10 two-digit identifying numbers (ignoring numbers over 20 and duplications). Write the 10 population members that correspond to the 10 selected identifying numbers. How does your mean compare to the mean of 8.50?

3. As in problem 2, the method, not the numbers, determines a correct answer. You should give each number an identifying number of 01 to 38, begin at some chance starting place, and select 12 numbers. The starting place for every random sample taken from Table B should be haphazardly chosen. If you start at the same place in the table every time, you will have the same identifying numbers every time.

4. Those with greater educational accomplishments are more likely to return the questionnaire. Such a biased sample will overestimate the accomplishments of the population.

5. The sample will be biased. Many of the general population will not even read the article and thus cannot be in the sample. Of those who read it, some will have no opinion and some will not bother to mark the ballot and return it. In short, the results of such a questionnaire cannot be generalized to the larger population.

6. **a.** Biased, although the bias is not considered serious. The sample is biased because every sample does not have an equal chance of being selected. For example, two brothers named Kotki, next to each other on the list, could not both be in one sample. Thus, samples with both Kotki brothers are not possible.

   b. Biased, because freshmen, sophomores, and seniors, although they are members of the population, would be in none of the samples.

   c. Random, assuming the names in the box were thoroughly mixed.

   d. Random; though small, this sample fits the definition of a random sample.

   e. Biased; some members of the population have completed the English course and no samples could include these people.

7. Standard error; expected value

8. **a.** Select a set of numbers as the population. Draw many samples of the same size, find the range for each sample, and arrange those ranges into a frequency distribution.

   b. The standard error of the range

   c. The expected value of the range, which is the mean of the sampling distribution of the range, will be *less* than the population range. Only a very few sample ranges (maybe only one) can have a value equal to the population range. Thus, almost all the sample ranges will be less than the population range. Statistics (such as the range) with expected values that are not equal to the corresponding population parameters are called *biased statistics* and are not popular among mathematical statisticians.

9. Please check your answer with the definition and explanation in the chapter.

10. If this standard error (whose name is the standard error of estimate) is very small, it means that the actual Y will be close to the predicted $\hat{Y}$. A very large standard error indicates that you can not put much confidence in a predicted $\hat{Y}$. (The actual Y is subject to other influences besides X.)

11.

		σ		
N	1	2	4	8
1	1	2	4	8
4	0.50	1	2	4
16	0.25	0.50	1	2
64	0.125	0.25	0.50	1

12. As N increases by a factor of 4, $\sigma_{\bar{x}}$ is reduced by one-half. More simply but less accurately, as N increases, $\sigma_{\bar{x}}$ decreases.

13. 16 times

14. $z = \dfrac{8.5 - 9}{2/\sqrt{8}} = \dfrac{-0.5}{0.707} = -.71$; proportion = .2389

15. For sample means of 8.5 or less,

$$z = \dfrac{8.5 - 9}{2/\sqrt{16}} = \dfrac{-0.5}{0.5} = -1.00; \text{ proportion} = .1587$$

For sample means of 10 or greater,

$$z = \dfrac{10 - 9}{2/\sqrt{16}} = \dfrac{1}{0.5} = 2.00; \text{ proportion} = .0228$$

16. For mean IQs of 105 or greater,

$$z = \dfrac{105 - 100}{15/\sqrt{25}} = \dfrac{5}{3} = 1.67; \text{ probability} = .0475$$

For mean IQs of 90 or less,

$$z = \dfrac{90 - 100}{15/\sqrt{25}} = \dfrac{-10}{3} = -3.33; \text{ probability} = .0004$$

Classrooms with 25 students will have a mean IQ fairly close to 100; departures of as much as 5 points will happen less than 5 percent of the time; departures of as much as 10 points will be very rare, occurring less than one time in a thousand. This analysis is based on the premise that no systematic factors are influencing a child's assignment to a classroom.

17. $\sigma_{\bar{X}} = \dfrac{15,000}{\sqrt{40}} = 2372.$ Therefore, $z = \dfrac{54,700 - 57,000}{2372} = -.97; p = .1660.$

Thus, the junior's suspicion that the $54,700 sample mean might be just a chance fluctuation from a true mean of $57,000 has some foundation. Such a sample mean (or one even smaller) would be expected in about 17 percent (one-sixth) of a large number of random samples if the true campus mean was $57,000.

18. a. Normal; t **b.** $t;$ normal   **c.** $t;$ normal

19. Degrees of freedom

20. W. S. Gosset, who wrote under the pseudonym "Student," invented the t distribution so that he could assess probabilities for small samples from populations for which σ was not known.

21. $\bar{X} = \dfrac{\Sigma X}{N} = \dfrac{15,600}{16} = 975$

$$\hat{s} = \sqrt{\dfrac{15,247,500 - \dfrac{(15,600)^2}{16}}{15}} = 50.0$$

$$s_{\bar{X}} = \dfrac{\hat{s}}{\sqrt{N}} = \dfrac{50}{\sqrt{16}} = 12.5$$

22. $\bar{X} = 27.00, \hat{s} = 3.863, s_{\bar{X}} = 1.032, df = 13$

The t value for 13 df and 95 percent confidence is 2.160.

$\text{LL} = \bar{X} - t(s_{\bar{X}}) = 27 - 2.160(1.032) = 24.77$

$\text{UL} = \bar{X} + t(s_{\bar{X}}) = 27 + 2.160(1.032) = 29.23$

The workshop is effective. A 95 percent confidence interval about the mean of all who might take the workshop, based on a sample of 14 who completed the workshop, gives a range of 24.77 to 29.23. Because the national mean for the test is *less* than the lower limit of the confidence interval of those taking the workshop, you may conclude that clients are more assertive than the national average after completing the workshop. (This conclusion assumes that the clients were not better than average in assertiveness before the workshop.)

23. $\bar{X} = 1.20, \hat{s} = 0.10, s_{\bar{X}} = 0.0577, t = 31.598$

$\text{LL} = 1.20 - 31.598(0.0577) = -0.62$ inches

$\text{UL} = 1.20 + 31.598(0.0577) = 3.02$ inches

The 120 Krups coffee pots meet the specifications. The company can have 99.9 percent confidence that the true mean burn rate for this box of pots is less than 4 inches per minute. (The lower limit is not of interest to Raytheon, and of course, there is no such thing as -0.62 inches.)

24. a. $\pm 3.182(2) = \pm 6.364$ (12.728 units wide)

   b. $\pm 2.131(1) = \pm 2.131$ (4.262 units wide). The fourfold increase in N produces a confidence interval about one-third as wide. The effect of changing the sample size depends on the relative change in N, not the absolute change.

c. $\pm 3.182(1) = \pm 3.182$ (6.364 units wide). Reducing $\hat{s}$ by half reduces the size of the confidence interval by half. This relationship is true regardless of the size of $\hat{s}$. Data gatherers reduce $\hat{s}$ by taking measurements carefully, using precise instruments, and increasing N.

25. The size of the standard deviation of each sample. If you happen to get a sample that produces a large standard deviation, a large standard error will result and the confidence interval will be wide.

26. a. 19

    b. Narrower; because N was increased by four, $s_{\bar{x}}$ is reduced to half of its size. In addition, increasing N decreases the t value from Table D. Therefore, the lines are less than half as long.

27. a. 18

    b. Narrower; the t value is smaller.

28. The standard error of the mean for each of the four trial means is 1.0; the t value (24 *df*) for the 95 percent confidence interval is 2.064. The lower and upper limits for each mean are ± 2.064.

    Look at the graph that accompanies this problem. The graph in problem 28 has been modified by adding 95 percent confidence intervals to each trial mean. The overlap of the trial 2 confidence interval with that of trial 1 indicates that the dip at trial 2 may not be reliable. That is, you are 95 percent confident that the population mean for trial 2 is in the range of about 3.94 to 8.06. Because the upper half of this range is also included in the trial 1 confidence interval, you aren't confident that a second sample of data will show the dip at trial 2. [Note that the figure that shows only the means invites you to conclude that performance (always) goes down on trial 2. The figure with the confidence intervals leads to a more cautious interpretation.]

    Looking at the *lack* of overlap between the confidence intervals of trial 3 and trial 2 leads you to conclude that the trial 3 performance is *reliably* better than that of trial 2; that is, you are more than 95 percent confident that a second sample of data will also show that the trial 3 mean is higher than the trial 2 mean.

chapter **eight**

1. .20 is the probability of data such as those actually observed, if the equality hypothesis is true.

2. The null hypothesis is a statement about a parameter.

3. The null hypothesis (symbolized H_0) is a tentative hypothesis about the value of some parameter (often the mean) of the population that the sample is taken from. The alternative hypothesis (symbolized H_1) specifies other values for the parameter.

4. Your outline should include the following points:

    a. Gather sample data from the population you are interested in and calculate a statistic.

    b. Recognize two logical possibilities for the population:
    H_0: a statement specifying an exact value for the parameter of the population
    H_1: a statement specifying all other values for the parameter

    c. Using a sampling distribution that assumes H_0 is correct, find the probability of the statistic you calculated.

    d. If the probability is very small, reject H_0 and accept H_1. If the probability is large, retain both H_0 and H_1, acknowledging that the data do not allow you to reject H_0.

5. Disagree

6. False. Although the researcher probably thinks that H_0 is false, the sampling distribution is based on the assumption that H_0 is true.

7. A significance level (α level) is chosen by the researcher as the point that separates "This is an unlikely event if H_0 is true and therefore reject H_0" from "This could reasonably happen if H_0 is true and therefore retain H_0." The largest acceptable significance level without justification is .05.

8. Phrases that should be circled are "p is small" and "Accept H_1."

9. **a.** $t_{.01}\,(17\ df) = 2.898$
    b. $t_{.001}\,(40\ df) = 3.551$
    c. $t_{.05}\,(\infty\ df) = 1.96$

10. A good interpretation explains the results using the terms and concepts used in the description of the problem. It also tells the direction of any difference that was observed.

11. α is the probability of a Type I error; a Type I error occurs when a researcher mistakenly rejects H_0. To reduce Type I errors, researchers set α at a low level, such as .05 or less. A statistical test on the data produces a p value, which is the probability of obtaining the sample statistic actually obtained, if H_0 is true.

12. A Type II error is retaining a false null hypothesis.

13. The probability of making a Type I error is zero. If, at first, you don't agree with this answer, keep thinking until you understand.

14. The probability of a Type I error decreases from .05 to .01, and the probability of a Type II error increases.

15. I suggest that you check each part of this problem as soon as you do it. If you have already worked them all, just look at each answer until you miss one. If an answer here surprises you, make a note of what surprised you. (Students, like researchers, have expectations and surprises. A surprise often means that you are about to learn something.)

Critical value	t-test value	Reject or retain H_0?	p value
a. 2.262	2.25	Retain	$p > .05$
b. 2.539	2.57	Reject	$p < .01$
c. 2.042	2.03	Retain	$p > .05$
d. 6.869	6.72	Retain	$p > .001$
e. 1.714	1.72	Reject	$p < .05$
f. 2.423	2.41	Retain	$p > .02$

16. $t_{.05} (23\ df) = 2.069$

   a. The mean paranoia score on the MMPI for the 24 police officers was 54.3, which is significantly greater than the population mean of 50 ($p < .05$). Police officers are significantly more paranoid than the general population.

   b. The mean paranoia score on the MMPI for the 24 police officers was 54.3, which is not significantly different from the population mean of 50 ($p > .05$). Police officers are not significantly more paranoid than the general population.

17. $t_{.05} (9\ df) = 2.262$

   a. The reduction in return rate from 5.3 percent to 4.9 percent was not statistically significant ($p > .05$). Including testimonials with orders does not reduce returns by a statistically significant amount.

   b. Including testimonials with orders reduced the return rate from 5.3 percent to 4.9 percent, a statistically significant reduction ($p < .05$).

18. The null hypothesis is $H_0: \mu_0 = 100$.

$$\overline{X} = \frac{5250}{50} = 105.00; \quad \hat{s} = \sqrt{\frac{560,854 - \dfrac{(5250)^2}{50}}{49}} = 14.00$$

$$s_{\overline{X}} = \frac{14.00}{\sqrt{50}} = 1.98; \quad t = \frac{\overline{X} - \mu_0}{s_{\overline{X}}} = \frac{105 - 100}{1.98} = 2.53; 49\ df$$

$t_{.02} (40\ df) = 2.423$

The null hypothesis can be rejected; $p < .02$. The conclusion is that the Head Start participants had IQs that were significantly above average.

   The actual studies that evaluated Head Start were more sophisticated than the one described here, but the conclusions were the same. The IQ differences, however, weren't lasting, but other benefits were.

19. The null hypothesis is H_0: $\mu_0 = 10$. The best alternative hypothesis is H_1: $\mu_1 \neq 10$. A two-tailed alternative hypothesis will allow you to conclude that odometers indicate too much or too little distance.

$$\bar{X} = \frac{121.2}{12} = 10.1$$

$$\hat{s} = \sqrt{\frac{1224.5 - \frac{(121.2)^2}{12}}{11}} = 0.186$$

$$s_{\bar{X}} = \frac{0.186}{\sqrt{12}} = 0.0537$$

$$t = \frac{10.1 - 10}{0.0537} = 1.86; 11\ df$$

$$t_{.05}\ (11\ df) = 2.201$$

Retain the null hypothesis. Although the sample mean (10.1 miles) is greater than the actual distance traveled (10 miles), the t test does not provide evidence that odometers record significantly more distance.

20. $t = \dfrac{\bar{X} - \mu_0}{s_{\bar{X}}} = \dfrac{56.10 - 50.00}{2.774} = 2.199; df = 12$

$$t_{.05}\ (12\ df) = 2.179$$

Reject the null hypothesis and conclude that people who were successful in used car sales have significantly higher extroversion scores than the general population.

21. $d = \dfrac{|\bar{X} - \mu|}{\sigma} = \dfrac{|270.675 - 269.3|}{0.523} = 2.63$

The effect size index, 2.63, is much larger than the conventional designation of "large." Even though the sample mean was only 1.375 grams more than the Frito-Lay company claimed, the effect was large because the bag-filling machinery was so precise (the standard deviation was small).

22. $d = \dfrac{|\bar{X} - \mu|}{\sigma} = \dfrac{|56.1 - 50.0|}{10} = 0.61$

The difference in extroversion between successful used car salespeople and the population mean has a medium effect size index.

23. $d = \dfrac{|\bar{X} - \mu|}{\sigma} = \dfrac{|98.2 - 98.6|}{0.7} = |-0.57| = 0.57$

An effect size index of .57 shows that a medium-size effect has existed undetected for a long time.

24. $H_0: \mu_0 = 58.7$ $\Sigma X = 360.8$ $\Sigma X^2 = 21{,}703.86$ $N = 6$

$\overline{X} = 60.1333$ $s_{\overline{X}} = 0.5084$

$$t = \frac{\overline{X} - \mu_0}{s_{\overline{X}}} = \frac{60.1333 - 58.7}{0.5084} = 2.82;\ 5\ df$$

$t_{.05}\ (5\ df) = 2.571$

The statistical decision is to reject the null hypothesis. Conclude that Snickers candy bars weigh more than the company claims, $p < .05$.

25. $t = (r)\sqrt{\dfrac{N-2}{1-r^2}}$

    a. $t = 2.24$; $df = 8$. Because $t_{.05}\ (8\ df) = 2.31$, $r = .62$ is not significantly different from .00.

    b. $t = -2.12$; $df = 120$. Because $t_{.05}\ (120\ df) = 1.98$, $r = -.19$ is significantly different from .00.

    c. $t = 2.08$; $df = 13$; $t_{.05}\ (13\ df) = 2.16$. Retain the null hypothesis.

    d. $t = -2.85$; $df = 62$; $t_{.05}\ (60\ df) = 2.00$. Reject the null hypothesis.

26. Here's my response: Chance is not a likely explanation for a correlation coefficient of .74 based on an N of 11 pairs. If there was really no relationship between cigarette smoking and lung cancer, chance would produce an $r = \pm.7348$ only 1 time in 100 (Table A, $df = 9$). An $r = .74$ is less likely than an $r = .7348$, so the probability that chance produced this result is less than 1 in 100. I don't believe chance did it.

27. $r = \dfrac{N\Sigma XY - (\Sigma X)(\Sigma Y)}{\sqrt{[N\Sigma X^2 - (\Sigma X)^2][N\Sigma Y^2 - (\Sigma Y)^2]}}$

$$= \frac{42(46{,}885) - (903)(2079)}{\sqrt{[42(25{,}585) - (903)^2][42(107{,}707) - (2079)^2]}} = .40$$

From Table A, $r_{.01}\ (40\ df) = .3932$. Because $.40 > .3932$, reject the null hypothesis and conclude that there is a positive correlation between the time a student works on a test and the grade, $p < .01$.

If, as you finished up this problem, you thought about waiting until the end of the period to turn in your next test (so that you would get a better grade), you drew a cause-and-effect conclusion from a correlation coefficient. Better go back and reread page 96.

chapter **nine**

    1. Your outline should include at least the following points:

      a. Two samples are selected using random assignment or random selection.

      b. The samples are treated the same except for one thing.

      c. The samples are measured.

      d. The difference in measurements either is attributed to the difference in treatments or is about what would be expected if the treatment does not have an effect.

2. Random assignment of participants to groups
3. **a.** *Independent variable:* personality type; Type A and Type B
   Dependent variable: satisfaction in life
   Null hypothesis: The mean satisfaction in life is the same for people with Type A personalities as for those with Type B personalities.
   b. *Independent variable:* type of crime; robbers and embezzlers
   Dependent variable: years served in prison
   Null hypothesis: Convicted robbers and convicted embezzlers spend, on the average, the same number of years in jail.
4. Your outline should include the following points:
   a. Recognize two logical possibilities:
   1. The treatment had no effect (null hypothesis, H_0).
   2. The treatment had an effect (alternative hypothesis, H_1).
   b. Assume H_0 to be correct and establish an α level.
   c. Calculate a test statistic from the sample data. Compare the test statistic to a sampling distribution (which is based on the assumption that H_0 is true).
   d. If the test statistic has a probability less than α, reject H_0, adopt H_1, and write a conclusion about the effects of the treatment. If the test statistic has a probability greater than α, retain H_0 and write a conclusion that says the data did not allow you to choose between H_0 and H_1.
5. Experiments are conducted so that researchers can write a conclusion about the effect of an independent variable on a dependent variable.
6. Degrees of freedom are always equal to the number of observations minus the (a) number of necessary relationships that exist among the observations or (b) number of parameters that are estimated from the observations.
7. **a.** Correlated samples. With twins divided between the two groups, there is a logical pairing (natural pairs); *independent variable:* environment; *levels:* good and bad; *dependent variable:* attitude toward education.
   b. Correlated samples. This is a before-and-after experiment. For each group, the amount of aggression before watching the other children is paired with the amount afterward (repeated measures); *independent variable:* whether children had watched others with better toys; *levels:* yes and no; *dependent variable:* number of aggressive encounters.
   c. Correlated samples. This is a repeated-measures design; *independent variable:* bright artificial light; *levels:* yes and no; *dependent variable:* depression score.
   d. Correlated samples (yoked control design). Participants are paired so that the amount of interrupted sleep is equal for the two. One, however, is deprived of REM sleep, the other is not; *independent variable:* REM deprivation; *levels:* yes and no; *dependent variable:* mood questionnaire score.
   e. Independent samples. The random assignment of individuals and the subsequent procedures give you no reason to pair two particular participants' mood questionnaire scores; *independent variable:* REM deprivation; *levels:* yes and no; *dependent variable:* mood questionnaire score.
   f. Independent samples. The dean randomly assigned individuals to one of the two groups; *independent variable:* sophomore honors course; *levels:* yes and no; *dependent variable:* grade point average.

This paragraph is really not about statistics, and you may skip it if you wish. Were you somewhat more anxious about your decisions on experiments c and d than on experiments a and b? Was this anxiety based on the expectation that surely it was time for an answer to be "independent samples"? Did you base your answer on your expectation rather than on the problem? If your answers are yes, you were exhibiting a *response bias*. A response bias occurs when you make a response on the basis of previous responses rather than on the basis of the current stimulus. Response biases often lead to a correct answer in textbooks, and you may learn to make some decisions (those you are not sure about) on the basis of irrelevant cues (such as what your response was on the last question). To the extent it rewards your response biases, a textbook is doing you a disservice. So, be forewarned; recognize response bias and resist it. (I'll try to do my part.)

8. A two-tailed test is appropriate; if one of the two kinds of animals is superior, you want to know which one.

$H_0: \mu_1 = \mu_2$
$H_1: \mu_1 \neq \mu_2$

	Fellow humans	Pigeons
ΣX	147	75
ΣX^2	7515	1937
N	3	3
$\overline{X}$	49	25
$\hat{s}$	12.4900	5.5678

$$t = \frac{\overline{X}_1 - \overline{X}_2}{\sqrt{\left(\frac{\hat{s}_1}{\sqrt{N_1}}\right)^2 + \left(\frac{\hat{s}_2}{\sqrt{N_2}}\right)^2}} = \frac{49 - 25}{\sqrt{\left(\frac{12.49}{\sqrt{3}}\right)^2 + \left(\frac{5.5678}{\sqrt{3}}\right)^2}} = \frac{24}{7.895} = 3.04$$

$df = 4$

Reject the null hypothesis and conclude that pigeons can spot a person in the ocean more quickly than humans can, $p < .05$.

9. *Independent variable:* percent of cortex removed; *levels:* 0 percent and 20 percent; *dependent variable:* number of errors

$$\hat{s}_{0 \text{ percent}} = \sqrt{\frac{1706 - \frac{(208)^2}{40}}{40 - 1}} = 4.00$$

$$\hat{s}_{20 \text{ percent}} = \sqrt{\frac{2212 - \frac{(252)^2}{40}}{40 - 1}} = 4.00$$

$$s_{\overline{X}_1 - \overline{X}_2} = \sqrt{\left(\frac{4.00}{\sqrt{40}}\right)^2 + \left(\frac{4.00}{\sqrt{40}}\right)^2} = 0.894$$

$$t = \frac{\overline{X}_0 - \overline{X}_{20}}{s_{\overline{X}_1 - \overline{X}_2}} = \frac{5.2 - 6.3}{0.894} = -1.23; \quad df = 78$$

$t_{.05}(60\ df) = 2.00$. Retain the null hypothesis and conclude that a 20 percent loss of cortex does not reduce significantly a rat's memory for a simple maze. *Note:* Some brain operations *improve* performance on some tasks, so a two-tailed test is appropriate.

10.

	New package	Old package
ΣX	45.9	52.8
ΣX^2	238.73	354.08
$\overline{X}$	5.1	6.6
N	9	8
$\hat{s}$	0.76	0.89

$$t = \frac{6.6 - 5.1}{\sqrt{\left(\dfrac{10.24}{15}\right)(0.24)}} = \frac{1.5}{0.40} = 3.75; \quad df = 15$$

$t_{.005}(15\ df) = 2.947$ (one-tailed test). Thus, $p < .005$; therefore, reject the null hypothesis and conclude that the search report can be compiled more quickly with the new package. A one-tailed test is appropriate here because the only interest is whether the new package is better than the one on hand. (See the section "One- and Two-Tailed Tests" in Chapter 8.)

11. First of all, you might point out to the sisters that if the populations are the two freshman classes, no statistics are necessary; you have the population data and there is no sampling error. State U. is one-tenth of a point higher than The U. If the question is not about those freshman classes but about the two schools, and the two freshman classes can be treated as representative samples, a two-tailed test is called for because superiority of either school would be of interest.

$$s_{\overline{X}_1 - \overline{X}_2} = \sqrt{\left(\frac{\hat{s}_1}{\sqrt{N_1}}\right)^2 + \left(\frac{\hat{s}_2}{\sqrt{N_2}}\right)^2} = 0.0474$$

$$t = \frac{\overline{X}_1 - \overline{X}_2}{s_{\overline{X}_1 - \overline{X}_2}} = \frac{23.4 - 23.5}{0.0474} = -2.11; \quad df = 120+$$

$t_{.05}(120\ df) = 1.980$. Students at State U. have statistically significantly higher ACT admission scores than do students at The U. You may have noted how small the difference actually is — only one-tenth of a point, an issue that is discussed in a later section in the text.

12. a. $\hat{s}_D = \sqrt{\dfrac{\Sigma D^2 - \dfrac{(\Sigma D)^2}{N}}{N - 1}}$

$\hat{s}_D$ is the standard deviation of the distribution of differences between correlated scores.

b. $D = X - Y$. D is the difference between two correlated scores.

c. $s_{\bar{D}} = \hat{s}_D/\sqrt{N}$. $s_{\bar{D}}$ is the standard error of the difference between means for a correlated set of scores.

d. $\bar{Y} = \Sigma Y/N$. $\bar{Y}$ is the mean of a set of scores that is correlated with another set.

13.

	X	Y	D	D^2
	16	18	−2	4
	10	11	−1	1
	17	19	−2	4
	4	6	−2	4
	9	10	−1	1
	12	14	−2	4
Σ	68	78	−10	18
Mean	11.3	13.0		

$$\hat{s}_D = \sqrt{\frac{\Sigma D^2 - \dfrac{(\Sigma D)^2}{N}}{N-1}} = \sqrt{\frac{18 - \dfrac{(-10)^2}{6}}{5}} = \sqrt{0.2667} = 0.5164$$

$$s_{\bar{D}} = \frac{\hat{s}_D}{\sqrt{N}} = \frac{0.5164}{\sqrt{6}} = 0.2108$$

$$t = \frac{11.3333 - 13.0000}{0.2108} = \frac{-1.6667}{0.2108} = -7.91; \quad df = 5$$

$t_{.001}$ (5 df) = 6.869. Therefore, $p < .001$. The conclusion reached by the researchers was that frustration (produced by seeing others treated better) leads to aggression.

Notice that the small difference between means (1.67) is highly significant even though the data consist of only six pairs of scores. This illustrates the power that a large correlation can have in reducing the standard error.

14. On this problem you cannot use the direct-difference method because you do not have the raw data. This leaves you with the definition formula, $t = (\bar{X} - \bar{Y})/\sqrt{s_{\bar{X}}^2 + s_{\bar{Y}}^2 - 2r_{XY}(s_{\bar{X}})(s_{\bar{Y}})}$, which requires a correlation coefficient. Fortunately, you have the data necessary to calculate r. (The initial clue for many students is the term ΣXY.)

Women	Men
$\bar{X} = \$12,782.25$	$\bar{Y} = \$15,733.25$
$\hat{s}_X = 2,358.99$	$\hat{s}_Y = 3,951.27$
$s_{\bar{X}} = 589.75$	$s_{\bar{Y}} = 987.82$
	$r = .693$

$$t = \frac{12{,}782.25 - 15{,}733.25}{\sqrt{(589.75)^2 + (987.82)^2 - (2)(.693)(589.75)(987.82)}} = -4.11$$

$t_{.001}$ (15 df) = 4.073. The null hypothesis can be rejected at the .001 level. Here is the conclusion written for the court.

> Chance is not a likely explanation for the $2951 annual difference in favor of men. Chance would be expected to produce such a difference less than one time in a thousand. In addition, the difference cannot be attributed to education because both groups were equally educated. Likewise, the two groups were equal in their work experience at the rehabilitation center. One explanation that has not been eliminated is discrimination, based on sex.

(These data are 1979 salary data submitted in Hartman and Hobgood *v.* Hot Springs Rehabilitation Center, HS-76-23-C.)

15. This is a correlated-samples study. It helps to carry four or five decimal places.

$\overline{X}$ = 0.1629 second $\overline{Y}$ = 0.1814 second

$\Sigma D = -0.1300$ $\Sigma D^2 = 0.0053$ $\overline{X} - \overline{Y} = \dfrac{\Sigma D}{N} = -0.0186$ $N = 7$

$$\hat{s}_D = \sqrt{\frac{\Sigma D^2 - \dfrac{(\Sigma D)^2}{N}}{N - 1}} = \sqrt{0.00048} = 0.02193$$

$$s_{\overline{D}} = \frac{0.02193}{\sqrt{7}} = 0.00829$$

$$t = \frac{-0.0186}{0.00829} = -2.24; \quad df = 6$$

$t_{.05}$ (6 df) = 2.447. The obtained t does not indicate statistical significance. (This is a case in which a small sample led to a Type II error. Auditory RT is faster than visual RT — approximately 0.16 second for auditory RT and 0.18 second for visual RT, using practiced subjects.)

16. I hope that you did not treat these data as correlated samples. If you did, you exhibited a response bias that led you astray. If you recognized the design as independent samples, you are on your (mental) toes.

Recency	Primacy
$\Sigma X_1 = 29$	$\Sigma X_2 = 29$
$\Sigma X_1^2 = 225$	$\Sigma X_2^2 = 225$
$N_1 = 5$	$N_2 = 5$

$$s_{\bar{X}_1 - \bar{X}_2} = \sqrt{\frac{\Sigma X_1^2 - \dfrac{(\Sigma X_1)^2}{N} + \Sigma X_2^2 - \dfrac{(\Sigma X_2)^2}{N}}{N(N-1)}} = 2.049$$

$$t = \frac{\bar{X}_1 - \bar{X}_2}{s_{\bar{X}_1 - \bar{X}_2}} = \frac{4.60 - 5.80}{2.049} = -0.59; \quad df = 8$$

$t_{.05}\,(8\ df) = 2.306$. Therefore, retain the null hypothesis. The conclusion is: This experiment does not provide evidence that primacy or recency is more powerful than the other.

17. $$t = \frac{\bar{X}_1 - \bar{X}_2}{s_{\bar{X}_1 - \bar{X}_2}} = \frac{60 - 36}{\sqrt{\left(\dfrac{12}{\sqrt{45}}\right)^2 + \left(\dfrac{6}{\sqrt{45}}\right)^2}} = \frac{24}{2} = 12.00; \quad df = 88$$

$t_{.001}\,(60\ df) = 3.460$. The meaning of very large t-test values is quite clear; they indicate a very, very small probability. Thus, conclude that children whose illness begins before age 3 exhibit symptoms longer than do those whose illness begins after age 6, $p < .001$.

18. The p is the probability of obtaining such a difference in samples (or a larger difference) if the two populations they come from (holistic scores and lecture – discussion scores) are identical. That is, if the populations are the same, then chance would produce such a difference (or one larger) less than 1 time in 100.

19. Here is one version: Significant differences are due to the independent variable and not to chance; important differences are ones that change our understanding about something. Important differences are significant, but significant differences may or may not be important.

20. $$d = \frac{|\bar{X}_1 - \bar{X}_2|}{\hat{s}} = \frac{|5.2 - 6.3|}{\sqrt{40}(0.894)} = |-0.19| = 0.19$$

An effect size index of 0.19 is small. Thus, d is small, and a t test using large samples did not find a significant difference. This combination leads to the conclusion that the effect (if any) of removal of 20 percent of the cortex is small. If there is an effect, it will be difficult to detect.

21. $$d = \frac{|\bar{X} - \bar{Y}|}{\hat{s}} = \frac{|28.36 - 34.14|}{5.74} = |-1.01| = 1.01$$

An effect size index of 1.01 is large; thus, the effect of the biracial camp on racial attitudes was both reliable (statistically significant) and sizable. Note that the effect size index gives you information that the means do not. That is, $d = 1.01$ tells you much more than does the fact that the girls scored an average of 5.78 points higher after the camp.

22. This is a correlated-samples design.

$$\overline{X}_{asleep} = 3.75 \qquad \overline{Y}_{awake} = 2.50 \qquad \Sigma D = 10 \qquad \Sigma D^2 = 28 \qquad \overline{D} = 1.25$$

$$\hat{s}_D = \sqrt{\dfrac{\Sigma D^2 - \dfrac{(\Sigma D)^2}{N}}{N - 1}} = \sqrt{\dfrac{28 - \dfrac{(10)^2}{8}}{7}} = 1.488$$

$$s_{\overline{D}} = \dfrac{\hat{s}_D}{\sqrt{N}} = \dfrac{1.488}{\sqrt{8}} = 0.5261$$

LL $= (\overline{X} - \overline{Y}) - t(s_{\overline{D}}) = (3.75 - 2.50) - 2.365(0.5261) = 1.25 - 1.24 = 0.01$

UL $= (\overline{X} - \overline{Y}) + t(s_{\overline{D}}) = (3.75 - 2.50) + 2.365(0.5261) = 1.25 + 1.24 = 2.49$

We are 95 percent confident that sleep improved memory by 0.01 to 2.49 words in this experiment. With $\alpha = .05$, the null hypothesis can be rejected with these data. (Note how much more informative "a 95 percent confidence interval of 0.01 to 2.49" is compared to "the null hypothesis was rejected.")

23. This is an independent-samples study. Note that problem 7c had the same independent and dependent variables. The procedures used, not the variables, determine the design.

LL $= (\overline{X}_1 - \overline{X}_2) - t(s_{\overline{X}_1 - \overline{X}_2}) = (11 - 6) - 2.878(1.20) = 5 - 3.45 = 1.55$

UL $= (\overline{X}_1 - \overline{X}_2) + t(s_{\overline{X}_1 - \overline{X}_2}) = (11 - 6) + 2.878(1.20) = 5 + 3.45 = 8.45$

These data provide 99 percent confidence that 2 hours a day in bright light reduces depression by 1.55 to 8.45 units on the scale used to measure depression. Also note that these data allow the null hypothesis to be rejected, $p < .01$.

24. You can be confident that the probabilities are accurate when the dependent-variable scores are normally distributed and have equal variances, and when the samples are randomly assigned or randomly selected from the populations.

25. Power is the probability of rejecting the null hypothesis when it is false.

26. My version of the list is (1) effect size, (2) sample size, (3) α level, and (4) preciseness of measuring the dependent variable.

27. The null hypothesis is that experience has no effect on performance. A two-tailed test would allow a conclusion about the advantage *or* the disadvantage of experience. The experimental design is independent samples.

$$t = \dfrac{7.40 - 5.05}{\sqrt{\left(\dfrac{2.13}{\sqrt{20}}\right)^2 + \left(\dfrac{2.31}{\sqrt{20}}\right)^2}} = \dfrac{2.35}{0.7026} = 3.34; \quad df = 38$$

$t_{.01}$ (30 df) $= 2.750$. Because the experienced group took longer, you should conclude that the previous experience with the switch *retarded* the subject's ability to recognize the solution.

$$d = \dfrac{|\overline{X}_1 - \overline{X}_2|}{\hat{s}} = \dfrac{|7.40 - 5.05|}{\sqrt{20}(0.7026)} = 0.75$$

An effect size index of 0.75 borders on being large.

28. I hope you recognized that these data could be arranged as correlated samples (matched pairs). Each pair represents a level of achievement in French II.

| | Listened to tape | | | | | |
| | In car | | In laboratory | | | |
	Rank	Errors	Rank	Errors	D	D^2
	1	7	2	7	0	0
	4	7	3	15	−8	64
	5	12	6	22	−10	100
	8	12	7	24	−12	144
	9	21	10	32	−11	121
Sum		59		100	−41	429
Mean		11.8		20.0	−8.2	

$$\hat{s}_D = \sqrt{\frac{\Sigma D^2 - \dfrac{(\Sigma D)^2}{N}}{N - 1}} = \sqrt{\frac{429 - \dfrac{(-41)^2}{5}}{4}} = 4.817$$

$$s_{\bar{D}} = \frac{\hat{s}_D}{\sqrt{N}} = \frac{4.817}{\sqrt{5}} = 2.154$$

$$t = \frac{\bar{X} - \bar{Y}}{s_{\bar{D}}} = \frac{11.8 - 20.0}{2.154} = \frac{-8.2}{2.154} = -3.81; \quad df = 4$$

$$d = \frac{|\bar{X} - \bar{Y}|}{\hat{s}} = \frac{|11.8 - 20.0|}{4.817} = |-1.70| = 1.70$$

$t_{.02}\,(4\;df) = 3.747$. Reject the null hypothesis at the .02 level. Because the car group made fewer errors, conclude that concrete, immediate experience facilitates learning vocabulary. A two-tailed test is appropriate because *disproving* a claim is always of interest to the scientifically minded. To use your data to support *or* refute a claim requires you to choose a two-tailed test (before the data are gathered). An effect size index of 1.70 is very large. (It is true that concrete, immediate experience has a great effect on learning foreign vocabulary words.)

Perhaps you analyzed these data with a confidence interval. Here is one for 95 percent confidence.

$$\text{LL} = (\bar{X} - \bar{Y}) - t(s_{\bar{D}}) = (11.8 - 20.0) - 2.776(2.154) = -8.2 - 5.98 = -14.18$$
$$\text{UL} = (\bar{X} - \bar{Y}) + t(s_{\bar{D}}) = (11.8 - 20.0) + 2.776(2.154) = -8.2 + 5.98 = -2.22$$

With 95 percent confidence, you can be sure that the effect of concrete, immediate experience is to reduce the number of errors on the vocabulary test by 2.22 to 14.18 items.

WHAT WOULD YOU RECOMMEND? chapters six–nine

a. A correlated-samples t test will allow you to determine if the number of smiles (dependent variable) is related to whether or not there is an audience (independent variable). The samples are correlated because each infant contributes data to both the audience and the no-audience condition.

b. A z-score analysis using the normal curve will give the percent of the population that is less gregarious than the social worker.

c. A one-sample t test will give the probability of obtaining a sample with a mean of 55 from a population with a mean of 50. If this probability is very low, you might conclude that the workshop increased gregariousness scores.

d. The lower and upper limits of a confidence interval will capture a population mean with a certain level of confidence (usually 95 or 99 percent).

e. The state's population mean is not captured in problem D because the sample is not a random one (or representative either) from the state. A better interpretation is that the interval captures the mean of "those students planning on going to college."

f. To determine the probability that an $r = .20$ could be a chance fluctuation, test the null hypothesis H_0: $\rho = 0$. If H_0 is rejected, conclude that there is a reliable relationship between the two variables. (The relationship between illness and stress is reliable though the r is small, about .20 to .30 — Cohen and Williamson, 1991.)

g. The probability that a student, chosen at random, comes from one of the six categories is found by dividing the number of students in that category by the total number of students. (See the most recent issue of *Statistical Abstract of the United States* for current data.)

h. A correlated-samples t test will allow you to determine if activity (the dependent variable) depends on whether or not a child ate sugar for breakfast (the independent variable). The samples are correlated because each child served in both conditions.

chapter**ten**

1. *Example 2:* The independent variable is the schedule of reinforcement; there are four levels. The dependent variable is persistance. The null hypothesis is that the mean persistance is the same for the four populations; that is, the schedule of reinforcement has no effect on persistance.

Example 3: The independent variable is degree of modernization; there are three levels. The dependent variable is suicide rate. The null hypothesis is that suicide rates are not dependent on degree of modernization, that the mean suicide rates for countries with low, medium, and high degrees of modernization are the same. In formal terms, the null hypothesis is H_0: $\mu_{low} = \mu_{med} = \mu_{high}$.

2. Larger

3. One estimate is obtained from variability among the sample means, and another estimate is obtained by pooling the sample variances.

4. F is a ratio of the two estimates of the population variance.

5. a. Retain the null hypothesis and conclude that the data do not provide evidence that social classes differ in their attitude toward religion.

b. Reject the null hypothesis and conclude that there is a relationship between degree of modernization and suicide rate.

6. There are several ways to craft a fine answer to this question. Some ways use graphs, some use algebra, and some use only words. A good answer includes such elements as: (a) F is a ratio of two variances — a between-treatments variance over an error variance; (b) when the null hypothesis is true, both variances are good estimators of the population variance, which produces a ratio of about 1.00; (c) when the null hypothesis is false, the between-treatments variance overestimates the population variance, resulting in a ratio that is greater than 1.00; (d) the F value calculated from the data is evaluated by comparing it to a critical value from the F distribution, a distribution that assumes the null hypothesis is true.

7.

	X_1	X_2	X_3	Σ
ΣX	12	12	24	48
ΣX^2	50	56	224	330
$\overline{X}$	4	4	8	

$$SS_{tot} = 330 - \frac{48^2}{9} = 330 - 256 = 74.00$$

$$SS_{treat} = \frac{12^2}{3} + \frac{12^2}{3} + \frac{24^2}{3} - \frac{48^2}{9} = 48 + 48 + 192 - 256$$

$$= 288 - 256 = 32.00$$

$$SS_{error} = \left(50 - \frac{12^2}{3}\right) + \left(56 - \frac{12^2}{3}\right) + \left(224 - \frac{24^2}{3}\right)$$

$$= 2 + 8 + 32 = 42.00$$

Check: $74.00 = 32.00 + 42.00$

8. *Independent variable:* degree of modernization, three levels; *dependent variable:* suicide rate.

	\multicolumn{3}{c}{Degree of modernization}	
	Low	Medium	High	Σ
ΣX	24	48	84	156
ΣX^2	154	614	1522	2290
$\overline{X}$	6.0	12.0	16.8	

$$SS_{tot} = 2290 - \frac{(156)^2}{13} = 418.00$$

$$SS_{mod} = \frac{(24)^2}{4} + \frac{(48)^2}{4} + \frac{(84)^2}{5} - \frac{(156)^2}{13} = 259.20$$

$$SS_{error} = \left(154 - \frac{(24)^2}{4} \right) + \left(614 - \frac{(48)^2}{4} \right) + \left(1522 - \frac{(84)^2}{5} \right) = 158.80$$

Check: $418.00 = 259.20 + 158.80$

Incidentally, the rate per 100,000 is about 18 in the United States and about 17 in Canada.

9. *Independent variable:* kind of therapy — model, film, desensitization, and control; *dependent variable:* snake-approach responses.

	Model	Film	Desensitization	Control	Σ
ΣX	104	72	68	40	284
ΣX^2	2740	1322	1182	430	5674
$\overline{X}$	26	18	17	10	
N	4	4	4	4	16

$$SS_{tot} = 5674 - \frac{284^2}{16} = 5674 - 5041 = 633$$

$$SS_{treat} = \frac{104^2}{4} + \frac{72^2}{4} + \frac{68^2}{4} + \frac{40^2}{4} - \frac{284^2}{16}$$

$$= 2704 + 1296 + 1156 + 400 - 5041$$

$$= 5556 - 5041 = 515$$

$$SS_{error} = \left(2740 - \frac{104^2}{4} \right) + \left(1322 - \frac{72^2}{4} \right)$$

$$+ \left(1182 - \frac{68^2}{4} \right) + \left(430 - \frac{40^2}{4} \right)$$

$$= 36 + 26 + 26 + 30 = 118$$

Check: $633 = 515 + 118$

10. Six groups. The critical values of *F* are based on 5, 70 *df* and are 2.35 and 3.29 at the .05 and .01 levels, respectively. If $\alpha = .01$, retain the null hypothesis; if $\alpha = .05$, reject the null hypothesis.

11. Data from problem 7:

a.

Source	SS	df	MS	F	p
Treatments	32.00	2	16.00	2.29	>.05
Error	42.00	6	7.00		
Total	74.00	8			

$F_{.05}(2, 6\ df) = 5.14$. Because 2.29 does not reach the critical value, retain the null hypothesis and conclude that you do not have evidence against the hypothesis that these three groups came from the same population.

b. $SS_{tot} = 280 - \dfrac{36^2}{6} = 64.00$

$SS_{treat} = \dfrac{12^2}{3} + \dfrac{24^2}{3} - \dfrac{36^2}{6} = 24.00$

$SS_{error} = \left(56 - \dfrac{12^2}{3}\right) + \left(224 - \dfrac{24^2}{3}\right) = 40.00$

Source	SS	df	MS	F	p
Treatments	24.00	1	24.00	2.40	>.05
Error	40.00	4	10.00		
Total	64.00	5			

$F_{.05}(1, 4\ df) = 7.71$. Retain the null hypothesis.

$t = \dfrac{\overline{X}_1 - \overline{X}_2}{s_{\overline{X}_1 - \overline{X}_2}} = \dfrac{8 - 4}{2.582} = 1.55; \quad df = 4$

Note that $(1.55)^2 = 2.40$; that is, $t^2 = F$.

12. Durkheim's modernization and suicide data from problem 8: For df, use

$df_{tot} = N_{tot} - 1 = 13 - 1 = 12$
$df_{mod} = K - 1 = 3 - 1 = 2$
$df_{error} = N_{tot} - K = 13 - 3 = 10$

Source	SS	df	MS	F	p
Modernization	259.20	2	129.60	8.16	<.01
Error	158.80	10	15.88		
Total	418.00	12			

Because $F_{.01}(2, 10\ df) = 7.56$, reject the null hypothesis and conclude that there is a relationship between suicide rate and a country's degree of modernization.

13.

Source	SS	df	MS	F	p
Treatments	515	3	171.667	17.46	<.01
Error	118	12	9.833		
Total	633	15			

$F_{.01}$ (3, 12 df) = 5.95. Because 17.46 > 5.95, reject the null hypothesis at the .01 level and conclude that the treatment method had an effect on the number of snake-approach responses. By the end of this chapter, you will be able to compare the means to determine which method is best (or worst).

14. *A priori* tests are those planned in advance. They are tests based on the logic of the design of an experiment. *Post hoc* tests are chosen after examining the data.

15. Recall that the mean suicide rates for the different degrees of modernization were: low — 6.0; medium — 12.0; high — 16.8. Comparing countries with low and medium degrees of modernization:

$$\text{HSD} = \frac{12.0 - 6.0}{\sqrt{15.88/4}} = \frac{6.0}{1.99} = 3.02; \quad \text{NS}; \quad \text{HSD}_{.05} = 3.88$$

Comparing countries with medium and high degrees of modernization:

$$\text{HSD} = \frac{16.8 - 12.0}{\sqrt{\dfrac{15.88}{2}\left(\dfrac{1}{4} + \dfrac{1}{5}\right)}} = \frac{4.8}{1.890} = 2.54; \quad \text{NS}; \quad \text{HSD}_{.05} = 3.88$$

Comparing countries with low and high degrees of modernization:

$$\text{HSD} = \frac{16.8 - 6.0}{1.890} = \frac{10.8}{1.890} = 5.71; \quad p < .01; \quad \text{HSD}_{.01} = 5.27$$

Suicide rates are significantly higher in countries with a high degree of modernization than they are in countries with a low degree of modernization. Although the other differences were not significant, the trend is in the direction predicted by Durkheim's hypothesis.

16. I calculated HSDs for each of the six pairwise differences. For some of these HSDs the equal-N formula was appropriate, and for some the unequal-N formula was necessary. The two values of $s_{\bar{x}}$ are

$$s_{\bar{x}} = \sqrt{\frac{MS_{\text{error}}}{N_t}} = \sqrt{\frac{3.93}{5}} = 0.887$$

$$s_{\bar{x}} = \sqrt{\frac{MS_{\text{error}}}{2}\left(\frac{1}{N_1} + \frac{1}{N_2}\right)} = \sqrt{\frac{3.93}{2}\left(\frac{1}{5} + \frac{1}{7}\right)} = 0.821$$

The table shows all six HSD values. The significance level of each comparison is indicated with one or two asterisks. $\text{HSD}_{.05} = 4.00$, $\text{HSD}_{.01} = 5.09$.

	crf	FR2	FR4
FR2	4.63*		
FR4	7.44**	3.41	
FR8	9.47**	5.60**	2.03

* $p < .05$
** $p < .01$

The trend of the means is consistent; the higher the ratio of response to reinforcement during learning, the greater the persistence during extinction. The crf schedule produced the least persistence, significantly less than any other schedule. The FR2 schedule produced less persistence than the FR4 schedule (NS) and significantly less than the FR8 schedule ($p < .01$).

17. Comparing the best treatment (model) with the second best (film):

$$\text{HSD} = \frac{\overline{X}_1 - \overline{X}_2}{\sqrt{MS_{\text{error}}/N_t}} = \frac{26 - 18}{1.568} = 5.10; \quad p < .05; \quad \text{HSD}_{.05} = 4.20$$

Comparing the poorest treatment (desensitization) with the control group:

$$\text{HSD} = \frac{\overline{X}_3 - \overline{X}_4}{s_{\overline{X}}} = \frac{17 - 10}{1.568} = 4.46; \quad p < .05; \quad \text{HSD}_{.05} = 4.20$$

For the treatment of phobias, using a live model and encouraging participation are significantly better than using a film or desensitization. All three of the treatments produced significant improvement compared to an untreated control group.

18. Normally distributed dependent variable, homogeneity of variance, and random assignment (or random sampling)

19. $f = \dfrac{\sqrt{\dfrac{K-1}{N_{\text{tot}}}(MS_{\text{treat}} - MS_{\text{error}})}}{\sqrt{MS_{\text{error}}}} = \dfrac{\sqrt{\dfrac{4-1}{16}(171.667 - 9.833)}}{\sqrt{9.833}} = 1.76$

An effect size index of 1.76 is much greater than the $f = .40$ that Cohen identifies as a large effect; thus, the type of therapy given to those suffering from a phobia has a large effect on improvement.

20.

	Normals	Schizophrenics	Depressives	Σ
ΣX	178	138	72	388
ΣX^2	4162	2796	890	7848
$\overline{X}$	22.25	19.71	12.00	
N	8	7	6	21

Source	SS	df	MS	F	p
Disorder categories	376.31	2	188.16	11.18	<.01
Error	302.93	18	16.83		
Total	679.24	20			

$F_{.01}(2, 18\ df) = 6.01$

Depressive and schizophrenic:

$$\text{HSD} = \frac{12.00 - 19.71}{\sqrt{\dfrac{16.83}{2}\left(\dfrac{1}{7} + \dfrac{1}{6}\right)}} = \frac{-7.71}{1.614} = -4.78;\quad p < .01;\quad \text{HSD}_{.01} = 4.70$$

Depressive and normal:

$$\text{HSD} = \frac{12.00 - 22.25}{\sqrt{\dfrac{16.83}{2}\left(\dfrac{1}{6} + \dfrac{1}{8}\right)}} = \frac{-10.25}{1.567} = -6.54;\quad p < .01;\quad \text{HSD}_{.01} = 4.70$$

Normal and schizophrenic:

$$\text{HSD} = \frac{22.25 - 19.71}{\sqrt{\dfrac{16.83}{2}\left(\dfrac{1}{8} + \dfrac{1}{7}\right)}} = \frac{2.54}{1.501} = 1.69;\quad \text{NS};\quad \text{HSD}_{.05} = 3.61$$

The R scores on the Rorschach test do not distinguish schizophrenics from normals. Both schizophrenics and normals score significantly higher than depressives.

21.

	ACS	4MAT	Control
$\overline{X}$	8.9	18.1	3.1

Source	SS	df	MS	F	p
Instruction methods	2288.533	2	1144.267	20.81	<.01
Error	3134.400	57	54.989		
Total	5422.933	59			

$F_{.01}(2, 55\ df) = 5.01$

HSD values calculated for the three pairwise comparisons are:

ACS and 4MAT:	HSD = 5.55**
ACS and control:	HSD = 3.50*
4MAT and control:	HSD = 9.05**

$* p < .05$ $** p < .01$ $\text{HSD}_{.01} = 4.37$
$\text{HSD}_{.05} = 3.44$

$$f = \frac{\sqrt{\dfrac{K-1}{N_{tot}}(MS_{treat} - MS_{error})}}{\sqrt{MS_{error}}} = \frac{\sqrt{\dfrac{3-1}{60}(1144.267 - 54.989)}}{\sqrt{54.989}} = 0.81$$

Three months after BSE training, women retained significantly more information if they had learned with the 4MAT method than if they were taught with the ACS method. Either method is better than no training. The effect size index of 0.81 shows that BSE training has a very large effect on information retained three months later.

22.

Source	SS	df	MS	F	p
Between embarrassments	141.733	2	70.867	5.76	<.05
Error	147.600	12	12.300		
Total	289.333	14			

$F_{.05}$ (2, 12 df) = 3.88

$$f = \frac{\sqrt{\dfrac{K-1}{N_{tot}}(MS_{treat} - MS_{error})}}{\sqrt{MS_{error}}} = \frac{\sqrt{\dfrac{3-1}{15}(70.867 - 12.300)}}{\sqrt{12.300}} = 0.80$$

HSD values are calculated from pairwise comparisons with

$$s_{\bar{X}} = \sqrt{\frac{12.300}{5}} = 1.568$$

Severe and mild embarrassment: HSD = 3.95*
Severe and no embarrassment: HSD = 4.34*
Mild and no embarrassment: HSD = 0.38

*$p < .05$, $HSD_{.05}$ = 3.77

The participants who were subjected to severe embarrassment in order to be in the experiment judged the uninteresting discussion as more favorable than did those who experienced mild embarrassment ($p < .05$) or no embarrassment ($p < .05$). The effect size index for the embarrassment procedures in this experiment is quite large ($f = 0.80$).

Although common sense says that a valuable goal is worth suffering for, these data (and many more) say that if we have suffered, then whatever we get is valuable. This balance that we create in our minds is called *effort justification* by social psychologists.

chapter **eleven**

1. **a.** 2×2 **b.** 3×2 **c.** 4×3
   d. $4 \times 3 \times 3$; this is a factorial design with three independent variables. If you are interested in factorial designs with more than two factors, see footnote 1 in Chapter 11.
2. a, b, and c all have one factor.
3. a, b, and d all have one independent variable with two levels.
4. An interaction occurs when the effect of changing levels of one factor depends on which level of the other factor you are looking at.
5. The effect on problem solving of time of day depends on gender. There are several ways to write a specific answer; here are two:
   a. Women solve problems better at 7:00 A.M.; men solve them better at 7:00 P.M.
   b. A 7:00 A.M. session in problem solving produces more correct answers from men; a 7:00 P.M. session produces more correct answers from women.
6. A main effect is the effect that the different levels of one factor have on the dependent variable.

7. **i. a.**

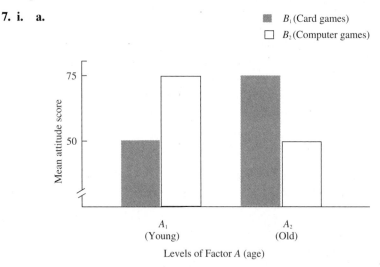

b. There appears to be an interaction.
c. Attitudes toward games depend on age. The young would rather play computer games, and the old would rather play cards.
d. There appear to be no main effects. Margin means for games are cards = 62.5 and computers = 62.5. Margin means for age are young = 62.5 and old = 62.5.
e. Later in this chapter you will learn that when there is a significant interaction, the interpretation of main effects isn't simple. Although all four means are exactly the same, there is a difference in attitudes toward the two games, but this difference depends on the age of the participant.

ii. a.

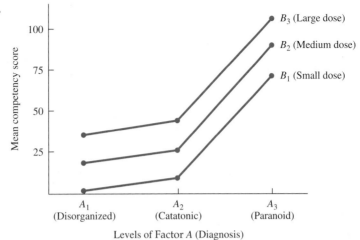

  b. There appears to be no interaction.
  c. The effect of the drug dose does not depend on the diagnosis of the patient.
  d. There appear to be main effects for both factors.
  e. For people with all types of schizophrenia, the larger the dose, the higher the competency score (Factor *B*, main effect). People with paranoia have higher competency scores than the others at all dose levels (Factor *A*, main effect).

iii. a.

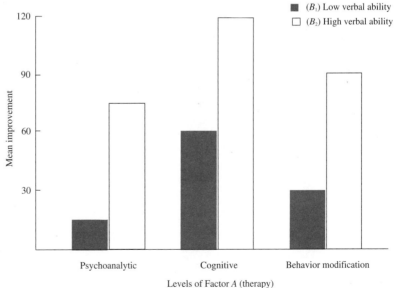

  b. There appears to be no interaction.
  c. The improvement observed with the three kinds of therapy does not depend on whether the patient is high or low in verbal ability.

    d. There appear to be main effects for both factors.

    e. High-verbal patients showed greater improvement than low-verbal patients. Patients who received cognitive therapy showed the most improvement. Psychoanalytic patients showed the least improvement; behavior modification patients were just above psychoanalytic patients.

iv. a.

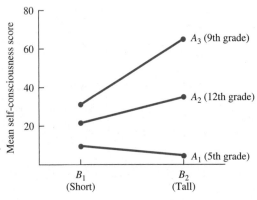

Levels of Factor B (Height)

    b. There appears to be an interaction.

    c. The effect of a person's height on self-consciousness depends on the person's grade in school. Height seems to make little difference in the 5th grade. In the 9th grade, tall students are much more self-conscious than short students, but this difference is much smaller in the 12th grade.

    d. The main effects appear to be significant. The margin means are 20 and 35 for height; 7.5, 47.5, and 27.5 for grade level.

    e. Here's an interpretation that is appropriate for the information you've been given so far. (More on interpretation of main effects when the interaction is significant will come later.) Height: tall students ($\overline{X} = 35$) are more self-conscious than short students ($\overline{X} = 20$). Grade level: 9th-graders ($\overline{X} = 47.5$) are more self-conscious than 12th-graders ($\overline{X} = 27.5$) or 5th-graders ($\overline{X} = 7.5$).

v. a.

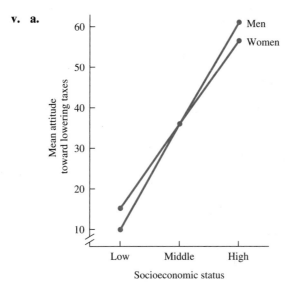

Socioeconomic status

b. There appears to be no interaction. Although the lines in the graph are not parallel — they even cross — the departure from parallel is slight and probably should be attributed to sampling variation.

c. Gender differences in attitudes toward lowering taxes on investments do not depend on socioeconomic status.

d. There appears to be a main effect for socioeconomic status but not for gender.

e. Men and women do not differ in their attitudes toward lowering taxes on investments (Factor B). The higher the socioeconomic status, the more positive are attitudes toward lowering taxes on investments.

8. a. 2×2

b. Independent variables: gender and handedness; dependent variable: age at death

c. $SS_{tot} = 61,455 - \dfrac{855^2}{12} = 61,455 - 60,918.75 = 536.25$

$$SS_{cells} = \dfrac{219^2}{3} + \dfrac{186^2}{3} + \dfrac{234^2}{3} + \dfrac{216^2}{3} - \dfrac{855^2}{12}$$

$$= 61,323 - 60,918.75 = 404.25$$

$$SS_{gender} = \dfrac{453^2}{6} + \dfrac{402^2}{6} - 60,918.75 = 61,135.50 - 60,918.75$$

$$= 216.75$$

$$SS_{handedness} = \dfrac{405^2}{6} + \dfrac{450^2}{6} - 60,918.75 = 61,087.50 - 60,918.75$$

$$= 168.75$$

$$SS_{AB} = 3[(73 - 75.5 - 67.5 + 71.25)^2 + (62 - 67.5 - 67 + 71.25)^2$$
$$+ (78 - 75 - 75.5 + 71.25)^2 + (72 - 75 - 67 + 71.25)^2]$$
$$= 3(6.25) = 18.75$$

Check: $SS_{AB} = 404.25 - 216.75 - 168.75 = 18.75$

$$SS_{error} = \left[16{,}013 - \frac{219^2}{3} \right] + \left[11{,}574 - \frac{186^2}{3} \right]$$
$$+ \left[18{,}284 - \frac{234^2}{3} \right] + \left[15{,}584 - \frac{216^2}{3} \right]$$
$$= 26.00 + 42.00 + 32.00 + 32.00 = 132.00$$

Check: $536.25 = 404.25 + 132.00$

9. a. 2×3

b. Teacher response and gender

c. Errors on an arithmetic test

d. $SS_{tot} = 13{,}281 - \dfrac{845^2}{60} = 1380.58$

$$SS_{cells} = \frac{(169)^2}{10} + \frac{(136)^2}{10} + \frac{(118)^2}{10} + \frac{(146)^2}{10} + \frac{(167)^2}{10} + \frac{(109)^2}{10} - \frac{(845)^2}{60}$$
$$= 306.28$$

$$SS_{response} = \frac{(315)^2}{20} + \frac{(303)^2}{20} + \frac{(227)^2}{20} - \frac{(845)^2}{60} = 227.73$$

$$SS_{gender} = \frac{(423)^2}{30} + \frac{(422)^2}{30} - \frac{(845)^2}{60} = 0.02$$

$$SS_{AB} = 10[(16.90 - 15.75 - 14.10 + 14.0833)^2$$
$$+ (13.60 - 15.15 - 14.10 + 14.0833)^2$$
$$+ (11.80 - 11.35 - 14.10 + 14.0833)^2$$
$$+ (14.60 - 15.75 - 14.0667 + 14.0833)^2$$
$$+ (16.70 - 15.15 - 14.0667 + 14.0833)^2$$
$$+ (10.90 - 11.35 - 14.0667 + 14.0833)^2]$$
$$= 78.53$$

Check: $SS_{AB} = 306.28 - 227.73 - 0.02 = 78.53$

$$SS_{error} = \left(3129 - \frac{(169)^2}{10} \right) + \left(2042 - \frac{(136)^2}{10} \right) + \left(1620 - \frac{(118)^2}{10} \right)$$
$$+ \left(2244 - \frac{(146)^2}{10} \right) + \left(2955 - \frac{(167)^2}{10} \right) + \left(1291 - \frac{(109)^2}{10} \right)$$
$$= 1074.30$$

Check: $1380.58 = 306.28 + 1074.30$

10. $SS_{tot} = 1405 - \dfrac{145^2}{20} = 353.75$

$SS_{cells} = \dfrac{25^2}{5} + \dfrac{57^2}{5} + \dfrac{43^2}{5} + \dfrac{20^2}{5} - \dfrac{145^2}{20} = 173.35$

$SS_{learn} = \dfrac{68^2}{10} + \dfrac{77^2}{10} - \dfrac{145^2}{20} = 4.05$

$SS_{recall} = \dfrac{63^2}{10} + \dfrac{82^2}{10} - \dfrac{145^2}{20} = 18.05$

$SS_{AB} = 5[(8.60 - 6.80 - 6.30 + 7.25)^2 + (4.00 - 7.70 - 6.30 + 7.25)^2$
$\qquad + (5.00 - 6.80 - 8.20 + 7.25)^2 + (11.40 - 7.70 - 8.20 + 7.25)^2]$
$\qquad = 151.25$

Check: $SS_{AB} = 173.35 - 4.05 - 18.05 = 151.25$

$SS_{error} = \left(400 - \dfrac{43^2}{5}\right) + \left(100 - \dfrac{20^2}{5}\right) + \left(155 - \dfrac{25^2}{5}\right) + \left(750 - \dfrac{57^2}{5}\right)$
$\qquad = 180.40$

Check: $353.75 = 173.35 + 180.40$

11.

Source	SS	df	MS	F	p
A (gender)	216.75	1	216.75	13.14	<.01
B (handedness)	168.75	1	168.75	10.23	<.05
A × B	18.75	1	18.75	1.14	>.05
Error	132.00	8	16.50		
Total	536.25	11			

$F_{.05}\,(1, 8\ df) = 5.32; \quad F_{.01}\,(1, 8\ df) = 11.26$

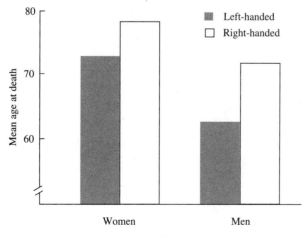

Interpretation: The interaction is not significant, so the main effects can be interpreted directly. Both gender and handedness produced significant differences. Right-handed people live longer than left-handed people ($p < .05$) and women live longer than men ($p < .01$).

12.

Source	SS	df	MS	F	p
A (response)	227.73	2	113.87	5.72	<.01
B (gender)	0.02	1	0.02	0.00	>.05
AB	78.53	2	39.27	1.97	>.05
Error	1074.30	54	19.89		
Total	1380.58	59			

$F_{.05} (2, 50\ df) = 3.18;\quad F_{.01} (2, 50\ df) = 5.06$

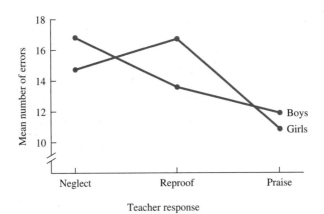

Teacher response

Interpretation: The interaction was not statistically significant. Gender was not statistically significant. Only the teacher-response variable produced results that were statistically significant ($p < .01$). The null hypothesis $\mu_{neglect} = \mu_{reproof} = \mu_{praise}$ is rejected. The three kinds of teacher response produced different numbers of errors.

13.

Source	SS	df	MS	F	p
A (learning)	4.05	1	4.05	0.36	>.05
B (recall)	18.05	1	18.05	1.60	>.05
AB	151.25	1	151.25	13.41	<.01
Error	180.40	16	11.28		
Total	353.75	19			

$F_{.05} (1, 16\ df) = 4.49;\quad F_{.01} (1, 16\ df) = 8.53$

■ Recall with drug
□ Recall without drug

14. i. a. *Independent variables:* (1) treatment the participant received (insulted or treated neutrally) and (2) administrator of test (same or different). *Dependent variable:* number of captions produced.

b. $F_{\text{treatments}} = 9.08, p < .01; F_{\text{experimenters}} = 6.82, p < .05; F_{AB} = 4.88, p < .05.$

c. The significant interaction means that the effect of the experimenter *depended* on whether the participant was insulted or treated neutrally. If the participant had been insulted, then taking the humor test from the person who had done the insulting caused a larger number of captions to be written — larger, that is, than if a new experimenter had administered the test. If the participant had not been insulted, there was little difference between the two kinds of experimenters.

Both main effects are statistically significant. For both main effects, the difference in margin means depends heavily on one cell, Cell A_1B_1. Interpreting the main effects for this problem isn't advisable.

ii. a. *Independent variables:* (1) name and (2) occupation of father. *Dependent variable:* score assigned to the theme.

b. $F_{\text{name}} = 0.44, p > .05; F_{\text{occupation}} = 0.05, p > .05; F_{AB} = 8.20, p < .01$

c. The significant interaction means that the effect of the name depends on the occupation of the father. The effect of the name David is good if his father is a research chemist (Cell A_1B_1) and bad if his father is unemployed (Cell A_1B_2). The effect of the name Elmer is good if his father is unemployed (Cell A_2B_2) and bad if his father is a research chemist (Cell A_2B_1).

The main effects are not significant. The average "David" theme is about equal to the average "Elmer" theme. Similarly, the themes do not differ significantly according to whether the father was unemployed or a research chemist. Notice that this interpretation of the main effects is misleading; that is, *names do have an effect,* but the effect depends on the occupation of the father. The appropriate response to these main effects is to ignore them.

iii. a. *Independent variables:* (1) questions (present and absent) and (2) demonstrations (present and absent). *Dependent variable:* score on a comprehension test.

b. $F_{questions} = 4.59, p < .05; F_{demos} = 9.80, p < .01; F_{AB} = 0.24, p > .05$

c. Because the interaction is not significant, the main effects can be interpreted in a straightforward manner. Questions are important; when questions are incorporated, students learn significantly more than when questions are not used. The demonstrations make a difference; presentations that have demonstrations produce comprehension scores that are significantly better than presentations without demonstrations.

15. *Interpretation:* Only the interaction is significant. The conclusion is that when conditions of learning and recall are the same, memory is good. When conditions of learning and recall are different, memory is poorer.

16. a. The number of scores in each group (cell) must be the same.

b. The samples in each cell must be independent.

c. The levels of both factors must be chosen by the experimenter and not left to chance.

d. The samples are drawn from normally distributed populations.

e. The variances of the populations are equal.

f. The participants are assigned randomly to the group, or the samples are drawn randomly from the populations.

17.

Source	SS	df	MS	F	p
A (rewards)	199.31	3	66.44	3.90	<.05
B (classes)	2.25	1	2.25	0.13	>.05
AB	0.13	3	0.04	0.00	>.05
Error	953.25	56	17.02		
Total	1154.94	63			

$F_{.05} (3, 55\ df) = 2.78$

There is no significant interaction between rewards and class standing, and freshmen and seniors do not differ significantly. There is a significant difference among the rewards; that is, rewards have an effect on attitudes toward the police. Pairwise comparisons are called for.

The table that follows shows the four treatment means, N's, and a matrix of HSD values calculated from the formula,

$$HSD = \frac{\overline{X}_1 - \overline{X}_2}{\sqrt{17.02/16}}$$

	$10	$5	$1	50¢
Means	8.938	10.250	12.125	13.563
N	16	16	16	16
$10		1.27	3.09	4.48*
5			1.82	3.21
1				1.39

* $HSD_{.05} = 3.79$

The less students were paid to defend the police, the more positive their attitude toward the police. Those paid 50 cents had significantly more positive attitudes toward the police than those paid ten dollars, $p < .05$.

18. $SS_{tot} = 480 - \dfrac{(84)^2}{18} = 88.00$

$SS_{cells} = \dfrac{(12)^2}{3} + \dfrac{(15)^2}{3} + \dfrac{(24)^2}{3} + \dfrac{(9)^2}{3} + \dfrac{(9)^2}{3}$

$\quad + \dfrac{(15)^2}{3} - \dfrac{(84)^2}{18} = 52.00$

$SS_{marital} = \dfrac{(51)^2}{9} + \dfrac{(33)^2}{9} - \dfrac{(84)^2}{18} = 18.00$

$SS_{religious} = \dfrac{(21)^2}{6} + \dfrac{(24)^2}{6} + \dfrac{(39)^2}{6} - \dfrac{(84)^2}{18} = 31.00$

$SS_{AB} = 3[(4.0 - 3.5 - 5.667 + 4.667)^2$

$\quad + (5.0 - 4.0 - 5.667 + 4.667)^2$

$\quad + (8.0 - 6.5 - 5.667 + 4.667)^2$

$\quad + (3.0 - 3.5 - 3.667 + 4.667)^2$

$\quad + (3.0 - 4.0 - 3.667 + 4.667)^2$

$\quad + (5.0 - 6.5 - 3.667 + 4.667)^2]$

$\quad = 3(1.00) = 3.00$

Check: $SS_{cells} = 18.00 + 31.00 + 3.00 = 52.00$

$SS_{error} = \left(56 - \dfrac{(12)^2}{3}\right) + \left(83 - \dfrac{(15)^2}{3}\right) + \left(194 - \dfrac{(24)^2}{3}\right) + \left(29 - \dfrac{(9)^2}{3}\right)$

$\quad + \left(35 - \dfrac{(9)^2}{3}\right) + \left(83 - \dfrac{(15)^2}{3}\right)$

$\quad = 36.00$

Check: $88.00 = 52.00 + 36.00$

Source	df	MS	F	p
Marital status	1	18.00	6.00	<.05
Religious behavior	2	15.50	5.17	<.05
AB	2	1.50	0.50	>.05
Error	12	3.00		
Total	17			

$F_{.05}$ (1, 12 df) = 4.75; $F_{.05}$ (2, 12 df) = 3.88

The interaction is not significant; HSD tests are appropriate for comparing the three levels of frequency of religious behavior. Using the HSD formula, you get

$$\text{HSD} = \frac{\overline{X}_1 - \overline{X}_2}{\sqrt{3.00/6}};\quad \text{critical value of HSD}_{.05} = 3.77$$

$\text{HSD}_{\text{often \& occasionally}}$ = 3.54;   NS

$\text{HSD}_{\text{never \& occasionally}}$ = 0.71;   NS

$\text{HSD}_{\text{often \& never}}$ = 4.24;   p < .05

Interpretation: Both marital status and frequency of religious behavior have an effect on reports of happiness. Married people reported being more happy than unmarried people (p < .05). Those frequently engaged in religious behavior reported higher happiness scores than those who never engaged in religious behavior (p < .05). Happiness was intermediate (and not significantly different) for those who occasionally practiced religious behavior. There was not an interaction between marital status and frequency of religious behavior.

chapter twelve

1. Obviously, there are several solutions to this problem. Before reading further, you might ask yourself if your solution is a practical one. If not, try to create a more practical solution.

   My answer is to select participants in groups of three. Within each group the three are matched in some way related to the material that is on the multiple-choice test. If the material is unfamiliar and the participants are college students, I would match on grade point average. Assignment of treatments within a group would be random. This is a three-group version of the matched-pairs design (page 187).

2.

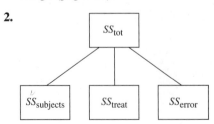

3.

Subject	Rest interval (weeks) 2	4	8	Σ
1	40	50	58	148
2	58	56	65	179
3	44	70	69	183
4	57	61	74	192
Σ	199	237	266	702
$\overline{X}$	49.75	59.25	66.50	

Source	SS	df	MS	F
Subjects	365.667	3		
Intervals	564.500	2	282.250	7.21
Error	234.833	6	39.139	
Total	1165.000	11		

Tukey HSD tests; $HSD_{.05} = 4.34$; $s_{\overline{X}} = \sqrt{\dfrac{39.139}{4}} = 3.128$

$HSD_{8v4} = 2.32$; NS

$HSD_{8v2} = 5.35$; $p < .05$

$HSD_{4v2} = 3.04$; NS

These data show that as the rest interval increases from 2 weeks to 4 weeks to 8 weeks, there is an increase in percent of recall. That is, the longer the rest interval, the better the percent of recall. Only the difference between 2 weeks and 8 weeks was statistically significant, $p < .05$.

4. $HSD_{1st\ v.\ cont} = \dfrac{74 - 75}{\sqrt{1.333/4}} = \dfrac{-1}{0.577} = -1.73$; $p > .05$

$HSD_{.05} = 4.34$. Although always staying with your first choice resulted in an average loss of one point, the difference was not statistically significant.

5. The comparison of the three therapy techniques does not lend itself to a correlated-measures design using the same participants. Clearly, it doesn't make sense to provide therapy for someone for a period sufficient to help and then use the next therapy on that same person.

6. 1. When the same participants serve at all treatment levels, you should be sure there are no carryover effects (unless you are looking for carryover effects).
   2. To use the analysis presented in this chapter, the levels of the independent variable should be chosen by the researcher and the participants randomly assigned to rows.

7. The assumptions for a repeated-measures ANOVA are (1) normal distribution of scores, (2) random assignment or random sampling, and (3) the covariance matrix is spherical.

8.

Source	SS	df	MS	F	p
Subjects	34,543.33	4			
Periods	5,890.00	2	2945.00	44.73	<.01
Error	526.67	8	65.83		
Total	40,960.00	14			

Tukey HSD tests; $\text{HSD}_{.01} = 5.64$; $s_{\bar{x}} = \sqrt{\dfrac{65.833}{5}} = 3.629$

$\text{HSD}_{\text{before v during}} = 11.30;\quad p < .01$

$\text{HSD}_{\text{before v after}} = 0.55;\quad \text{NS}$

$\text{HSD}_{\text{after v during}} = 11.85;\quad p < .01$

During meditation, oxygen consumption drops to about 80 percent of the level observed before and after meditation. The drop is statistically significant, $p < .01$. The oxygen consumption rates before and after meditation are not significantly different.

9. If the conclusion that meditation lowers oxygen consumption is wrong, it is a Type I error. If the conclusion that oxygen consumption before and after meditation is not significantly different is wrong, it is a Type II error.

10.

Matching characteristic	No therapy	Behavior therapy	Cognitive therapy	Drug therapy	ΣX_s
Women, at age 20	18	11	8	10	47
Men, at age 30	21	9	15	10	55
Women, at age 45	24	16	10	16	66
ΣX_t	63	36	33	36	168
$\bar{X}_t$	21	12	11	12	

Source	SS	df	MS	F
Subjects	45.500	2		
Therapies	198.000	3	66.000	8.16*
Error	48.500	6	8.083	
Total	292.000	11		

$F_{.05}\,(3, 6\,df) = 4.76 \qquad * p < .05$

Tukey HSD tests; $\text{HSD}_{.05} = 4.90$; $s_{\bar{x}} = \sqrt{\dfrac{8.083}{3}} = 1.641$

$\text{HSD}_{\text{no v behavior}} = 5.48;\quad p < .05$

$\text{HSD}_{\text{no v drug}} = 5.48;\quad p < .05$

$\text{HSD}_{\text{no v cognitive}} = 6.09;\quad p < .05$

$\text{HSD}_{\text{drug v cognitive}} = 0.61;\quad \text{NS}$

Each of the three therapies is significantly better than the no-therapy control group. There are no significant differences among the three therapy groups.

chapter**thirteen**

1.

	Received program	Control	Total
Police record	114	101	215
No police record	211	224	435
Total	325	325	650

The expected frequencies are:

$$\frac{(215)(325)}{650} = 107.5 \qquad \frac{(215)(325)}{650} = 107.5$$

$$\frac{(435)(325)}{650} = 217.5 \qquad \frac{(435)(325)}{650} = 217.5$$

O	E	$O - E$	$(O - E)^2$	$\dfrac{(O - E)^2}{E}$
114	107.5	6.5	42.25	0.393
101	107.5	−6.5	42.25	0.393
211	217.5	−6.5	42.25	0.194
224	217.5	6.5	42.25	0.194
Σ 650	650.0			$\chi^2 = 1.174$
				$df = 1$

By the shortcut method:

$$\chi^2 = \frac{650[(114)(224) - (101)(211)]^2}{(215)(435)(325)(325)} = 1.17$$

$\chi^2_{.05}\,(1\ df) = 3.84$. Therefore, retain the null hypothesis and conclude that the two groups did not differ significantly in the number who had police records. The Cambridge–Somerville Youth Study, discussed in many sociology texts, was first reported by Powers and Witmer (1951).

Did you notice that the proportion of those with police records was *higher* for the boys who participated in the program?

When a statistical test does not show a statistically reliable effect, it probably isn't useful to calculate an effect size index.

2. $\chi^2 = \dfrac{100[(9)(34) - (26)(31)]^2}{(35)(65)(40)(60)} = 4.579$

Calculations of the expected frequencies are as follows:

$$\frac{(35)(40)}{100} = 14 \qquad \frac{(35)(60)}{100} = 21$$

$$\frac{(65)(40)}{100} = 26 \qquad \frac{(65)(60)}{100} = 39$$

O	E	$O - E$	$(O - E)^2$	$\dfrac{(O - E)^2}{E}$
9	14	−5	25	1.786
26	21	5	25	1.191
31	26	5	25	0.962
34	39	−5	25	0.641
Σ 100	100			$x^2 = 4.58$
				$df = 1$

$x^2_{.05}$ (1 df) = 3.84. Reject the hypothesis that group size and joining are independent. Conclude that passersby are more likely to join a group of five than a group of two.

$$\phi = \sqrt{\frac{x^2}{N}} = \sqrt{\frac{4.58}{100}} = \sqrt{.0458} = 0.21$$

The effect that group size has on joining is midway between a small effect and a medium effect.

3. a.

Recapture site

		Issaquah	East Fork
	Issaquah	46 (34.027)	0 (11.973)
Capture site			
	East Fork	8 (19.973)	19 (7.027)

$$x^2 = \frac{73[(46)(19) - (0)(8)]^2}{(46)(27)(54)(19)} = 43.76$$

$$\phi = \sqrt{\frac{x^2}{N}} = \sqrt{\frac{43.76}{73}} = \sqrt{.5995} = 0.77$$

$x^2_{.001}$ (1 df) = 10.83. Reject the null hypothesis (which is that the second choice is independent of the first) and conclude that choices are very consistent; salmon tend to choose the same stream each time. The effect size index is 0.77, a value much greater than .50, which is labeled *large*.

b.

Recapture site

		Issaquah	East Fork
	Issaquah	39 (40.071)	12 (10.929)
Capture site			
	East Fork	16 (14.929)	3 (4.071)

$$\chi^2 = \frac{70[(39)(3) - (12)(16)]^2}{(51)(19)(55)(15)} = 0.49$$

$\chi^2_{.05}(1\ df) = 3.84$. Hasler's hypothesis is supported; fish with plugged nasal openings do not make a consistent choice of streams, but those that get olfactory cues consistently choose the same stream. For further confirmation and a short summary of Hasler's work, see Hasler, Scholz, and Horrall (1978).

4. The work that needs to be done on the data is to divide each set of applicants into two independent categories, hired and not hired. The following table results. Expected values are in parentheses.

	White	Black	Total
Hired	390 (339.327)	18 (68.673)	408
Not hired	3810 (3860.673)	832 (781.327)	4642
Total	4200	850	5050

$$\chi^2 = \frac{5050[(390)(832) - (3810)(18)]^2}{(4200)(850)(408)(4642)} = \frac{3.3070 \times 10^{14}}{6.7514 \times 10^{12}} = 48.91$$

Given that $\chi^2_{.001}(1\ df) = 10.83$, the null hypothesis can be rejected. Because 9.29 percent of the white applicants were hired compared to 2.12 percent of the black applicants, the statistical consultant and the company in question concluded that discrimination had occurred.

$$\phi = \sqrt{\frac{\chi^2}{N}} = \sqrt{\frac{48.91}{5050}} = \sqrt{.00969} = 0.10$$

The effect size index is 0.10, which is a small effect. The lesson here is that a small effect practiced on a large sample produces a reliable difference between the two groups.

5. $\chi^2 = \Sigma\left[\frac{(O - E)^2}{E}\right] = \frac{(316 - 300)^2}{300} + \frac{(84 - 100)^2}{100}$

$= 0.8533 + 2.5600$

$= 3.4133;\quad df = 1$

$\chi^2_{.05}(1\ df) = 3.84$. Retain the null hypothesis and conclude that these data are consistent with a 3:1 hypothesis.

6. The expected frequencies are 20 correct and 40 incorrect: $(\frac{1}{3})(60) = 20$ and $(\frac{2}{3})(60) = 40$.

$$\chi^2 = \frac{(32 - 20)^2}{20} + \frac{(28 - 40)^2}{40} = 10.80;\quad df = 1;\quad p < .01$$

Thus, these three emotions can be distinguished from each other. Observers did not respond in a chance fashion.

There is an interesting sequel to this experiment. Subsequent researchers did not find the simple, clear-cut results that Watson reported. One experimenter (Sherman, 1927) found that if only the infants' reactions were observed, there was a great deal of disagreement. However, if the observers also knew the stimuli (dropping, stroking, and so forth), they agreed with each other. This seems to be a case in which Watson's design did not permit him to separate the effect of the infant's reaction (the independent variable) from the effect of knowing what caused the reaction (an extraneous variable).

7. To get the expected frequencies, multiply the percentages given by Professor Stickler by the 340 students. Then enter these expected frequencies in the usual table.

O	E	$O - E$	$(O - E)^2$	$\dfrac{(O - E)^2}{E}$
20	23.8	−3.8	14.44	0.607
74	81.6	−7.6	57.76	0.708
120	129.2	−9.2	84.64	0.655
88	81.6	6.4	40.96	0.502
38	23.8	14.2	201.64	8.472
Σ 340	340.0			$\chi^2 = 10.94$
				$df = 5 - 1 = 4$

(In this problem, the only restriction on the theoretical frequencies is that $\Sigma E = \Sigma O$.) $\chi^2_{.05}$ (4 df) = 9.49. Therefore, reject the contention of the professor that his grades conform to "the curve." By examining the data, you can also reject the contention of the colleague that the professor is too soft. The primary reason the data do not fit the curve is that there were too many "flunks."

8. Goodness of fit

9. Each of the six sides of a die is equally likely. With the total number of throws as 1200, the expected value for any one of them is $\frac{1}{6}$ times 1200, or 200. With $\chi^2 = 5.75$, retain the null hypothesis. The results of the evening do not differ significantly from the "unbiased dice" model.

O	E	$O - E$	$(O - E)^2$	$\dfrac{(O - E)^2}{E}$
195	200	−5	25	0.125
200	200	0	0	0.000
220	200	20	400	2.000
215	200	15	225	1.125
190	200	−10	100	0.500
180	200	−200	400	2.000
Σ 1200	1200			$\chi^2 = 5.75$
$\chi^2_{.05}$ (5 df) = 11.07				$df = 6 - 1 = 5$

10. Goodness of fit

11.

O	E	$O - E$	$(O - E)^2$	$\dfrac{(O - E)^2}{E}$
21	12.735	8.265	68.310	5.365
5	13.265	−8.265	68.310	5.150
27	28.408	−1.408	1.982	0.070
31	29.592	1.408	1.982	0.067
0	6.857	−6.857	47.018	6.857
14	7.143	6.857	47.018	6.583
Σ 98	98.000			$\chi^2 = 24.09$
$\chi^2_{.001}(2\,df) = 13.82$			$df = (3 - 1)(2 - 1) = 2$	

Reject the null hypothesis and conclude that Madison used the word *by* significantly more frequently than did Hamilton. The stage is now set to examine the 12 disputed papers for their use of *by,* an examination that Mosteller and Wallace carried out. They describe the rates they found as "Madisonian." In addition, other words that distinguished between Madison and Hamilton (*upon, also,* and others) were found in Madisonian rates in the 12 papers.

12. Independence

13. To convert the information you have into frequencies that can be analyzed with χ^2, multiply the percentages (as proportions) by the number of observations. Thus, for *observed frequencies:*
.193(140) = 27 schizophrenic offspring; .807(140) = 113 nonschizophrenic offspring
For the *expected frequencies:*
.25(140) = 35 schizophrenic offspring; .75(140) = 105 nonschizophrenic offspring

O	E	$O - E$	$(O - E)^2$	$\dfrac{(O - E)^2}{E}$
27	35	−8	64	1.829
113	105	−8	64	0.610
Σ 140	140			$\chi^2 = 2.439$
				$df = 2 - 1 = 1$

$\chi^2_{.05}(1\,df) = 3.84$. Thus, retain the null hypothesis and conclude that these data are consistent with the 1:3 ratio predicted by the theory.

This problem illustrates that data that are consistent with the theory do not prove that the theory is true. Current explanations of schizophrenia include a genetic component, but not the simple one envisioned by the early hypothesis.

14. The difference between the tests for goodness of fit and for independence is in how the expected frequencies are obtained. In the goodness-of-fit test, expected frequencies are predicted by a theory, whereas in the independence test, they are obtained from the data.

15. The first task in this problem is to determine the expected frequencies. To do this, multiply each color's population percentage by the total number of M&Ms, 23. Here is the summary table with the steps for calculating χ^2.

Color	O	E	$O - E$	$(O - E)^2$	$\dfrac{(O - E)^2}{E}$
Brown	4	6.9	−2.9	8.41	1.219
Red	6	4.6	1.4	1.96	0.426
Yellow	8	4.6	3.4	11.56	2.513
Green	2	2.3	−0.3	0.09	0.039
Orange	3	2.3	0.7	0.48	0.213
Blue	0	2.3	−2.3	5.29	2.300
Σ	23	23.0			$\chi^2 = 6.71$
					$df = 6 - 1 = 5$

Degrees of freedom is number of categories minus one; $df = 6 - 1 = 5$. The critical value is $\chi^2_{.05} (5\ df) = 11.07$. Retain the null hypothesis. The observed frequencies are not significantly different from the expected frequencies, which are based on the company's claims.

Perhaps you have some complaints about this problem. (If not, reread the problem and its answer before reading further.)

Perhaps your complaint is that an N of 23 isn't large enough for a goodness-of-fit problem with six categories. A more sophisticated version of the complaint is that five of the six categories have expected values that are less than 5. The most sophisticated version is that such a small sample carries a very high probability of making a Type II error.

Fortunately, this is the kind of problem you can gather your own data on. Of course, you *must* use a larger sample (and having a research partner would be nice, too). Should you and your partner find yourselves having to test peanut M&Ms, the percentages are the same except that brown is 20 percent and blue is 20 percent. Finally, I want to acknowledge my debt to Randolph A. Smith, who explained to me the value of M&Ms as a statistical tool.

16. You should be gentle with your friend but explain that the observations in his study are not independent and cannot be analyzed with χ^2. Explain that, because one person is making five observations, the observations are not independent; the choice of one female candidate may cause the subject to pick a male candidate next (or vice versa).

17.

Candidates

		Hill	Dale	Σ
	Yes	57 (59.248)	31 (28.752)	88
Signaled turn	No	11 (8.752)	2 (4.248)	13
	Σ	68	33	101

$$\chi^2 = \frac{101[(57)(2) - (31)(11)]^2}{(88)(13)(68)(33)} = 2.03$$

$\chi^2_{.05}$ (1 df) = 3.84. Cars with bumper stickers for Hill and for Dale were observed as they turned left at a busy intersection. Sixteen percent of the 68 Hill cars failed to signal their left turn. Six percent of the 33 Dale cars failed to signal. This difference is not statistically significant; that is, a difference this big is expected more than 5 times in 100 if there is really no difference in percentages among the supporters of the two candidates.

18. I hope you would say something like, "Gee, I've been able to analyze data like these since I learned the t test. I will need to know the standard deviations for those means, though." If you attempted to analyze these data using χ^2, you erred (which, according to Alexander Pope, is human). If you erred, forgive yourself and reread pages 279 and 296.

19.

Houses

		Brick	Frame	Σ
	Hill	17 (37.116)	59 (38.884)	76
Candidates	Dale	88 (67.884)	51 (71.116)	139
	Σ	105	110	215

$$\chi^2 = \frac{N(AD - BC)^2}{(A + B)(C + D)(A + C)(B + D)}$$

$$= \frac{215(867 - 5192)^2}{(76)(139)(105)(110)} = 32.96; \quad df = 1$$

$$\phi = \sqrt{\frac{\chi^2}{N}} = \sqrt{\frac{32.96}{215}} = \sqrt{.1533} = 0.39$$

$\chi^2_{.001}$ (1 df) = 10.83; $p < .001$. Houses with yard signs for Hill and for Dale were classified as brick (affluent) or one-story frame (less affluent). Eighty-four percent of the 105 affluent houses had Dale signs; 16 percent had Hill signs. The relationship was in the other direction at the 110 less affluent houses: 46 percent had Dale signs and 54 percent had Hill signs. These differences, which are statistically significant, support the conclusion that Dale's supporters are more affluent than Hill's supporters. The effect size index, ϕ, is 0.39, a value that indicates that the degree of relationship is between medium and large.

20. $df = (R - 1)(C - 1)$, except for a table with only one row.

    a. 3   **b.** 12   **c.** 3   **d.** 10

chapter**fourteen**

1. The relative of the independent-samples t test is the Mann–Whitney U test. Corresponding to a correlated-samples t test is the Wilcoxon matched-pairs signed-ranks T test.

2. There are two differences: (a) Parametric tests give accurate probabilities if the populations the samples are drawn from have the characteristics that parametric tests assume (e.g., ANOVA assumes the populations are normally distributed and have equal variances). Nonparametric tests do not have assumptions about the populations built in. (b) The null hypothesis for parametric tests is that the population means are equal; for nonparametric tests, the null hypothesis is that the population distributions are the same.

3. Compare your answer with the material in the section "The Rationale of Nonparametric Tests."

4. Scale of measurement and power

5. Sum of ranks of those given estrogen = 76
Sum of ranks of the control animals = 44

$$\text{Smaller } U = (7)(8) + \frac{(7)(8)}{2} - 76 = 8; \quad \text{larger } U = 48$$

For the critical value, look in Table H for a two-tailed test with $\alpha = .05$ (the second page). At the intersection of $N_1 = 7$, $N_2 = 8$, the U value is 10. The obtained U of 8 is less than the critical value, so reject the null hypothesis. By examining the data, you see that the estrogen-injected animals ranked lower. Conclude that estrogen causes rats to be less dominant.

If you gave the rat that won zero bouts a rank of 1, the sum of ranks for the estrogen-injected group is 36; the sum for the sesame oil-injected group is 84. However, the same U values result.

6. With $\alpha = .01$, there are *no* possible results that would allow H_0 to be rejected. The friend must find more cars.

7. $\Sigma R_Y = 13$, $\Sigma R_Z = 32$

$$\text{Smaller } U = (4)(5) + \frac{(5)(6)}{2} - 32 = 3$$

Even with $\alpha = .05$ and a two-tailed test, a U of 1 or less is required to reject H_0. With an obtained $U = 3$, you must conclude that the quietness test did not produce evidence that Y cars are quieter than Z cars.

8. Giving 0 parasites a rank of 2 and 110 parasites a rank of 40, you find the sum of the ranks for Present day is 409.5; the sum of the ranks for Ten years earlier is 410.5. If you ranked in reverse order, the sum of the ranks for Present day is 574.5 and the sum for Ten years earlier is 245.5. Either way, the smaller U is 109.5 and the larger U is 274.5.

$$U_1 = (24)(16) + \frac{(24)(25)}{2} - 409.5 = 274.5$$

$$U_2 = (24)(16) + \frac{(16)(17)}{2} - 410.5 = 109.5$$

$$\mu_U = \frac{(24)(16)}{2} = 192$$

$$\sigma_U = \sqrt{\frac{(24)(16)(41)}{12}} = 36.22$$

$$z = \frac{(109.5 + 0.5) - 192}{36.22} = -2.26$$

The critical value of z is $|1.96|$, so reject the null hypothesis that the distributions are the same. By examining the average ranks (23.9 for present-day birds and 15.3 for 10-years-ago birds), you can conclude that present-day birds have significantly fewer brain parasites. (If the 0 birds rank 2, the average ranks are 17.1 for present-day birds and 25.7 for 10-years-ago birds. The same conclusion results.)

9. $$\mu_T = \frac{N(N + 1)}{4} = \frac{112(113)}{4} = 3164$$

$$\sigma_T = \sqrt{\frac{N(N + 1)(2N + 1)}{24}} = \sqrt{\frac{112(113)(225)}{24}} = 344.46$$

$$z = \frac{(4077 + 0.5) - 3164}{344.46} = 2.65$$

Because 2.65 is greater than 1.96, reject H_0 and conclude that incomes were significantly different after the program. Because you do not have the actual data, you cannot tell whether the incomes were higher or lower than before.

10.

Worker	Without rests	With rests	D	Signed ranks
1	2240	2421	181	2
2	2069	2260	191	4
3	2132	2333	201	5
4	2095	2314	219	6
5	2162	2297	135	1
6	2203	2389	186	3

Check: $21 + 0 = 21$

$$\frac{(6)(7)}{2} = 21$$

$\Sigma \text{(positive)} = 21$
$\Sigma \text{(negative)} = 0$
$T = 0$
$N = 6$

The critical value of T at the .05 level for a two-tailed test is 0. Because 0 (obtained) is equal to or less than 0 (table), reject the null hypothesis and conclude that the output with rests is greater than the output without rests. Here is the story behind this study.

From 1927 to 1932 the Western Electric Company conducted a study on a group of workers at the Hawthorne Works near Chicago. The final conclusion reached by this study has come to be called the Hawthorne effect. In the study, workers who assembled telephone relays were separated from the rest of the workers. A variety of changes in their daily routine followed, one at a time, with the following results: 5-minute rest periods morning and afternoon increased output; 10-minute rest periods increased output; company-provided snacks during rest periods increased output; and quitting 30 minutes early increased output. Going back to the original no-rest schedule increased output again, as did the reintroduction of the rest periods. Finally, management concluded that it was the special attention paid to the workers, rather than the changes, that increased output. Thus, the Hawthorne effect is an improvement in performance that is due to being in an experiment (getting special attention) rather than to the specific manipulation in the experiment. For a summary of this study, see Mayo (1946).

11. The experimental design has two independent samples. A Mann–Whitney U test is needed. If high scores are given high rank (that is, 39 ranks 1):

$\Sigma \text{(Canadians)} = 208; \quad \Sigma \text{(U.S.)} = 257$

(If low scores are given high ranks, the sums are reversed.)

$$U_1 = (15)(15) + \frac{(15)(16)}{2} - 257 = 88; \quad U_2 = 137$$

Consulting the second page of Table H (boldface type), you find a U value of 64 is required to reject the null hypothesis ($\alpha = .05$, two-tailed test). The lower of the two calculated U values for this problem is 88. Because 88 is greater than the table value of 64, the null hypothesis must be retained. Thus, conclude that there is no strong evidence that Canadians and people from the United States have different attitudes toward the regulation of business.

Once again you must be cautious in writing the conclusion when the null hypothesis is retained. You have not demonstrated that the two groups are the same; you have shown only that the groups are not significantly different.

12.

Student	Before	After	D	Signed rank
1	18	4	14	13
2	14	14	0	1.5
3	20	10	10	9
4	6	9	−3	−4
5	15	10	5	6
6	17	5	12	11
7	29	16	13	12
8	5	4	1	3
9	8	8	0	−1.5
10	10	4	6	7
11	26	15	11	10
12	17	9	8	8
13	14	10	4	5
14	12	12	0	Eliminated
Check:	85.5 + 5.5 = 91		Σ (positive) = 85.5	
	$\dfrac{(13)(14)}{2} = 91$		Σ (negative) = −5.5	
			$T = 5.5$	
			$N = 13$	

The tabled value for T for a two-tailed test with $\alpha = .01$ is 9, so reject H_0 and conclude that the after-distribution is from a different population than the before-distribution. Except for one person, the number of misconceptions stayed the same or decreased, so conclude that the course *reduced* the number of misconceptions.

I would like to remind you here of the distinction made in Chapter 9 between statistically significant and important. There is a statistically significant decrease in the number of misconceptions, but a professor might be quite dismayed at the number of misconceptions that remain.

13. Σ (positive) = 81.5   *Check:* 81.5 + 54.5 = 136

Σ (negative) = −54.5 $\dfrac{(16)(17)}{2} = 136$

$T = 54.5$

$N = 16$

A $T \leq 29$ is required for rejection at the .05 level. Thus, there is no significant difference in the weight 10 months later. Put in the most positive language, there is no statistically significant evidence that the participants gained back the weight lost during the workshop.

14. The tables show the raw data converted to ranks. The table on the left shows the lowest score ranked 1. With the highest score ranked 1, the table on the right results. Although the sums at the bottom are not the same, the difference between any two pairs is the same.

Friends	Romans	Countrymen
15	9	2
18	13	10
12	4.5	6
14	11	1
16	3	8
17	4.5	7
Σ = 92	45	34

Friends	Romans	Countrymen
4	10	17
1	6	9
7	14.5	13
5	8	18
3	16	11
2	14.5	12
Σ = 22	69	80

	Friends	Romans
Romans	47*	
Countrymen	58**	11

* $p < .05$ ** $p < .01$

For $K = 3$, $N = 6$, the critical values are 43.3 at the .05 level and 53.9 at the .01 level. Romans and countrymen did not differ from each other, but both groups offered significantly less than friends did. So, when you have a plea, ask a friend to lend an ear.

15. The table shows the raw scores converted to ranks.

	NO	RE	SE	GE
	6	9	8	15.5
	4	6	18	20
	2	14	17	12.5
	1	6	12.5	19
	3	10	11	15.5
Σ =	16	45	66.5	82.5
Sum with ranks reversed	89	60	38.5	22.5

	NO	RE	SE
RE	29		
SE	50.5*	21.5	
GE	66.5**	37.5	16

* $p < .05$ ** $p < .01$

For $K = 4$, $N = 5$, the critical values are 48.1 at the .05 level and 58.2 at the .01 level. Although group exercise is better than solitary exercise, the difference is not significant. Both of these groups improved significantly when compared to the no-treatment group. The relaxation group was intermediate between no treatment and solitary exercise, but not significantly different from the other three groups. Take-home message: Exercise with others.

16.

	Authoritarian		Democratic		Laissez-faire	
	X	Rank	X	Rank	X	Rank
	77	5	90	10.5	50	1
	86	9	92	12	62	2
	90	10.5	100	16	69	3
	97	13	105	21	76	4
	100	16	107	22	79	6
	102	19	108	23	82	7
	120	27.5	110	24	84	8
	121	29	118	26	99	14
	128	32	125	30	100	16
	130	33	131	34	101	18
	135	37	132	35.5	103	20
	137	38	132	35.5	114	25
	141	39.5	146	41	120	27.5
	147	42	156	44	126	31
	153	43	161	45	141	39.5
Σ		393.5		419.5		222.0
Sum with ranks reversed		296.5		270.5		468.0

	Authoritarian	Democratic
Democratic	26	
Laissez-faire	171.5*	197.5*

$*p < .05$

For $K = 3$, $N = 15$, a difference in the sum of the ranks of 168.6 is required to reject H_0 at the .05 level. Therefore, both the authoritarian leadership and the democratic leadership resulted in higher personal satisfaction scores than laissez-faire leadership, but the authoritarian and democratic types of leadership did not differ significantly from each other.

17. Arranging the data into a summary table, you have:

Groups

		1	2	3	4	5
	2	85				
	3	31	54			
Groups	4	18	67	13		
	5	103	188*	134	121	
	6	31	116	62	49	72

* $p < .05$

For $K = 6$, $N = 8$, differences of 159.6 and 188.4 are required to reject the null hypothesis at the .05 and .01 level, respectively. Thus, the only significant difference is between the means of groups 2 and 5 at the .05 level. (With data such as these, the novice investigator may be tempted to report "almost significant at the .01 level." Resist that temptation.)

18. The Mann–Whitney U test is used for ranked data from two independent samples with equal or unequal N's. The Wilcoxon matched-pairs signed-ranks T test is used for ranked data from two correlated samples. The Wilcoxon–Wilcox multiple-comparisons test makes all possible pairwise comparisons among K independent equal-N groups of ranked data.

19. a. Wilcoxon matched-pairs signed-ranks test; this is a before-and-after study.
b. Wilcoxon matched-pairs signed-ranks test; pairs are formed by family.
c. Spearman's r_s; the degree of relationship is desired.
d. Wilcoxon–Wilcox multiple-comparisons test; the experiment has three independent groups.
e. Wilcoxon matched-pairs signed-ranks test; again, this is a before-and-after design.

20.

	Symbol of statistic	Appropriate for what design?	Reject H_0 when statistic is (greater, less) than critical value?
Mann–Whitney test	U	Two independent samples	Less
Wilcoxon matched-pairs signed-ranks test	T	Two correlated samples	Less
Wilcoxon–Wilcox multiple-comparisons test	None	More than two independent samples, equal N's	Greater

21. A low correlation means the philosophers have different sets of criteria of what is important. On the other hand, a high correlation means they have about the same set of criteria.

Candidates	Locke	Kant	D	D^2
A	7	8	−1	1
B	10	10	0	0
C	3	5	−2	4
D	9	9	0	0
E	1	1	0	0
F	8	7	1	1
G	5	3	2	4
H	2	4	−2	4
I	6	6	0	0
J	4	2	2	4
			$\Sigma D^2 = 18$	

$$r_s = 1 - \frac{6\Sigma D^2}{N(N^2 - 1)} = 1 - \frac{6(18)}{10(99)} = 1 - .109 = .891$$

The two professors seem to be in pretty close agreement on the selection criteria.

22.

Year	Rank in marriages X	X^2	Rank in grain Y	Y^2	$(X)(Y)$
1930	1	1.00	3	9.00	3.0
1931	2	4.00	1	1.00	2.0
1927	3	9.00	4	16.00	12.0
1925	4	16.00	2	4.00	8.0
1926	5	25.00	5	25.00	25.0
1923	6	36.00	6	36.00	36.0
1922	7	49.00	7	49.00	49.0
1919	8.5	72.25	11	121.00	93.5
1924	8.5	72.25	9	81.00	76.5
1928	10	100.00	9	81.00	90.0
1929	11	121.00	9	81.00	99.0
1918	12	144.00	12.5	156.25	150.0
1921	13	169.00	12.5	156.25	162.5
1920	14	196.00	14	196.00	196.0
Σ	105	1014.5	105	1012.5	1002.5

This is a case of tied ranks, so you should compute a Pearson r on the ranks:

$$r = \frac{N\Sigma XY - (\Sigma X)(\Sigma Y)}{\sqrt{[N\Sigma X^2 - (\Sigma Y)^2][N\Sigma Y^2 - (\Sigma Y)^2]}}$$

$$= \frac{14(1002.5) - (105)(105)}{\sqrt{[(14)(1014.5) - (105)^2][(14)(1012.5) - (105)^2]}} = .95$$

Table A shows that for 12 df, a coefficient of .78 is significant when $\alpha = .001$. The sentence of explanation might be: There is a highly significant relationship between the number of marriages and the value of the grain crop; as one goes up, the other does, too.

23. $r_s = 1 - \dfrac{6\Sigma D^2}{N(N^2 - 1)} = 1 - \dfrac{6(308)}{16(16^2 - 1)} = .55$

From Table L, a coefficient of .503 is required for significance (two-tailed test). Thus, an $r_s = .55$ is significantly different from .00.

24. With only four pairs of scores there is no possible way to reject the hypothesis that the population correlation is .00 (see Table L). Advise your friend that more data must be obtained before any inference can be made about the population.

WHAT WOULD YOU RECOMMEND? chapters ten–fourteen

a. A Mann–Whitney U test should be used to compare flooding and systematic desensitization. The two groups are independent samples — there is no reason to pair the score of a trainee who experienced flooding with the score of a trainee who experienced systematic desensitization. A nonparametric test is called for because the data were skewed.

b. The proper analysis for these data is a chi square test of independence. The score that a participant received was counted as a frequency in a category (comply or not comply).

c. These data should be analyzed with a factorial ANOVA. The two independent variables are the looks and intentionality of the defendant. The dependent variable is the likelihood estimate (a quantitative measure). (The factorial ANOVA of these data produced a significant interaction. Baby-faced defendants were less likely to be convicted of an intentional offense than a mature-faced defendant, but more likely to be convicted of a negligent offense.)

d. A Spearman r_s gives the degree of relationship between two variables. A Spearman r_s is more appropriate than a Pearson r because one variable is rank scores.

e. The theory, which predicts the number of participants on subsequent occasions of the exercise, can be evaluated using a chi square goodness-of-fit test. (The expected values are 50, 25, and 12.5.)

f. This experiment has one independent variable that has three levels. Because the variance of the "no cues" condition was much greater than that of the other two conditions, a Wilcoxon–Wilcox multiple-comparisons test is appropriate.

g. A Wilcoxon matched-pairs signed-ranks T test is appropriate for these skewed weight data. This is a correlated-samples design because a before-meal score is paired with an after-meal score. (Both scores were obtained from the same dog.)

h. To compare each pair of recipes, a Tukey HSD test is appropriate. (No evidence in the problem suggests that the data are not normally distributed or have unequal variances.)

i. A one-factor correlated-measures ANOVA will help evaluate the program. Tukey HSD tests can determine the immediate effect of the program (post-test vs. pretest), the long-term effect (follow-up vs. pretest), and the effect of time on retention (follow-up vs. post-test).

chapter**fifteen**

Set A

1. A mean of a set of means is needed, and to find it you need to know the number of dollars invested in each division.

2. Wilcoxon matched-pairs signed-ranks T test. You have evidence that the population of dependent-variable scores (reaction time) is not normally distributed.

3. A one-factor correlated-measures ANOVA

4. A median is appropriate because the scores are skewed.

5. Assume that these measures are normally distributed and use a normal curve to find the proportion (0.0228 is the answer).

6. Either an independent-measures t test or a Mann–Whitney U test

7. A 95 or 99 percent confidence interval about the mean reading achievement of those 50 sixth-graders

8. χ^2 goodness-of-fit test

9. Mode

10. Neither will do. The relationship described is nonlinear — one that first increases and then decreases. A statistic, *eta,* is appropriate for curved relationships.

11. Correlated-samples t test or Wilcoxon matched-pairs signed-ranks T test

12. A regression equation will provide the predictions, and a correlation coefficient will indicate how accurate the predictions will be.

13. A 2 × 2 factorial ANOVA

14. The student's wondering may be translated into a question of whether a correlation of .20 is statistically significant. Use a t test to find whether $r = .20$ for that class is significantly different from $r = .00$. Or, look in Table A in Appendix C.

15. This is a χ^2 problem but it cannot be worked using the techniques in this text because these before-and-after data are correlated, not independent. Intermediate texts describe appropriate techniques for correlated χ^2 problems.

16. A confidence interval about the sample mean. Does the CI include the manufacturer's claim of 1000 hours?

17. Pearson product-moment correlation coefficient

18. Wilcoxon–Wilcox multiple-comparisons test (if the reaction-time scores are skewed) or one-way analysis of variance

19. A χ^2 test of independence will determine whether a decision to report shoplifting is influenced by gender (one χ^2 test) and by the dress of the shoplifter (a second χ^2 test).

20. A line graph with serial position on the X axis and number of errors on the Y axis will illustrate this relationship.

21. A Mann–Whitney U test is preferred because the dependent variable is skewed. The question is whether there is a relationship between diet and cancer. (Note that if the two groups differ, two interpretations are possible. For example, if the incidence is higher among the red meat cultures, it might be because of the red meat or because of the lack of cereals.)

22. An r_s will give the degree of relationship for these two ranked variables.

23. For each species of fish, a set of z scores may be calculated. The fish with the largest z score should be declared the overall winner.

24. Calculating an effect size for the experiments mentioned in the first part of this question will tell you how big a difference psychotherapy makes.

Set B

25. The critical values are 377.6 at the .05 level and 457.6 at the .01 level. You can conclude that Herbicide B is significantly better than A, C, or D at the .01 level and that D is better than C at the .05 level.

26. An effect size of .10 is quite small. Using Cohen's (1969) conventions, effect sizes of .20, .50, and .80 are classified as small, medium, and large. In the case of a company offering instruction, a large number of clients provides a large N, which can ensure a "statistically significant improvement," even if the effect size is quite small.

27. The screening process is clearly worthwhile in making each batch profitable. Of course, the cost of screening will have to be taken into account.

28. $t_{.001}$ (40 df) = 3.55; therefore, reject the null hypothesis and conclude that extended practice *improved* performance. Because this is just the opposite of the theory's prediction, you can conclude that the theory is not adequate to explain the serial position effect.

29. Because $\chi^2_{.05}$ (2 df) = 5.99, the null hypothesis for this goodness-of-fit test is retained. The data do fit the theory; the theory is adequate.

30. The overall F is significant ($p < .01$), so conclude that the differences should not be attributed to chance. The low dose is significantly better than the placebo ($HSD_{.01} = 4.70$) and significantly better than the high dose (which has a smaller mean than the placebo). The placebo and high dose are not significantly different.

31. The critical value of U for a two-tailed test with $\alpha = .01$ is 70. You may conclude that the difference between the two methods is statistically significant at the .01 level. You cannot, from the information supplied, tell which of the two methods is superior.

32. The critical value at the .001 level of an r_s with $N = 18$ is .708 (Table A). Attitudes of college students and people in business toward the 18 groups are similar. There is little likelihood that chance would produce the similarity of attitudes found in the sample.

33. $t_{.05}$ (60 df) = 2.00; $F_{.01}$ (30, 40 df) = 2.20; the poetry unit appears to have no significant effect on mean attitudes toward poetry. There is a very significant effect on the variability of the attitudes of those who studied poetry. It appears that the poetry unit turned some students on and some students off, thus causing a great deal of variability in the attitude scores.

34. $F_{.05}$ (1, 44 df) = 4.06; $F_{.01}$ (1, 44 df) = 7.24; only the interaction is significant. A graph, as always, will help in the interpretation. The question of whether to present one or both sides to get the most attitude change depends on the level of education of the audience. If the members of a group have less than a high school education, present one side. If they have some college education, present both sides.

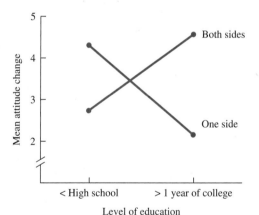

35. The tabled value of $F_{.05}$ (2, 60 df) = 3.15, so the null hypothesis is retained. There are no significant differences in the verbal ability scores of third-graders who are left- or right-handed or mixed.

36. Because $\chi^2_{.02}$ (2 df) = 7.82, the null hypothesis for this test of independence is rejected at the .02 level. Alcoholism and toilet-training are not independent; they are related. Experiments such as this do not allow you to make cause-and-effect statements, however.

appendix**A**

Part 1 (Pretest)

1. About 16	**16.** 0.83	**31.** 33%	**46.** 4.00
2. About 15	**17.** 0.1875	**32.** 22.68	**47.** 4.50
3. About 15	**18.** 0.056	**33.** 0.19	**48.** 0.45
4. 6.1	**19.** 0.60	**34.** 60 birds	**49.** 15.25
5. 0.4	**20.** 0.10	**35.** 5	**50.** 8.00
6. 10.3	**21.** 1.50	**36.** 4	**51.** 16.25
7. 11.424	**22.** 3.11	**37.** 6, 10	**52.** 1.02
8. 27.141	**23.** 6	**38.** 4, 22	**53.** 12.00
9. 19.85	**24.** −22	**39.** 16	**54.** 10, 30
10. 2.065	**25.** −7	**40.** 6.25	**55.** 12, 56
11. 49.972	**26.** −6	**41.** 0.1225	**56.** 13.00
12. 0.0567	**27.** 25	**42.** 3.05	**57.** 11.00
13. 3.02	**28.** −24	**43.** 0.96	**58.** 2.50
14. 17.11	**29.** 3.33	**44.** 0.30	**59.** −2.33
15. 1.375	**30.** −5.25	**45.** 4.20	

Part 2 (Review)

1. a. A sum is the answer to an addition problem.
   b. A quotient is the answer to a division problem.
   c. A product is the answer to a multiplication problem.
   d. A difference is the answer to a subtraction problem.

2. a. About 10   **b.** About 7   **c.** About 100   **d.** About 0.2
   e. About 0.16   **f.** About 60   **g.** Just less than 1   **h.** About 0.1

3. a. 14   **b.** 126   **c.** 9   **d.** 0   **e.** 128
   f. 13   **g.** 12   **h.** 13   **i.** 9

4. a. 6.33   **b.** 13.00   **c.** 0.05   **d.** 0.97   **e.** 2.61
   f. 0.34   **g.** 0.00   **h.** 0.02   **i.** 0.99

5.
$$
\begin{array}{r}
0.001 \\
10.000 \\
3.652 \\
2.500 \\
\hline
16.153
\end{array}
$$

6.
$$
\begin{array}{r}
14.20 \\
-7.31 \\
\hline
6.89
\end{array}
$$

7.
$$
\begin{array}{r}
1.26 \\
\times\, 0.04 \\
\hline
0.0504
\end{array}
$$

8.
$$
\begin{array}{r}
143.300 \\
16.920 \\
2.307 \\
8.100 \\
\hline
170.627
\end{array}
$$

9.
$$
\begin{array}{r}
76.5 \\
0.04\overline{)3.06} \\
2\,8 \\
\hline
26 \\
24 \\
\hline
20 \\
20 \\
\hline
\end{array}
$$

10.
$$
\begin{array}{r}
2.04 \\
11.75\overline{)24.00} \\
23\,50 \\
\hline
5000 \\
4700 \\
\hline
300
\end{array}
$$

11.
$$
\begin{array}{r}
152.12 \\
-127.40 \\
\hline
24.72
\end{array}
$$

12.
$$
\begin{array}{r}
0.07 \\
\times\, 0.5 \\
\hline
0.035
\end{array}
$$

13. $\dfrac{9}{10} + \dfrac{1}{2} + \dfrac{2}{5} = 0.90 + 0.50 + 0.40 = 1.80$

14. $\dfrac{9}{20} \div \dfrac{19}{20} = 0.45 \div 0.95 = 0.47$

15. $\left(\dfrac{1}{3}\right)\left(\dfrac{5}{6}\right) = 0.333 \times 0.833 = 0.28$ **16.** $\dfrac{4}{5} - \dfrac{1}{6} = 0.80 - 0.167 = 0.63$

17. $\dfrac{1}{3} \div \dfrac{5}{6} = 0.333 \div 0.833 = 0.40$ **18.** $\dfrac{3}{4} \times \dfrac{5}{6} = 0.75 \times 0.833 = 0.62$

19. $18 \div \dfrac{1}{3} = 18 \div 0.333 = 54$

20. a. $(24) + (-28) = -4$ **b.** -23
    c. $(-8) + (11) = 3$ **d.** $(-15) + (8) = -7$

21. -40 **22.** 24

23. 48 **24.** -33

25. $(-18) - (-9) = -9$ **26.** $14 \div (-6) = -2.33$

27. $12 - (-3) = 15$ **28.** $(-6) - (-7) = 1$

29. $(-9) \div (-3) = 3$ **30.** $(-10) \div 5 = -2$

31. $4 \div (-12) = -0.33$ **32.** $(-7) - 5 = -12$

33. $6 \div 13 = 0.46$ **34.** $6/13 = .46; .46 \times 100 = 46$ percent

35. $18 \div 25 = 0.72$ **36.** $85 \div 115 = 0.74 \times 100 = 74$ percent

37. 25.92 **38.** 0.31 **39.** 128 **40.** 187 coupons

41. $|-31| = 31$ **42.** $|21 - 25| = |-4| = 4$

43. $12 \pm (2)(5) = 12 \pm 10 = 2, 22$ **44.** $\pm(5)(6) + 10 = \pm 30 + 10 = -20, 40$

45. $\pm(2)(2) - 6 = \pm 4 - 6 = -10, -2$ **46.** $(2.5)^2 = 2.5 \times 2.5 = 6.25$

47. $9^2 = 9 \times 9 = 81$ **48.** $0.3 \times 0.3 = 0.09$

49. $(\tfrac{1}{4})^2 = (0.25)(0.25) = 0.0625$

50. a. 25.00 **b.** 2.50 **c.** 0.79 **d.** 0.25 **e.** 4.10
    f. 0.0548 **g.** 426.00 **h.** 0.50 **i.** 4.74

51. $\dfrac{(4 - 2)^2 + (0 - 2)^2}{6} = \dfrac{2^2 + (-2)^2}{6} = \dfrac{4 + 4}{6} = \dfrac{8}{6} = 1.33$

52. $\dfrac{(12 - 8)^2 + (8 - 8)^2 + (5 - 8)^2 + (7 - 8)^2}{4 - 1}$

$= \dfrac{4^2 + 0^2 + (-3)^2 + (-1)^2}{3}$

$= \dfrac{16 + 0 + 9 + 1}{3} = \dfrac{26}{3} = 8.67$

53. $\left(\dfrac{5 + 6}{3 + 2 - 2}\right)\left(\dfrac{1}{3} + \dfrac{1}{2}\right) = \left(\dfrac{11}{3}\right)(0.333 + 0.500) = (3.667)(0.833) = 3.05$

54. $\left(\dfrac{13 + 18}{6 + 8 - 2}\right)\left(\dfrac{1}{6} + \dfrac{1}{8}\right) = \left(\dfrac{31}{12}\right)(0.167 + 0.125) = (2.583)(0.292) = 0.75$

55. $\dfrac{8[(6 - 2)^2 - 5]}{(3)(2)(4)} = \dfrac{8[4^2 - 5]}{24} = \dfrac{8[16 - 5]}{24} = \dfrac{8[11]}{24} = \dfrac{88}{24} = 3.67$

56. $\dfrac{[(8 - 2)(5 - 1)]^2}{5(10 - 7)} = \dfrac{[(6)(4)]^2}{5(3)} = \dfrac{[24]^2}{15} = \dfrac{576}{15} = 38.40$

57. $\dfrac{6}{1/2} + \dfrac{8}{1/3} = (6 \div 0.5) + (8 \div 0.333) = 12 + 24 = 36$

58. $\left(\dfrac{9}{2/3}\right)^2 + \left(\dfrac{8}{3/4}\right)^2 = (9 \div 0.667)^2 + (8 \div 0.75)^2$

$= 13.5^2 + 10.667^2 = 182.25 + 113.785 = 296.03$

59. $\dfrac{10 - (6^2/9)}{8} = \dfrac{10 - (36/9)}{8} = \dfrac{10 - 4}{8} = \dfrac{6}{8} = 0.75$

60. $\dfrac{104 - (12^2/6)}{5} = \dfrac{104 - (144/6)}{5} = \dfrac{104 - 24}{5} = \dfrac{80}{5} = 16.00$

61. $\dfrac{x - 4}{2} = 2.58, \quad x - 4 = 5.16, \quad x = 9.16$

62. $\dfrac{x - 21}{6.1} = 1.04, \quad x - 21 = 6.344, \quad x = 27.34$

63. $x = \dfrac{14 - 11}{2.5} = \dfrac{3}{2.5} = 1.20$

64. $x = \dfrac{36 - 41}{8.2} = \dfrac{-5}{8.2} = -0.61$

appendix B

1.

Test scores (class interval)	Tally marks	f
63–67	/	1
58–62	/	1
53–57	///	3
48–52	////	4
43–47	////	4
38–42	卌 //	7
33–37	卌 //	7
28–32	卌 //	7
23–27	卌 //	7
18–22	////	4
13–17	///	3
8–12	/	1
3–7	/	1

The data are positively skewed.

2.

Weights	Tally marks	f
189–191	/	1
186–188		0
183–185		0
180–182	/	1
177–179		0
174–176	//	2
171–173	/	1
168–170	//	2
165–167	/	1
162–164	/	1
159–161	///	3
156–158	卌	5
153–155	////	4
150–152	卌 /	6
147–149	////	4
		$\Sigma = 31$

3. $\bar{X} = \dfrac{\Sigma X}{N} = \dfrac{1720}{50} = 34.40$

Median location $= \dfrac{N+1}{2} = \dfrac{50+1}{2} = 25.5$. There are 23 frequencies in the intervals below 33–37 and 7 scores in the interval. The median is 35, the midpoint of the interval 33–37.

Because four intervals share the largest number of scores, 7, the mode is the midpoint of these four intervals. Mode $= 32.5$.

4. $\bar{X} = \dfrac{\Sigma X}{N} = \dfrac{4939}{31} = 159.32$ pounds

Median location $= \dfrac{N+1}{2} = \dfrac{31+1}{2} = 16$. There are 14 frequencies below the interval 156–158, which has 5 measurements. The median is 157 pounds.

The mode is 151 pounds because the interval 150–152 has 6 frequencies, more than any other interval.

Note that the mean is larger than the median, confirming your judgment that the distribution is positively skewed.

references

Aron, A., & Aron, E. N. (1999). *Statistics for psychology.* (2nd ed.). Englewood Cliffs, NJ: Prentice Hall.

Aronson, E., & Mills, J. (1959). The effect of severity of initiation on liking for a group. *Journal of Abnormal and Social Psychology, 59,* 177–181.

Bahrick, H. P., Bahrick, L. E., Bahrick, A. S., & Bahrick, P. E. (1993). Maintenance of foreign language vocabulary and the spacing effect. *Psychological Science, 4,* 316–321.

Bandura, A., Blanchard, E. B., & Ritter, B. (1969). The relative efficacy of desensitization and modeling approaches for inducing behavioral, affective, and attitudinal changes. *Journal of Personality and Social Psychology, 64,* 173–199.

Barber, T. X. (1976). Suggested ("hypnotic") behavior: The trance paradigm versus an alternative paradigm. In T. X. Barber (Ed.), *Advances in altered states of consciousness & human potentialities* (Vol. 1, pp. 175–259). New York: Psychological Dimensions.

Barker, R. B., Dembo, T., & Lewin, K. (1941). Frustration and regression: An experiment with young children. *University of Iowa Studies in Child Welfare, 18,* No. 1.

Benjamin, L. T., Cavell, T. A., & Shallenberger, W. R. (1984). Staying with initial answers on objective tests: Is it a myth? *Teaching of Psychology, 11,* 133–141.

Berger, R. W., & Hart, T. H. (1986). *Statistical process control.* New York: Marcel Dekker.

Berk, L. E. (1998). *Development through the lifespan.* Boston: Allyn & Bacon.

Berry, D. S., & McArthur, L. Z. (1986). Perceiving character in faces: The impact of age-related craniofacial changes on social perception. *Psychological Bulletin, 100,* 3–18.

Biffen, R. H. (1905). Mendel's laws of inheritance and wheat breeding. *Journal of Agricultural Science, 1,* 4–48.

Birch, H. G., & Rabinowitz, H. S. (1951). The negative effect of previous experience on productive thinking. *Journal of Experimental Psychology, 41,* 121–125.

Blair, R. C., & Higgins, J. J. (1985). Comparison of the power of the paired samples t test to that of Wilcoxon's signed-ranks test under various population shapes. *Psychological Bulletin, 97,* 119–128.

Blair, R. C., Higgins, J. J., & Smitley, W. D. S. (1980). On the relative power of the *U* and *t* tests. *British Journal of Mathematical and Statistical Psychology, 33,* 114–120.

Boehner, C., & Howe, S. (1996, June). *Statistical graphics in psychological research.* Poster session presented at the annual meeting of the American Psychological Society, San Francisco, CA.

Box, J. F. (1981). Gosset, Fisher, and the *t* distribution. *American Statistician, 35,* 61–66.

Bradley, D. R., Bradley, T. D., McGrath, S. G., & Cutcomb, S. D. (1979). Type I error rate of the chi-square test of independence in R × C tables that have small expected frequencies. *Psychological Bulletin, 86,* 1290–1297.

Bradley, J. V. (1978). Robustness? *British Journal of Mathematical Statistical Psychology, 31,* 144–152.

Bransford, J. D., & Franks, J. J. (1971). The abstraction of linguistic ideas. *Cognitive Psychology, 2,* 331–350.

Brehm, J. W., & Cohen, A. R. (1962). *Explorations in cognitive dissonance.* New York: Wiley.

Bureau of Labor Statistics. (December 1944). Union wages and hours of motor truck drivers and helpers. July 1, 1944. *Monthly Labor Review.*

Camilli, G., & Hopkins, K. D. (1978). Applicability of chi-square to 2×2 contingency tables with small expected cell frequencies. *Psychological Bulletin, 85,* 163–167.

Campbell, S. K. (1974). *Flaws and fallacies in statistical thinking.* Englewood Cliffs, NJ: Prentice-Hall.

Cohen, I. B. (March 1984). Florence Nightingale. *Scientific American, 250,* 128–137.

Cohen, J. (1969). *Statistical power analysis for the behavioral sciences.* New York: Academic.

Cohen, J. (1988). *Statistical power analysis for the behavioral sciences* (2d ed.). Hillsdale, NJ: Erlbaum.

Cohen, J. (1992). A power primer. *Psychological Bulletin, 112,* 155–159.

Cohen, S., & Williamson, G. M. (1991). Stress and infectious disease in humans. *Psychological Bulletin, 109,* 5–24.

Coren, S., & Halpern, D. F. (1991). Left-handedness: A marker for decreased survival fitness. *Psychological Bulletin, 109,* 90–106.

David, F. N. (1968). Francis Galton. In D. L. Sills (Ed.), *International encyclopedia of the social sciences* (Vol. 6, pp. 48–53). New York: Macmillan.

Diener, E., Suh, E. M., Lucas, R. E., & Smith, H. L. (1999). Subjective well-being: Three decades of progress. *Psychological Bulletin, 125,* 276–302.

Dillon, K. M. (1999). I am 95% confident that the world is round: An interview about statistics with Chris Spatz. *Teaching of Psychology, 26,* 232–234.

Doll, R. (1955). Etiology of lung cancer. *Advances in Cancer Research, 3,* 1–50.

Dunkling, L. (1993). *The Guinness book of names.* Enfield, UK: Guinness.

Durkheim, E. (1951, reprint). *Suicide.* New York: Free Press.

Ellis, D. (1938). *A source book of Gestalt psychology.* London: Routledge & Kegan Paul.

Everitt, B. S., & Hay, D. F. (1992). *Talking statistics: A psychologist's guide to design & analysis.* New York: Halstead.

Ferguson, G. A., & Takane, Y. (1989). *Statistical analysis in psychology and education* (6th ed.). New York: McGraw-Hill.

Fisher, R. A. (1925). *Statistical methods for research workers.* London: Oliver and Boyd.

Fisher, R. A., & Yates, F. (1963). *Statistical tables for biological, agricultural, and medical research* (6th ed.). Edinburgh: Oliver and Boyd.

Flynn, J. R. (1987). Massive IQ gains in 14 nations: What IQ tests really measure. *Psychological Bulletin, 101,* 171–191.

Forbs, R., & Meyer, A. B. (1955). *Forestry handbook.* New York: Ronald.

Galton, F. (1869). *Hereditary genius.* London: Macmillan.

Galton, F. (1889). *Natural inheritance.* London: Macmillan.

Galton, F. (1901). Biometry. *Biometrika, 1,* 7–10.

Greene, J. E. (Ed.-in-chief). (1966). *McGraw-Hill modern men of science.* New York: McGraw-Hill.

Guinness World Records 2000. (2000). Enfield, UK: Guinness World Records.

Hacking, I. (1984). Trial by number. *Science 84, 5,* 69–70.

Hald, A. (1998). *A history of mathematical statistics from 1750 to 1930.* New York: Wiley.

Harter, H. L. (1960). Tables of range and studentized range. *Annals of Mathematical Statistics, 31,* 1122–1147.

Hasler, A. D. (1966). *Underwater guideposts.* Madison: University of Wisconsin Press.

Hasler, A. D., & Scholz, A. T. (1983). *Olfactory imprinting and homing in salmon.* New York: Springer-Verlag.

Hasler, A. D., Scholz, A. T., & Horrall, R. M. (1978). Olfactory imprinting and homing in salmon. *American Scientist, 66,* 347–355.

Hedges, L. V., & Nowell, A. (1995). Sex differences in mental test scores, variability, and numbers of high-scoring individuals. *Science, 269,* 41–45.

Hellekson, C. J., Kline, J. A., & Rosenthal, N. E. (1986). Phototherapy for seasonal affective disorder in Alaska. *American Journal of Psychiatry, 143,* 1035–1037.

Henry, G. T. (1994). *Graphing data.* Thousand Oaks, CA: Sage.

Howell, D. C. (1997). *Statistical methods for psychology* (4th ed.). Boston: Duxbury.

Howell, D. C. (1999). *Fundamental statistics for the behavioral sciences* (4th ed.). Belmont, CA: Duxbury.

Huff, D. (1954). *How to lie with statistics.* New York: Norton.

Irion, A. L. (1976). A survey of the introductory course in psychology. *Teaching of Psychology, 3,* 3–8.

Jenkins, J. G., & Dallenbach, K. M. (1924). Obliviscence during sleep and waking. *American Journal of Psychology, 35,* 605–612.

Johnson, R. C., McClearn, G. E., Yuen, S., Nagoshi, C. T., Ahern, F. M., & Cole, R. E. (1985). Galton's data a century later. *American Psychologist, 40,* 875–892.

Jones, S. S., Collins, K., & Hong, H. (1991). An audience effect on smile production in 10-month-old infants. *Psychological Science, 2,* 45–49.

Keppel, G. (1982). *Design and analysis: A researcher's handbook.* Englewood Cliffs, NJ: Prentice-Hall.

Kirk, R. E. (1984). *Elementary statistics* (2d ed.). Pacific Grove, CA: Brooks/Cole.

Kirk, R. E. (1995). *Experimental design: Procedures for the behavioral sciences* (3d ed.). Pacific Grove, CA: Brooks/Cole.

Kirk, R. E. (1999). *Statistics: An introduction* (4th ed.). Fort Worth, TX: Harcourt Brace.

Kramer, C. Y. (1956). Extension of multiple range tests to group means with unequal numbers of replications. *Biometrics, 12,* 307–310.

Landau, D., & Lazarfeld, P. F. (1968). Quételet, Adolphe. In D. L. Sills (Ed.), *International encyclopedia of the social sciences* (Vol. 13, pp. 247–257). New York: Macmillan.

Lewin, K. (1958). Group decision and social change. In E. E. Maccoby, T. M. Newcomb, & E. L. Hartley (Eds.), *Readings in social psychology* (3d ed.). New York: Holt, Rinehart & Winston.

Loftus, E. F. (1979). *Eyewitness testimony.* Cambridge, MA: Harvard University Press.

Loftus, G. R., & Loftus, E. F. (1988). *Essentials of statistics* (2d ed.). New York: Random House.

Mackowiak, P. A., Wasserman, S. S., & Levine, M. M. (1992). A critical appraisal of 98.6°F, the upper limit of the normal body temperature, and other legacies of Carl Reinhold August Wunderlich. *Journal of the American Medical Association, 268,* 1578–1580.

Mann, H. B., & Whitney, D. R. (1947). On a test of whether one or two random variables is stochastically larger than the other. *Annals of Mathematical Statistics, 18,* 50–60.

Mayo, E. (1946). *The human problems of an industrial civilization.* Boston: Harvard University Press.

McCall, R. B. (1998). *Fundamental statistics for behavioral sciences* (7th ed.). Pacific Grove, CA: Brooks/Cole.

McCrae, R. R., & Costa, P. T. (1990). *Personality in adulthood.* New York: Guilford.

McKeachie, W. J., Pollie, D., & Speisman, J. (1955). Relieving anxiety in classroom examinations. *Journal of Abnormal and Social Psychology, 50,* 93–98.

McKeown, T., & Gibson, J. R. (1951). Observation on all births (23,970) in Birmingham, 1947: IV. "Premature birth." *British Medical Journal, 2,* 513–517.

McMullen, L., & Pearson, E. S. (1939). William Sealy Gosset, 1876–1937. *Biometrika, 30,* 205–253.

Micceri, T. (1989). The unicorn, the normal curve, and other improbable creatures. *Psychological Bulletin, 105,* 156–166.

Milgram, S. (1969). Note on the drawing power of crowds of different size. *Journal of Personality and Social Psychology, 13,* 79–82.

Milnor, W. R. (1990). *Cardiovascular physiology.* New York: Oxford.

Minium, E. W., Clarke, R. B., & Coladarci, T. (1999). *Elements of statistical reasoning* (2d ed.). New York: Wiley.

Minium, E. W., King, B. M., & Bear, G. (1993). *Statistical reasoning in psychology and education* (3d ed.). New York: Wiley.

Mischel, H. N. (1974). Sex bias in the evaluation of professional achievements. *Journal of Educational Psychology, 66,* 157–166.

M&M/Mars. (1993). *A little illustrated encyclopedia of M&M/Mars.* Hackettstown, NJ: Author.

Moore, D. S., & McCabe, G. P. (1993). *Introduction to the practice of statistics* (2d ed.). New York: Freeman.

Mosteller, F., & Wallace, D. L. (1989). Deciding authorship. In J. M. Tanur, F. Mosteller, W. H. Kruskal, R. F. Link, R. S. Picters, G. R. Rising, & E. L. Lehmann (Eds.), *Statistics: A guide to the unknown* (3d ed.). Pacific Grove, CA: Wadsworth & Brooks/Cole.

Nelson, N., Rosenthal, R., & Rosnow, R. L. (1986). Interpretation of significance levels and effect sizes by psychological researchers. *American Psychologist, 41,* 1299–1301.

Nijsse, M. (1988). Testing the significance of Kendall's τ and Spearman's r_s. *Psychological Bulletin, 103,* 235–237.

Overall, J. E. (1980). Power of chi-square tests for 2×2 contingency tables with small expected frequencies. *Psychological Bulletin, 87,* 132–135.

Paulhus, D. L., & Van Selst, M. (1990). The spheres of control scale: 10 years of research. *Personality & Individual Differences, 11,* 1029–1036.

Pearson, E. S. (1949). W. S. Gosset. *Dictionary of national biography: 1931–1940.* London: Oxford University Press.

Pearson, K., & Lee, A. (1903). Inheritance of physical characters. *Biometrika, 2,* 357–462.

Peden, B. F., & Hausmann, S. E. (2000). Data graphs in introductory and upper-level psychology textbooks: A content analysis. *Teaching of Psychology, 27,* 93–97.

Porter, T. M. (1986). *The rise of statistical thinking.* Princeton, NJ: Princeton.

Powers, E., & Witmer, H. (1951). *An experiment in the prevention of delinquency: The Cambridge-Somerville Youth Study.* New York: Columbia University Press.

Robinson, L. A., Berman, J. S., & Neimeyer, R. A. (1990). Psychotherapy for the treatment of depression: A comprehensive review of controlled outcome research. *Psychological Bulletin, 108,* 30–49.

Rokeach, M., Homant, R., & Penner, L. (1970). A value analysis of the disputed Federalist papers. *Journal of Personality and Social Psychology, 16,* 245–250.

Rosen, L. A., Booth, S. R., Bender, M. E., McGrath, M. L., Sorrell, S., & Drabman, R. S. (1988). Effects of sugar (sucrose) on children's behavior. *Journal of Consulting and Clinical Psychology, 56,* 583–589.

Rosenthal, R., & Rosnow, R. L. (1991). *Essentials of behavioral research* (2d ed.). New York: McGraw-Hill.

Rosnow, R. L., & Rosenthal, R. (1999). *Beginning behavioral research: A conceptual primer* (3d ed.). Upper Saddle River, NJ: Prentice-Hall.

Runyon, R. P. (1981). *How numbers lie.* Lexington, MA: Lewis.

Runyon, R. P., Coleman, K. A., & Pittenger, D. J. (2000). *Fundamentals of behavioral statistics* (9th ed.). Boston: McGraw-Hill.

Sax, L. J., Astin, A. W., Korn, W. S., & Mahoney, K. M. (1999). *The American freshman: National norms for Fall, 1999.* Los Angeles: Higher Education Research Institute.

Schachter, S., & Gross, L. P. (1968). Manipulated time and eating behavior. *Journal of Personality and Social Psychology, 10,* 98–106.

Schumacher, E. F. (1979). *Good work.* New York: Harper & Row.

Shedler, J., & Block, J. (1990). Adolescent drug use and psychological health. *American Psychologist, 45,* 612–630.

Sherif, M. (1935). A study of some social factors in perception. *Archives of Psychology.* No. 187.

Sherman, M. (1927). The differentiation of emotional responses in infants: The ability of observers to judge the emotional characteristics of the crying infants, and of the voice of an adult. *Journal of Comparative Psychology, 7,* 335–351.

Smith, M. L., & Glass, G. V. (1977). Meta-analysis of psychotherapy outcome studies. *American Psychologist, 32,* 752–760.

Smith, R. A. (1971). The effect of unequal group size on Tukey's HSD procedure. *Psychometrika, 36,* 31–34.

Snedecor, G. W., & Cochran, W. G. (1980). *Statistical methods* (7th ed.). Ames: Iowa State University Press.

Spatz, T. S. (1991). Improving breast self-examination training by using the 4MAT instructional model. *Journal of Cancer Education, 6,* 179–183.

Spearman, C. (1930). Autobiography. In C. Murchison (Ed.), *History of psychology in autobiography.* New York: Russell & Russell.

Sprent, P. (1989). *Applied nonparametric statistical methods.* London: Chapman and Hall.

Statistical abstract of the United States: 1999 (119th ed.). (1999). Washington, D.C.: U.S. Bureau of the Census.

Sternberg, R. J., & Williams, W. M. (1997). Does the Graduate Record Examination predict meaningful success in the graduate training of psychologists? A case study. *American Psychologist, 52,* 630–641.

Stevens, S. S. (1946). On the theory of scales of measurement. *Science, 103,* 677–680.

Stewart, I. (July 1977). Gauss. *Scientific American, 237,* 123–131.

Tanur, J. M., Mosteller, F., Kruskal, W. H., Lehmann, E. L., Link, R. F., Pieters, R. S., & Rising, G. R. (Eds.). (1989). *Statistics: A guide to the unknown* (3d ed.). Pacific Grove, CA: Wadsworth & Brooks/Cole.

Tufte, E. R. (1983). *The visual display of quantitative information.* Cheshire, CT: Graphics.

Tufte, E. R. (1990). Envisioning information. Cheshire, CT: Graphics.

Tufte, E. R. (1997). *Visual explanations.* Cheshire, CT: Graphics.

Wainer, H. (1984). How to display data badly. *American Statistician, 38,* 137–147.

Walker, H. M. (1929). *Studies in the history of statistical method.* Baltimore: Williams & Wilkins.

Walker, H. M. (1940). Degrees of freedom. *Journal of Educational Psychology, 31,* 253–269.

Walker, H. M. (1968). Karl Pearson. In D. L. Sills (Ed.), *International encyclopedia of the social sciences.* New York: Macmillan and Free Press.

Wallace, R. K., & Benson, H. (1972). The physiology of meditation. *Scientific American, 226,* 84–90.

Wallace, W. L. (1972). College Board Scholastic Aptitude Test. In O. K. Buros (Ed.), *The seventh mental measurements yearbook.* Highland Park, NJ: Gryphon.

Wallgren, A., Wallgren, B., Persson, R., Jorner, U., & Haaland, J. (1996). *Graphing statistics and data: Creating better charts.* Thousand Oaks, CA: Sage.

Watson, J. B. (1924). *Psychology from the standpoint of a behaviorist* (2d ed.). Philadelphia: Lippincott.

White, E. B. (1970). *The trumpet of the swan.* New York: Harper & Row.

Wilcox, R. R. (1987). New designs in analysis of variance. *Annual Review of Psychology, 38,* 29–60.

Wilcoxon, F. (1945). Individual comparisons by ranking methods. *Biometrics, 1,* 80–83.

Wilcoxon, F., & Wilcox, R. A. (1964). *Some rapid approximate statistical procedures* (Rev. ed.). Pearl River, NY: Lederle Laboratories.

Wilde, G. J. S. (1991). Economics and accidents: A commentary. *Journal of Applied Behavior Analysis, 24,* 81–84.

Winer, B. J., Brown, D. R., & Michels, K. M. (1991). *Statistical principles in experimental design* (3d ed.). New York: McGraw-Hill.

Winter, L., Uleman, J. S., & Cunniff, C. (1985). How automatic are social judgments? *Journal of Personality and Social Psychology, 49,* 904–917.

Wittrock, M. C. (1986). Students' thought processes. In M. C. Wittrock (Ed.), *Handbook of research on teaching* (3d ed.). New York: Macmillan.

Wittrock, M. C. (1991). Generative teaching of comprehension. *The Elementary School Journal, 92,* 169–184.

Wood, T. B., & Stratton, F. J. M. (1910). The interpretation of experimental results. *Journal of Agricultural Science, 3,* 417–440.

Woodworth, R. S. (1926). Introduction. In H. E. Garrett (Ed.), *Statistics in psychology and education.* New York: Longmans, Green.

Work, M. S., & Rogers, H. (1972). Effects of estrogen level on food-seeking dominance among male rats. *Journal of Comparative and Physiological Psychology, 79,* 414–418.

Yates, F. (1981). Sir Ronald Aylmer Fisher. In E. T. Williams & C. S. Nichols (Eds.), *Dictionary of national biography 1961–1970.* Oxford: Oxford University.

Youden, W. J. (1962). *Experimentation and measurement.* Washington, D.C.: National Science Teachers Association.

Zajonc, R. B., & Bargh, J. (1980). Birth order, family size, and decline of SAT scores. *American Psychologist, 35,* 662–668.

Zwick, R. (1993). Pairwise comparison procedures for one-way analysis of variance designs. In G. Keren & C. Lewis (Eds.), *A handbook for data analysis in the behavioral sciences: Statistical issues* (pp. 43–71). Hillsdale, NJ: Erlbaum.

index